Propagation of Horticultural Plants

Propagation of Horticultural Plants

Arid and Semi-Arid Regions

Edited by

R. S. Singh

Principal Scientist (Horticulture)
Central Institute for Arid Horticulture
Bikaner – 334 006, Rajasthan

and

R. Bhargava

Principal Scientist (Plant Physiology)
Central Institute for Arid Horticulture
Bikaner – 334 006, Rajasthan

NEW INDIA PUBLISHING AGENCY

New Delhi – 110 034

NEW INDIA PUBLISHING AGENCY

101, Vikas Surya Plaza, CU Block, LSC Market
Pitam Pura, New Delhi 110 034, India
Phone: + 91 (11)27 34 17 17
Fax: + 91(11) 27 34 16 16
Email: info@nipabooks.com
Web: www.nipabooks.com

Feedback at feedbacks@nipabooks.com

ISBN: 978-93-83305-25-4

Composed, Designed and Printed at Jai Bharat Press, Delhi

Dr. T. A. More
Vice Chancellor

Mahatma Phule Krishi Vidyapeeth
Rahuri - 413 722,
Ahmednagar, Maharashtra

Foreword

To be competitive in global market, quality productions of fruits are essential for which availability of quality planting material has the key role. Horticulture has invariably improved the socio-economic status of the farmers. In the recent years, there has been phenomenal increase in area and production of fruit crops in India, which has brought 'Golden Revolution' in our country. The credit of this achievement of Revolution goes to untiring efforts and the hard work of our scientists, horticulturists, extension workers and adoption of the standardized technologies by the farmers/growers of the country. It has also played an important role in woman empowerment, employment opportunity through floriculture, nursery husbandry, seed production, bee keeping, etc.

India is blessed with different types of agro-climatic conditions, among which the extreme dry cold climate of Leh and Ladakh and dry hot desertic climate of western Rajasthan are unique. In order to develop agriculture in these regions, special technologies are required for crop production which can give sustainable production under such extreme conditions. In this pursuit, production of planting material requires special attention and the authors have addressed this issue in the present book. In their present book, the authors have given comprehensive information about the botany, uses, propagation methodology and quality standards of planting material of each crop selected by them. This was the demand of the hour and the information compiled by them is extremely useful which will go a long way in development of strategies for horticultural development of cold/ hot arid regions.

The efforts of Dr. R.S. Singh and Dr. R. Bhargava in bringing out a comprehensive compilation of information in the form of a valuable book entitled "Propagation of Horticultural Plants" is praiseworthy. The book has covered the different aspects of propagation, use of plant growth regulators, micro-propagation, quality standards of planting material and disease management in nursery plants of arid and semi arid regions. The major feature of this book is that the problems confronting the arid horticultural production and their solutions have been specifically elaborated and discussed with full justification and scientific support.

I hope that this book will be useful for the students, teachers, researchers, extension workers and Nurserymen and growers of our country, not only because of its latest reference material but also because of its simple language.

(T. A. More)

Rahuri (Maharashtra)

Dated : September, 2013

Preface

Plant propagation is an important aspect of agriculture to enhance production in general and horticulture in particular. Quite a few treatises on this important field of plant science have been written but the needs of the post-graduate students and research workers in this field have not been adequately considered. The availability of genuine planting material to growers at right time on affordable cost is also a constraint responsible for lower productivity of horticultural crops. Numerous results of investigation have been published dealing with the propagation of horticultural plants of semi-arid and arid regions in many journals over a long period of time. This book compiles an up-to-date account of the researches on different aspects of propagation of semi arid and arid fruits, economic plants, spices, medicinal and aromatic plants and ornamentals, under a single book.

The book also contains chapters on principles and practices of seed and vegetative propagation, physiology of regeneration of adventitious roots on cuttings as well as viability and invigoration of seeds, use of plant growth substances, micro-propagation, and management of diseases in nursery plants and quality standards of seed and planting materials. An extensive bibliography on propagation of major semi- arid and arid fruit crops should prove useful to the students in horticulture as well as the researchers working on plant propagation, policy makers and nursery management.

We are highly grateful to the contributors of the articles for their sincere efforts and cooperation to bringing out the book. We are thankful to Dr. S.K. Sharma, Director, CIAH, Bikaner; Dr. S.K.Malhotra, ADG (Hort), ICAR, New Delhi; Dr. Vishal Nath, Director, NRC on Litchi, Muzaffarpur, Dr. B. D. Sharma, Dr. J. P. Singh, Dr. Sanjay Singh, Dr. N. D. Yadav, Dr. J. C. Tewari , Dr. D. B. Singh and Prof. P. K. Yadav, Dr. G. Pandey, Dr. A.K. Singh, Dr. Hare Krishna and Dr. S. K. Maheshwari for the encouragement and help in editing of the chapters. Our sincere thanks are due to Er. B.R. Khatri (Sr. Computer) for type setting of manuscript and designing the cover of the book and also to the staff of M/s New India Publishing Agency, New Delhi for timely publishing the book.

We are happy to record the appreciation for the help of Mrs. Garima Pal, Res. Assoc.; Ms. Poornima, SRF, Dr. Vinita Singh, BAMS, in preparation of the manuscript including arrangement of references of several crops. We are extremely thankful to family members and friends for the cooperation and help for bringing the publication.

Our sincere thanks to Scientists and staff members of CIAH, Bikaner, Rajasthan and CHES, Godhra, Gujarat for rendering help and cooperation for preparation of this publication.

Bikaner   R. S. Singh
Rajasthan   R. Bhargava

Contents

Foreword .... *v*

Preface .... *vii*

1. **Introduction** **1**

2. **Propagation through Seeds** **9**
   R. Bhargava

3. **Vegetative Propagation** **29**
   R. Bhargava

4. **Micro Propagation of Horticultural Crops** **41**
   Hare Krishna

5. **Physiology of Rooting and Growth in Fruit Plants** **55**
   Hare Krishna and S.K. Maheshwari

6. **Role of Plant Growth Regulators in Propagation of Arid Fruit Plants** **73**
   N.K.Gupta and Sunita Gupta

7. **Nursery Management for Production of Quality Planting Materials** **91**
   R. S. Singh and Hare Krishna

8. **Ber (*Ziziphus mauritiana* Lamk.)** **113**
   R.S. Singh

9. **Pomegranate (*Punica granatum* L.)** **135**
   R. S. Singh and B.D. Sharma

10. **Date palm (*Phoenix dactylifera* L.)** **147**
    R.S. Singh

11. **Aonla (*Emblica officinalis* Gaertn.)** **177**
    A.K. Singh and R.S. Singh

12. **Bael (*Aegle marmelos* Correa.)** 189
D. Pandey, S.K. Shukla and G. Pandey

13. **Karonda (*Carissa carandas* L.)** 205
K.K. Misra and Subhash Chandra Singh

14. **Fig (*Ficus carica* L.)** 217
R.S. Singh

15. **Manila Tamarind (*Pithecellobium dulce* Roxb.)** 229
R.S. Singh and S.K.Maheshwari

16. **Custard Apple (*Annona squamosa* L .)** 237
R.S. Singh and A.K. Singh

17. **Tamarind (*Tamarindus indica* L.)** 247
J. Suresh, D. Sarla Devi, N. Kumar and S. Anbu

18. **Wood Apple (*Feronia limonia* L.)** 267
S.S. Hiwale

19. **Phalsa (*Grewia subinaequalis* DC.)** 277
D. B. Singh

20. **Gangana [*Grewia tenax* (Forsk.) Fiori.]** 285
J.P. Singh and R.S. Singh

21. **Mulberry (*Morus* sp.)** 291
R.S. Singh

22. **Kair (*Capparis decidua* Forsk.)** 299
R. Bhargava and R. S. Singh

23. **Lasoda (*Cordia myxa* L.)** 307
P.K.Yadav and Vartika Srivastava

24. **Jamun (*Syzygium cuminii* Skeels.)** 315
Sanjay Singh and A. K. Singh

25. **Khirni (*Manilkara hexandra* (Roxb.) Dub.)** 325
Sanjay Singh and R. S. Singh

26. **Citrus Fruits** 329
Atul Chandra and M. K.Kaul

27. **Kamrakh (*Averrhoa carambola* L.)** 349
Vishal Nath and Bikash Das

28. **Chironji (*Buchanania lanzan* Spreng.)** 363
Sanjay Singh and A. K. Singh

29. **Salvadora (*Salvadora oleoides*)** 371
J.C. Tewari and R. S. Singh

30. **Noni (*Morinda citrifolia* L.)** 379
A.K. Singh

31. **West Indian Cherry (*Malphigia puncifolia* L.)** 389
Jasmine Jaya, Swaminathan, A. A. and S. Krishnasamy,

32. **Mahua (*Bassia latifolia* Roxb.)** 395
Sanjay Singh and A.K. Singh

33. **Jojoba [*Simmondsia chinensis* (Link) Schn.]** 403
J.C. Tewari and R. S. Singh

34. **Cactus pear [*Opuntia ficus indica* (L.)Mill.]** 413
R.S. Singh

35. **Marula nut (*Sclerocarya birrea sub* sp. caffra)** 433
R.S. Singh

36. **Propagation of Less Known Shrubs of Hot Arid Region** .. 441
J. P. Singh and V.S. Rathore

37. **Propagation of Seed Spice Crops** 451
S.K. Malhotra

38. **Propagation of Ornamental Plants of Cold Arid Region**... 467
B.P. Sharma

39. **Guggul [*Commiphora wightii* (Arnott.) Bhand.]** 483
J. P. Singh and R. S. Singh

40. **Senna (*Cassia angustifolia* Vahl.)** 491
R.S. Singh and J. P. Singh

41. **Henna (*Lawsonia inermis* L.)** 499
R. S. Singh

42. **Khejri (*Prosopis cineraria*)** 507
J.C. Tewari and R.S. Singh

43. **Indian Aloe (*Aloe barbadensis* Mill.)** 517
R. S. Singh

44. Drumstick (*Moringa oleifera* Lam.) 527
K. Krishnakumary and K. V. Suresh Babu

45. Management of Diseases and Insects in Nursery Plants ... 535
S. K Maheshwari and Hare Krishna

Chapter-1

Introduction

Even since, the man adopted plant cultivation; he attempted various methods of propagation. In ancient time, the only source of plant multiplication was by seeds. Hence, techniques for higher seed germination, improved seed vigour were developed by giving pre-treatment. However, as the knowledge advanced and demands for having true to type plants was realized to man shifted his cultivation practices for some plants from seed to vegetative propagation such as cutting, layering and grafting. The use of cuttings, corm, bulbs, tubers for multiplication were known to ancient peoples as mentioned in old literature. This method also gave an edge to conserve the genetic gain once developed in a cultivar, since all vegetatively multiplied plants will have some genome and will produce true to type plants. Thus, the development of vegetative propagation method because it is easy method to multiply plants on large scale and develop uniform quality produce.

Many new methods of budding and grafting have been standardized with the progress in the field of horticulture. For example, invention of glass house, green house, mist chambers, micro-propagation, micro-grafting, application of rooting co-factors in rooting of cuttings difficult to root epicotyls grafting, stooling in guava, bottom heat technique in rooting of cutting are some of new invention in the field of propagation of horticultural plants. In semi arid and arid region environmental condition is very harsh and difficult for plant multiplication but advancement in methodology and growing structures for the plants, many horticultural plant species earlier which were multiplied through seeds, are now being propagated through cuttings, micro-grafting and micro-propagation. Soft wood cutting in pomegranate under mist condition has been found successful in arid region. For example, Kair (*Capparis decidua*) is an under exploited fruit tree of arid region, was earlier multiplied through seeds, now it can be propagated through semi hard wood cuttings by giving some hormonal treatments as well appropriate season for cutting planting. Khejri (*Prosopis cineraria*) a multipurpose tree of hot ard region, is commonly propagated by seeds but now vegetative propagation technique has been developed for large scale multiplication of plants. In case of date palm, it was commonly multiplied through suckers or offshoots which is now multiplied through micro-propagation and large scale planting material are produced in different tissue culture laboratory in the country. Pomegranate is commonly multiplied through cuttings and layering now, it is propagated through tissue culture technique to raise disease free material.

Methods of Propagation of Horticultural Plants

There are different propagation methods, which can be used for commercial multiplication of horticultural plants of semi arid and arid regions. These include cutting, layering, stooling, budding, micro-grafting and grafting. Because of advancement in techniques, different vegetative methods are used for multiplication of fruits, medicinal and aromatic plants, flowering and plants of other economic value like olive, jojoba, henna, on large scale.

In case of kair, bael, *Salvadora* plants emerged through root suckers near to main stem can be removed carefully during rainy season and transplanted in the field for establishment.

Sexual propagation

Propagation by seed is considered as the most important, simple method of multiplication of most of the annuals, biennials and many perennial fruit plants; vegetables crops; plantation, aromatic and medicinal plants; ornamental, flowering and shade trees, shrubs and forestry plants. In many seed spice crops, medicinal and aromatic plants, seed is commonly used for multiplication. Even at present time, seed propagation is simple and useful method for multiplication and is in practice by the farmers of the region.

Vegetative propagation

It is the method by which true- to- type plants can be raised under nursery conditions. Vegetative method is being used since ancient time but certain improvement has been made. Standardization of time for multiplication of plants is also essential for different crops region wise.

The vegetative methods for multiplication of several horticultural plants are given below.

Cutting

It is the method of propagating fruit plants in which the part of a plant (generally stem) having at least few buds, when detached from mother plant and placed under favorable conditions develop into a complete plant resembling in all characteristics to the parent from which it was taken. This method is commonly used in plants, which root easily and readily, hence, multiplication of plants by cutting is very fast and cheap. The fruit plants like phalsa, lemon, pomegranate, marula nut, mulberry, Argan, Fig, grape, etc. are commercially propagated by cuttings. Plant growth substances such as IBA, NAA, have used for root initiation in hard wood cuttings. In some plant species, rooting is very difficult which can be enhanced by treating cuttings with plant growth substances like auxins and other chemicals. In case of olive, twigs are suitable for cuttings which are growing straight to upper direction. Selection of cutting and time of planting besides growing environment play an important role in multiplication of quality plants.

a. Stem cuttings

- Herbaceous cuttings: Dahlia, begonia etc.
- Soft wood cuttings: Pomegranate, Juniper, Duranta, *Jasminum*, etc.
- Semi hardwood cuttings: Pomegranate, guava, ker, lasoda, etc.
- Hardwood cuttings: Fig, Olive, Marula nut, Argan, Moringa, Guggul, etc.

b. Leaf cuttings

- Bigonia, Bryophyllum, etc.

c. Leaf bud cuttings

- Blackberry, lemon, etc.

d. Root cuttings

- Fig, cherry, etc.

e. Cladodes/ pads

- Cactus pear (*Opuntia ficus indica*)

Layering

Layering is a method of vegetative propagation, in which roots are initiated on the shoots while they are still attached to the mother plants. This is an alternate method of propagation in fruit plants which do not root easily when detached from the mother plants. Most commonly used methods of layering are air layering, ground and mound layering for multiplication of fruit plants. This method is being used by nursery men in the country.

Air layering: It is very common method for multiplication of litchi, pomegranate, guava, etc. during February–March and July-August.

Ground layering: In this method, a branch of plant, which is near the ground, is chosen and a ring of bark about 2.5 cm is diameter is removed just below the bud. This branch is then bended and buried in soil when still attached to the mother plant. The soil is regularly watered to keep it moist. These new plants should preferably be planted in pots or nursery beds for development of better roots and shoot system before planting in the fields.

Mound layering or stooling: In this method, plant is headed back either in February or in July. The new shoots come out during April and September, from ground level. A ring of bark is removed from these shoots and they are covered with moist soil. This method is also known as stool layering and is used for propagation of guava and apple rootstocks. For this, tropical, sub-tropical humid climate is required. Availability of moisture is main factor in callusing and root development.

- Tip layering: raspberry etc.
- Simple layering: Ornamental climbers etc.
- Stooling/mound layering: mango, litchi, bael, guava, etc.
- Serpentine layering: Clematis etc.
- Trench layering: Cherry,

Budding

Budding is a method in which only one bud is inserted in the rootstock. This method is very easy and suitable for fast multiplication.

This method saves bud wood as compared to grafting. This method is generally employed during spring and rainy season. The common methods of budding are T- or I- budding, patch budding, and chip budding.

T- or I - budding: This is also known as shield budding and it is commonly used for *in situ* establishment of ber orchard in hot arid region. It is successful method for propagation of ber during June to August month in hot arid region.

Patch budding: This type of budding is quite successful in guava and it gives 60 to 70 percent successes during May and June. To have better success, a patch having two buds is used as scion instead of a single bud. This method is termed as improved patch budding method and it is used in lasoda, aonla, mulberry, khejri, etc,

Chip budding: This method is usually employed when the stock and scion are still dormant, that is just before the start of new growth.

- T-budding: Citrus, ber, etc.
- Patch budding: Lasoda, aonla, khejri, citrus,
- Chip budding: mango
- Ring budding: ber, etc.

Grafting

Grafting is most popular method of vegetative propagation, where two plant parts are joined together in such a manner that they unite and continue their growth as one plant. The different methods of grafting are tongue grafting, cleft grafting, approach grafting, side grafting and veneer grafting. In grafting technique, age, type of rootstock, season play important role for success of graft union.

Tongue Grafting: This method is commonly used when the stock and scion are of equal diameter. This is then wrapped with polythene strip.

Cleft grafting: This is also known as wedge grafting. This method is useful in the nursery where the rootstock is quite thicker than scion and tongue grafting cannot be employed successfully. Cleft grafting is done during dormant phase of plant.

Approach Grafting: This method of grafting is termed approach grafting, as the rootstock is approached to the scion, while it is still attached to the mother plant. It is used in mango.

Side Grafting: A three sided rectangular cut about 4 x 1.25 cm is made on the rootstock at a height of about 15-20 cm from the ground level and the bark of the demarcated portion is lifted away from the rootstock. A matching cut is also made on the base of the scion to expose cambium. The scion should be prepared well before the actual grafting is done. When the scion has sprouted and its leaves turned green, the root stock portion above the graft union is cut away. It is used for multiplication of ornamental plants.

Veneer Grafting: It is very common method for multiplication of mango and improved method of grafting.

- Whip grafting: many fruit plants
- Cleft grafting: avocado, pear etc.
- Bark grafting: different fruit plants
- Side grafting:' Mango, avocado etc.
- Inarching: Mango, litchi etc.
- Veneer grafting: Mango
- Stone/epicotyls grafting: mango
- Top working: mango, ber etc.
- Soft wood grafting: Jamun, Mahua, chironji, wood apple, bael,
- Bridge grafting: For repairing of damaged fruit plants.

Other Specialized Methods of Vegetative Propagation

- **Propagation by division**
- Division of tuberous roots: Sweet potato, turmeric, bigonia etc.
- Division of stem tubers: Potato,
- Division of rhizomes: Ginger, Banana, etc.
- **Propagation by separation**

  Bulbs: Onion, tulip etc

  Corms: Gladiolus etc.
- **Propagation by suckers**

  Pineapple, banana, date palm, etc.
- **Propagation by runners**

  Strawberry

- **Propagation by stolon**

  Lawn grasses
- **Other vegetative structures**

  Bulbils: Lilies

  Offsets: Pineapple

Micro propagation

Micro-propagation refers to the production of plants from very small plant parts, tissues or cells, grown aseptically in a test tube or containers under controlled nutritional, environmental and aseptic conditions. It is also called as *in vitro* propagation, which implies clonal, true-to-type propagation of plants by a variety of tissue and cell culture methods and is considered as the most efficient and commercial method of plant . Micro-propagation is suitable method for the rapid and large-scale clonal multiplication of elite materials. The meristem explant used for micro- propagation is about 0.1-0.5 mm size having only one or two leaf primordial which multiply thousands number plants. Tissue culture or *in vitro* culture is two broadly used terms for micro-propagation, which basically include aseptic culture of various plant parts. At present, many horticultural plants *viz.* banana, date palm, pomegranate, flowers, is being multiplied through micro-propagation techniques in our country. The following plants parts are used for micro-propagation as under:

- Cell culture: Carrot, sunflower etc.
- Meristem culture: Banana, guava, etc.
- Shoot tip culture: Papaya, grape, date palm, ker, etc.
- Anther culture: Dhatura
- Embryo culture and embryo rescue: peach, plum etc.

Nursery and Production of Quality Planting Material

Nursery plays an important role in multiplication of plants of any types. The role of quality seeds and planting materials are very important for increasing productivity of horticultural crops in semi arid and arid region of the country. Further, genuine and improved variety plant is needed for quality production of fruits and vegetable crops. The importance of nursery is more viable in that area where new orchards are to be established. The availability of quality planting material at affordable price is important for the farmers/growers. The initial input of orchard establishment is comparatively higher than other crop cultivation, which can be minimized by establishing a nursery to provide quality planting material at nearest site of plantations. Use of growth substances also improves the rooting and growth of plants under nursery conditions. Treatment of seedlings and plants is also necessary to minimize the spread of diseases. Planting material of improved varieties should be purchased from any registered nursery or recognized sources.

Chapter-2

Propagation through Seeds

R. Bhargava
Central Institute for Arid Horticulture, Bikaner-334 006

Plant propagation is the science of generating new plants. Basically, in plant kingdom, the plants multiply themselves either asexually or by sexual means. In asexual propagation methods, plants use any of the vegetative part of the plant to produce offspring such as stem, root, buds, etc. The plants thus produced are having exactly the same genetic blueprint as that of the parent plant. However, in the sexual method of propagation, the specialized parts of the plants which have undergone meiosis takes parts such as pollen and ovule to produce seed which on germination produces the plants having characters encoded on its genome. Such plants may have a blend of characters of both the parents. The seed production is a complex phenomenon which takes place after fertilization (Bhatnagar and Johri, 1972).

Seed Formation

Bhatnagar and Johri (1972) gave a detail account of development of angiosperm seeds. In general, seed formation in angiosperm involves the following events:

(i) Formation of stamens and pistils in the flower bud.

(ii) The anthesis or opening of flowers which signals the sexual maturity of essential organs.

(iii) Pollination which involves the transfer of pollen grains from the stamens (anthers) to the pistils (stigma), germination of the pollen and formation of the pollen tube.

(iv) Fertilization of the egg cell in the ovule with one male nucleus and fusion of the polar nuclei with the sperm nucleus from the pollen tube.

(v) Growth of the fertilized egg cell and its differentiation into an embryo, plus developments of seed coat (testa) from one or two integuments of the ovule, which forms a covering to protect the embryo from drying out, mechanical injury and other unfavorable situation.

The development of seeds in the fruit are characterized by two distinct stages (i) ripening stage during which the fresh and dry weight of the seed increases rapidly and (ii) maturation stage- in which the seeds are the major food storage organs which store food either in form of carbohydrate, protein or fat, which is consumed for the growth and development of embryo.

Advantages

Propagating plants from seed is essentially nurturing baby plants, the genetic offspring of at least one parent plant. There are several advantages to propagate from seeds:

1. It is a relatively cheap way of producing lots of new plants.
2. It is a way to obtain new gene combinations which may have novel characters.
3. It is a convenient way to share the bounty of your garden with friends.

Disadvantages

1. It takes longer period to produce mature plants from seed than it does from cutting or other horticultural means.
2. Propagation from seed takes a lot of care and attention, and in many cases, a good deal of experience and knowledge.
3. New plants from seed won't necessarily come true to type. They will be genetically distinct from their parent(s).

Seed Dormancy

Seeds of some species of plants will germinate almost immediately after harvest, if placed under suitable conditions of moisture, temperature and aeration. In contrast, there are many species especially woody perennials, in which the freshly extracted seeds do not germinate even when the environmental conditions are most favorable. This phenomenon of delay in germination after harvest is called seed dormancy. This is an important survival mechanism for the species, because these species do not germinate unless the adverse weather condition ends. For certain group of species chilling temperature for certain length of the time helps in the termination of dormancy. For certain other kinds of seeds (e.g. certain desert plants), germination is withheld unless the seeds experience certain amount of rainfall. Often, the dormancy is due to several factors and may persist indefinitely, and unless certain problems for the propagators to sow the seeds at a proper time for germination. A detail knowledge about dormancy type, dormancy mechanism and dormancy termination methods is essential to handle the seeds effectively and in a useful manner.

Type of dormancy

1. **Seed coat dormancy-** In certain species, seed coats or other related structures covering the embryo are impermeable to water and gases, particularly oxygen, and therefore, the physiological process leading to germination are prevented. Consequently, such freshly harvested seeds remain dormant until the impermeable barrier has been removed or weakened. The seeds which do not imbibe water after being placed under normal germination conditions for a specific period are called 'hard seeds' (Watson, 1948).

   In certain species the seed coat is apparently permeable to water and gases, but is so hard and thick that it prevents embryo expansion during germination. In these species, germination does not take place unless the mechanically resistant seed coat is removed or weakened due to mechanical aberrations, alternate freezing and thawing, passage through digestive tracts of animals or birds, or artificially by mechanical or acid scarification (Varner, 1965).

2. **Due to rudimentary embryo-** In some plants species, the fruits are dropped much before the seed matures and embryo is physiological in mature or rudimentary. In such cases, the seeds do not germinate when planted immediately after harvest.

3. **Due to physiologically dormant embryos-** In some plant species, the seeds do not germinate immediately after harvest even if all conditions for germination are provided to them.

However, they require some storage period for germination and growth. In such plant species, the dormancy is because of endogenous factors. This type of dormancy may also be due to the requirement of chilling treatment. There are evidences to show that during stratification the level of endogenous growth -promoting substances increases, while that of growth inhibiters decreases, thus removing the germination block and permitting germination (Martin *et. al.*, 1969; Ross and Bradbeer, 1968).

Double dormancy

In some species, seed coat is impervious and at the same the embryo is also dormant. Such seeds require 2 years for breaking of dormancy in nature. In the first spring, the micro-organisms act upon the seed coat making it weak and soft, and then the embryo dormancy is broken by receiving chilling temperature in the winter. The seedling emerges in the second spring. These seeds are known as 2 year seeds. In autumn olive (*Elaeagnus umbellata*), both mechanically resistant seed coat and dormant embryo are responsible for seed dormancy. Combination of two or more types of seed dormancy is known as double dormancy.

Secondary dormancy

Seeds that are ready to germinate after removal of all germination blocks may become dormant again, if exposed to unfavourable environmental conditions. For example, seeds of some woody perennials, after undergoing stratification to overcome embryo dormancy, become dormant again if the germination temperature reaches to 26 to 32° C. This type of dormancy is known as secondary dormancy. Exposure to high temperature or to low moisture may induce secondary dormancy.

Methods to Break Seed Dormancy

1. **Natural breaking of dormancy:** In nature dormancy terminates when embryo gets suitable environment such as adequate moisture, aeration and temperature. The impermeable seed coat present in many species became permeable due to the rupturing of softening action of natural agents like micro organism, high or low temperature, humidity fiber and abrasion due to wind or digestive tracts of birds and animals which feed on these seeds. For e.g.,

Rhizoctonia damages seed coat. In case of *Prosopis* spp natural seed germination takes place if pods are eaten by animals in grazing lands.

2. **Treatments to break Dormancy:** The various treatments for overcoming dormancy may be divided into the following three groups

    1. **Seed coat treatments:** These treatments aim at making hard seed coat permeable to water or gases either cracking or softening them. The process is usually referred as scarification. These treatments are either physical or chemical in nature. It has been observed that seed soaking enhanced germination and seedling growth by pre-treatments of seeds with cow dung for 24 hours in *Phoenix* species (Singh and Bhargava, 2009). *Grewia* sp. seeds treated with 50% Sulphuric acid solution for 5 minute exhibited maximum germination in a span of 65 days (Singh *et al.*, 1993). Chemical treated and water soaked seeds shown superiority over control but in no case germination could exceed 60% in *Grewia tenax*. It has been observed that if seeds of *lasoda are* buried in compost pits for shorter period for seed treatment. After treatment seeds germinate easily in the nursery.

        a. **Scarification**

            i. Acid scarification: treating seeds with concentrated acids like sulphuric acid, Hydrochloric acid etc.,

            ii. Thermal scarification: the seeds are treated with different temperatures and gases

            iii. Mechanical scarification: The seed coat is damaged using mechanical means. viz.,

            - Rubbing seeds on sand paper or by using mechanical scarifier as in subabul
            - Making small incision by piercing a needle as in bitter gourd. Cutting of operculum in marula nut seed is necessary for germination.
            - Removing of entire seed coat as in rubber

b. **Embryo treatments:**

i. Stratification: the incubation of seeds at a suitable low temperature (Usually 0-5 °C) over a moist substratum before transferring them to a temperature optimum for germination. Ex. Cherry (*Prunus cerasus*), Mustard and rape seeds

ii. High temperature treatment: in some species, incubation at 40-50 °C for few hours to 1-5 days may be effective in overcoming dormancy.

iii. Chemical treatments: Alternatively growth regulators or other chemicals may be applied to induce germination. Growth regulators commonly used are GA_3 (100 ppm), kinetin (10-15 ppm) and thio-urea (0.5-3%)

3. **Miscellaneous approaches:**

i. Exposing seeds to light

ii. Pressure treatment

iii. Infra red radiation treatment

iv. Magnetic treatment

Seed Germination

In seed germination process, the seed's role is that of reproductive unit; it is the thread of life that assures the survival of all plant species. Furthermore, because of its role in stand establishment, seed germination remains a key to modern agriculture.

Seed would normally germinate only after they have undergone a predefined period of growth and development accumulating food reserves and finally becoming air dry.

Seed germination depends on a favourable combination of several external and internal factors; in nature, seed must wait for this combination to occur for their germination.

Types of germination

1. **Hypogeal germination:** the cotyledons or storage organs do not emerges above the soil surface; only plumule emerges above the ground. Ex: Most of the monocots and pea

2. **Epigeal Germination:** The cotyledon or storage organs emerge above the soil surface. Ex: Most of the dicots and pine.

Phases of seed germination

i. Imbibition: Rapid water uptake

ii. Active Metabolism: Major metabolic events begin

iii. Cell expansion: Seedling protrusion

Factors affecting seed germination

1. Internal factor: Seed Maturity, Mechanical damages
2. External factors: i) Water ii) Oxygen iii) Temperature iv) Light and v) Soil Factors

   i. Water: Water is a basic requirement for germination. It is essential for enzyme activation, breakdown, translocation and use of reserve storage materials.

   ii. Oxygen: Atmospheric air is composed of 79.9 % Nitrogen, 20% oxygen and 0.03% carbon dioxide. Oxygen is required for germination of most of species. If CO_2 concentration is higher than 0.03% it retards germination. Respiration increases sharply during seed germination. Since respiration is essentially an oxidative process, an adequate supply of oxygen is a must.

   iii. Temperature: Seed germination is a complex process involving many individual reactions and phases, each of which is affected by temperature. The effect on germination can be expressed in terms of cardial temperature i.e., minimum, optimum and maximum temperature. The optimum temperature for most of the seeds is between 15 to 30°C maximum temperature is between 30 to 40°C. Some species will germinate even at freezing point also.

   iv. Light: Some species required light for seed germination. Both light intensity (lux) and light quality (colour and wavelength) influence seed germination

   v. Soil factor: Soil structure, soil texture and soil temperature influences on seed germination.

Seed Viability

A viable seed is one which is capable of germination under suitable conditions. The definition includes dormant but viable seeds, in which case the dormancy must be broken before viability can be measured by germination. A non-viable seed, therefore, is one which fails to germinate even under optimal conditions, including treatments for the removal of dormancy. The practical definition of viability depends upon the context in which it is used; for example, to the ecologist, viability implies the ability of the seed to germinate and the ability of the seedling to establish itself in the environment in which the seed finds itself. However, when seeds are utilized, to produce a crop, for example, then viability is a measure of the suitability of the seed batch to produce a satisfactory crop.

Legal and commercial requirements demand that there must be rigrous consistency in the determination of seed viability. Under the International Rules for seed Testing (International Seed Testing Association, 1985), precise criteria are defined for testing the viability of all of the main species of crop plants and of many trees. They include the criteria necessary for making a representative sampling of a seed batch, whether it be a bag of seed or a lot, for the number of test replication, and for the precise condition of the test for each species. These conditions specify the amount of water necessary, the air temperature, and through the specification of the substrate, the aeration, the amount of contact between seed and moisture, the seed temperature and the illumination. The germination tests may be conducted in dishes with the seeds placed on a specified grade of germination paper. Some seeds should be placed between sheets of paper; larger seeds may be placed on sand under sheets of paper, pressed into sand or buried in sand. The dishes of seed may be placed in incubators or plant growth cabinets and are normally subjected to continuous illumination.

Seed Vigour

Seed vigour is an important quality parameter which needs to be assessed to supplement germination and viability tests to gain insight into the performance of a seed lot in the field or in storage. Several definitions have been offered to explain seed vigour. Looking into the complexity of the situation, the ISTA congress in 1977 adopted the definition of seed vigour as "the sum total of those properties of the seed which determine the level of activity and performance of the seed or seed lot during germination and seedling emergence". Although, differences in physiological attributes of seed lots can be demonstrated in the laboratory, it was recommended that the term should be used to describe the performance of seeds when sown in the field (Perry, 1984a).

As the germination test is conducted in an optimum condition specific to different species, it is not always possible to get an idea of the performance of a seed lot in the field on the basis of germination test in the laboratory. It is mainly because of the reason that field conditions are seldom optimum and the emerging seeding suffers from one or the other kind of stress. In many cases seed lots having similar laboratory germinations may give widely differing field emergence values.

Seed vigour is still a concept rather than a specific property of a seed or seed lot. Several factors like; genetic constitution, environment and nutrition of mother plant, maturity at harvest, seed weight and size, mechanical integrity, deterioration and ageing and pathogens are known to influence seed vigour (Perry, 1984a). Therefore, care has to be exercised in selecting a seed vigour test to do the job. Two criteria have been employed by the ISTA seed vigour committee to evaluate, the performance of seed vigour test methods for different crops:

(i) Reproducibility of vigour method

(ii) The relationship between vigour test results and seedling emergence in field soil.

There is no universally accepted vigour test for all kinds of seeds. Determination of following vigour tests will be useful in gaining additional information on seed quality.

1. Growth tests

(a) **First count:** The test is done along with the regular germination test. The number of normal seedlings germinated on the first count day, as specified in the germination test for each species are counted. The number of normal seedlings gives an idea of the level of seed vigour in the sample. Higher the number of normal seedlings greater is the seed vigour.

(b) **Seedling growth rate and dry weight:** The seedlings are grown either in laboratory, green house or field. In laboratory, in between rolled towel paper method should be followed. Ten seeds are planted in the centre of the moist towel papers in such a way that the micropyles are oriented towards bottom to avoid root twisting. The rolled towel papers are kept in the germinator maintained at a temperature recommended for crop in reference. After a specified period of time (5-10 days) towel papers are removed and five seedlings are selected, their lengths is measured and mean seedling length is calculated. Seed lots producing the taller seedlings are considered more vigorous than the seed lots producing shorter seedlings. For dry weight determination, the seedlings are removed and dried in an air oven at 100°C temperature for 24 hours. The seedling dry weight provides additional information for assessing seed vigour.

(c) **Period of germination:** One hundred seeds, each in four replications, are planted in recommended substratum for germination. The substratum is kept in a germinator maintained at recommended temperature for the crop in reference. Numbers of seedlings emerging daily are counted from day of planting the seeds in the medium till the time germination is complete. Thereafter a germination index (G.I.) is computed by using the following formula:

$G=n/d$

where, n =number of seedlings emerging on day 'd'

d = day after planting

The seed lot having greater germination index is considered to be more vigorous.

(d) **Seed vigour index (SVI):** This is calculated by determining the germination percentage and seedling length of the same seed lot. Fifty seeds each in four replications are germinated in towel papers as prescribed for the crop species in germination test. While evaluating the number of normal seedlings at the time of final count, the seedling length of 5 randomly selected seedlings are measured.

Seed vigour index is calculated by multiplying germination (%) and seedling length.The seed lot showing the higher seed vigour index is considered to be morevigorous (Abdul-Baki and Anderson, 1973).

2. Conductivity test

A seed sample of 2-5 g. is weighed and surface sterilized with 0.1% $HgCl_2$ for 5-10 minutes. The sample is washed thoroughly in distilled water. The clean seeds are immersed in 100 ml of water at 25 ± °C temperature for 1-12 hours. After this the seeds are removed with a clean forcep. The steep water left is decanted and is termed as leachate.

First the conductance of distilled water is measured in a beaker. The electrode is then cleaned with a tissue paper and conductance of the leachate is read. To get the EC of leachate the reading of distilled water is substracted from the sample reading. The value is then corrected for the temperature and multiplied by the cell constant factor. The reading is expressed as mm mhos/cm/g of seed. Lower the value of EC greater is the seed vigour.

3. Hiltner test (Brick Gravel Test)

The sand is seived, moistured and filled in the germination box leaving about 3 cm empty at the top. One hundred seeds are placed in each box in the impressions made by a sand marker. After this 2-2.5 cm of porous brick gravel is spread over the seeds. The box is kept in the germinator at appropriate temperature. After the period required for germination, the box is removed and the seedlings which have emerged through the brick gravel layer are counted. The percentage of emerged seedlings is used to compare seed vigour of different lots. The test should be repeated 3-4 times to get authentic value.

4. Paper piercing test

The seeds are placed on 1.5 cm moist sand in a tray or sand box. The seeds are covered with specially selected dry filter paper which is then covered with 2 cm of moist sand. After this, the sand boxes/trays are kept in a germinator maintained at 20°C temperature for 8 days. After 8 days sand boxes/trays are taken out and seedlings emerging above the paper are counted. A seed lot having maximum number of seedlings coming out of paper is considered to be most vigorous. The test is highly dependent on the quality of paper and should be used when such papers are available.

5. Cold test

After grinding and properly sieving the soil is filled in tray upto 2 cm depth. Fifty seeds were placed over the sand and covered with another 2 cm thick layer of soil. The soil is compacted and enough water is added to make the soil about 70% of its water holding capacity. The temperature of the water should be 10 °C. After watering the trays are covered with polythene bags and placed in the refrigerator maintained at 10°C temperature for one week. After one week the trays are removed and placed in the germinator at 25°C temperature. The seedlings emerged after 4 days are counted. The germination percentage is computed by counting the number of normal seedlings as in germination test. Higher the germination percentage greater is the vigour.

6. Accelerated ageing test

One hundred seeds each in four replications are tied in a fine muslin cloth. The tied seeds are placed in jar on a wire mesh. The lower part of the jar is filled with water. There should not be a direct contact between water and the seed. The jar is covered with the lid and sealed with paraffin wax to make it air tight. The jar is then placed in the accelerated aging chamber maintained at 45 ± 2°C temperature for 3-5 days. The jar is removed after this period and the seeds are cooled in desiccators. The seeds are then tested in a normal germination test specific to different crops. The percent germination gives level of seed vigour. Higher the germination percentage greater is the vigour of the seed.

Future Role of Seed Vigour Testing

Seed vigour is an important component of seed quality and satisfactory levels are necessary in addition to traditional quality criteria of moisture, purity, germination and seed health to obtain optimum plant stand and high production of crops. As agricultural and horticultural techniques become progressively more sophisticated, the need for high vigour seed will increase and testing standards, similar to those recognized for germination will be required (Perry, 1984b). The technology of seed vigour testing has not been perfected so far, so much so that there is not a single universally accepted seed vigour test method. Research is needed to further refine the current seed vigour test methods and to develop new methods which are more related to field/storage conditions.

Seed Certification

The General Seed Certification Standards are applicable to all crops which are eligible for certification, and with field and seed standards for the individual crops, shall constitute the Minimum Seed Certification Standards. The word 'seed' or 'seeds' as used in these standards shall include all propagating materials.

Purpose of Seed Certification

The purpose of seed certification is to maintain and make available to the public, through certification, high quality seeds and propagating materials of notified kind and varieties so grown and distributed as to ensure genetic identity and genetic purity. Seed certification is also designed to achieve prescribed standards.

Seed Certification Agencies

Certification shall be conducted by the Certification Agencies notified under section 8 of the Seeds Act, 1966. In every state seed certification agencies are working and these are governed by rule.

Certified Seed Producer

Certified seed producer means a person/organization who grows or distributes certified seed in accordance with the procedures and standards of the certification.

Eligibility Requirements for Certification of Crop Varieties

Seed of only those varieties which are notified under section 5 of the Seeds Act, 1966 shall be eligible for certification.

Classes and Sources of Seed

A. Breeder seed

Breeder seed is seed or vegetative propagating material directly controlled by the originating or sponsoring-plant breeder of the breeding programme or institution and/or seed whose production is personally supervised by a qualified plant breeder and which provides the source for the initial and recurring increase of Foundation seed.

Breeder seed shall be genetically so pure as to guarantee that in the subsequent generation *i.e.* certified Foundation seed class shall conform to the prescribed standards of genetic purity. The other quality factors of Breeder seed such as physical purity, inert matter, germination *etc.* shall be indicated on the label on actual basis. The Breeder seed shall be packed and supplied by the breeders in the form and manner indicated in.

B. Certified seed

Certified seed shall be the seed certified by Certification Agency notified under section 8 of the Seeds Act, 1966 or seed certified by any Certification Agency established in any foreign country provided the Certification Agency has been recognised by the Central Government through notification in the Official Gazette. Certified seed shall consist of two classes, namely, Foundation and certified seed and each class shall conform to the following description:

1. Certified Foundation seed shall be the progeny of Breeder seed, or be produced from Foundation seed which can be clearly traced to Breeder seed. Thus, Foundation seed can even be produced from Foundation seed. During the production of certified Foundation seed, the following guidelines shall be observed:

   (a) Certified Foundation seed produced directly from Breeder seed shall be designated as Foundation seed stage-I;

   (b) Certified Foundation seed produced from Foundation seed stage-I shall be designated as Foundation seed stage-II;

   (c) Certified Foundation seed stage-II will not be used for further increase of Foundation seed and shall be used only for production of certified seed class;

   (d) Minimum Seed Certification Standards shall be the same for both Foundation seed stage-I and II unless otherwise prescribed;

   (e) Production of Foundation seed stage-II shall ordinarily be adopted in respect of such crop varieties provided, when it is expressly felt by the Certification Agency that Breeder seed is in short supply;

   (f) Production of Foundation seed stage-II may be adopted for the following group of crops:

   - vegetatively propagated crops
   - apodictically reproduced crops
   - self- po1linated crops

2. Production of Foundation seed stage-I and II shall be supervised and approved by the Certification Agency and be so handled as to maintain specific genetic identity and genetic purity and shall be required to conform to certification standards specified for the crop variety being certified.

3. (a) Certified seed shall be the progeny of Foundation seed and its production shall be so handled as to maintain specific genetic identity and purity according to standards prescribed for the crop being certified

(b) Certified seed may be the progeny of Certified seed provided this reproduction does not exceed three generations beyond Foundation seed stage- I and;

- It is determined by the Certification Agency that genetic identity and genetic purity will not be significantly altered;

(c) Certified seed produced from certified seed shall not be eligible for further seed increase under certification. Certification tags for such production which is not eligible for further seed increase under certification shall be superscribed with, "not eligible for further seed increase under certification."

Phases of seed certification

Certification shall be completed in six broad phases listed as under:

(a) receipt and scrutiny of application;

(b) verification of seed source, class and other requirements of the seed used for raising the seed crop;

(c) field inspections to verify conformity to the prescribed field standards;

(d) supervision at post-harvest stages including processing and packing;

(e) seed sampling and analysis, including genetic purity test and/or seed health test, if any, in order to verity conformity to the prescribed standards; and

(f) grant of certificate and certification tags, tagging and sealing.

Establishing source of seed

The individual intending to produce seed under certification shall submit to the Certification Agency, one or more relevant evidence such as certification tags, seals, labels, seed containers, purchase records, sale records *etc.*, as may be demanded by the Certification Agency during submission of the application, its scrutiny and/or during first inspection of the seed crop, in order to confirm if the seed used for raising the crop has been obtained from the source approved by it and conforms to the provisions contains in para- V. This requirement also applies to both parents in seed production involving two parental lines.

Field Area for Certification

There is' no minimum or maximum limit for the area offered by a person for certification, provided the certified seed production meets all the prescribed requirements.

Unit of certification

For the purpose of field inspections, the entire area planted under seed production by an individual shall constitute one unit provided:

(a) it is all under one variety;

(b) it does not exceed ten hectares;

(c) it is not divided into fields separated by more than fifty meters between them;

(d) it is planted with or is meant to produce seed belonging to the same class and stage in the generation chain;

(e) the crop over the entire area is more or less of the same stage of growth so that observations made are representative of the entire crop;

(f) the total area planted, by and large, corresponds to the quantity of seed reported to have been used; and the Certification Agency's permission had been obtained to sow a larger area by economizing on seed rate; if that be the case;

(g) raised strictly as a single crop and never as mixed;

(h) not so heavily and uniformly lodged that more than one third of the plant population is trailing on the ground leaving no scope for it to stand up again thus making it impossible for

the Certification Agency to inspect the seed crop at the appropriate growth stage in the prescribed manner;

(*i*) as far as possible, so maintained as to show adequate evidence of good crop husbandry thereby improving the'reputation *for* certified seeds; and

(*j*) not grown as inter, companion or ratoon crop unless otherwise specified

Field Inspection

(*a*) the field inspection work which requires technically-trained personnel, shall be performed by the persons who have been so authorized by the Certification Agency;

(*b*) field inspection meant to verify those factors which can cause irreversible damage to the genetic purity or seed health shall be conducted without prior notice *to* the seed producer.

(*c*) soon after the completion of the field inspection, a copy of the report shall be handed over to the seed producer or his representative.

Harvesting, Threshing and Transportation

Seed crop meeting field standards for certification shall be harvested, threshed and transported to the seed processing plant in accordance with the guidelines issued by the Certification Agency. During these operations, seed producer will take all precautions to safeguard the seed from admixture and other causes of seed deterioration.

Bulking

Bulking of unprocessed seed stocks to obtain larger homogeneous seed stocks may be permitted by the Certification Agency provided the stocks to be bulked meet the following requirements:

- belong to the same certified seed producer;
- belong to the same crop, variety, class of seed and stage in the generation chain;
- were produced in the same season and under similar agro-climatic conditions;
- were subjected to certification by the same Certification Agency;

- have more or less similar physical appearance and levels of moisture;
- are adequately homogeneous in composition.

Seed Processing and Packing Schedule

The Certification Agency shall prepare and communicate seed processing and packing schedule to all certified seed producers soon after the certification of seed crops at field stage. The seed producers shall adhere to the schedule specified by thd Certification Agency. However, re-scheduling may be accepted by the Certification Agency on the request of seed producer on genuine grounds.

Seed Processing

Seed processing means cleaning, drying, treating, grading and other operations which will improve the quality of seeds. Seed from fields which conformed to the standards of certification at field stage shall, as soon as possible after the harvest will be brought at processing plant for processing. The screen aperture of specific size shall be used for cleaning and grading of seeds so that typical contaminants such as weed seeds, small seeds, damaged seeds, broken and shriveled seeds, straw, chaff, leaves, twigs, stones, soil particles *etc.* are removed. However, the Certification Agency is authorized to deviate under exigencies to use the screen of small aperture size than specified. In such cases, the Certification Agency shall record the reasons for reduction in the aperture size of the screen. Processed seed shall not have seed of the size lower than the bottom screen used beyond 5.0% (by weight).

Seed Treatment

When a variety, seed of which is under certification is susceptible to a seed borne disease organism or when seed under certification is carrying a seed borne pathogen and a seed treatment is available which may control the disease or pathogen when properly applied, the Certification Agency may require such seed to undergo such treatment before Certification. In case seed in required to be treated before sowing by the user, the chemical calculated at the recommended dose shall be kept in a plastic packet and placed inside the seed container with complete direction and precautions required for treating of the seed. The information about the treatment shall also be displayed on seed' containers. If the seeds have been treated, the following instructions shall also be complied with:

(*a*) a statement indicating that the seed has been treated;

(*b*) the commonly accepted chemical or abbreviated chemical name of the applied substance and

(*c*) if the substance of the chemical used for treatment and present with the seed is harmful to human beings or other vertebrate animals, a caution statement such as "Do not Use for Food Feed or Oil purposes".

Samples and Sampling of Seeds

Soon after completion of the seed processing or after seed treatment as the case may be, the Certification Agency shall draw a representative composite sample as per procedure specified in Seed Testing Manual. The quantity of seed samples so drawn shall be sufficient to provide three samples of the size of submitted sample. The composite sample will be divided into three equal parts, and one shall be sent for analysis to a notified Seed Testing Laboratory, the second part to the seed producer and retain the third part as a guard sample.

Seed Analysis Report

The Seed Testing Laboratory shall analyze the seed samples in accordance with the prescribed procedure and deliver the Seed Analysis Report to the Certification Agency as soon as may be, but not later than 30 days from the date of receipt of the samples unless the seed is subjected to such tests which require more than 30 days for completion of the test.

Validity Period of the Certificate

The validity period shall be nine months from the date of test at the time of initial certification. The validity period could be further extended for six months provided on retesting seed conforms to the prescribed standards in respect of physical purity, germination and insect damage for all seeds except vegetative propagating material for which lot shall be reexamined for seed standards specified for respective crop. A seed lot will be eligible for extension of the validity period as long as it conforms to the prescribed standards.

Revocation of Certificate

If the Certification Agency is satisfied, either on reference made to it in this behalf or otherwise that:

(*a*) the certificate granted by it under section 9(3) of the Act has been obtained by misrepresentation as to an essential fact; or

(*b*) the holder of the certificate has, without reasonable cause, failed to comply with the conditions subject to which the certificate has been granted or has contravened any of the provisions of the Act or the Rules made there under, then, without prejudice to any other penalty to which the holder of the certificate may be liable under the Act, the Certification Agency may, after giving the holder of the certificate an opportunity of showing cause revoke the certificate, under the provisions of section 10 of the Act.

Retention of Certification Records

The Certification Agency shall preserve in order all the documents including the guard samples pertaining to certification of each seed lot for two years from the date of grant/extension of the certificate and four years in respect of rejected seed crops or lots from the date of communication of rejection unless and otherwise required for longer period.

References

Abdul-Baki, A.A. and Anderson, J.D. 1973. Vigour determination in soyabean seed by multiple criteria. Crop Science, 13: 630-633.

Bhatnagar, S.P and Johri, B.M. 1972. In, Seeds Biology (Kozlowski, T.T. *ed*) Vol.1, p.77. New York: Academic Press.

Martin, G.C., Iona, M., Mason, R and Forde, H.I. 1969. Changes in endogenous growth substances in the embryo of *Juglans regia* during stratification. J. Amer. Soc. Hort. Sci., 94: 13-17.

Perry, D.A. 1984a. Factors influencing the establishment of cereal crops. Aspects of Applied Biology, 7: 65-83.

Perry, D.D. 1984b. Commentary in ISTA Vigour test committee collaborative trials. Seed Science and Technology, 12: 301-308.

Ross, J.D. and Bradbeer, J.W. 1968. Concentration of gibberellins in Chilled Hazel Seeds. Nature, London, 220: 85.

Singh, R. S. and Bhargava, R. 2009. Effect of seed treatment on germination and growth behavior of date palm (*Phoenix* spp.) under hot arid conditions. Jour. Trop. Forestry, 25: (1) 42-48.

Singh, R.S., Tewari, J. C. and Sharma, S. K. 1993. Effect of different seed treatment on germination and growth behaviour in *Grewia tenax* Forsk. under nursery conditions. My Forest, 29 (2):151-154.

Varner, J.E. 1965. Seed development and germination. In, J. Bonner and J.E. Varner (*eds.*) Plant Biochemistry Academic Press, NY pp 763-792.

Watson, D.P. 1948. Structure of the testa and its relation to germination in the Papilionaceae tribes Trifolae and Loteae. Annals of Botany, 12: 385-409.

Chapter-3

Vegetative Propagation

R. Bhargava
Central Institute for Arid Horticulture, Bikaner-334 006

In global competitive agriculture market, the need of the hour is to provide uniform and quality agricultural produce which can compete in market and give remunerative returns. For this, the agriculture production needs to be addressed on more scientific grounds and full package of practices are to be adopted to produce quality produce for global market. Earlier, the orchards were established using seedling material where uniformity of the produce was not guaranteed. Thus, in recent years more emphasis is being paid on scientific establishment of orchards which requires true to type planting material from which uniform produce is expected. This true to type planting material can only be produced by vegetative methods. Thus, standardization of vegetative propagation in horticultural crops, particularly in perennial crops has gained much importance. Some references are available on propagation of plants which shows that vegetative propagation methods were also practiced in ancient time as mentioned in old treaties literature *viz.* Upvan vinod, Ain-e-akbari, etc.

In recent year's crop production, sexual and asexual methods of propagation complement to each other. The trees usually consist of usually two portions, joined together by budding or grafting, the scion having been multiplied vegetatively and stock frequently, though not always, grown from seed. Though, there are many fruit species, plantation crops and forestry trees, which are propagated in a way that rootstock is not used, due to the readiness with which they may be propagated from stem or root or specialized organs or their division. A large number of plants is now propagated by asexual or vegetative methods (Brison and Aldriance, 2000). Asexual propagation involves reproduction from vegetative parts of plants. In general, a large number of plants are propagated commercially through asexual or vegetative means. In asexual propagation, vegetative parts of the plants are used for multiplication. It is possible because all the living cells of a plant have a capacity to regenerate into a full plant under favorable environmental conditions. The plants produced through vegetative means are therefore genetically identical and similar to the mother plant, possessing all the characteristics of the mother plant (Sharma, 2002). Vegetative methods of propagation are important because they permit large-scale multiplication of a genetically identical individual plant. Vegetative methods of propagation have certain advantages and initiation over sexual (seed) method of propagation is given below:

Advantages of Vegetative Propagation

In some respects, different methods of vegetative propagation are advantageous over sexual methods of propagation as under:

1. Plants raised through vegetative methods are true-to-type, having genotype and phenotype similar to the mother plant.
2. Vegetatively propagated plants have uniform growth and give consistent yield of better fruit quality.
3. Vegetatively propagated plants are less vigorous and thus their management is easy and less costly than seedling plants.
4. It is the only suitable method of multiplication of plants, which produce seedless fruit and have not viable seeds. For example, banana, pineapple, fig and varieties of gape, guava and lemon. Thus, such plants can only be perpetuated through vegetative means.
5. Vegetatively propagated plants have less juvenility than that of seedling trees and thus they come into bearing at early age that the seedlings plants.
6. With vegetative means, a large plant is obtained within a short period.
7. The tree size, fruit quality, and precocity in bearing can be easily regulated in vegetative propagation be exploiting the desirable effects of different rootstocks.
8. The desirable abiotic effects or rootstock on the scion cultivar can be utilized effectively budding or grafting them on desirable rootstock.

9. Self-incompatibility exists in a number of fruit plants like mango, aonla, apple, almond and cherry etc. Provision for adequate pollinizer in an established fruit orchard can be obtained by top working with the desirable pollinizer/cultivar.
10. Viral indexing is a useful technique for studying viruses in citrus, for which indicator plants are required. It can be done by grafting or budding the desirable scion on susceptible rootstock. For example, Kagzi lime can be used as a rootstock (indicator plant) for detecting the present of Tristeza virus in citrus.
11. Early assessment of hybrids/varieties can be done by grafting or budding the desirable hybrid/variety on a large established tree or on certain dwarfing rootstock.
12. Damaged part of the tree trunk or root can be repaired by bridge grafting or inarching.

Limitations in Vegetative Propagation

Besides advantages, certain limitations of vegetative methods of propagation are as given below:

1. Vegetative propagated plants are usually shortlived.
2. For raising plants through vegetative means, some technical expertise is required.
3. Because the plants raised through vegetative means don't show much variation and hence no new varieties can be evolved by vegetative means.
4. Vegetatively propagated plants are usually more susceptible to insect-pests diseases.
5. Vegetatively propagated plants are usually prone to suckering.
6. Vegetatively propagated plants are usually less hardy to abiotic stresses as compared the plants raised through seeds.
7. Vegetative propagation more expensive than the seed propagation.
8. Vegetative propagation is failure in some fruit plants and forestry trees in which no method of vegetative propagation has been found successful.

Methods of Vegetative Propagation

There are different methods, which can be used for commercial multiplication of various horticultural plants. These include cutting, layering (goottee), stooling, budding and grafting (Bose *et al.*, 1997). Because of various advantages, different vegetative methods are employed for large scale multiplication of plants.

Cutting: It is the method of propagating fruit plants in which the part of a plant (generally stem) having at least few buds, when detached from parent plant and placed under favorable conditions develop into a complete plant resembling in all characteristics to the parent from which it was taken. This method is commonly used in plants, which root easily and readily, thus, multiplication of plants is very quick and cheap. The fruit plants like *phalsa,* baramasi lemon, pomegranate, marula nut and grapes are commercially propagated by cuttings.

The hardwood cuttings are the common method of propagation, which are prepared from fully mature tissues. Round cuttings are preferred over angular immature cuttings. The shoots of about one year old or more can easily be used for preparing hardwood cuttings. In case of deciduous fruit plants such as grape, pomegranate, phalsa and fig the cuttings are made after pruning. While in evergreen fruit plants like baramasi lemon, the cuttings can be prepared during the spring (February - March) and rainy season (August - September). Generally the cuttings of 15-20 cm length and having 3-5 buds are made. The lower cut is given in a slanting manner just below the bud to increase the absorption of nutrients. The upper cut is given at a right angle to reduce the size of the wound and as far as possible away from the upper bud to avoid its drying. After the cuttings are prepared they should be allowed to dry. These cuttings are usually tied in small bundles (20-25 cuttings) and buried in moist soil/ sand for a certain period for healing of wounds, which is known as callusing. In some fruit and flowering plants, soft wood cuttings are also used for multiplication. Plant growth substances such as IBA, NAA, have used for root initiation in hard wood cuttings. In some plant species, rooting is very difficult which can be enhanced by treating cuttings with auxins (Hartman *et al.*,1993).

Layering

Layering is a method of vegetative propagation, in which roots are induced on the shoots while they are still attached to the mother plants. This is a alternate method of propagation in fruit plants which do not root easily when detached from the mother plants. Most commonly used methods of layering are air layering, ground and mound layering.

Air layering

As name refers, in this method the layering is done in air. To be more precise the rooting is done on the shoot itself when it is still attached to the mother plant. In this method one year old, healthy and straight shoot is selected and ring of bark measuring about 2.5 cm just below a bud is removed. Moist sphagnum moss is placed around this portion and is wrapped with a polythene strip. It is light in weight, and has a very high water holding capacity. If sphagnum moss is not available, any other material, which can retain moisture for long period of time, can be used for this purpose. The polythene covering does not allow the moisture to come out but permit gas exchange. Moreover the layers need not be watered afterwards, which saves a considerable labour. This method of layering is also known as goottee method. This is very common method for multiplication of litchi, pomegranate etc.

Air layering can be practiced during February–March and July-August in guava, litchi, sapota, loquat etc. After a few weeks the roots are developed which are visible through the polythene covering. Then a half way cut should be given to the rooted layers on the parent branch at least 15 days prior to their permanent removal from the mother plant. At the time of separation, a few leaves or small shoot is retained. It is also advisable to plant these rooted layers in nursery for close attention than to plant them directly in field. These layers can be planted in the fields during the following year in February or September–October.

Ground layering

In this method, a branch of plant, which is near the ground, is chosen and a ring of bark about 2.5 cm is diameter is removed just below the bud. This branch is then bended and buried in soil when still attached to the mother plant. The soil is regularly watered to keep it moist. Within a few weeks, the roots are formed and new plant is separated from the mother plant. Separation should be done in such a way that the roots formed also go with the detached plant. These new plants should preferably be planted in pots or nursery rows for development of better root system and shoot system before planting in the fields.

Mound layering

In this method, plant is headed back either in February or in July. The new shoots come out during April and September, from ground level. A ring of bark is removed from these shoots and they are covered with moist soil. The rooted stools of April stooling are separated during rainy season and those of August are removed in the following spring. These stools, after separating from the parent plant are planted in the nursery fields. This method is also known as stool layering and is used for propagation of guava and apple rootstocks.

Budding

Budding is a method in which only one bud is inserted in the rootstock. This method is very easy and fast. This method saves budwood as compared to grafting. As soon as the bark starts slipping both on the stock and scion, this is considered to be the optimum time for budding. This shows that the cambium, which is the tissue responsible for union, is active. This method is generally employed during spring and rainy season. The common methods of budding are T-budding, patch budding and chip budding.

T-budding

This is also known as shield budding. A horizontal cut about 1/3rd the distance around the stock is given on the stock 15-20 cm above the ground level. Another vertical cut 2-3 cm in length is made down from the middle of the horizontal cut and flaps of the bark are loosened with ivory end of the budding knife to receive the bud. After the 'T' has been made in the stock the bud is removed from the bud stick. To remove the shield of bark containing the bud, a slicing cut is started at a point on the bud stick about 1.25 cm below the bud, continuing underneath about 2.5 cm above the bud. A second horizontal cut is then made 1.25 to 2 cm above the bud, thus permitting the removal of the shield piece. The shield is removed along with a very thin slice of wood. The shield is then pushed under the two raised flaps of bark until its upper horizontal cut matches the same cut on the stock. The shield should fix properly in place, well covered by the two flaps of bark, but the bud itself exposed. The bud union should be wrapped with polythene strip to hold the two components firmly together until the union is completed. T budding can be performed at any time of the year provided cell sap flows freely. In *in-situ* orchard establishment of ber, I- budding was found successful. However, July to September months was suitable for budding in the field under arid conditions (Vishal Nath *et al.*, 2000).

Patch budding

This type of budding is quite successful in guava and it gives 60 to 70 percent successes during May and June. Freshly cut angular bud-wood from current season's growth should be used as scion. A rectangular or square patch or piece of bark about 1.0-1.5 cm broad and 2.5 cm long is removed from the rootstock at about 15 to 20 cm from ground level. A similar patch with a bud on it is removed from the bud stick taking care not to split the bark beneath the bud. This patch is then transferred to rootstock and fixed smoothly at its new position and tied immediately with polythene strip. To have better success, a patch having two buds is used as scion instead of a single bud. This method is termed as improved patch budding method.

Chip budding

This method is usually employed when the stock and scion are still dormant, that is just before the start of new growth. In this method, one about 2.5 cm long slanting cut is given into the stock followed by another cut at lower end of this first cut, in such a way that a chip of bark is removed from the stock. The bud from the scion wood is removed in the same way so that it matches the cuts given in the rootstock. This chip with a bud on it is fitted smoothly into the cut made in the rootstock taking care that the cambium layers of the stock and scion unite at least on one side. The bud is then tied and wrapped with polythene strip, to prevent drying up of the bud.

Grafting

Grafting is another method of vegetative propagation, where two plant parts are joined together in such a manner that they unite and continue their growth as one plant. In this method, the scion twig has more than two buds on it. Grafting is commonly done in pear, peach, plum, almond, mango etc. In temperate fruits like peach, plum and almond grafting is done when the plants are dormant while, in mango it is done when the trees are in active growth. The different methods of grafting are tongue grafting, cleft grafting, approach grafting, side grafting and veneer grafting.

Tongue grafting

This method is commonly used when the stock and scion are of equal diameter. First, a long, smooth, slanting cut of about 4 to 5 cm long is made on the rootstock. Another downward cut is given starting approximately 1/3rd from the top and about centimeter in length. Similar cuts are made in the scion wood exactly matching the cut given in the rootstock. The scion having 2 to 3 buds is then tightly fitted with the rootstock taking care that the cambium layer of at least one side of the stock and scion unites together. This is then wrapped with polythene strip.

Cleft grafting

This is also known as wedge grafting. This method is useful in the nursery where the rootstock is quite thicker than scion and tongue grafting cannot be employed successfully. The stock up to 8 cm in thickness can be grafted with this method. The rootstock to be grafted is cut smoothly with a secateur or saw. It is then split in the middle down to about 4 cm. The bud stick having 3 to 4 buds is trimmed like a wedge at the lower end with outer side slightly broader than the inner side. The lower bud on the scion should be located just well in to the stock making sure that the cambium layers of both the stock and scion are perfectly matched. Cleft grafting is done during dormant period.

Approach grafting

This method of grafting is termed approach grafting, as the rootstock is approached to the scion, while it is still attached to the mother plant. Alternatively, the mother plants are trained to be low headed and the stock is sown under their canopy. Last week of July or the first week of August is the best period for approach grafting. In this method the diameter of rootstock and scion should be approximately the same. A slice of bark along with a thin piece of wood about 4 cm long is removed from matching portions of both the stock and the scion. They are then brought together making sure that their cambium layers make contact at least on one side. These grafts are then tied firmly with polythene strip or any other tying material. The stock and scion plants are watered regularly to hasten the union. The union is complete in about 2 to 3 months.

Side grafting

A three sided rectangular cut about 4 x 1.25 cm is made on the rootstock at a height of about 15-20 cm from the ground level and the bark of the demarcated portion is lifted away from the rootstock. A matching cut is also made on the base of the scion to expose cambium. The scion should be prepared well before the actual grafting is done. The healthy scion shoots from the last mature flush are selected for this purpose. The selected scion shoots should have plump terminal buds. After the selection of the scion shoots, remove the leaf blades, leaving petioles intact. In about 7 to 10 days the petioles shall drop and terminal buds become swollen. At this stage the scion stick should be detached from the mother tree and grafted on the stock. The prepared scion is inserted under the bark flap of the rootstock so that the exposed cambia of the two components are in close contact with each other. The bark flap of the rootstock is resorted in its position. The graft union is then tied firmly with polythene strip. After the completion of the grafting operation, a part of the top of rootstock is removed to encourage growth of the scion. When the scion has sprouted and its leaves turned green, the root stock portion above the graft union is cut away.

Veneer grafting

In this method, a shallow downward cut of about 4 cm long is given on the rootstock at a height of about 15-20 cm from the ground level. At the base of this cut, a second short downward and inward cut is made to join the first cut, so as to remove a piece of wood and bark. The scion is prepared exactly as in side grafting. The cuts on the rootstock and scion shoot should be of the same length and width so that the cambial layers of both components match each other. Then, the prepared scion is inserted into the rootstock and tied security with polythene strip. After the union is complete the stock is cut back, leaving time for doing veneer grafting. It is very common method for multiplication of mango.

D. Propagation through Specialized Organs

1. **Runners:** A runner is a specialized stem that develops from the axil of a leaf at the crown of a plant. It grows horizontally along the ground and forms a new plant at one of the nodes e.g. strawberry. The runner production is favoured by long day and high temperature. The daughter plants are separated and used as new planting material.

2. **Suckers:** A shoot arising on an old stem or underground part of the stem is known as suckers. In other words, a sucker is a shoot, which arises on a plant below the ground. These shoots, when separated from the mother plant and transplanted produce adventitious roots. The capacity of a plant to form suckers varies from plant to plant, variety to variety and is even climate dependent. The sucker formation is common in fruit plants like pear and banana. In banana, sword suckers

are commonly used for propagation of plants. In case of date palm, suckers/offshoots emerged from female palm of appropriate size and weight are used for new plantations.

3. **Slips:** Slips are shoots just arising below the crown but above the ground. Pineapple is commercially propagated through this method of propagation.

Special Techniques of Propagation

Apart from conventional methods of propagation such as sexual and asexual methods, there are special techniques of propagation, which have played an important role in enhancing efficiency plant production. One of such methods is *in vitro*, also commonly referred as micropropagation.

Micropropagation

Recently, the tissue culture technique *i.e* micropropagation has expanded their scope and potential on commercial scale. Micropropagation is suitable for the rapid and large-scale clonal multiplication of elite germplasm. The technique has been referred as micropropagation because the size of the tissue in culture is very minute as compared to conventional vegetative cutting or any other plant part. The meristem explant used for micropropagation is about 0.1-0.5 mm size having only one or two leaf primordia. Morel and Martin (1952) for the first time demonstrated that virus free plants can be obtained by culturing shoot meristems. Later on with the discovery of the hormonal control of organogenesis by Shoog and Miller (1957) and finding of most commonly used tissue culture media by Murashigue and Skoog (1962), the scope of micropropagation was further extended to vast range of plant species, including fruit and plantation crops. With the advancement in science and technology, micropropagation technique has also been standardized for many plants, and it, is now widely used for multiplication of many horticultural plants.

Advantages: Micropropagation has the following advantages over conventional propagation techniques:

1. Year around production of plants irrespective of seasonal constraints.
2. Small space is required to maintain and multiply large number of plants.
3. Small tissue is required as an explant, hence saves the scion wood to a great extent.

4. Speedy international exchange of germplasm, requiring minimum quarantine checks is possible.
5. Micropropagated plants exhibit vigorous growth, and higher yields.
6. Micropropagated plants are usually free from viruses.
7. Micropropagation is highly beneficial in dioecious fruit plant species (date palm and papaya), where large-scale production of female plants is possible.
8. It helps in reducing the breeding cycle.
9. Production of homozygous plants is possible under *in vitro* conditions.
10. It is highly beneficial in plants in which vegetative propagation is not possible or the propagation rate is very slow (papaya and date palm).
11. *In vitro* systems have the potential for long-term transportation or shipment of propagation material.

Besides above advantages, Micropropagation technique has some limitations such as the facilities required for tissue culture is very costly as highly skilled persons are required. But year round plants can be produced and hardened planted can be used for plantation in the field.

Green Houses

To propagate plants around the year and to produce off season fruits and vegetables the green houses are used. The use of green houses is also essential to mass multiply the plants through tissue culture. The main cause of promoting optimum growth in green houses is high relative humidity and temperature control, adequate day length and light intensity. Good light conditions are essential for the sturdy growth of the seedlings. Experiments have shown that a greenhouse that runs from east to west is the best for better light penetration in winter and early spring. Greenhouses may be plastic polyethylene covered or made from fiberglass or other appropriate covering material. Modern greenhouses are well equipped with elaborate structures and have precise control on temperature, light intensity and humidity. The size and type of greenhouse, primarily depends upon the need of the plant propagator. Glass covered greenhouses are expensive but they have long life. However, for short-term benefits plastic covered greenhouses can also be made. Several types of sheets are available for their construction like UV stabilized polyethylene, polycarbonate and fiberglass. These are lightweight and inexpensive as compared to glass. Plastic covered greenhouses are lighter than the glass covered ones but there is very high increase in humidity in such houses, especially in winters which results undesirable water drops on the plants. It can, however be, overcome by making adequate ventilation (Whitehead, 1970).

Polyvinyl chloride (PVC) has also been used in construction of greenhouses but it tends to darken pre-maturely in sunlight. In India, construction of temporarily low cost poly houses is in fashion for raising nursery of fruit plant in off-season. Such low cost greenhouses are constructed either on wood or metal framework and are covered with Polyethylene sheet of 0.10 to 0.15mm thickness, which is resistant to ultra-violet rays. These houses are equipped with thermostat, cooler or an air conditioner or humidifier etc for rigid control on temperature and humidity. In greenhouses, two types of beds are used for growing seedlings. The beds may either be prepared on ground itself or raised beds or bench type beds are use.

References

Bose, T.K., Mitra S.K., Sadhu M.K. and Das, P. 1997. Propagation of Tropical and Sub tropical Horticultural Crops. IInd Edition, Naya Prakash, Calcutta.

Brison, F.R. and Aldriance, G.W. 2000. Propagation of Horticultural Plants. Mc Graw Hill Book Company. Inc, New York.

Hartman, H.T., Kester, D.E. and Davies, F.T. 1993. Plant propagation: Principles and Practices. 5th Edition. Prentice Hall of India, New Delhi..

Morel, G. and Martin, C. 1952. Guerison de dahlias attaints d'ume Maladie a virus. C. R. Academy. Sci. Paris, 235: 1324-1325.

Murashige, T. and Skoog, F. 1962. A revised medium for rapid growth and bioassay with tobacco tissue culture. Physiol. Plant, 15: 473-497.

Sharma, R.R. 2002. Propagation of Horticultural Crops: Principles and Practices. Kalyani Publishers, New Delhi.

Skoog, F. and Miller, C.O. 1957. Chemical regulation of growth and organ formation in plant tissue culture *in vitro*. Symp. Soc. Exp. Biol. 11: 118-131.

Vishal Nath, Saroj, P.L., Singh, R.S., Bhargava, R. and Pareek, O.P. 2000. *In situ* establishment of ber orchard under hot arid ecosystem of Rajasthan. Indian Jour.Hort. 57(1):21-26.

Whitehead, G.E. 1970. Grow fruit in your green house. Faber and Faber, London..

Chapter-4

Micro Propagation of Horticultural Crops

Hare Krishna
Central Institute for Arid Horticulture, Bikaner-334 006

Micropropagation refers to the production of plants from very small plant parts, tissues or cells, grown aseptically in a test tube or containers under controlled nutritional, environmental and aseptic conditions. It is also called as *in vitro* propagation, which implies clonal, true-to-type propagation of plants by a variety of tissue and cell culture methods and is considered as the most efficient and commercial method of plant propagation. Tissue culture or *in vitro* culture is two broadly used terms for micropropagation, which basically include aseptic culture of various plant parts. Micropropagation was developed as the result of accidental findings of French Professor G. Morel in 1958, when he was attempting to obtain virus-free plants through meristem culture in dahlia and orchids. All the biological principles of micropropagation technique are based on the phenomenon of totipotency of cell, in which a cell of a plant has the capacity to regenerate in to a full-fledged plant having different organs. German plant physiologist, Haberlandt (1902) is known as the father of tissue culture technique, who for the first time coined the term *totipotency*.

Commercial status in micropropagation has been achieved in over 500 species and genotypes including herbaceous species, foliage plants and fruit crops like banana, strawberry, papaya etc. and vegetables like potato, asparagus, etc. In fruit crops different woody species, such as rootstocks of temperate fruit crops like apple, pear, peach and plum is also under multiplication in some western countries. In India there are over 130 commercial tissue culture laboratories, which mostly multiply fruit crops like banana, papaya, strawberry, sugarcane, medicinal and aromatic plants, spices, floricultural crops etc. At present, India produces about 80 millions plants per year; however, the installed capacity of such laboratories is 300 million plants per annum, which accounts to about twenty five to thirty percent of the installed capacity. About seventy five per cent share is from the private sectors and joint ventures.

Advantages and Disadvantages of *in vitro* Propagation

In vitro propagation offers some significant advantages in quality, quantity and economics over vegetative and seed propagation for many plant species. However, is has certain disadvantages also that can't be ignored if maximum potential of this technology has to be taken.

Advantages

1. **Production of large propagating material in shorter time:** Micropropagation has the advantage of producing large number of planting material. Due to higher multiplication rate, millions of plants can rapidly be produced within a short time as compared to asexual and sexual methods of propagation.
2. **Production of disease free material:** Many diseases and the contamination of the source plant material are the limiting factors in conventional propagation methods. *In vitro* propagation techniques address these problems to minimize infection and produce disease free planting material in many horticultural crops.
3. **Production of true-to-type plant:** In micropropagation, there is gurantee of production of true-to-type propagation material having characters similar to mother plant. Now, methods have been developed and designed to provide the maximum number of true-to-type plants from a minimum number of selected mother plants.
4. **Ability to Air-Ship large quantity of planting material:** As many as 30,000-50,0000 micropropagated plants can be packed in to a cubic meter of shipping space. It has been found as the most efficient, quick and inexpensive method of transport of plants.
5. **Newly bred varieties/hybrids can be brought to markets quickly:** There is a great demand for the propagating material of the newly released varieties/hybrids, which can't be met

with traditional methods of propagation. However, it is quite easy to meet this demand by exploiting *in vitro* propagation techniques.

6. **Plant production throughout the year:** In conventional propagation, seeds or other propagation material is available at a particular time of the year. However, under *in vitro* techniques, the process of multiplication continues throughout the year because of controlled environmental conditions.

7. **Useful in propagation of dioecious plants:** Under field conditions, male and female plants of dioecious species (date palm, papaya etc.) are to planted in a definite ratio and in general, male are required in a very less proportion, whereas these are produced in large amount under natural conditions. In micropropagation, only female or male plants can be produced in desired number by taking explant from the wanted ones.

8. **Production of secondary metabolites:** In tissue culture systems, many secondary metabolites are produced, which are used for the production of medicines by pharmaceutical industries.

9. **Breeding cycle is reduced:** Using tissue culture techniques, brceding cycle of the crops is reduced.

10. **Production of seedless hybrids:** By the use of embryo rescue technique, it is possible to develop seedless varieties as in grape.

11. **Useful in hard-to propagate plant species:** *In vitro* propagation is highly useful for those plants which are hard-to-propagate by vegetative methods or their propagation rate is very slow as in datc palm, coconut ctc.

12. **Production of homozygous plants:** By using protoplast culture, it is possible to raise homozygous plants easily, which takes years together through back crossing method under conventional breeding programme.

13. **To obtain rare hybrids:** In many interspecific and intergeneric cross, fertilization occurs normally, but poor and abnormal development of endosperm causes pre-mature death of the hybrid embryos. The embryo culture technique is extensively used to rescue such hybrid embryos from wide crosses, which otherwise would abort.

14. **To overcome seed dormancy:** Dormancy in the seeds of deciduous plants can be reduced by growing excised embryos in the culture medium.
15. **Seed viability can be tested more rapidly:** The possibility of breaking seed dormancy by embro culture allows the use of this technique for testing the seed viability of a particular seeds more rapidly.

Disadvantages

1. **Contamination:** One of the greatest economic losses in commercial micropropagation is caused by endogenous and environment induced contamination of plant cultures. Contaminant once introduced in controlled environment, multiplies at a faster rate due to favourable environmental conditions. It is almost impossible to get rid of contaminant, if introduced in controlled conditions.
2. **Higher level of somatic variation:** In micropropagation, there are always chances of high level of variation in the plantlets. Variation may be due to excision of explant from diseased mother plant or incomplete monitoring of production methods and plant material during micropropagation.
3. **Poor establishment of the plantlets in field:** One of the major concerns in raising material by micropropagation has been the poor establishment of the propagating material in the field.
4. **High production costs:** Expensive technology and facilities, and the labour intensive nature of tissue culture, offer render micropropagation economically unfeasible to many propagators.
5. **Poor response of woody perennials:** It is very difficult to raise most of the woody perennials through tissue culture techniques because of exudation of very high amount of phenolics from the explants, which hinder the establishment in the culture media.
6. **Technical skill is required:** For raising plants through *in vitro* techniques, a technical expert is required, whereas for other methods of propagation, even illiterate peoples or farmers can raise plants.

Other Implications of Micropropagation

In addition to major use of tissue culture techniques for rapid clonal multiplication of plants, this technique is highly important for several purposes as under:

1. Production and maintenance of pathogen free stock plants.
2. Long-term *in vitro* conservation of germplasm.
3. Selection and regeneration of transgenic plants.
4. Conservation of pollen grains.

Principles of Micropropagation

In raising plants through micropropagation some fundamental principles are involved, which are discussed below;

1. Totipotency and plant regeneration

All *in vitro* methods of plant propagation rely on the phenomenon of totipotency of plant cells. Totipotency may be defined as the capacity of plant cell, tissue or organ to have the capacity to regenerate in to a full plant. This term was given a German Plant Physiologist, Haberlandt in 1902, who is considered as the father of tissue culture systems. This can be achieved by excising the explant and placing it on culture medium.

2. Axillary bud proliferation

It has been observed by many biotechnologists that a very large number of axillary shoot buds grow extensively when shoot tips, or even smaller apical meristems are cultured in an appropriate culture medium, containing higher concentration of cytokinins. Both the excision of the explant and the cytokinin rich medium activate the bud, leading to proliferation of many side shoots, which can later be separated for further culture.

3. Organogenesis: development of organs

Organogenesis from explants results in the formation of both shoot and roots. Formation of these organs takes place simultaneously, but either shoots or roots are formed first. However, in general, shoots are formed first than the roots.

4. Somatic embryogenesis

Somatic embryogenesis is different from organogenesis in that the regeneration and organization are bipolar i.e. the shoot and root meristems in the explants are formed simultaneously from proembryonic masses.

Basic Requirements for Micropropagation

For successful *in* vitro multiplication, the following pre-requisites are required:

i) A well equipped laboratory with all required equipment (balance, autoclave, laminar flow, culture trolley, glass wares, chemicals etc) and facilities (media preparation room, washing and drying room, inoculation room and culture room).

ii) Asepsis i.e. absence of any microbial contamination.

iii) A suitable culture medium, which depend upon plant requirement. The common medium which are used are MS (Murashige and Skoog) medium, B5 or Gamborg medium, Woody plant medium etc.

iv) Controlled environment i.e. temperatures between 22° and 27°C, 60-80% humidity and light intensity of 20 to 200 μmol $m^{-2}s^{-1}$.

Stages of micropropagation

Micropropagation takes place in a series of morphogenetic processes, i.e. culture initiation to acclimatization of plants designated as Stage 0 to IV, which vary with plant species. The standardized technique is often referred to as a protocol in which the entire steps along with the media and cultural conditions are well defined and also reproducible when commercial production is practiced. Micropropagation takes place in the following stages;

1. Stage 0: This is a preparatory stage wherein the mother/stock plants are maintained under extreme hygienic conditions and prepared for establishing cultures.
2. Stage I: This called the culture initiation stage where any explant excised from the mother plant(s) is cultured *in vitro* to initiate healthy cultures.

3. Stage II: In this stage, shoot proliferation is achieved to produce multiple shoots, i.e. it is the most critical stage which decides the commercial fate of any protocol. The number and rate of multiple shoot proliferation vary with species and variety.

4. Stage III: During this stage, *in vitro* rooting of the regenerated shoots is achieved to produce complete plantlets.

5. Stage IV: It is the last stage where the *in vitro* plants are subjected to acclimatization treatment so that they survive better in the field conditions.

Hardening or Stage-IV of micropropagation is considered as the most critical stage of micropropagation as it is the beginning of the biochemical and physiological processes in plants, which are necessary for survival and making them functional. During hardening, the *in vitro* raised plantlets are subjected to *ex vitro* conditions and thus they resort to develop functional root system, increased absorption of water and mineral nutrients, increased photosynthesis, etc.

Owing to huge requirement of planting material Government of India through Department of Biotechnology has initiated the programme of establishing big sizes hardening facilities for the large-scale multiplication of forest trees and horticultural plants through tissue culture in the different agro climatic regions. Six regional hardening units have been established at:

1. Tata Energy and Resource Institute (TERI), Gual Pahari, Gurgaon
2. Jai Narayan Vyas University, Jodhpur
3. Haryana Council of S&T, Hisar, Haryana
4. G. B. Pant Institute of Himalayan Environment and Development, Almora
5. West Bengal Council of Science & Technology, Kolkata
6. Regional Research Laboratory, Jammu (J&K)
7. Tata Energy and Resource Institute (TERI), Guwahati.

Synthetic Seeds

To avoid the cumbersome hardening processes, plant propagators are now shifting toward alternative options such as use of synthetic seeds or synseeds, which need no hardening and can be planted even directly under field conditions. Synseeds are defined as artificially encapsulated somatic embryos, shoot buds, cell aggregates or any other tissue that can be used for sowing as seeds and posses the ability to convert into a plant under *in vitro* or *ex vitro* conditions (Fig. 1). The encapsulated seed also contains nutrients, growth regulators, pesticides, antibiotics etc. needed for the development of a complete plantlet from the enclosed somatic embryo (embryos developed from vegetative cells and tissues within *in vitro* system) or shoot bud.

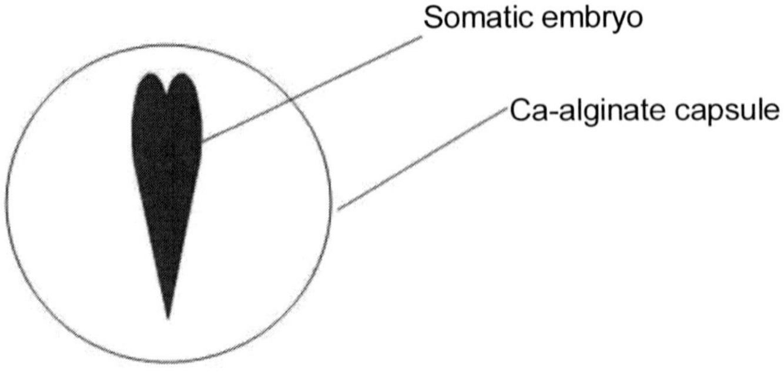

Fig. 1 : Synthetic seed containing a somatic embryo.

Technique of Synseed Production

Shoot bud/ somatic embryos are allowed to disintegrate into individual seeds from the differentiated callus. These are then dropped in a solution of sodium alginate and added into the fresh medium along with somatic embryos/shoot buds. With the help of micropipette, individual somatic embryo/shoot bud is dropped into calcium chloride solution. The outer layer of each drop immediately forms biodegradable layer of calcium alginate. Later, these synseeds can be used for multiplication like any other sexually propagated seeds. This technique has been standardized in many horticultural crops including fruits, vegetables, ornamentals and aromatic and medicinal plants (Table 1).

Advantages

- Combined advantages of clonal and seed propagation
- Introduction of agricultural chemicals and microorganisms
- Yield uniformity
- Advantages mainly with high-valued germplasm conservation and exchange
- Elimination of rooting and hardening phase

Table 1 : Horticultural crops in which synseed technology has been standardized.

Plant	Propagules used
Fruits and plantation crops	
Apple	Shoot apices
Banana	Shoot apices
Tea	Somatic embryos
Coffee	Somatic embryos
Mulberry	Axillary buds
Mango	Somatic embryos
Vegetable crops	
Potato	Somatic embryos
Carrot	Somatic embryos
Tomato	Axillary buds
Brinjal	Somatic embryos
Ornamental crops	
Cymbidium pendulum	Protocorms
Cymbidium giganteum	Protocorms
Geranium	Somatic embryos
Lilium	Somatic embryos
Geodorum densiflorum	Protocorm like bodies
Spices, medicinal and aromatic plants	
Black pepper	Shoot buds
Cardamom	Somatic embryos
Cinnamon	Shoot buds
Basil	Nodal segments
Vanilla	Shoot buds

Major Micropropagation Systems for Fruit Crops

Based on the type of explant, the different *in vitro* techniques have been employed in micropropagation of different plant species. A few of them are briefly discussed hereunder:

Meristem tip culture: This technique is widely used in horticultural plants like potato, dahlia, carnation and orchids. In this method, the meristem tip consisting of one or two pairs of leaf primordia are cultured in a medium. After a few weeks, the plantlets are re-generated and after hardening of the plantlets, these are transplanted in the soil under natural environmental conditions. Meristem tip cultured plants give rise to polyploid plants instead of diploid plants. Moreover, meristem tip culture is very useful for the elimination of viruses from infected plant material. Rapid multiplication of the plants, which are otherwise not easily propagated by vegetative means, is also possible through meristem culture. Plants produced are free from pathogens and can be stored for longer period and in smaller space.

Embryo culture: Aseptic culture of embryo has been standardized in many plant species. The embryo is rescued while it is immature and before the degeneration is initiated. The excised embryo is cultured on the semi solid or solid culture medium as the case may be. However, in embryo culture, the addition of some plant extracts like coconut milk or yeast extract in the culture medium enhances growth and development of the embryos. Embryo techniques are primarily useful in case of making distant crosses in hybrid production. It is also useful technique to overcome seed dormancy. Many embryological problems can be studied through embryo culture. In grape, embryo rescue technique has been used world wide to produce seedless varieties by excising immature embryo in crosses between Seedless x Seedless, which is otherwise not feasible through conventional propagation or breeding techniques.

Callus culture: A callus is an unorganized mass of parenchymatous cells. Callus culture has been induced artificially in a wide range of plant species using some explants. Callus formation is basically a natural phenomenon and is essential in the healing of wounds, graft and bud union and rooting of cuttings. In callus culture, the stimulation of cell division and multiplication is required and thus culture medium is accordingly supplemented with necessary growth regulators. So, for callus culture, auxins are required frequently followed by cytokinins. In callus culture, organogenesis occurs in two stages. In first stage, formation of meristems takes place and in the second stage, active growth of stem buds and roots takes place. As the nutritional requirement of these two stages is different, transferring of first stage to the second stage with culture medium having different supplementations (especially hormones) is a must.

Anther culture: Anther culture is used to produce haploid plants. Haploid plants are very useful in genetical and biochemical studies and for producing homozygous lines (within six months) of F_1 hybrids which, otherwise may take six generations or more with conventional methods of breeding. Guha and Maheshwari reported the first anther culture in Datura in 1959. In this technique, anthers are excised and cultured on to required and standardized culture medium. The stage, at which the anther is excised, depends on different species. However, in most cases, it is excised at or near microscopic mitosis, when the pollen grains passes from uninucleate to the binucleate condition. As a result of cell division, multicellular masses are developed, which may develop in to embroids or callus. Plantlets are developed from callus by organogenesis. Haploid plants are treated with colchicine for chromosome doubling and for producing the true breeding diploid plants.

Cell culture: Cell suspension cultures are obtained by homogenizing a piece of callus into liquid medium and shaking with shaker until medium becomes cloudy with suspended cells. Due to recent advances in plant Biotechnology, it is now possible to derive full plant from a single cell. This technique is primarily useful to plant breeders because now it is possible to induce desirable variability in individual cell, which can be regenerated into a full plant.

Protoplast culture: In protoplast culture, the rigid cell walls are first removed either mechanically or enzymatically so as to expose the protoplast. The exposed protoplast is then cultured onto a suitable culture medium. Individual protoplast of the plants has been regenerated into full plant as in Nicotiana, Brassica and Aparagus etc. In plant breeding systems, protoplast culture has many advantages i). A large homogenous population (10^6 to 10^7) of plants can be obtained from a small sample (1 g leaf), ii). Fusion of two protoplasts of two different plant species is possible, which may bring a greater variability in plants.

Ovule culture: In this method, unfertilized ovules are excised and cultured into medium. This technique has potential application to produce hybrid seeds in wide genetic crosses, which is rather difficult through conventional means. It is also efficient method to obtain plants in self-incompatible species. The technique of growing ovules in sterile culture is not a difficult job. However, one should have sound knowledge of the flowering and fertilization timing for isolation of the ovule at proper time.

Status of Micropropagation Industry in India

At present, an estimated 80 millions tissue cultured plants, from 130 tissue culture companies, are produced every year in the country. The major share of produced plants is occupied by banana (41%) followed by sugarcane (31%), ornamental plants (14%), spices (6%) and medicinal plants (4%). To bridge the gap between research and field testing, and to check the clonal fidelity in such planting material, the Department of Bio-Technology, Government of India, has set up pilot-scale facilities including Micropropagation Technology Parks (MTPs) at the The Tata Energy and Research Institute, New Delhi; National Chemical Laboratory, Pune, and National Facility for Virus Diagnostics and Quality Control of Tissue Culture Raised Plants (NFVDQCTP) at IARI, Pusa, New Delhi, with 5 satellite centers NCL, Pune; TERI, New Delhi; IHBT, Palampur; IIHR, Bangalore and SPIC, Chennai.

Commercial status in micropropagation has been achieved in over 500 species and genotypes including herbaceous species foliage plants and fruit crops like banana, strawberry, papaya etc. and vegetables like potato, asparagus, etc. In fruit crops different woody species, such as rootstocks of temperate fruit crops like apple, pear, peach and plum is also under multiplication in some western countries. In India there are over 130 commercial tissue laboratories, which mostly multiply fruit crops like banana, papaya, strawberry etc. At present, India produces about 80 millions plants per year; however, the installed capacity of such laboratories is 300 million plants per annum, which accounts to about twenty five to thirty percent of the installed capacity. About seventy five per cent share is from the private sectors and joint ventures.

Micropropagation techniques have been successfully adopted in several agricultural and horticultural crops. Among fruit crops, strawberry was the first fruit crop to be propagated commercially through micropropagation. Micropropagation techniques have been standardized for many temperate, tropical and sub-tropical fruit crops (Table 2). In India, tissue culture technique has also been perfected in banana, grape and papaya. In banana, shoot tip excised from rhizomes of sword suckers are suitable explants and MS medium suppleme nted with sucrose (3 per cent), gelite (0.25 per cent), is the best. Shoot tips and two nodal micro cuttings are highly suitable explants for faster and disease free production of grape plants through tissue culture. Salt tolerant rootstock of grape (Salt Creek and St. George) has great demand and *in vitro* propagation has been successfully used. Shoot tip culture technique has been demonstrated in papaya to produce female and male plants in the desired ratio. Similarly, pineapple is now being commercially propagated through micropropagation. In hard-to-root fruit species like, date palm and coconut, *in vitro* techniques have been perfected only recently. Scientists in India and different parts of the world have been trying hard for the micropropagation of mango and perennial forestry plants through tissue culture but the success has been limited due to recalcitrance and browning of explants.

Table 2 : Micropropagation technology developed for some fruit crops in India.

Sl. No.	Fruit crop	Institution responsible for the development of technology
1.	Almond	Punjab University, Chandigarh
2.	Annona	IIHR, Bangalore; BARC, Mumbai; NCL, Pune
3.	Apple	UHF, Solan
4.	Banana	IIHR, Bangalore; TNAU, Coimbatore; NCL, Pune
5.	Ber	Pune University, Pune
6.	Citrus	NBRI, Lucknow; NRC, Citrus, Nagpur; University of Jodhpur, Jodhpur
7.	Grape	IARI, New Delhi
8.	Guava	GBPUAT, Pantnagar
9.	Papaya	IARI, New Delhi
10.	Pear	Punjab University, Chandigarh
11.	Pineapple	BARC, Mumbai
12.	Pomegranate	NCL, Pune
13.	Strawberry	TERI, New Delhi
14.	Date palm	AAU, Anand

References

Bhojwani, S. S. and Bhatnagar, S.P. 2000. The embryology of angiosperm. 4th revised edition, Vikash Publishing House Pvt. Ltd., New Delhi-110 014.

Bose, T.K., Sanyal, D. and Sadhu, M.K. 1998. Propagation of Horticultural Crops. Naya Prokash Publishers, Calcutta.

Brison, F.R. and Adriance, G.W. 2000. Propagation of horticultural plants. Biotech Books, New Delhi.

Rao, P.S., Heble, M.R. and Chadha, M.S. 1980. Plant tissue culture, genetic manipulations and somatic technology of plant cells. BARC, Bombay.

Reinert, J. and Bajaj, Y.P.S. 1977. Application and Fundamental aspects of plant cell, tissue and organ culture. Springer-Verlag, Berlin, Germany.

Reinert, J. and Bajaj, Y.P.S. 1977. Plant cell, tissue and organ culture, Springer-Verlag, Berlin, London.

Sharma R.R. and Krishna Hare 2013. Textbook of Plant Propagation and Nursery Management. International Book Distributing Co., Lucknow.

Thorpe, T.A. 1981. Plant tissue culture : Methods and applications in agriculture. Academic Press, New York.

Vasil, I. K. 1980. Perspectives in plant cell and tissue culture. Intl. Rev. Cytol. Suppl., 11B, Academic Press, New York.

Chapter-5

Physiology of Rooting and Growth in Fruit Plants

Hare Krishna and S.K. Maheshwari
Central Institute for Arid Horticulture, Bikaner-334 006

Many horticultural and forestry plants don't come true-to-type when propagated by seed owing to cross pollination and highly heterogeneous nature of fruit crops, in general. As a result, when such plants are grown, the individuals differ in morphological as well as fruit characteristics as compared to mother plant. Though, it was a common practice in old days to establish new orchards with seedling plants, but now with the development and standardization of different vegetative propagation techniques, such plantings are avoided. In modern system of crop production, instead of a single tree derived from seed, use of composite plants (consist of usually scion and rootstock, joined together by budding or grating) is in vogue. This could have been possible by growing interest of mankind in vegetative means of plant propagation. In asexual propagation, vegetative parts of the plant are used for propagation. It is practicable because all the living cells of a plant have a capacity to regenerate into a full plant under favourable environmental conditions. The plants produced through vegetative means are therefore genetically identical and similar to the mother plant, possessing all the characteristics of the mother plant. Vegetative methods of propagation are important because they permit large-scale multiplication of a genetically identical individual plant. Principally, plant can be asexually reproduced either by using part of two or more plants in a union (grafting) or parts of the same plant (rooting).

1. Rooting of Cuttings

Adventitious root formation is a key step in vegetative propagation of woody or horticultural species, and it is a complex process known to be affected by multiple factors. Roots, which arise in the plant from sites other than their normal sites in the embryo or the primary roots, are defined as adventitious, as are roots that arise on explants from stem or leaf tissues cultured *in vitro*. In propagation by cutting, a portion of root, stem or leaf is detached from the mother plants and several chemical, environmental, mechanical treatments are given to induce the formation of root and shoot. A cutting, when put in suitable propagation medium becomes an individual plant in its own root, similar to the mother plant. A key stage in adventitious rooting is the formation of root meristemoids (a clump of small cells growing through synchronous mitotic divisions). Histologically, the formation of adventitious roots within tissues, which usually do not produce roots, requires that a cell, or group of cells to begin some developmental process (such as competence, determination and meristemoid formation) to form roots. In general, the rooting competence of some cell types remains unexpressed in plants; however, it may be induced by culturing them *in vitro*. This could be attributed to wounding reaction, as such or in combination with the action of specific growth promoters/hormones, for instance, auxin. In genera such as citron, preformed adventitious root primordia may already be present in plants. These preformed primordia normally lie dormant until the stem is made into cuttings and placed under conditions favourable for emergence as adventitious roots. Competence for root formation is defined as the ability of specific cells in a tissue to respond to specific root-inducing stimuli for example auxin. Once competent cells have been exposed to a root inducer, they may become determined for root formation. These root meristemoid later develops into root primordia. Subsequent primordium development leads to root emergence.

1.1. Anatomical basis of rooting of cuttings

The roots to develop (adventitious roots), either in stem, root or leaf cuttings, have internal origin in parent structures of most plant species. However, in some species (e.g. *Tamarix*) they may develop exogenously and then connect to the internal tissues of the stem. In the process of rooting of cuttings, generally four anatomical changes are observed. These are i) Formation of place for root initials by de-differentiation of specific mature cells, ii) Formation of root initials from certain cells near vascular bundles, iii) Development of root primordial and formation of vascular connection between root primordia and conducting tissues of cuttings and emerging through the cortex and epidermis, iv) Emergence of roots outside the cuttings. The herbaceous stem has four major areas i) a large pith in centre ii) a ring of vascular bundles outside the pith, iii) a cortex outside vascular bundles and iv) a thin epidermis. In herbaceous plants, adventitious roots are formed just outside and between the vascular bundles, though it may vary depending upon kind of plant. e.g. roots arise in phloem parenchyma region in tomato, pumpkin and from epidermis in crassula and roots arise from vascular bundles in castor. In woody plants, one or more layers of xylem and phloem are present and roots are formed in stem cuttings from living parenchyma cells. Earlier it was thought that callus formation is necessary for rooting of cuttings but it has been established that callusing and root initiation are two independent phenomenons and can occur simultaneously. The excess callus formation may hinder root initiation in some species, while it is a precursor for root initiation in others.

1.1.1. Stem structure

The development of continuous sclerenchyma ring between phloem and cortex of the stem is generally considered as anatomical barrier to rooting as can be seen in olive. Presence of lignified tissue in stem may act as mechanical barrier to rooting. In some species, some structures within the stem favour initiation of root primordia than others. For example, Citron (*C. medica*) produces root profusely from preformed root initials within short time compared to sour orange (*C. aurantium*) which form only few roots after several weeks.

1.1.2. Preformed root initials

Adventitious root initials are developed in the intact stem of certain woody plants even before cuttings are made. These root initials develop to some extent, become dormant and get embedded within the stem. Such dormant root initials are called as preformed or latent root initials. When the cuttings are made from such stems are placed in favourable environmental conditions, the dormant or latent root initials become active and roots are developed rapidly and easily. Occurrence of root initials is quite common in willow, hydrangea, poplar, jasmine, currant, citron and old plants apple, cherry, and quince.

2. Physiological Basis of Rooting of Cuttings

2.1. Hormones

The first naturally occurring auxin, Indole-3-acetic acid (IAA) was identified in 1934 to induce root formation in cuttings. Later, it was found that indole butyric acid (IBA) and Naphthalene acetic acid (NAA) are more effective in inducing rooting in cuttings. Cytokinin group of hormones is responsible for growth and differentiation of cells. Cuttings having high cytokinin levels are difficult to root. Similarly, if cytokinins are applied artificially, root initiation process is inhibited. Gibberellins and ethylene at low concentration promote root initiation and *vice versa.*

2.2. Vitamins

Endogenous optimum level of vitamins or exogenous application of vitamins gives beneficial effects in rooting initiation in cuttings. Thiamine, Pyridoxine, Niacin and biotin vitamins, if applied with some auxin (indoleacetic acid), stimulate root initiation.

2.3. Endogenous rooting inhibitors

Cuttings of some plants may not root mainly because of presence of natural occurring inhibitors. For example, the cuttings of *Vitis berlandieri* do not root easily as compared to *V. vinifera* due to presence of higher amount of inhibitors. The leaching of inhibitors by placing the cuttings in water, before planting, enhances the root initiation and subsequent growth and development.

2.4. Presence of buds and leaves

The presence of leaves on the cutting exerts a strong stimulating influence on the root initiation process in cuttings. The leaves produce carbohydrates, besides leaves and buds are powerful auxin producers, which are undoubtedly translocated downward to the base of stem, where it promotes root formation.

2.5. Rooting co-factors

Rooting co-factors are natural occurring substances that appear to act synergistically with auxins, primarily indoleacetic acid (IAA), for root formation in cuttings. Bouillenne and Went were the first to name the root-forming factor of leaf, cotyledon and buds as rhizocaline in 1933. Later, a new hypothesis was proposed by Bouillenne and Bouillenne-Walrand in 1955, highlighting that rhizocaline is a complex of three components *viz.* i) a specific rooting factor, which is translocated from buds, leaves, characterized chemically as ortho-dihydroxy phenol, ii) a non specific factor (auxin) in low concentration, which is also translocated from leaves and the buds and iii) a specific enzyme, probably polyphenol oxidase, present in pericycle, phloem or cambium. According to this hypothetically scheme, ortho-hydroxyphenol reacts with auxin, giving rise to rhizocaline. The formation of rhizocaline leads to the reactions involved in the root initiation process (Figure 1). Another hypothesis was proposed by Hess (1968) indicating that a number of co-factors interact with IAA, resulting in IAA co-factor complex, which subsequently initiates rooting. Usually, the easily rooted plants have a large number of co-factors and in higher concentration than in difficult to root plants. One of these co-factors (No 4) consists of highly active substance, characterized as oxygenated terpenoids. Another factor, No 3 was identified as chlorogenic acid, Later, it was postulated that phenolic compound more especially catechol, reacts synergistically with indole acetic acid in root production. This oxidation of ortho-dihydroxy phenol is one of the first steps leading to root initiation process. Similarly, abscisic acid was also found to be associated as one of the co-factors in root initiation as it promotes root initiation. The action of phenolic compounds in root initiation is mainly in protecting the natural occurring auxin-indoleacetic acid from destruction by indole acid oxidase. Rooting co-factors have been reported to occur in many plant species. They have been found in *Hedera helix*, Crab C, MM_{106} and M_{26} apple rootstocks. These were also found in easy to root Old Hume pear cultivar but not in difficult to root Bartlett.

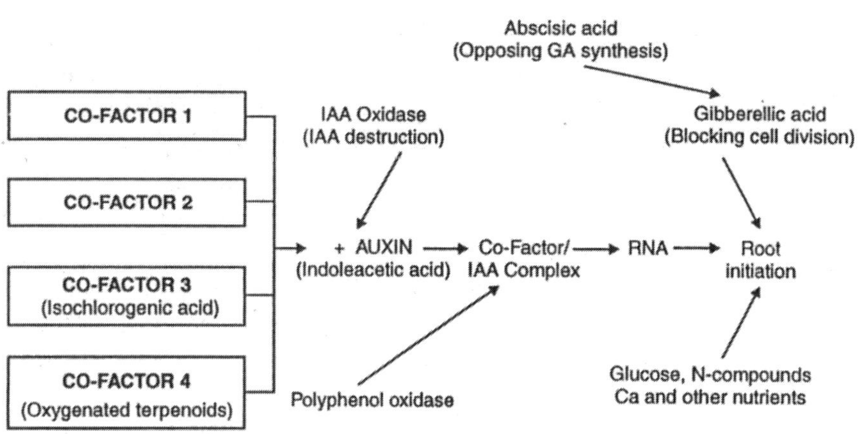

Fig.1 : Schematic representation of involvement of rooting co-factors in rooting process of cutting.

2.6. Nutritional factors

The nutrient status of plant from which cuttings are taken also plays vital role in root initiation process of the cuttings. However, among different, nitrogen and carbohydrate reserve is important. Low nitrogen and high carbohydrate reserve usually favour root initiation process compared to high nitrogen and low carbohydrate reserve. High N content usually favours luxuriant and hinder root initiation process but extreme N deficiency also hinders the rooting process. Thus, high C:N ratio should be maintained in the stock plant before taking cuttings from them.

3. Phases of Rooting

Four phases of rooting have been recognized, which are discussed as under;

i) a) **Physiological Inception:** During this phase, no visual change is noticed in cuttings. This entails internal physiological changes that are a prerequisite for the subsequent developmental changes. Probably, this phase involves hormonal changes, gene deregulation, etc. This is likely to be the stage of cuttings, where hormonal treatments (rooting powders) have their effect.

b) **De-differentiation:** At this stage, cytoplasm becomes dense and nucleus enlarges. Further partially differentiated cells, usually parenchyma, become less specialized and de-differentiate to a meristematic state, which is capable of cell division. Some researchers consider that at this stage, cells become competent to respond to the rhizogenic signal of auxin. Formation of starch grains can also be seen during this phase. Growth regulator like gibberellin is inhibitory to the process of de-differentiation.

ii) **Induction or Initiation:** The initiation last from the first cell division to formation of a primordia. By the action of auxin, some cells become determined to root formation at this stage. This begins with the first cell division, the formation of a group of unorganized and non-differentiated cells, followed by the development in to root primordia. The starch grain can be seen degraded, when observed under a microscope. Hormones like cytokinins and ethylene are inhibitory to induction of rooting.

iii) **Differentiation:** During differentiation, emerging primordia develops apical zonation and vascular differentiates. The emerging primordia take on the morphological characteristics of a root. Application of auxin at this stage has been found inhibitory.

iv) **Elongation:** The emerging root elongates through the parent plant body towards the exterior.

v) **Maturation:** The root emerges and continues growth and development characteristic of a typical root for that particular plant. For all intents and purposes, it behaves like a primary root apex.

4. Molecular Biology of Rooting

The different stages of root development are controlled and regulated by various phytohormones with auxin playing a major role. In roots, auxin is involved in lateral root formation, maintenance of apical dominance and adventitious root formation. All these developmental events require correct auxin transport and signaling. During embryonic RAM (root apical meristem) formation many auxin related mutants, such as those involved in its biosynthesis *yuc1 yuc4 yuc10 yuc11* quadruple mutant, or mutants affected in transport or signaling (monopteros, bodenlos, auxin transport inhibitor resistant 1 (*tir1*) and related *tir1/afb1–3* quadruple mutant are unable to form the embryonic RAM. The auxin flux coming from the apical region of the embryo into the hypophysis leads to *auxin transport inhibitor resistant*, 1TIR1 (and related redundant AFBs) pathway activation and induction of auxin-response genes such as PIN genes (coding for auxin efflux carriers), whose products will increase auxin transport and accumulation to further differentiate the hypophysis. In addition, over-expression of the DFL1/GH3-6 or the IAMT1 genes, which encode enzymes modulating free IAA levels, results in a reduction in lateral root formation. Auxin transport into the regions where lateral root initiate also seems crucial for the regulation of root branching. Functional analyses of mutants impaired in auxin transport, such as the *aux1, axr4*, and *pin* multiple mutants, have demonstrated the crucial role of auxin transport in lateral root formation. Studies using auxin transport inhibitors have shown that the polarity of auxin movement provides an important developmental signal during lateral root initiation and root gravitropism. The strong bias in the direction of auxin transport within a tissue results from the asymmetrical cellular localization of an influx system involving the AUX1 protein and an efflux apparatus with PIN-type efflux Facilitators. When auxin reaches the target tissue it induces a transcriptional response. Several auxin-responsive genes and gene families involved in auxin signalling have been identified, of which the Aux/IAAs and ARFs are the best studied. Auxin induces expression of Aux/IAA proteins, which in many cases reduces the sensitivity of cells toward auxin. This induction is mediated by ARF proteins that bind to the auxin responsive elements (AuxREs) in the promoters of auxin-responsive genes, and activate or repress transcription through interaction with specific Aux/ IAA proteins. These loops represent extensive feedback as well as feed-forward regulation. Alterations in these control systems affect several auxin-dependent plant pathways including root development. For example, normal differentiation of a root cap from the distal initials depends on a pair of redundantly acting auxin response factors, ARF10 and ARF16. In double mutants, the root cap initials over proliferate, thereby generating a mass of undifferentiated progeny. Another example are the gain-of-function solitary root (*slr*) mutants which carries a stabilizing mutation in the IAA14 negative regulator of auxin signaling and cannot form lateral root primordia. Detailed analyses of the *slr* mutant demonstrated that the *slr* mutation blocks early cell divisions during initiation.

Histological studies indicate that cytokinin inhibits the differentiation of primordia at an early stage in development. The cytokinins type-B response regulator PtRR13 (DDDKPtRR13) have been found to delay rooting phenotype and cause misregulation of CONTINUOUS VASCULAR RING1, a negative regulator of vascularization; PLEIOTROPIC DRUG RESISTANCE TRANSPORTER9, an auxin efflux transporter; and two APETALA2/ETHYLENE RESPONSE FACTOR genes with sequence similarity to TINY. Inappropriate cytokinin action *via* DDDKPtRR13 expression appeared to disrupt adventitious root development, when root founder cells are hypothesized to be sensitive to the negative effects of cytokinin. PtRR13 acts downstream of cytokinin to repress adventitious root formation in intact plants, and that reduced cytokinin signaling after shoot excision enables coordinated expression of ethylene, auxin, and vascularization pathways leading to adventitious root development. Besides, it has been found that ethylene also plays an important role in induction of rooting as evidenced by NR (*never ripe* mutant of tomato) cuttings produced fewer adventitious roots than wild-type cuttings. This suggests that the promotive effect of auxin on adventitious rooting is influenced by ethylene responsiveness. Seedling root growth of tomato in response to mechanical impedance was also influenced by ethylene sensitivity. Apart from ethlylene, another gas, nitric oxide (NO) also mediates the auxin response during adventitious root formation. NO induces the activation of soluble guanylate cyclase (GC) by binding to the heme group of the enzyme, which results in a transient increase of the cellular messenger cGMP. Results indicate that NO operates downstream of IAA promoting adventitious root development through the GC-catalyzed synthesis of cGMP. There are molecular evidences that auxin rapidly activates HO (heme oxygenase) activity and that the product of HO action, CO (carbon monoxide), then triggers the signal transduction events that lead to the auxin responses of adventitious root formation.

5. *Agrobacterium rhizogenes:* for Improved Rooting Ability

Many woody plants, economically important for fruit production, are often difficult to root, both in conventional and *in vitro* propagation. In some cases, it was possible to improve *in vitro* rooting with hormone application, etiolation etc. However, the difficulty of rooting is still one of the major obstacles to successful micropropagation. Researchers, now, aim at the production of genetically modified fruit rootstocks with an improved root system. For this purpose the introduction of the root loci, the ROL genes from *Agrobacterium rhizogenes* (a gram negative soil bacterium) has been chosen. These genes are well-known for their phenotypical effects in the hairy-root disease. The ability of *A. rhizogenes* to incite hairy root disease is determined by a virulence plasmid. *Agrobacterium rhizogenes* transfers a segment of its plasmid to the plant genome. The transferred DNA contains genes which are involved in the synthesis of plant hormones. These genes express in the plant cell and give rise to rooty-tumors at the infection site. Transgenic plants can be readily regenerated from the rooty-tumors and the transferred DNA is transmitted to progeny plants. High regeneration potential and sustained maintenance of transferred DNA makes the bacterium a suitable vector for plant genetic engineering. Four genes (ROLA, B, C and D) are involved in the observed changes in root formation and root morphology.

The *rolB* gene, derived from the *Ri* plasmid of *A. rhizogenes,* is effective in promoting rooting in different plant species through increased auxin sensitivity and/or changes in the auxin perception pathway. This gene was successfully used for transformation of apple rootstocks M26, M9/29, Jork 9, and pear rootstock BP10030 in order to increase their rooting abilities.

A plant nuclear gene specifically binding to domain *B* promoter of *rolB* has been identified, which is characterized by a single zinc finger domain of the *Dof* (DNA binding with one finger) type, and was called '*NtBBF1*' (*N. tabacum rolB* domain *B* factor 1). It has been shown that the pattern of expression of *NtBBF1* in tobacco plants is remarkably similar to that of *rolB*. The Dof protein *NtBBF1* causes the induction of *rolB* mediating the auxin-inducible expression of the gene. Auxin has been hypothesized to affect the activity and/or stability of *NtBBF1*. The protein RolB causes increased auxin sensitivity in the target cells. This increased sensitivity is the possible trigger of the meristematic condition (i.e., formation of indeterminate meristems). The specification of the meristem into a definite organ is affected by the local hormonal balance. Thus, for example, in the TCL system, auxin favours the formation of roots, cytokinin favours the formation of shoots, and auxin plus cytokinin (with a possible prevalent role of auxin, favour the formation of flower buds, but only in plants already determined to flower.

6. Factors Affecting Rooting of Cuttings

The ability of the rooting of cuttings in different plant species varies greatly and is influenced by various factors as under:

i. **Physiological condition of the stock plant:** It can exert a significant influence on rooting of the cuttings. In many plant species the cuttings taken in the early morning hours, when plants are in turgid condition, usually root better than the cuttings taken from water deficient plants. Cuttings taken from nutrient deficient plants rarely root better. Restriction of downward movement of carbohydrates, auxins and other possible root promoting factors, also result in better rooting of cuttings. Thus the girdling and etiolation of shoots prior to their use as cuttings is useful for stimulation of rooting.

ii. **Age of the plant:** The ability of cuttings to form adventitious roots, decreases with the increase in the age of the plant due to presence of more rooting inhibitors and decreased level of polyphenols in the older plants. In fruit plants like mango, apple, citrus, pear, peach, etc., the cuttings often root better if taken in juvenile period (pre-bearing period) and the cuttings taken from old, non bearing plants rarely root better. The juvenility is induced artificially in mature old plants by heading back, spraying GA_3, stooling, disbudding, etc.

iii. **Type of wood :** Cuttings taken from lateral shoots root better than those taken from terminal shoots because lateral shoots have more stored food, facilitating better rooting in the cuttings. In hard wood, rooting is more in basal portion of the shoot than on the middle or top portion because accumulation of food and root promoting substances are more in the basal parts. Further, cuttings taken from a vegetative shoot, root better than those taken from flowering shoot. Similarly, cuttings are made with a heel (a small piece of old wood) or mallet (a short segment), usually initiate rooting faster due to presence of preformed root initials in them.

iv. **Presence of leaves and buds:** The presence of leaves and buds is essential for the rooting of cuttings as they are the primary source of carbohydrate and auxin synthesis in plants which are transported to the basal portion of the cutting for root initiation. Usually, growing buds promote rooting and dormant buds inhibit it. However, in some cases, the leaves are removed to some extent to reduce the loss of water due to transpiration

v. **Position of the basal cut:** In some plants, the rooting is better when cut is made below the node, however in some rooting is much better if cut is made at the node.

vi. **Time of year:** In deciduous plants, the hard wood cuttings should be made in dormant season while in evergreen plants semi-hard wood or soft wood cuttings should be made in growing season or in spring or later fall, depending upon the species.

vii. **Treatment of cuttings**

a. **Growth regulators:** Among growth regulators, indole butyric acetic acid (IBA) has been found to be the most ideal compound for promoting rooting in cutting in most of the plant species. The concentration varies from plant to plant and type of the cuttings used. Naphthalene acetic acid (NAA) and 2,4-D (2,4-dichlorophenoxy acetic acid) are other root promoting auxins with varying degree of success.

b. **Mineral nutrients:** Treatment of cuttings with nitrogenous compounds (organic and inorganic) usually promotes rooting in cuttings. Boron also stimulates the root production in cuttings. The combination of nutrients with auxins is most effective.

c. **Fungicides:** Treatment of cuttings with fungicides like captan and benomyl, oftenly gives better results. The fungicides can also be mixed with IBA before treatment.

d. **Wounding:** Basal wounding of the cuttings is helpful for inducing rooting in cuttings as it promotes better absorption of water and growth substances from the rooting medium. Wounded portion also releases ethylene, which is known as rooting promoter in cuttings.

e. **Association with mycorrhiza:** The addition of mycorrhizal fungi into rooting substrate during cutting propagation can increase rooting. This has been observed in fruit crops like blue berry, huckle berry, grape, pomegranate etc.; besides, several ornamentals. Therefore, addition of AM fungal inoculum to rooting substrate could be an effective strategy for the nursery production of mycorrhizal plants.

viii. **Environmental conditions:** Environmental conditions during rooting, like water relations, light, temperature and rooting medium, may also affect the rootings in cuttings. For better rooting in the leaf cuttings, the turgidity of the leaves is to be maintained. If rooting medium is water deficient and the leaves continue transpiration, it results in loss of turgidity. A day temperature of 21-24^{0}C and a night temperature of 13-15^{0}C are considered quite satisfactory for rooting of cuttings. Excessively high or low temperature inhibits root initiation in cuttings. Light is the major source of energy in photosynthesis. In rooting cuttings, the formation of carbohydrates is most important. Light intensity, day length or light quality affect rooting of cuttings Usually, low light intensity gives better rooting in cutting than obtained from stock plants grown at high light intensity, thus etiolated shoots root better than non-etiolated shoots. Some plants manufacture good carbohydrate under short day conditions and others at long day or neutral day conditions and thus rooting in the cuttings is influenced. An ideal rooting medium must provide sufficient porosity to allow free passage of air, good water holding capacity, with capacity to easy draining and must be free from harmful pathogens. The pH of the rooting medium above and below 7 may be harmful in root initiation in cuttings of most of plant species. The level of exchangeable calcium should not be high. Similarly availability of oxygen (1-10ppm) is essential for root initiation.

7. Grafting

Grafting is an asexual propagation method in which parts of two different plants are joined so that they continue their growth as one plant. To accomplish this, one of the two parts serves as the bottom part, which is in contact with the soil and is called the rootstock and the other top part or scion. In this union, the plant material used as the scion is being propagated and is usually the only one allowed to grow. In effect the scion becomes the new shoot of the plant union and stock serve as the root, conducting nutrients across the graft junction and into the shoot. Many horticultural plants are propagated by grafting. In grafting, the desired cultivar can be raised on other plants for achieving the desirable benefits.

7.1. Graft union

The successful graft union is the basis of grafting. The formation of graft union is basically a process of healing of wounds. A graft is successful if healing is complete and vascular transport restored between the two parts (scion and rootstock). In this process, the parts of graft that are originally prepared and placed in close contact don't themselves move about or grow together. The union is accomplished entirely by cells that grow and develop after the actual grafting operation has been done. Further, in a graft union there is no intermingling of the cell contents of both stock and scion and cells maintain their own identity.

7.1.1. Formation of graft union

The main steps involved in the formation of graft union include, i) wounding response and the formation of callus tissue consisting of parenchymatous cells by the exposed cells of cambium of both the partners i.e. stock and scion, ii) inter locking of cells and callus cells of stock and scion, iii) differentiation of certain parenchymatous cells into new cambium cells, iv) formation of new vascular tissues by the new cambium for establishing a vascular connection for the translocation of water, nutrients and other metabolites.

7.1.2. Steps in graft union formation

i. **Wound response and callus formation**

In response to wounding (result of cuts made on scion and stock by knife) a necrotic layer 1 cell deep forms on both scion and stock, which is followed by the production of undifferentiated callus tissue from uninjured parenchyma cells below the necrotic layer. Thereafter, callus forms a wound periderm (outer "bark"), which becomes suberized to prevent entry of pathogens. Later, the necrotic layer dissolves.

ii. **Interlocking of parenchymatous cells**

Callus, which mostly comes from scion (due to basal movement of auxins and carbohydrates etc.) proliferates for 1-7 days followed by adhesion of scion and stock cells with a mix of pectins, carbohydrates and proteins, which is probably secreted by dictyosomes, a part of the Golgi bodies in cells. An exception to callus formation mostly taking from scion could be observed on established rootstock, which can develop more callus than the scion. Callus is mainly comprised of spongy tissue, which penetrates the thin necrotic layer and soon fills up the gap between the stock and the scion and become intimately interlocked. This interlocking of the cells provides a mechanical support to the graftage and also allows the passage of nutrients and water between the stock and the scion. Initially, there is development of a brown line between the callus of stock and scion but it disappears gradually. Similarly, the old vascular tissue of both the components is sealed off with a deposit of gum and thereby accelerates the healing process.

iii. **Production of new cambial cells**

In the callus bridge of the stock and the scion, there are certain parenchymatous cells, which differentiate into new cambial cells after 2-3 weeks of grafting and development of these cells result in the formation of continuous cambial connections between the stock and scion. This is good indication of formation of successful graftage.

iv. **Formation of new vascular tissues**

After the formation of continuous connection between the stock and the scion through the callus bridge, new cambial activity results in the formation of new xylem towards inside and phloem outside the original vascular cambium of the stock and the scion. The new xylem tissues generally originate from the activities of the scion tissues rather than from the stock. Thus, the production of new xylem and phloem tissues permits vascular connection between the stock and the scion, continuing throughout the life of a grafted plant. Ii is important that this stage should be completed before much new leaf development on scion otherwise the leaves may wilt and the scion may die. Though, some water can be translocated through callus cells but it is not enough to support leaves. The cell-to-cell transport takes place *via* plasmodesmata (symplastic transport), which links cells membranes. On the other hand, apoplastic transport occurs between adhering cells.

8. Molecular Biology of Vascular Development

The earliest event of vascular bundle formation is establishment of the procambium. Polar auxin transport (PAT), which is established by the polar localization of PIN proteins, is required to determine the zone of procambium formation. Both auxin and sterols modulate PAT and influence the earliest step of vascular tissue formation. The two conducting vascular tissues phloem and xylem are arranged in an organ-specific pattern. In the vascular bundles of the shoot, HD-ZIPIII genes (class III homeodomain-leucine-zipper-containing) and KANADI genes determine the abaxial and the adaxial position of phloem or xylem, respectively. In *phb phv rev* [PHABULOSA (PHB), REVOLUTA (REV), PHAVOLUTA (PHV)] triple mutants phloem surrounds xylem and in gain-of-function HD-ZIPIII mutants and kan1 kan2 kan3 triple mutants xylem surrounds phloem. Thus, HD-ZIPIII and KANADI genes antagonize each other. Further regulation of the HD-ZIPIII genes occurs via miR165/166, auxin, and brassinosteroids (BR). AtHB8 expression is stimulated by auxin. The vascular meristem gives rise to phloem and xylem cells. Auxin and cytokinins have been implicated in both specification and activity of the cambium. Also gibberellins (GA) are positive regulators of cambial activity. GA appears to promote PAT, as GA induces PIN1 expression in cambium. Furthermore, auxin stimulates GA biosynthesis genes and inhibits GA degradation genes. An as yet unknown flowering-related signal from the shoot promotes the late stage of cambial activity. The last phase in vascular tissue development is differentiation of the conductive tissues. To obtain the ring-like secondary wall pattern of the xylem at protoxylem position, cytokinin signaling has to be inhibited by AHP6 (histidine phospho transfer proteins). Xylogen controls xylem differentiation and is positively regulated by both auxin and cytokinin. VND6 and VND7 determine differentiation into meta- and protoxylem, respectively. Both transcription factors are positively regulated by auxin, cytokinin, and brassinosteroids.

9. Factors Affecting Healing of Graft Union

The formation of graft union is basically a process of wound healing. This process is influenced by many factors thereby affecting the success or failure of the graftage. These are as under:

i. **Kind of plant:** The ability of two kinds of plants to form a successful graft union is related largely to their natural relationship. Distantly related plants show incompatibility or very slow healing of graft union. The plants like apple, pear, mango, sapota etc can easily be grafted, whereas, walnut, hickories, oaks are difficult to form successful graft union.

ii. **Environmental conditions:** Optimum temperature, moisture and O_2 supply is necessary during and after grafting, for formation and healing of the graft union. Optimum conditions usually vary from plant to plant, For example optimum temperature for grafting of apple is 21-24^{0}C whereas, it is 25-30^{0}C for walnut. For better healing, high humidity is essential to avoid drying of graft union and better growth of the callus. Similarly, optimum supply of O_2 helps in rapid cell division, which is necessary for callus enlargement and proper moisture (irrigation) to the stock plants favours high graft success.

iii. **Growth activity of the stock and scion plant:** The condition of the stock and scion during grafting also plays an important role in healing process. If plants are in active growth, they heal soon. However, certain plants (walnut, maple and grape) bleed heavily when cuts are made and such grafts do not heal properly. In mango, defoliation of scion is done before grafting, for encouraging bud break.

iv. **Method of grafting:** Grafting technique also influences the healing process of a graft union and it varies in different plant species. A highly successful method of grafting for one plant species may be a complete failure for other species.

v. **Infestation by insect-pests and diseases:** Infection of bud wood / scion sticks by viruses, insect-pests (borer) and diseases may result in poor healing of graft union. Control of these pathogens promotes healing of graft union.

vi. **Application of plant bio-regulators and growth promoting microbes:** Plant bio-regulators such as auxin, cytokinins, albumin and ascorbic acid and plant growth promoting rhizobia (PGPR) have been found to promote graft-take in various plants like walnut, grape and sweet pepper. Foliar application of bio-regulators/ PGPR promotes/induce the initiation and proliferation of callus and new vascular tissue by promoting cell division and/or cell development at the cut surface of scion stem and improves the survival rate after grafting.

vii. **Polarity:** Maintenance of proper polarity is important for proper healing and successful bud or graft union. In budding and grafting, the proximal end of the scion should be inserted into the distal end of the stock (Fig. 2). However, in root grafting the proximal end of the scion should be inserted in to the proximal end of the root pieces. If the order is reversed i.e. if the scion is inserted upside down, the union may stay alive for some time but don't grow afterwards. Similarly, in nurse root grafting, the rootstock is knowingly grafted to scion with reverse polarity.

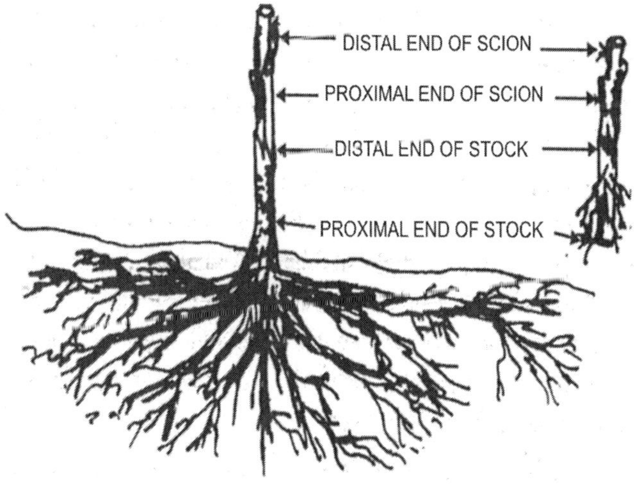

Fig. 2 : Polarity in grafting.

'Each generation will reap what the former generation has sown'.

- Chinese Proverb

References

Finch-Savage, W.E. and Gerhard L.M. 2006. Seed dormancy and the control of germination. New Phytologist, 171: 501–523.

Hilhorst, H.W.M. and Koornneef, M. 2007. Dormancy in Plants. Encyclopedia of Life Sciences. 1-4.

Kucera B. Cohn M.A., Leubner-Metzger G. 2005. Plant hormone interactions during seed dormancy release and germination. Seed Science Research 15: 281–307.

Nickell, L.G. 1982. Plant growth regulators: agricultural uses. Springer Verlag, New York.

Pandey, S. N. and Chadha, A. 2000. A text book of botany: Plant anatomy and economic botany, Vol. III, Vikas Publishing House, New Delhi-110 002.

Reddy, K. K. 1992. Vegetative propagation and biotechnology for tree improvement, Natraj Publisher, Dehradun, India.

Sharma R.R. and Krishna H. 2013. Textbook of Plant Propagation and Nursery Management. International Book Distributing Co., Lucknow, India (ISBN: 978-81-8189-198-3).

Sharma, R.R. 2002. Propagation of Horticultural Crops. Kalyani Publishers, Ludhiana, India.

Tewari, J. P. and Mishra, K.K. 2003. Nursery Production and Management of Sub-Tropical Fruits, Training Manual, Department of Horticulture, G.B.P.U.A.&T., Pantnagar, Uttarakhand.

Tilney-Basset, R.A. 1963. The structure of periclinal chimeras. Heredity, 18: 265-285.

Villiers, T.A. 1972. Seed dormancy. *In*: Seed Biology, Vol. 2, (T.T. Kozlowski, Ed), Academic Press, New York, pp.220-282.

Chapter-6

Role of Plant Growth Regulators in Propagation of Arid Fruit Plants

N.K.Gupta and Sunita Gupta
SK Rajasthan Agricultural University, Bikaner-334 006

The classical idea of the existence of a hormone as a chemical messenger was first reported by Darwin and Darwin (1880). However, Boysen-Jensen (1911) was first who actually demonstrated their prescence by a grafting experiment. Evidence exist to show that early mankinds often used these phytohormones for rooting of cuttings, flower induction, control of flower drop etc. The modern age of phytohormones began in 1926 when F.W.Went isolated an active substance from oat coleoptile tip which later on proved as IAA (Indole Acetic Acid) or auxin. A dramatic revolution in this field started after the discovery that a chemical different in structure to IAA could exhibit auxin like activity such as rooting of cuttings (Zimmermann and Wilcoxon,1935) and by 1940,the use of synthetic auxins was well established with the progress in this field, the first generation phytohormones have been replaced with superior product and new crops are being added to take the advantages of its beneficial effects.

Plant propagation is an integral part of horticultural development. It can be done sexually or asexually.The sexual reproduction or propagation means propagation of the individual plant through seed /seedling, which is obtained from fusion of two gametes i.e. male and female sex organs. Asexual propagation means the development of new individual plant through vegetative plant part, without function of sex-organs, which have compatibility to regenerate new individuals with same genetically true characters of parent materials. Both are equally important in horticulture development ,particularly under arid ecosystem as root stock s are mostly multiplied by seeds and improved varieties are perpetuated by vegetative means.It has been observed that a large number of seeds fail to germinate under traditional methods of raising the rootstocks. Physical treatments and certain growth regulators have been found helpful in germination by way of reducing moisture requirement of the tissue or alternatively an increase in water uptake by exerting an effect on membrane permeability (Looney,1998). Similarly ,the vegetative propagation has become more feasible with the development of knowledge about auxins,particularly IBA and IAA. The various methods of propagation which employ bioregulators in horticultural production include traditional methods of rooting of cuttings of shoots and newer techniques of tissue culture. While the first method of propagation has been in practice for last several decades, the tissue culture techniques in the production of horticultural crops have taken a lead and are emerging as an important tool in the rescent past. Initiation and or acceleration of rooting of cuttings, produce a large root system, increase the per cent rooting and make it possible to root cutting of some species that normally root poorly (Reighard *et al.*,1990; Gupta *et al.*, 2004). A large number of horticultural crops especially fruit crops are propagated by this method.

The most commonly used plant growth regulators in the rooting of cuttings are auxins, especially IBA, that has low degradability by auxin destroying enzyme. Other auxins of importance are NAA; 2,4-D; 2,4,5,-T and IAA. In a few cases, growth retardants such as SADH have also proved beneficial in root induction. GA has antagonistic effect on rooting of cuttings. There are many methods of applying auxins to stem cuttings, among which the quick dip,prolonged soaking and powder methods are widely used. In the quick deep method, a high concentration of auxin dissolved in a volatile solvent like ethyl alcohol is used for a quick treatment of cuttings for a few seconds. In the other methods, a low concentration of plant bioregulators is employed for effective and profuse rooting of cuttings.

Several fruit crops are being propagated by cuttings, however, cuttings of many tropical / subtropical fruit trees poorly responded to growth regulator treatments. Techniques of shooting and layering have been successful in the crops like apple, mango, guava,lemon, etc. both of these techniques have offer excellent results by auxin treatments. Auxin such as NAA and IAA have also been found promising in increasing the graft intake in fruit crops especially in grapes by stimulating cambium activity. Root initiation in cuttings depend on the the presence of relative concentration of auxin, cytokinin and cytokinin like substances. Application of plant growth substances enhances rooting in stem cuttings especially those who are hard to root. IBA is used commercially for rooting in many plant species including litchi, mango, guava, pomegranate, etc. Indole Butyric Acid (250-10,000 ppm) and NAA applied in lanoline paste are helpful in inducing good rooting in air-layering (Pandey, 2010).

Plant tissue culture or aseptic culture of plant cells and organs has gained considerable prominence in research and commercial applications. This method of propagation has a potential of being the rapid and cheaper method of producing the horticultural plants.The procedure for manipulation through tissue culture techniques rely on mass production of shoots, either adventitiously from explants or by enhanced auxillary branching of excised shoot tips and their rootings. All the success with tissue culture has been accomplished with nutrient media supplemented with nutrient suitable gowth regulators. The micropropagation shoot tip cultures of a large number of fruitscrops such as banana is done commercially.

Dormancy is observed in various plant parts such as seeds, buds and bulb. Dormancy of seed plants is a natural phenomenon for regulation of plant development. It limits the time at which seeds germinate and helps in adapting a species to seasonal vagaries and unfavourable environment. The studies on the levels endogenous hormones in dormant and non dormant seeds/buds and the responce obtained from their exogenous application indicate that dormancy is regulated by hormones. Incrase in endogenous growth promoter hormones. GA_3 and cytokinins and decrease in ABA at dormancy break and vice versa during the period in seed /bud led the hypotheis that inhibitor-promoter balance bdetermines the onset and termination of rest. Use of bioregulators for the control of dormancy has important applications in rapid plant propagation .

Seeds of many plant species often require an extended period of long days after ripening at low temperature before germination. Breaking of rest is of great significance in accelerating the breeding programme through rapid propagation. Since the discovery of role of GA in dormancy,rapid progress has been made in this direction. GA applications are now being made for accelerated seed germination in citrus, cherry, grapes, apple, and peach. The treatment with GA significantly reduced /substituted the period of seed stratification. Pre –sowing soaking of seeds in GA solution was found to stimulate the germination of trifoliate orange, sweet orange, cleoptra mandarin and acid lime. GA was found to be also effective in breaking seed dormancy in grapes, ber and in scarified seed of ber and anonna.

In nutshell the plant growth regulators may be used in fruit tree propagation to identify (1) superior genotypes in terms of production, quality and environmental responses and (2) stages and tissues component for an effective large scale propagation of elite material. The plant growth regulators are basically classified into five major groups i.e., auxins, gibberellins, cytokinins, ethylene and abscisic acid. In addition, certain chemicals-natural as well as synthetic also show hormonal effect to plants. In this review attempts have been made to summarise the effect of growth regulators on propagation of different fruit plants grown under arid and semi-arid conditions.

Ber (*Ziziphus mauritiana* Lamk.)

Ber is an important fruit crop of arid and semi arid regions of India. Germination of ber seeds has been found to be promoted by breaking hard shell by soaking the extacted seeds in chemical solutions like gibberellins (Krishnan and Kulshekhran,1984).Treatment of seed kernels by soaking in 0.1% thiourea enhanced the germination percentage (Murthy and Reddy,1990). Kulkarni *et al.* (1992) obtained over 45 per cent germination with seeds soaked in 100 ppm GA for 12 hours, with better seedling vigour in terms of root and shoot growth. Nema *et al.*(1994) observed that use of GA_3 300 ppm in ber seed promoted germination. Increased seedling height,stem diameter and survival percentage. These results further indicated that after 90 days of seed sowing, these seedlings attained satisfactory height and diameter for budding operation. Thus, grafts may be raised within a short period as compared to traditional methods(Nema *et al.*, 1994). The germination of freshly harvested ber seeds can also be improved by soaking in sulphuric acid for six minutes (Kajal,1983). Hore and Sen (1994) studied the role of pre sowing seed treatment on germination and seedling growth. Higher germination rates were obtained from seeds treated with GA_3 200 ppm. The incidence of polyembryony was highest with GA_3 400 ppm and cycocel 1000 ppm.GA treatment has been found to enhance shoot growth of seedlings.(Shamugavelu,1966).

In Chinese jujube,treatment of seeds without endocarps or of seeds with acid scarified endocarps with 400 ppm GA, interrupted dormancy and increased germination (Casini and Salvadori,1975;1980). Application of 10 ppm GA or benzyladenine induced germination in immature seeds of Chinese jujube (Kim and Kim,1983). Mortality of young plants at the time of transplanting is a major difficulty in establishing a ber orchard,

Singh *et al.*(2003) reported that treatment of the roots of defoliated budling with 200 ppm IBA increases the translanting success, number of leaves, budlling diameter just above the bud union and budling height (Table 1).

Table 1: Effect of various transplanting treatments on growth parameters of bare rooted budling of ber after 120 days .

Treatments	No,of leaves /plant	Budling diameter (cm)	Budling Diameter (cm)	Stem diameter (cm)	Transplanting success (cm)
Control defoliated budlings	50.20	0.76	50.85	1.00	93.50
Defoliated +200 ppm IBA	75.35	0.91	65.20	1.12	100.00
Defoliated	56.26	0.64	61.65	1.10	100.00
Defoliated+200 ppm IBA+12% Waxol	50.70	0.84	50.20	1.17	100.00
CD at 5%	9.51	0.09	5.59	NS	NS

(*Source :* Singh *et al.*, 2003)

Attempts have been made to propagate ber by stooling (Singh *et al.*,1973) and air layering (Chatterjee and Rao,1978; Ughreja and Chauhan, 1983). In stool beds, nearly 100 per cent rooting has been obtained within a month in juvenile shoots induced by heading back one month old ber plants,and application of either 800to 1200 ppm (Singh *et al.*,1973) or 1500 ppm IBA (Ughreja and Chauhan 1983) immediately after ringing at 10 cm height during the monsoon.In air layers also, nearly 100 per cent rooting has been obtained by treatment with 7500 ppm IBA along with 100 ppm boron (Misra and Jauhari,1970) or 1000 ppm IBA (Chatterjee and Rao,1978)

Attempts to propagate ber by cuttings have been unsuccessful even with the use of growth regulators (Misra and Jauhari,1970) although root initiation has been observed on treatment with 200 to 4000 ppm NAA (Jauhari,1960).Treatment of Chinese jujube cuttings with 50 ppm IBA for 18 hours (Tarasenko and Shcherbakova,1977) or 10 hours (Shcherbakova and Kulikov,1972) induce 88.8 and 96 per cent rootimg , respectively .Shen *et al.*(1992) obtained best results by aquick dip in 250 -500 ppm IBA or NAA or in a mixture of 250 ppm IBA and 500 ppm NAA.

Aonla (*Emblica officinalis* Garten.)

Aonla can be successfully cultivated in marginal Soils and various kinds of wastelands,particularly in salt affected soils.The poor establishment of budlings after transplanting and heavy mortality (20-50%) creates problem in expansion of area under aonla cultivation .This is due to the fact that aonla has strong tap root system which get disturbed during digging of plants from nursery.Seeds soaked in 500 ppm GA solution for 24 hhours quicken germination and increase root system .Singh and Pandey (2003) reported that the transplanting of leaves detached bare rooted budlings treated with 200ppm IBA gave maximum transplanting success and bud sprouting .Dipping roots in IBA futther improved the success and growth of aonla budlings (Table 2). Air layering with the use of Seradix and stooling with the use of 5000 ppm IBA+ ethrel gave encouraging results in aonla.The micro propagation studies in aonla showed that addition of Kinetin (0.4mg/litre)to MS medium produced the highest rate of shoot formation (37.5%) and number of leaves per explant(3.66).Microshoots in MS media supplemented with IAA and NAA exhibited callusing near the cut end, while paclobutrazol treated plants showed slight sweeling at the shoot bottom followed by splitting of outer tissue (Mishra *et al.,* 1999).

Table 2 : Effect of IBA on different parameters of plant growth in aonla.

Treatments	Percent establishment		Per cent sprouting		Shoot height (cm)		No.of branches	
	Leaves attached +200 ppm IBA	Leaves detached +200 ppm IBA	Leaves attached +200 ppm IBA	Leaves detached +200 ppm IBA	Leaves attached +200 ppm IBA	Leaves detached +200 ppm IBA	Leaves attached +200 ppm IBA	Leaves detached +200 ppm IBA
15th Dec	46.92	59.00	40.03	47.63	73.64	84.50	98.2	113.2
15th Jan	52.78	63.93	41.24	49.70	85.20	97.40	119.0	137.6
15th Feb	52.86	75.00	42.88	52.86	107.30	120.70	148.5	173.0
15th March	48.85	56.79	33.17	39.55	132.90	144.52	196.8	218.4
15th April	43.08	50.85	20.91	23.05	152.08	165.80	235.0	260.9
15th July	46.92	59.00	28.12	35.19	42.03	46.30	14.5	18.0
15th August	45.00	55.07	23.14	29.52	37.00	40.10	12.1	15.8
CD at 5%	Transplanting	4.48		1.69		4.11		5.54
	Transplanting time	3.39		2.24		5.43		7.33
	Interaction	NS		NS		NS		NS

(*Source :* Singh and Pandey, 2003)

Pomegranate (*Punica granatum* L.)

Pomegranate is a favourite fruit crop of dryland areas of the country. Its propagation by seeds is easy but true type plants are not obtained. Beside, the vegetatively propagated plants are precocious and high yielding. However, the rooting percentage of cutting is very low, particularly in some cultivars like Ganesh (Panwar *et al.*,2001). Etiolation and ringing of source shoot, treatments with auxin and non auxinic compounds ars found effective in root initiation of hard to root plant species. When pomegranate stem cuttings were placed in media containing moisture levels of 5.80, 11.85,15.95 or 20.14 per cent, the corresponding successful rooting were 39.23,67.21,56.79 and 90.0 per cent. Pre treatment with IBA at 50 or 100 ppm improved the rooting percentage only at 15.95 per cent moisture level although at increased the number and weight of primary roots at the the two higher moisture.

Purohit and Shekharappa (1985) reported that basal stem cutting with a diameter of 1.0 to 1.25 cm, treated with IBA at 5000 ppm showed the best rooting and shoot number. It is reported that IBA was more effective than NAA for inducing rooting in hard wood , semi hardwood and softwood stem cuttings of pomegranate. Treatment with IBA at 500 ppm resulted in maximum rooting (83.33%). Singh (1994) found that semi hard wood cuttings of pomegranate treated with IBA 1000ppm gave maximum success (51.9%) followed by Seradex -B (46%) as compared to control (26.9%).

Hore and Sen (1993) reported maximum rooting success in pomegranate cutting with PHB 1000 ppm + NAA 2500 ppm.These workers further reported that phenolic compounds were effective in promoting auxin action to success air layering that phenolic compounds were effective in promoting auxin action to success air layering of pomegranate (Hore and Sen,1997). It is now a days increasingly assumed that phenol auxin complex influence root formation through production of ethylene.In an another experimrnt, Panwar *et al.*(2001) took hard wood cutting of pomegranate var. Ganesh and dipped (basal end of cutting)for one minute in different concentrations of indole butyric acid before planting in Nursery. It was found that IBA 750 ppm produced maximum number of roots, average length of root and percentage of rooting (Table 3) Varma *et al.*(2005) recorded the maximum number of roots per cutting and percentage of rooted cuttings(49.69) and establishment of rooted cuttings(65.70%) due to 1000 ppm IBA.Among the different combinations of growth regulators IBA 2000 ppm+ NAA 500 ppm was found to be better than other combination studied .Saroj *et al.*(2005) studied the effect of various concentrations of IBA for rooting in pomegranate cuttings. Among them 2500 ppm IBA had given best responses in terms of rooting and number of roots per shoot in semi hardwood cutting regarding percentage rooting. The numbers of roots reduced with increasing concentrations of IBA. IBA 2500 ppm also induces fibrous root system,which is essential for better establishment of plants under field conditions.

Table 3 : Effect of IBA on rooting and shoot growth of Pomegranate hard wood cuttings cv. Ganesh.

IBA(ppm)	No.of roots/ cutting	Average length of roots(cm)	% of rooting	No.of shoots/ cutting	Av.length of shoot (cm)	No.of leaves / plant
250	10.6	23.9	65.2	4.2	47.1	351.7
500	14.5	27.7	70.6	5.9	51.4	439.6
750	15.4	27.7	73.3	5.6	59.0	469.8
1000	10.9	18.6	60.6	5.1	38.7	421.2
2000	7.3	16.4	58.8	4.3	32.2	397.9
Control	5.8	9.0	43.6	3.2	21.0	104.2
CD at 5%	1.29	2.12	3.87	1.41	2.16	4.52

(*Source:* Panwar *et al.*,2001)

The effect of growth regulators on *in vitro* regeneration in pomegranate cv. Ganesh has been reported by Murkute *et al.*(2004). These researches noticed the root differentiation of half strength MS+NAA 0.5mg/li induced maximum rooting (66.4%) within 8 to 9 days. However, IAA 0.5mg/l induced higher root length (3.2cm) as compared to NAA.The percentage planted survival was observed upto 50%.The role of auxin in rooting of pomegranate has also been earlier discussed by Jadika and Mehra (1986).Umrao(1999) reported that IBA at 7.389 m moles m^{-1} concentrations enhanced rooting and plant growth of pomegranate and thus produced taller and healthy plants.Further, a combination of IBA and NAA (2500 ppm each) as well as 1000ppm p-hydroxybenzoic acid plus 10000 ppm IBA also enhanced rooting of cuttings and plant survival of pomegranate (Reddy and Reddy,1990).

Citrus (*Citrus* species)

Lemon plants are generally raised from cuttings and layerings. Role of growth regulators in propagation of fruit plants have been reported by Arora and Yamdagni (1985) in Banarasi lemon and obtained better success in rooting and final survival of hardwood cuttings. Similarly,

Kumar *et al.*(1995) reported that IBA 2000 ppm in combination with phenolic compounds like p-cumric acid ,p-hydroxybenzoic acid and ferulic acid exhibits synergism in root initiation of cuttings and layers of hard to root species (Reddy and Majumdar,1978;Rakesh *et al.* 1997) Similarly,Baghel *et al* (1993) reported that root and shoot characters of seedless lemon cuttings were significantly influenced by different growth regulators like Rootex-B, Surootax and Veradix. However, Rootex-B was found best root promoting growth regulator for maximum rooting,growth and success of seedless lemon cuttings. Experiments conducted on different citrus species (Kagzi lime, Rangpur lime, sweetlime and seedless lemon) showed that IBA at 2000 ppm can be used for successful rooting of cuttings and their cent per cent survival (Pandey *et al.,* 2003)

Table 3 : Effect of IBA and PHA (p-hydroxybenzoic acid) treatments on various parameters after 120 days of planting of cuttings in Lemon cv. Banarasi

Treatments	Per cent cutting sprouted		No.of sprout per cutting		No.of leaves per cuttings	
	July	Feb.	July	Feb	July	Feb
Control	41.20	27.08	0.22	0.18	5.16	4.58
IBA 1000ppm	42.72	26.07	0.34	0.21	7.58	4.83
IBA 1500ppm	49.47	36.69	0.41	0.22	11.76	4.88
IBA 2000ppm	50.53	38.33	0.44	0.23	12.87	4.91
IBA1000ppm+ PHA1000 ppm	42.81	35.28	0.37	0.24	10.99	5.10
IBA1500ppm+ PHA1000ppm	50.51	37.17	0.44	0.30	12.29	4.26
IBA2000ppm+ PHA 1000 ppm	51.72	39.39	0.46	0.32	13.44	6.00
IBA2000ppm+ PHA 2000 ppm	46.22	38.15	0.41	0.32	13.00	8.15
IBA1500ppm+ PHA2000 ppm	55.70	41.15	0.50	0.38	13.26	9.00
IBA 2000ppm+ PHA 1000ppm	57.38	43.56	0.55	0.40	15.50	9.06
CD at 5%		4.98	0.02		2.02	
	Season	2.09		0.02		0.89

(*Source:* Kumar *et al.*,1995)

Guava (*Psidium guajava* L.)

Guava (*Psidium guajava* L.) can be raised by open pollinated seeds but vegetative propagation is preferred for getting desired characteristics and genetic uniformity. Secondly,the problem of wilt can be better controlled by searching the resistant / tolerant rootstocks. Air layering is the popular method of vegetative propagation in guava. Bhagvat *et al.* (1999) reported that treatment of guava air layers with 4500 ppm IBA resulted in the highest percentage rooting (91.97%) and survival (88%) over control. However, a combination IBA and NAA gave poor results.In another study,Tomar *et al.* (1999) found that the survival rates of guava air layers were highest (71.65%) with NAA 15000 ppm in first year of experience whereas in second year they were highest (70.29%) with IBA 15000 ppm. Dashora *et al.* (2008) reported that the treatment of IBA 10,000 ppm exhibited significantly highest rooting per centage of air layers (63.40%),maximum number of secondary roots per air layer (22.33%) and longest root per air layer (6.13cm)as compared to control and all other treatments..

Dutta *et al.* (1992) found 100 per cent rooting as well as survivality of the stool layers with the etiolation of shoots for 30 days and treatment with IBA at 5000ppm,IBA 3000 ppm+PHB 200 ppm and IBA 5000 +PHB 200 ppm caused. Singh (1998) stooled three etiolated plants of guava cultivars i.e. Seedless, Sardar and Allahabad Safed and aided with paclobutrazol.It was found that application of paclobutrazol 2500 ppm significantly increase rooting percentage, number of primary roots,number of secondary roots and survival per centage of rooted stools compared to the control treatment for all three cultivars.

Guava can also be propagated by semi hardwood cuttings. Sadhu and Bose(1980) reported that cuttings by pre-soaking treatment of Ethephon at 50 ppm or acetylene at 100 ppm followed by quick dip in 2500 ppm of IBA or IAA.Guava cuttings failed to r oot in the absence of auxin ,the best combination being acetylene+ IBA (50% rooting) . Kilany and Gabr (1986) recorded 1.67 -4.67 per cent rooting of hardwood cuttings, 18.3-57.5 per cent rooting of leaf bud cuttings and 81.4 per cent rooting of semi-hardwood cuttings by treatment with IBA at 2500 ppm and á-napthol at 10 ppm. Ahmed *et al.* (1998) found that maximum rooting (93%) in softwood cutting of guava was obtained with IBA 4000 ppm.The rooting response to NAA was very poor.

The seed germination of guava is also affected by soaking in growth regulators. Rodriguez *et al.* (1983) obtained higher rate of sed germination by soaking the seeds in water for 4 days followed by 24 hour in 1000 ppm $GA_{3.}$ However,Vijaykumar *et al.*(1991) reported that persowing seed treatment with GA_3 (for 8 hours in 100 ppm) had no significant effect on percentage of germination or shoot length but increased seedling vigour index and root length. Chandra and Govind (1993), reported that germination of guava seeds cv. Allahabad Safeda could enhance to 83.2% when soaked in GA_3 3000 ppm for 12 h as compared to control (50.0%).Seed soaking in GA_3 3000 ppm also resulted in significant increase in plant height, number of leaves per plant and leaf size, recorded 5 months after sowing.

Bael (*Aegle marmelos* Correa.)

Bael (*Aegle marmelos*) is reproduced either by seed or vegetatively. It can successfully be propagated by etiolated and invigorated cuttings with the use of 5000 ppm IBA(Ray and Chatterjee,1996) .Hazarika *et al.* (1998) observed that benzylaminopurine treated cuttings of bael rooted better than untreated cuttings.

Clonal propagation using shoot tip culture showed that strength Knop's medium supplemented with 15 mg /l IBA produced maximum rooting (43.89%)and maximum mean number of roots per regenerated shoot (3.0) in bael cv.Gonda Selection (Lal *et al.*, 2003).

Kair or Ker (*Capparis deciduas* Forsk)

Ker (*Capparis deciduas* Forsk.) grows wild on barren soil and waste lands of arid and semi arid parts of India. Its natural propagation is through seeds and root suckers.The seed germination of ker can be enhanced by soaking in 500 ppm GA_3 for 24 hours Sharma,(2005). Ker can also be propagated through stem cuttings with little success. Out of three types (hardwood, semi-hardwood and softwood) only semi-hardwood cuttings showed sprouting (40%) and rooting (20%). In round the year study, at monthly, interval to evaluate the effect of IBA and/or NAA, it was observed that January, March and August to November are the suitable months for taking cuttings and IBA (7500 ppm) + Thymine (1000 ppm), prepared in DMSO medium, was found to be the optimum concentration (Bhargava *et al*, 2006). Certain plant growth regulators can influence the per cent success of propagation by cuttings of ker. Meghwal and Vashishtha (1998) found that IBA failed to cause any significant influence on number and length of roots and shoots (Table 4).

Table 4 : Effect of auxins on rooting and shoot growth of cuttings of ker.

Treatment	Sprouting	No.of shoots	Shoot length	Percent rooting	No.of roots	Root length (cm)
Indole butyric acid						
0ppm	21.7	3.9	15.4	20.12	8.29	12.5
2500ppm	23.4	4.3	21.3	20.92	4.4	12.3
5000ppm	22.6	4.5	18.0		8.6	12.7
CD at 5%	NS	0.5	1.8	NS	NS	NS
f"-Napthoxy acetic acid						
O ppm	22.6	4.4	16.0	21.15	8.8	11.7
1000ppm	22.3	3.8	17.3	20.79	6.9	12.0
2000 ppm	22.9	4.5	18.9	21.39	10.5	3.8
CD at 5%	NS	0.5	1.8	NS	1.2	1.6

Phalsa (*Grewia subinaequalis*)

Phalsa (*Grewia subinaequalis*) is propagated by seeds but for quality production it can also be propagated through cuttings.However,per cent success is very low.Plant growth regulators can enhance the rooting of phalsa cuttings to certain extent. Baghel *et al.*, 2004 reported that biofertilizers can be beneficial for increasing the rooting and growth of cutting by producing growth regulators.These workers found that PSB (phosphorus solublizing bacteria) in slurry form increase the percent sprouting number of roots per cutting,root length,shoot length and other phenological characters of phalsa.

Date palm (*Phoenix dactylifera* L.)

In Date palm (*Phoenix dactylifera*), the availability of quality planting material is always problem. Actually, the survival of its small aerial offshoots/suckers without providing special environmental conditions ,growth regulators like NAA and or catecol are found to increase the rooting of ground offshoots (Al- Mana *et al.*,1996).The per cent germination of seed in date palm can also be enhanced by their treatment with GA_3 200 ppm (Hore and Sen,1996).

Karonda (*Carissa carandus* L.)

Karonda(*Carissa carandus* L.) is generally propagated by seeds. Among the vegetative propagation methods, air layering has been found quite successful with the aid of rooting hormones particularly IBA and NAA (Mishra and Singh,1990).Karonda produces many lateral shoots in the nursery during early periods of its growth and is most suitable plant for stooling .Mishra and Jaiswal (1993) showed that the application of 10,000 ppm IBA significantly increased the rooting of stool layered shoots of karonda over control (Table 5). It is suggested that the highest survival with IBA 10,000 ppm might be due to the presence of more number and length of roots ,thus causing better absorption of water and nutrients from soil and ultimately retaining higher percentage of survival of stool layers.Verma *et al.* (2005) reported that karonda cuttings dipped in the solution of cane sugar (4%) with IAA (100%) at room temperature for 24 hours gave 84 per cent success as against 24 per cent in control.

Table 5 : Effect of IBA on rooting ,root characteristics and survival of shoot layers of Karonda

Treatment	Rooted layers(%)	Number of roots/layer	Length of longest root(cm)	Survival
Control	37.22	4.25	8.06	26.56
IBA 2500 ppm	52.78	9.0	8.63	39.23
IBA 5000 ppm	57.00	12.90	10.48	43.08
IBA 7500 ppm	61.22	15.98	14.75	61.22
IBA 10,000 ppm	77.71	18.26	15.34	66.64
IBA 15000 ppm	71.56	17.85	15.60	59.01
IBA 20000 ppm	71.56	17.22	15.73	59.01
CD at 5%	12.63	1,44	1.76	7.86

(*Source:* Mishra and Jaiswal,1993)

Thrust Areas for Research

- Because the fundamental knowledge of plant propagation, subsequent growth and development is limited, only non-specific targets for controlling these have been identified.Thus, research is needed to understand how to alter the regulating factors rather than simply influencing propagation process and success rate.Potential for coupling the bioregulator treatments with other propagation practices,technologies and combined effects should be explored in future.

- In the recent years,with the discovery of new bioregulators , the hormonal picture has become rather complex .There are now five groups of growth regulators but there may well be many more in future,Researchers in the recent past have isolated new family of plantbioregulators.Thus, the research for identification of bioregulators with well defined objectives of propagation for particular crop,season and environmental conditions should be initiated with due emphasis on environmental protection and sustainability.
- More emphasis should be given on the use of natural PGR's rather than synthetic PGR's.
- Better methods of PGR application which have high efficiency for target should be evolved.
- Standardization of the protocol for micropropagation techniques and plant regeneration by cutting for rapid multiplication.

References

Ahmed,M., Ahmed, I., Laghare, M. H., and Hidayatullah, 1998. Effect of growth regulators on rooting in soft wood cuttings of guava under mist condition. Sarhad J.Agri.14:423-425

Al-Mana,F.A. Ed-Hamaday, M.A. Bacha, M.A. and Abedarahman, A.O: 1996, Improvimg root development on ground and aerial datepalm offshoots. Principles 40:179-181

Arora, R.K. and Yamdagni, R. 1985. Effect of growth regulators on rooting of lemo cuttings with and without leaves. Haryana Agric.Res.J.15:77-81

Baghel,B.S, Kirar,D.S; Nema , B.K.and Pandey,S.K.1993. Effect of mechanical treatments and growth regulators o rooting and growth of seedless lemon (*Citrus lemon* Burn.) cuttings. Indian J.Hort., 61:89-91

Bhagat ,B.K., Jain,B.P.; Singh ,C and Choudhary,B.M 1999. Studies on the propagation of guava (*Psidium guajava* L. *)* cv.Sardar by ground layering in polybags. Orissa J. Hort.27:19-21.

Bhargava,R., Verma,P., Saroj,P.L.and Chauhan,Neelam 2006.Propagationof ker (*Capparis deciduas*) through cuttings. Ind.Forester, 132 (6):737- 745.

Boysen-Jensen, *P* 1991. *La transmission de l'mitatiom phototropic I,Avena.* Bull ACad.Roy,Denmaru,1:3.

Casini,E. and Salvadori, S. 1980. The germination of seeds of jujube (Ziziphus sativa)I.The effect of gibberelic acid on seeds with and without endocarp. Rivista di Agricultura Subtropicale e Tropicale.74:39-74.

Casini,E.and Salvadori, S. 1975. Observation and research on the use of various growth regulators in the germination of seeds of fruit trees.La Nuova A.O.P.I.University di Firenze.

Chandra,R.and Govind, S. 1993. Gibberelic acid,thiourea, ethrel and acid treatments in relation to seed germination and seedling growth in guava (*Psidium guajava* L.) Progressive Hort., 22:40-43.

Chatterjee,B.K.and Rao ,D.P. 1978. Propagation of jujube(*Ziziphus jujube*) by air layering, Indian Agriculturist, 22:40-43.

Darwin,C.and Darwin, F. 1880. The Power of Movements in Plants.Murray,London.

Dashora,L.K, Kavita Arvindakshan. Lakhawat, S.S and Garina Sharma 2005. Effect of IBA and NAA on Rhizogenesis ofairlayers of guava (*Psidium guajava*) cv. L-49. In. National Seminar on Commercialization of Horticulture in Non-traditional Areas. Central Institute for Arid Horticulture,Bikaner.pp78

Dutta,P., Mitra, S.K. and Bose,T.K.1992. Indian Agric. 35:83-86.

Gupta,N.K, Jain ,S.K, Gupta, S. And Sen N.L.2004. Plant Bioregulators:Potential tools to enhance fruit production in India.In (Eds.K.K.Bora;Karan Singh and A.Kumar), Pointer publication, Jaipur,pp 1-24 .

Hazarika,B.N.,Nagaraya, V. And Parhasarthy, V.A. 1998. *In vitro* acclimatization of microshoots of Aegel marmelos L. International J.Tropical Agri.14:251-254.

Hore,J.K. and Sen, S.K. 1993 Root formation in pomegranate (*Punica granatum* L.)stem cuttings with NAA and auxins synegists under intermittent mist.Crop Res. 6:252-257.

Hore, J.K. and Sen, S.K. 1994.Role of presomwing treatment on germination ,seedling growth and longevity of ber (*Zizyphus mauritiana* Lamk) seds. Indian J.Agril. Res.28:285-289.

Hore,J.K. and Sen,S.K. 1996. Viability of datepalm seeds under different pre storage treatments. Curr. Res, 25:54-56.

Hore,J.K. and Sen, S.K. 1997. Effect of etiolation and plant growth regulators on the success of air layering of pomegranate (*Punica granatum* L.). Haryana J.Hort. Sci 26:63-66.

Jadika,k and Mehra,P.N.1986 .Morphogenesis in pomegranate (*Punica granatum*). Canadian J.Bot.64:1644-1653.

Jauhari,O. S. 1960. Some observations on vegetative propagation of Ziziphus mauritiana Lamk.for gooties and cuttings with the aid of growth regulators. Current Science. 29:30-31.

Kajal,R.S.1983.Studies in the effect of sowing depth,seed and budding treatments on germination of *Ziziphus mauritiana* Lamk..Thesis Abstract,CCSHAU,Hisar .9:176.

Kim ,Y.S.and Kim,W.S. 1983.Factors influencing seed germination of *Ziziphus jujube* Mill. South Korea Hort. 25:125-130.

Krishnan,B.M. and Kulsheharan,M 1984 .Studies on seed germination in wild ber (*Ziziphus rotundifolia*). South Indian Hort.32:153-154.

Kulkarni,A.R, Bhatt,N.R;Supe ,V.S. and Kale P.N. 1992 .Influence of storage conditions and presowing treatments on germination and seedling growth in ber. J.Maharashtra Agric.Univ.17:132-133.

Kumar,R., Gill,D.S. and Kaushik, R.A. 1995. Effect of indole butyric acid ,p-hydroxybenzoic acid and season on the propagation of lenmon cv. Baramasi from cuttings. Haryana J.Hort. Sci.24:13-18.

Lal,M.;Sindhu,S.S and Sharma ,S. 2003 .Rapid clonal propagation of bael through shoot tip culture, Haryana J. Hort.Sci 24:13-18.

Looney. L.N. 1998 .Plant bio-regulators in fruit production: An overview and outlook J.Korean Soc.Hort.Sci 39 : 125-128.

Meghwal, P.R.and Vashishtha, B.B. 1998. Effect of time of planting and auxins on rooting of cuttings of ker (*Capparis deciduas* Forsk.). Annals of Arid Zone,37:401-404.

Mishra,K.K. and Jaiswal ,H.R. 1993. A note on the effect of indole butyric acid on rooting and survival and survival of shoot layers of karonda(*Carissa carandas*.L) Haryana J.Hort.Sci.22:209-210.

Mishra,K.K. and Singh ,R 1990 .Effect of growth regulators and survival of air layers of karonda (*Carissa carandas*.L). Ann.Agric.Res.11:208-210.

Mishra,M;Saxena,R.P.; Pathak ,R.K.;Srivastaiva,A. And minimum M. 1999.Studies on micropropagation of aonla (*Emblica officinalis* Gaertn.). Progressive Hort. 31:116-122.

Misra,A.K.and Jauhari,O.S. 1970 .Root initiation in layers and stem cuttings of *Morus alba* L. and *Ziziphs mauritiana* Lamk. With special reference to plant growth from mature trees. Indian J.Hort.61:206-208.

Murkute,A.A.;Patil,S and Singh,S.K 2004 .*In vitro* regeneration in pomegranate cv.Ganesh from mature trees. Indian J.Hort.61:206-208.

Murthy,B.N.S and Reddy,Y.N. 1990. Temperature dependence of seed germination and seedlings growth in ber and their modification by pre –showing treatments.Seed Sci.Technol.18:621-627.

Nema, S; Jain ,P,K.; Nema, B.K. and Nema, N.K. 1994 .Effect of gibberelic acid,sucrose and date of sowing on seed germination and seedling growth in ber. Indian J.Hort.51:154-156.

Pandey, S.N. 2010. Role of Plant Growth Regulators in Production of Quality Planting material of Horticultural crops, *In*. National Conference on Production of Quality Seed and Planting material held at New Delhi 11-14th March,2010,.pp.13-14.

Pandey,A.,Patel, R.M.,Agrawal,S. and Sharma,H,G. 2003.Effect of plant growth regulators on rooting and survival of different speces of citrus cuttings. Orissa J.Hort.31:42-44

Panwar, R.D.;Kaushik,R.A.;Singh,S.and Gupta,R.B. 2001 .Effect of Indole Butyric Acid (IBA) on rooting of hardwood cuttings in pomegranate (*Punica granatum* L.). Haryana J.Hort.Sci.30:72.

Purohit,A.G. and K.E.Shekherappa, 1985. Indian Jour.Hort.,42:30-36.

Rakesh ,Gill,D.S.;Beniwal ,B.S and Dhankar ,R.S. 1997 .A note on rooting behaviour of banarasi lemon (Citrus limon L.) as affected by indole buteryic acid, p-hydrozybenzoic acid and season. Haryana J.Hor.Sci.26:217-219.

Ray,D.P. and Chatterjee,B.K. 1996. Effect of different concentration of growth regulators,etiolation and vigoration treatments on the rooting of stem cutting of bael (Aegal marmelous Corse). Orissa J.Hort. 24:36-41.

Reddy ,Y.N. and Majumdar,P.K 1978 . Synergism of phenols and flavanoids with IBA in regeneration of mango*(Mangifera indica L.)*and guava *(Psidium gujava L.)*cuttings. Vatika.1:37-44.

Reddy,Y.N.and Reddy, Y.T.N. 1990. Effect of basal bounding,growth regulator and polythene covering on rooting of pomegranate cuttings. J.Maharashtra Agril.Univ. 15:153-155.

Reighard, G.L. .Dawid, W.C. and Newall,W.C.Jr 1990. Rooting and survival of hardwood cuttings of 406 species,cultivars and hybrids of Prunas. Hort.Sci.25:517-518.

Rodriguez,N.Gorollilo,M.and Caballevo,S. 1983.Ciencia yTecnica en la Agricultura.6:39-48.

Sadhu,M.K. and Bose, S .1980. Indian J,Agric. Sci.50:448-450.

Saroj,P.L., Awasthi, O.P.; Bhargava,R. and Singh, U.V. 2005 .Propagation of pomegranate by cutting under mist system. National Seminar on Commercialization of Horticulture in Non-traditional Areas. Central Instittute for Arid Horticulture, Bikaner.pp.78.

Shanmugavelu,K.G. 1996. Studies on the effect of plant growth regulators on the seedlings of some tree plant species. South Ind.Hort. 14:24-35

Shcherbakova, L.T. and Kulikov,G.V. 1972 Vegetative propagation of jujube under mist. Subtropicheskie Kul'tury,No.1:94-98.

Shen,X.D.;Gao,F,G.;Chen,B.X.and Wang,Y.1992 .An experiment of propagation of Ziziphus jujube by green wood cuttings.Ningxia. J.Agroforestry Sci.Technol.,2:32-34

Singh,D.K. 1998.Regeneration of guava(Psidium gujava L.) cultivars by stooling with the aid of paclobutrazol. Ann,Agril.Res.19:317-320.

Singh,H.K.and Pandey,C.S. 2003. Effect of transplanting treatments and time of transplanting on bare rooted budling of aonla(*Emblica officinalis* Garten.). Haryana J.Hort.Sci.32:176-179.

Singh,R.S. 1994. Efffect of growth substances on rooting of pomegranate cuttings. Curr. Agri.,18:87-90.

Singh, S; Signrot,R.S.; Bhatia, S.K. and Joon.M.S. 2003 .Effect of transplanting treatment on the success of bare rooted budlings of ber (*Z.mauritiana* Lamk). Haryana J.Hort.Sci.32:174-175.

Tarasenko,M.T. and Shaumarow,K.B.1977. Raising of Zizyphus jujube transplants by means of softwood cuttings.Izvestiya Timiryazevskoi Sel'skokhozyaistvennoi Akademiya.3:150-163.

Tomar, S.K., Gurjar,B.S. and Tomar ,R.S.S. 1999. Response of different concentrations of IBA and NAA on rootage and growth of air layers of guava (*Psidium gujava.*L) cv.Gwaliar-27. Adv.Plant. Sci.12:535-538.

Ughreja, P.P and Chauhan,K.S 1983. .Effect of plant growth regulators on rooting in airlayers of ber (*Ziziphus mauritiana* Lamk.). Haryana J. Hort.Sci.12:160-164.

Umrao,V.K. 1990. IBA enhances rooting in Pomegranate cuttings. Ann.Arid Zone, 38:87-88.

Verma,L.R;Joshi,V.J.;P.P and Patel,B.S 2005. Effect of growth regulators on rooting of cutting in pomegranate (*Punica granatum* L.).In. National Seminar on Commercialization of Horticulture in non-traditional areas. Central Insitute for Arid Horticulture,Bikaner.pp-78.

Vijayakumar, A; Palanisammy, V., Jayraj,T. and Arumugam,R. 1991. South Ind.Hort.39:315-316.

Zimmermann, P.W. and Wilcoxon,F. 1935. Several chemical growth substances which cause initiation of roots and other responces in plants.Contr.Boyce Thompson Inst.7

Chapter-7

Nursery Management for Production of Quality Planting Materials

R. S. Singh and Hare Krishna
Central Institute for Arid Horticulture, Bikaner-334 006

Besides edapho-climatic constraints in the arid region, availability of genuine planting materials at the doorsteps of the farmer's is one of the major challenges that come in the way of horticultural development at commercial venture. Though, establishment of nursery is considered as remunerative farming enterprise if managed properly, in arid region it is still confined in the vicinity of big cities. It is also interesting to reveal that in the state like Rajasthan, most of the private nursery men are not producing planting materials in their nurseries except for the fruits like *ber, lasoda* and kinnow instead they are purchasing planting materials from the other states in bulk and selling them on higher rate to the farmers of the state. The main reasons for poor development of nursery in the arid region are; unfavorable edapho-climatic conditions and poor knowledge about innovations made in recent past for nursery management. Keeping these facts in view, there is a need to develop a long chain of hitech-nurseries coupled with Human Resource Development for mass multiplication of quality planting materials for fulfilling the demand of end users. Production of quality planting material is directly link with productivity of fruit crops (Bose *et al.*, 1998; Brison and Adriance, 2000).

Nursery

Nursery is place where planting materials are raised either by seeds or by vegetative means with care before transplanting at desired site. Often we consider the nursery management and plant propagation are same, though they are altogether different but interrelated. Planting material is costly input in development of orchards of fruit crops because of long time investment. In fact, multiplication of quality planting materials is the central theme of nursery management but management is a trade oriented dynamic process refers to efficient utilization of resources for economic return (Reddy *et al.*, 2001). Always use genuine planting material for development of orchards. Plants should be purchased from any registered/recognized nursery. About 50,000 nurseries propagating different ornamental and fruit plants in India spread over in 26 states (Prakash, 2010). Most of nurseries are located near by cities and producing ornamental plants. The demand of planting material has been increasing during last decade as the area in fruit crops has expanded. In case of establishment of nursery, mother plants, scion and root stock are very important for raising of plants. A right step in the direction is preparation of a Manual of Nursery which should serve as a reference document to both entrepreneur and grower. The main phases of management are (i) planning, (ii) implementation, (iii) monitoring and evaluation and (iv) feedback for further refinement, while the key elements of nursery management are (i) the place, (ii) the plant and (iii) the person behind.

Importance of nursery

1) The young plants can be easily maintained in the nursery. Similarly, it is easy to look after the young and tender seedlings in the nursery rather than in the field.
2) Propagation of plants by asexual means require special skill and care of plants before transplanting in the field, which can be easily done in nursery but not at consumer's level.
3) Seasoning of the seedlings against natural calamities is only possible in the nursery.
4) Many plants do not respond to direct sowing e.g. cabbage, tomato, papaya and conifers, in the field as compared to transplanting of seedling raised in the nursery.

5) Cuttings of different horticultural and forestry plants for rooting are first planted in nursery for better care and management.
6) For hardening of seedlings/grafts/layers, nursery is a pre-treatment place.

Pre-requisites for nursery establishment

Following pre-requisites are considered important for nursery establishment.

(i) Nursery should preferably be established in a locality where fruit cultivation is being practiced.

(ii) As far as possible, the nursery should be located in the important production areas.

(iii) It should be ensured that the site should be well connected by different means of communication and should be easily accessible.

(iv) Climatic conditions should be favourable for the plants to be propagated.

(v) The nursery should have its own resources to provide parent material (mother plants) for propagation.

(vi) The locality should have adequate supply of sweet water.

(vii) The soil should be deep, fertile, well drained and free from soil borne pathogens.

(viii) There should be sufficient provision for different propagation structures like glass house, net house, lath house etc.

(ix) Sufficient labour, budder and grafters should be available to handle different operations.

(x) The materials like fertilizers, pesticides, growth regulators, grafting wax, lanolin paste etc and equipments etc should be readily available in the locality.

Selection of site: In our country less emphasis is given on selection of site for the nursery, though there are several negative consequences of improper site and an unsatisfactory site will increase the cost of operations and poor stock production. Therefore, setting up of a nursery is a long term venture require careful planning, because mistake committed initially cannot be rectified easily at later stage and may adversely affect the return from the investment. For making job easier a site selection team comprising of 5-7 members should visit all the potential sites by using a common checklist as mentioned below and then finalize by considering merits and demerits of all the sites visited.

Check list for selection of potential nursery site

Location	:			
Site	:			
Soil	:	Top soil	:	Texture
			:	Depth
			:	pH
Depth of water table	:			
Drainage	:			
Water supply	:		:	Well
			:	Canal
			:	Storage tank
			:	Quality of water
			:	pH
Topography :			:	Level
			:	Rolling
			:	Slope
			:	Erosion
Cover	:		:	Principal tree species
			:	Ground flora
Climatic data	:		:	Annual precipitation
			:	Distribution
			:	Frost free months
			:	High temperature months
Approximate acreage				
Access	:			
Ownership	:			
Others	:			

Selection of Location and Site

- The nursery area should be near to the water source.
- Generally, the location should be partially shaded i.e. under the trees if it is not possible then artificial shade like shade net structure is to be provided.

- There should be strong living or non-living fencing so that plants can be protected from loss due to animals and hot winds during summer.
- Proper drainage facilities should be provided otherwise water logging problem would arise.

Preparation of Nursery Bed

Selected soil should be worked well to break the clods and weeds stones and stubbles should be removed. Height of the raised bed should be 10-15 cm with a width of 1.25 m and length may be according to requirement and convenience but generally, it is to be kept 6.25 m long. Two part of fine soil, one part of FYM can be incorporated to each bed to improve aeration and fertility of the soil. Before preparing the bed soil should be drenched with 0.4% formaldehyde or 0.3% Copper Oxychloride to kill pathogen/pests in the soil.

Types of nursery

Temporary nursery: It consists of raising nursery beds. It can be changed from one place to another depending upon requirements.

Permanent nursery: Sidewalls with drainage holes are constructed with concrete to a height of 75 cm. Seeds are sown in soil inside the concrete structure. After removal of each type of seedling the soil is enriched with fertilizers and manures.

Media for Propagating Nursery Plants

Saplings are raised in nurseries through different means of propagation. Therefore, various kinds of media and mixes are employed for the propagation of the different plants either through seeds, cuttings or other means of propagation. However, one medium may be suitable for seed sowing and the other for cuttage or layered or other propagules. Similarly, plants of one species may do well in a particular medium but may be a totally failure for the other. Hence, the choice of the use of a medium depends on the plant species, method of propagation and availability of a medium. However, an ideal medium should meet the following criteria;

1) The selected medium should be sufficiently firm and dense so that it holds the seeds or other propagating material (cuttings, layered plants, stools etc.) firmly in place during the process of seed germination or rooting.

2) Its volume of the medium should remain constant upon drying or wetting.

3) It should have enough ability to retain and supply ample amount of moisture to the seeds or other propagating materials.

4) It should be porous enough to permit adequate aeration and to drain out excessive moisture.

5) The medium must be free from seeds of weeds, nematodes and pathogens, which affect the germination of seeds or rooting of the propagules adversely.

6) Its pH should range from 5.5 to 6.5.

7) It should be of consistent quality from batch to batch and reproducible.

8) The selected medium must not have higher concentrations of salts.

9) It should be highly decomposed, preferably with 20:1 C:N ratio to prevent nitrogen immobilization and excessive shrinkage.

10) It should have high cation exchange capacity so that it retains the mineral nutrient.

11) It should be easily available.

12) It should not be costly.

13) It should be able to provide adequate nutrition to the germinating seeds or cuttings.

14) It must be capable of being pasteurized with steam or chemicals without harmful effects.

15) It must be easy to wet and should have the ability to retain sufficient moisture to reduce frequent watering.

A. Soil: Soil is composed of material in the solid, liquid and gaseous states. For satisfactory plant growth these materials must be existed in the proper proportions. Solid portion of a soil is comprised of both inorganic and organic forms. The liquid part of the soil, the soil solution is made up of water containing dissolve minerals in various quantities. In poorly drained, water logged soils, water replaces the air in soil thus depriving plants roots as well as certain desirable aerobic microorganisms of the oxygen necessary for their existence.

B. **Sand:** Sand consists of small rock grains. Quartz sand is generally, used for propagation purposes consisting chiefly of a silica complex. The type used in plastering is most satisfactory for rooting cuttings. Sand is the heaviest of all rooting media used, a cubic foot of dry sand weighing about 100 lb. It should be preferably be fumigated or heat treated before use as it may contain weed seeds and various harmful fungi.

C. **Peat:** Peat consists of the remains of aquatic, marsh, beg or swamp vegetation, which has been preserved under water in partially decomposed state. There are three types of peat i.e. moss peat, reed peat and peat humus. Out of these, moss peat is the least decomposed type, which derived from sphagnum or other mosses. It has high moisture holding capacity is high in acidity and contain small amount of nitrogen.

D. **Sphagnum Moss:** Commercial sphagnum moss in the dehydrated young residue or living portion of acid bog plants in the genus *Sphagnum* such as *S. papillosum, S. capillaceum* and *S. palustre.* It is relatively sterile, light in weight and has a very high water holding capacity being able to absorb 10 to 20 times its weight of water. The stem and leaf tissues of sphagnum moss consist largely of groups of water holding cells. Sphagnum moss has a pH of about 3.5.

E. **Vermiculite:** This is a micaceous mineral, which expands markedly when heated. Chemically it is a hydrated magnesium aluminum iron silicate when expanded, it is very light in weight (6 to 10 lb per cubic feet), neutral in reaction with good buffering property and insoluble in water, it is able to absorb large quantity of water 3 to 4 gallon per cubic ft. Vermiculite has a relatively high cata-ion exchange capacity and thus can hold nutrients in reserve and later release them. It contains enough Mg and K to apply most plant. Expanded vermiculite should not be pressed or compacted when wet, as this will destroy its desirable porous structure.

F. **Perlite:** This gray white silicaceous material is of volcanic origin mined from lava flows. A particle size of 1/16 to 1/8 inch in diameter is usually used in horticultural application. It is essentially neutral with a pH of 6.0 to 8.0 but with non-buffering capacity, unlike vermiculite, it has no cation exchange capacity and contains no mineral nutrients. It is most useful in increasing aeration in a mixture.

G. **Compost:** Humus content is limiting factor in arid region therefore, compost mixture may be useful as moisture holding humus material having small amount of plant nutrient. Grass clippings, leaves and garden refuse are accumulated and allowed to decompose preferably in a 4 to 6 feet bin with stated sides to give good aeration. Moisture should be added from time to time during dry seasons, decomposition will be hastened if some nitrogenous fertilizer is sprinkled through each batch of newly added material. Compost may contain weed seeds and nematodes as well as noxious insects and diseases, so preferably it should be sterilized before use.

H. **Shredded bark, Saw dust and Wood shavings:** These materials, consisting of redwood, cedar, fir or pine, can be used in soil mixes, serving much the same purposes as peat moss except that their rate decomposition is slower. Nitrogen is added in an amount sufficient to take care of the decomposition requirements of the sawdust plus an additional amount for use by the plants. Due to its low cost it is widely used as a soil amendment, although some types, especially when fresh, may contain material, toxic to plants.

I. **Leaf mould:** Leaf mould is prepared by decomposing fallen leaves of the plants. Sometimes, chemical fertilizers, farmyard manure or cow dung are also added to it. Its colour and composition depends mainly on the original material. It contains almost all the essential nutrient elements in it. Leaf mould should be properly treated with fungicides and insecticides before use to kill the pathogens.

Soil and its Management

Soil is medium plant growth therefore, it should have sufficient amount of plant nutrients which are essential for growth and development for raising seedling, soil should be friable and should absorb water readily and dry quickly at the surface, but it should not become dry too early. Loam and Sandy loam soils well supplied with organic matter are good, provided that they are free from pathogen. To make the soil free from soil borne diseases especially damping off it may be disinfected by shell soil fumigant, formaldehyde, steam and chloropicrin. Mehta (1959) recommends the following methods for sterilizing the soil. One part of formalin mixed with 100 parts of water should be applied at the rate of 4.55 liter per 0.09 square meter of surface soil so as to saturate it up to depth of 152.40 mm. Then the soil mixture is covered with gunny bags or a thick paper for a day or so that fumes may penetrate the soil and kill all fungi and insects. The soil mixture is then uncovered and left for a few days till the smell has disappeared. The sterilized soil mixture is then put through a 6 mm mesh screen.

Though under Indian conditions, nursery is not raised generally in wooden boxes, clay pans and flowerpots except for the flowers. But it is convenient to raise vegetable seedlings also in flats or plant boxes due to following advantages:

It is very easy to sow and transplant comfortably.

1. Sowing or transplanting may be continued whatever the weather may be outside.
2. It is convenient to check the plant growth during abnormal temporary weather condition.
3. Plants raised in flats or boxes can easily be removed for transplanting without any damaged to their roots.
4. There is an economy of labour and space if flats or boxes of standard size are used for this. It is recommended that the size of boxes made of deodar wood may be 40x30 cm and 7.5 cm deep.
5. For uniform and high percentage of germination the soil must be fine, moist and firm. Basic organic manures like well rotten compost or FYM should be mixed well in it.

Environmental Factors

In arid zone of India, rainfall is by far the most important factor limiting crop production and sustenance of life is at a stress level. A significant feature of the distribution of rainfall over the Indian arid zone is that a major part of the rainfall is received during the south-west monsoon period, June-September. The contribution of this seasonal rainfall to annual rainfall is quite high over the south and central parts of western Rajasthan.

A prime characteristics of the arid zone of India of which a major portion is land locked is the extreme variation in the diurnal and annual temperatures. The temperatures begin to rise from March onwards over the whole arid zone. However, the desert region in NW India records the highest day temperatures and May-June are the hottest months when mean temperatures are above 40°C. However, on individual days temperatures of the order of 47°C to 50°C are also recorded in these regions when heat waves occur. With the onset of the monsoon season, the mean daily maximum temperatures fall sharply from 42°C in June to 38°C in July and 36°C in August-September over the arid zone of NW India. In the Indian arid zone, the wind regime is strongest over the Gujarat region followed by western Rajasthan where wind speed of more than 20 km/ha are recorded during April, May, June and July months.

Growing Structures

There are a number of structures used for growing of plants in nursery (Sharma and Hare Krishna, 2013).These structure have important role in faster multiplication and growth of nursery plants. Agro-shade net (50%) structure of green shed is very common in hot arid conditions for nursery plants. Black shade net is also available for different areas.

a. **Green houses:** Commercial green houses are usually independent structures of even span gable roof construction, proportioned so that the space is well utilized for convenient walkways and propagating benches. In green houses construction, a wood or metal frame work is build to which are fastened wood is metal sash bars to support panes of glass embedded in putty. All metal prefabricated aluminum green houses are also widely used. A means of providing air movement and air exchange is necessary in all green houses to aid in controlling temperature and humidity. A ridge ventilator is almost always used and in some cases side ventilators are also used. Automatic, thermostatically operated controls are available to open and close the ventilators as the temperature fluctuates during the day and night. Forced air ventilator is usually used in large installations. Green houses can be mechanically cooled in the summer at low cost by the use of large evaporative cooling units. It is best if possible, to have the green house heating self-opening ventilators and evaporative cooling systems controlled by thermostat. As a general rule, a minimum night temperature of 60°F is suggested for which the heating thermostat should be set. Ventilator thermostats should be set to open at about 72°F and the evaporator cooler thermostat set to start operating the blower sat 75°F.

Plastic covered green houses- Lightweight frames covered with various types of plastic film are popular for small home garden structures as well as for large commercial installations. Plastic houses are usually of temporary construction except when the more permanent high cost covering such as fiberglass panels are used.

b. **Polyhouse**: This is the least expensive covering material but has shortest life. It breaks down in summer and must be replaced once a year or oftener this is generally done in the fall for use during the winter. Ultraviolet ray resisting polythene lasts longer but costs somewhat more. A thickness of 4-6 mils is recommended.

c. **Fiberglass**: Rigid panels corrugated or flat of fiberglass sheets embedded in plastic are widely used for greenhouse construction. Fiberglass is strong, long lasting, light weight and easily applied coming in a variety of width, lengths and thicknesses. Only the clean material especially, made for green house and in a thickness of 0.038 inch and weighing 4 to 5 oz. per sq. ft. should be used.

d. **Hot beds:** Hot bed is often used for the same purpose as a green house. Seedlings can be started and leafy cuttings rooted in such structures early in the season. Heat is provided below the propagating medium by electric heating cables, hot water, steam pipe, or hot air flues. In hot bed close attention must be paid to shading and ventilation as well as temperature and humidity control. The hot bed may consist of a large wood box or frame with slopping tight fitting lid made of window sash or, preferably, regular hot bed sash. It should be placed in a sunny but protected and well-drained location. A standard size of frame is 3′x6'. The hot bed is filled with 4 to 6 inch of the rooting or seed-germinating medium over the heating cables. Alternatively, flats containing the medium can be used there are placed directly on a thin layer of sand covering wire netting placed over the heating cables for protection from tools.

e. **Cold frames:** Cold frame construction is almost identical with that of hotbeds, except that no provision is made for supplying bottom heat. The standard glass 3′x6' sash is often used as a covering for the frame although light weight, less expensive covering can be made, utilizing frames covered with one of the plastic materials. Primary use of these structures is in conditioning or hardening rooted cuttings or young seedlings preceding field, nursery row or container planting. In cold frame only the heat of the sun retained by the transparent covering is utilized.

f. **Lath house:** These structures are useful in providing protection for container grown nursery stock especially in areas of high summer temperatures and high light intensity. For tender species the lath house is used as an intermediate step between the cold frame and field planting. More commonly, wood or pipe supports are used set in concrete with the necessary supporting cross members. Shade is provided by thin wood strips about 2 inch wide, placed to give 1/3 to 2/3 covers, depending on the need. The sides as well as the top are usually covered.

g. **Mist propagation unit:** The rooting of soft wood leafy cutting under spray or mist is a technique now widely used by nurserymen and other plant propagators throughout the world. Mist propagation units are used for propagation of "difficult to root" cuttings in most advanced countries. Mist beds are constructed within a greenhouse. A fine mist is sprayed intermittently over the cuttings at regular intervals during day and night. The mist unit is controlled by a time clock, operating a magnetic solenoid valve and is set in a way to turn on the mist for 3-5 seconds to wet the leaves and turn off for some time and when the leaves are dry, the mist is again turned on.

h. **Plastic tunnels:** A simple but effective method of protecting cuttings, which comprises of three feet wide wire hoops, which support the polyethylene are made from 0.2-inch diameter wire and are erected at 30-inch intervals. A white translucent polyethylene sheet 6 feet wide is then stretched over the hoops and is held in position by fastening two lines of polyethylene bailer twin to alternate hoops on either side of the tunnel. The seeds or cuttings or other propagating material is sown/ planted inside the tunnels during winter. Nowadays, perforated tunnels are used because in non-perforated tunnels, temperature and relative humidity rises sharply, which may affect rooting of cuttings or further growth of the propagating material.

Propagation methods

Plant propagation is the increase in number or perpetuation of the species by reproduction. This reproduction is achieved through sexual and asexual methods. Sexual reproduction is the reproduction of plants through a sexual process involving meiosis. By this method of propagation true to type plants can not be obtained. Asexual propagation is defined as the duplication of a whole plant from any cell, tissue or organ of that plant. Since the product of asexual method of propagation is true to type, it is considered to be the most desirable method of multiplication (Sharma, 2002). Marula nut is an exotic fruit species and stem cutting was found successful method for multiplication under hot arid conditions (Singh and Bhargava, 2008).

The propagation practice of various horticultural crop plants under arid ecosystem should be such that it can overcome the adverse natural conditions so that establishment and growth of the plants can be proper. The commercial methods of plant propagation in arid fruits in Table 1. Besides this, there are some standard of quality planting materials (Table 2) which should be taken into consideration while purchasing of plants from the nursery.

Table 1 : Propagation of arid and semi- arid fruit plants.

Sl.No	Name of crops	Method of propagation	Time	Important cultivars
1.	*Ziziphus mauritiana* (Ber)	Budding	June-September	Gola, Seb, Kaithli, Mundia, Umran,Thar Sevika, Banarasi Karaka
2.	*Punica granatum* (Pomegranate)	Cutting, Air layering	June-July	Ganesh, Jalore Seedless, Mridula, G-137, Bhagava, Phule Arakta,
3.	*Emblica officinalis* (Aonla)	Budding	June-July	Kanchan, Krishna, Chakaiya, NA 6, NA 7, NA 10, Laxmi-52,
4.	*Annona squmosa* (Custard apple)	Seed, Budding	June-July	Red Sitaphal, Mammoth, Balanagar, APK(Ca)-1,Arka Sahan, Local var.
5.	*Psidium guajava* (Guava)	Seed, Budding, Air layering,	June-July	Allahabad Safeda, L-49, Red flesh,

Contd.

				Chittidar, Lalit, Hisar Surkha,
6.	*Phoenix dactylifera* (Date palm)	Suckers, Tissue culture	Feb-March, Aug.-Sept.	Medjool, Zahidi, Barhee, Halawy, Chipchap, Khalas, Khadrawy, Khuneizi, Shamran,
7.	*Ficus carica* (Fig)	Hardwood cutting	Dec.-Feb., June-July	Puna fig, Brown Turdey, Conadria, Deanna, Excel,
8.	*Grewia subinaequalis* (Phalsa)	Seed, Hard wood cutting	March-April	Sharbati and local
9.	*Tamarindus indica* (Tamarind)	Seed, Budding	Feb.-March, June-July	PKM-1, Yogeshwari, Pratisthan, Goma Prateek,
10.	*Cordia myxa* (Lasoda)	Seed, Budding	April-May	CIAH Sel.-1, Big fruit type,
11.	*Syzygium cumminii* (Jamun)	Seed, grafting	June-July	Goma Priyanka, Seedless,, CISH Sel.-1
12.	*Carissa carandas* (Karonda)	Seed	Aug.-Sept.	Pant Manhar, Pant Sudershan, CHES K-1, Konkan Bold, Local var.
13.	*Aegle marmelos* (Bael)	Budding	June-July	NB 5, NB 6, NB 9, Pant Shivani, Pant Aparna, Goma Yashi, CISH B-1
14.	*Feronia lemonia* (Wood apple)	Seed, Budding, Grafting	Feb-March, June-July	Local
15.	*Morus* sp. (Mulberry)	Cutting	June-July	Local
16.	*Capparis decidua* (Ker)	Seed, cuttings	July-Sept.,	Local type,
17.	*Salvadora oleoides* (Pilu).	Seed, Suckers	June-July	Local
18.	Cactus pear *(Opuntia ficus indica)*	Cladodes	February-March	Clone 1269, 1270, 1271, 1308
19.	Manila tamarind (*Pithecelobium dulce*)	Seed	Feb.-March	PKM-1, Local,
20.	Marula nut (*S. birrea* sub sp. *caffra*)	Seed, stem cuttings	Feb.-March	Local

Table 2 : Standards for quality planting materials of arid horticultural crops.

Plant Type	Propagation methods	Considerations
Date palm	Sucker	8-10 Kg, 1-1.5 m ht., with roots, smaller base cut
Indian Aloe	Sucker	250 g wt., 30 cm ht., 2.5-3.0 cm basal dia., 3-5 leaves, with root
Tamarind	Budding	50-100 cm ht., 0.75 – 1.25 cm budling girth
Ber	Budding	30-50 cm ht., 0.75 –1.25 cm budling girth
Aonla	Budding	50-75 cm ht., 1-1.5 cm budling girth
Bael	Budding	30-50 cm ht., 0.75 cm budling girth
Guava	Budding	50-75 cm ht., 1-1.5 cm budling girth
Pomegranate	Cutting	25-40 cm ht., 0.75-1.25 cm girth
Karonda	Seeds	20-35 cm ht., 0.35-0.75 cm girth

Advantages of sexual propagation

1. Compared to vegetative propagation, propagation through seed is relatively a cheaper and simple method of obtaining a large number of plants. Even, in some fruit plants like papaya, this is the most popular method of propagation.
2. In general, seed raised plants have long life and better root system (tap root); thereby, provide better anchorage as compared to vegetatively propagated plants, which have adventitious roots.
3. Usually, seedling plants are resistant to insect-pests and diseases than the vegetatively propagated plants. In addition, they can tolerate vagaries of adverse climatic conditions in a better way than the vegetatively propagated plants.
4. Generally, rootstocks for budding and grafting are raised through seeds, except clonal rootstocks.
5. The hybrids, derived from controlled hybridization programme, are first raised through seeds. Likewise, sexual propagation sometimes leads to the production of chance seedlings, which may be superior to the parent plants.
6. Nucellear seedlings can be utilized to raise uniform plants. The mangosteen (*Garcinia mangostana*) fruit is developed through parthenogenesis. Seedlings obtained from the seeds of such fruits are similar to its mother plant and thus

propagated commercially through seed. Polyembryony is a rule in some fruit *viz.*, citrus (except pummelo, Tahiti lime and Turanj) and mango (varieties like Kurukkan, Chandrakaran, Carabao, Goa and Vellaicolumban etc.)

7. It is easy to store seeds for longer time than the vegetative parts and can be transported easily to distant places. Moreover, storage of seeds involves limited space and time than vegetative parts of a plant.
8. Handling of seeds is easy as compared to vegetative parts of the plant.
9. Usually, not much technical skill is required for raising plants through seeds.
10. Seed propagation is useful in producing virus free plants as most viruses are not transmitted through seeds.
11. With small quantity of seed, a large population can be raised.

Disadvantages

Seed propagation is traditional method for those plant species which are very difficult to multiply by other means. Germination and seedling growth was enhanced by pre-treatment of seed in date palm (Singh and Bhargava, 2009). Propagation through seeds has certain limitations over asexual methods, which are discussed as under;

1. In majority of instances, seedling plants are highly heterozygous in nature. Thus, the seedling plants are not uniform in growth, yield and fruit quality parameters than vegetatively propagated plants.
2. Seedling plants are usually tall, vigrous and spreading types and pose problems for efficient management. Thus, the management cost involved in operations like harvesting, pruning and plant protection etc. is comparatively higher than the vegetatively propagated plants.
3. In general, seedling plants have longer juvenile phase and take more time to come into bearing than vegetatively propagated plants.
4. Usually, seeds in most of fruit plants loose viability very soon after extraction and thus have very low germination rate.
5. It is not possible to derive the beneficial effects of rootstock on scion variety with seeds.

6. Some fruit plants like pineapple, banana does not produce viable seeds and propagation through seeds is not possible in them.
7. Seed borne viruses exist in a number of fruit plants as psorosis in citrus and mosaic in peach, cherry and almond. Thus, multiplication of these plants by seed is not feasible.
8. It is not possible to perpetuate the exact character of any superior selection through seeds.
9. Seed often require various treatment such as cold chilling, warm stratification, mechanical or chemical seed coat scarification, or some combination of these or other treatments as pre-requisite to germination to take place. Lengthy dormancy requirements may result in a more costly product or a longer construction schedule. Sometimes, unknown dormancy requirements may prevent propagation by seed entirely.

Advantages of asexual propagation

1. Fruit plants are true to type and are identical to mother plant.
2. Vegetatively propagated plants have uniform growth, and give consistent yield of better fruit quality.
3. Fruit trees come into bearing earlier than the seedling trees.
4. Vegetative propagation is the only choice in plants, the seeds of which are more prone to dormancy (peach and olive etc.).
5. Some fruit plants like banana, pineapple, fig and varieties of grape, guava and lemon have seedless fruit and produce little or no viable seeds. Such plants are easily perpetuated through vegetative means.
6. Vegetatively propagated plants are less vigorous and thus their management is easy and less costly than seedling plants.
7. With vegetative means, a large plant is obtained within short time.
8. The desirable abiotic effects of rootstocks on the scion cultivar can be utilized effectively by budding or grafting them on desirable rootstocks. Regulation of the tree size, fruit quality, precocity etc. by using different rootstocks or interstock is possible.

9. A 'clone' can only be perpetuated through vegetative means.
10. Self-incompatibility exists in a number of fruit plants like mango, aonla, apple, almond and cherry etc. Provision for adequate cross-pollination in an established fruit orchard can be obtained by top working in scattered trees throughout the orchard with the proper pollinising variety.
11. Indicator plants can be used for detecting various viral diseases through Viral indexing technique in which the desirable scion is grafted or budded on to a susceptible rootstock. For example Kagzi lime can be used as rootstock (indicator plant) for detecting the presence of tristeza virus in citrus.
12. Early assessment of hybrids/varieties can be done by grafting or budding the desirable hybrid/variety on a large established tree or on certain dwarfing rootstock.
13. Damaged part of the tree trunk or root can be repaired by bridge grafting or inarching.
14. A number of varieties can be grown on a single plant, which is a novelty.

Limitations

Vegetative propagation has certain limitations as under:

1. The vegetatively propagated plants, particularly budded and grafted are short lived.
2. No new varieties can be evolved by vegetative mean of propagation.
3. It is a specialized task and requires technical skill.
4. Vegetatively propagated plants are usually more susceptible to insect-pests and diseases.
5. Vegetatively propagated plants are usually prone to suckering.
6. Vegetatively propagated plants are usually less hardy to abiotic stresses as compared the plants raised through seeds.
7. Vegetative propagation is more expensive than the seed propagation.
8. Vegetative propagation is failure in some fruit plants and forestry trees in which no method of vegetative propagation has been found successful for instance, papaya and phalsa.

Irrigation and Nutrient Management

Water is limiting factor under arid ecosystem. In the times of scarcity supplementary irrigation helps in over coming mortality of plant in initial stage. Sprinkler system of irrigation is suitable for nursery plants under arid ecosystem where water supply is not adequate. The water is pumped with pressure through the sprinklers attached to pipes and these sprinklers are adjusted in such a manner to overlap up to one-fourth aqua covered by the other sprinklers. These are then moved to the next point after sufficient percolation has taken place.

Nutrients are supplied through different fertilizers and manures. Important organic manure, which is used for nursery bed are Farm Yard Manure (FYM), Compost, Leaf moulds, Oil cakes and Wood ash. This organic manure is incorporated in to the soil before sowing of seeds. Application of inorganic fertilizers in nursery is not so common but initially small amount of N, P and K can be given for fair growth and development of nursery plants. These fertilizers are applied only after irrigation because due to lack of moisture they will become ineffective.

Hardening of Nursery Plants

Hardening or hardening off is a term applied to any treatment that makes firm the tissues of plants to endure better the unfavorable environmental conditions such as freezing hot sunshine or hot drying winds. A well-hardened plant shows dark green colour than an unhardened plant and some may even develop reddish or purplish colour. Any treatment that checks growth increases the hardness in plants.

Technique of Hardening

- By watering the plant sparingly till it is not wilted and providing increased ventilation
- Lowering of temperature also retards growth and adds to the hardening process.

Above mentioned methods are employed according to the kind of the plants. Tomato, egg plants, watermelon, chilies plants are less hardy and they should not be hardened severely. Vegetables like cabbage and lettuce can withstand low temperature up to 18-29° F and are hardy. Very free air circulation hardening both tender and hardy plants. In warm seasons plants hardening is done to withstand possible burning due to sunshine, hot winds and dry soil at their planting season, while in cold season plants hardening may develop protection against freezing. Over hardening should be avoided as it delays renewal of growth after setting out. Hardening should be gradual in order to prevent a severe check to growth or the possible killing of plant. It is better to maintain a moderate rate of growth through out the plant-growing period than to have rapid growth up to the time of hardening and then check it suddenly.

It is very essential that plants should be hardened according their kind so that there is an assurance of high survival and slow growth under the conditions to be expected at the time of transplanting. Hardening should be severe in crops like tomato, brinjal, however, period and degree is varying according to method of hardening. In Indian conditions hardening by allowing the soil of the plant bed to become dry is the best and practical method. Hardening by keeping the plants cool by exposing them to low temperature is difficult to control and as such is not possible (Singh *et al.*, 2009). Hardening of cactus pear plant raised through tissue culture was done before transplanting under field conditions for better establishment and survival of plants.

Effect of Hardening

- Hardening increases the quantity and modifies the nature of colloids in the plant cells enabling them to resist the loss of water.
- Hardening increases the percentage of dry matter and sugars in the plant but decrease the percentage of freezable water and transpiration per unit area of leaf.
- The principal effect of hardening cabbage plant is a change in the constituents of protoplasm, which present their precipitation. It is also found that the proteins are changed to less easily precipitated forms as indicated by an increase with amino acid content on hardening.
- Hardening also decreases the rate of growth in the plants.
- In hardened plants there is greater resistance to plasmolysis and deplasmolysis due to greater resistance to the injurious effects of the protoplasmic strain.
- Hardened plants can withstand better the unfavorable conditions of transplanting such as less intake of water, hot dry winds and low temperature. They tend to be hard and stiff rather than succulent.

- Development of a pink colour in hardened plants especially in the stems, petioles and leaf veins is an indication of high sugar content and other changes in chemical composition. The leaves of hardened plants are of deeper colour and smaller in size.
- Hardening of plants increase the waxy covering on leaves of cabbage and some other plants.

After care

- Plants must be protected from hot wind and sunshine
- For controlling damping off disease in nursery beds drenching of 4% formaldehyde or 0.3% copper oxychloride.
- Plants should be irrigated time to time so that solubility of nutrients present in to the soil is increased.
- In initial stage seedling should be covered with polythene or straw in order to protects from direct sunlight.
- There should not be any weed population.

Constraints

There are some constraints such as low rainfall, high temperature in summer, high wind velocity, low water holding capacity, very low temperature in winter and high potential evapo-transpiration (PET) rate in hot arid regions. However, impact of such edapho-climatic constraints can be lessened by adopting suitable measures such as raising plants under controlled environment conditions, use of fortified rooting or growth medium, provision of temporary thatch or heating of nursery beds during extreme weathers etc.

Registration of Nursery

Availability of true -to- type quality planting material is crucial for success of commercial horticulture. States Govt. ensure supply of quality planting material for fruit crops by enactment of Fruit Plant Nursery (Regulation) Act and enforcement of its provisions through licencing of nurseries. NHB has made guidelines for recognition of horticultural nurseries. In case of approval, initially recognition shall be granted for 02 year. Nursery farms licenced under state Act by competent authority. Registration of fruit plants nurseries is done by State Horticulture department/ State Govt. and also recognition of nursery is done by NHB for a specific period. It is very essential for supply of quality of plants to the growers. Renewal of registration/ recognition of nursery is essential. Registration of nursery should be strictly followed in each state by implementing the Act.

References

Bose, T.K., Sanyal, D. and Sadhu, M.K. 1998. Propagation of Horticultural Crops. Naya Prokash Publishers, Calcutta.

Brison, F.R. and Adriance, G.W. 2000. Propagation of horticultural plants. Biotech Books, New Delhi.

Prakash, G. S. 2010. Demand and supply of quality planting material-National scenario, *In.* National Conference on Production of Quality Seed and Planting material-Health Management in Horticultural crops, New Delhi, pp.6-7.

Reddy, Y.T.N., Jankiram, T. and Reddy, D. S. 2001. Scientific nursery management. The House of Sapran (Media), Bangalore.

Sadhu, M.K. 1986. Plant Propagation. New Age International Publishers, New Delhi.

Sharma, R.R. 2002. Propagation of horticultural crops. Kalyani Publishers, Ludhiana, India.

Sharma, R.R. and Krishna, H. 2013. Textbook of Plant Propagation and Nursery Management. International Book Distributing Co., Lucknow, India (ISBN: 978-81-8189-198-3).

Singh, D., Sivalingam ,P.N., Bhargava, R. and Singh, R. S. 2009. Micro-propagation and *in vitro* conservation of elite genotypes of Cactus pear (*Opuntia ficus indica* Mill), Ind. J. Arid Hort., 4 (1):9-12.

Singh, R. S. and. Bhargava, R. 2009. Effect of seed treatment on germination and growth behavior of date palm (*Phoenix* species) under hot arid conditions. J. Trop. Forestry, 25(1): 42-48.

Singh, R. S. and Bhargava, R. 2008. Propagation of Marula nut through stem cuttings in arid region. Haryana J. Hort. Sci., 37 (3&4): 263-264.

Chapter-8

Ber (*Ziziphus mauritiana* Lamk.)

R.S. Singh
Central Institute for Arid Horticulture, Bikaner-334 006

Ber (*Ziziphus mauritiana* Lamk.) is one of the ancient and common fruits of the country. It is a drought hardy fruit tree and can grow successfully even under harsh climatic conditions. Ber is well known for its ability to thrive under adverse conditions of salinity, drought and poor marginal soils. However, for good tree's growth and yield, deep sandy loam soils with neutral or slightly alkaline reaction and good drainage are most suitable. It is grown throughout the country especially in arid and semi- arid parts. However, it is mainly grown in western parts of Rajasthan where rainfall is less and erratic (Pareek *et al*, 2002). In addition to Rajasthan state, it is cultivated in other states like U.P., Karnataka, Maharashtra, Tamil Nadu, A.P., Gujarat, M.P., Chhatisgarh, Haryana, Punjab and Bihar.

From 1984 to 1995, the area under ber cultivation has been doubled in India. The total area is about 88,000 ha with an estimated production of 0.9 million tonnes. However, the exact figure of area and production is not available. In Rajasthan state, the area under ber cultivation is about 1165 ha giving a production of 10,116 MT (2009-10). Ber is also grown in other south Asian countries viz. Pakistan and Bangladesh .In African countries like Botswana, Mali, Zimbabave, Niger, Burkinofaso, Senegal, Ethiopia, where ber are harvested from seedling plantations. The quality of fruits obtained from seedlings is not as good as improved varieties. In China, *Ziziphus jujuba* (Chinese jujube) is grown in cold desert areas for fresh fruit and processed products.

In North- India, fruits are available from December to March while in South India, it is available from September to November. Ber plants are regular and heavy bearer and the mature trees of different budded varieties can bear up to quintals of fruit per tree. Ber trees are relatively easy and inexpensive to grown in an orchard, yet the returns from it can exceed the profits from several other fruit trees.

Nutritive Value and Uses

The ber fruits are commonly used for fresh eating, dehydration and for making processed products. A number of value added products are prepared from ber fruits. Ber pulp is used for making drinks like ready to serve (R.T.S.), beverage and squash, etc. Preservation of fruits under arid condition is more important in rural areas due to non-availability of fresh produce throughout the year. Practice of preserving fruits and vegetable is in vogue with the inhabitants of desert region since ages due to frequent drought and famine conditions. It not only provides out of season food but also makes proper utilization of the surplus which other wise would have gone waste.

The ber fruits have been relished as dessert as well as used for making various products from the Vedic age is clear from its mention in the ancient Indian literature. Dehydrated ber fruit are exported from China under the name "Chinese dates". Even in India, fruits of the wild as well as cultivated ber are dehydrated in the sun by the villagers and are sold at remunerative price in the market. Work done at Hisar and Jodhpur has shown immense potential for the utilization of ber fruits for making excellent quality preserve, candy, jam, beverages, etc. The technique for making this value added products are now available. The feasibility of adoption of these techniques for commercial exploitation of cottage industries as well as on large scale should now be evaluated (Vashishtha, 2001). The value added products of ber should be commercialized.

Ber is one of the most nutritious fruits. It is also familiar as poor men's apple. It is one of richest source of vitamin C, next only to aonla and guava. The fruit of different varieties contains 12 to 19 per cent total soluble solids and 0.2 to 0.6 per cent acidity at the fully ripe stage. It is also a good source of minerals, protein and carbohydrates (Gopalan *et al*, 1985). Ber is richer than apple in protein, phosphorus, calcium, carotene and vitamin C. Ripe fruits provide 20.9 K calories per 100 g pulp.

The composition of *ber* fruits has been quite widely reported and much of the information is cultivar specific. Fresh, mature *ber* fruits contain 81 to 97 % pulp according to Pareek (1983) and 91.6 to 92.9 % according to Jawanda *et al.* (1981).

Botanical Description

The Ber (*Ziziphus* sp.) belongs to family Rhamnaceae having more than 600 species (Baily, 1947). Out of these, only few species yield edible fruits. Ber is a hardy, much branched, erect, semi-erect, droopping and spreading type growth habit with regular fruit bearing tree. Thorns are available on the stem/shoots. Leaves are green and in different shape. Watt (1908) distinguished between these two main groups of jujube *i.e.* Chinese jujube (*Ziziphus jujuba* Mill.) and Indian jujube (*Z. mauritiana* Lamk.). In most of the flora, Indian jujube is named as *Ziziphus jujuba* Lam. However, the nomenclature of entire genus *Ziziphus* is unclear. There are at least 200 varieties of ber in cultivation in India. Plants are susceptible to low temperature during winter season.

The climatic conditions are more responsible in plant growth, flowering and fruiting in *Ziziphus* species. Summer pruning is required for proper fruiting in arid regions. Fruits availability depends on agro-climatic regions of the country and varies from September to March. Immature fruit drop have been reported in ber which may be due to several factors. Fruit is drup and sweet in taste. The size of stone varies among cultivars. In Northern India, fruits matured during December to February according to the growing cultivars. Under south India conditions, fruit matures early.

Species and Cultivars

Seven species of *Ziziphus i.e. Z. mauritiana,* boradi (*Z. mauritiana* var. *rotundifolia*), *Z. nummularia, Z. oenoplia , Z. spina-christi , Ziziphus jujuba and Ziziphus spp.*(exotic) and around 338 ber cultivars/genotypes are being maintained in the field gene bank at CIAH, Bikaner. The cultivars are normally differentiated on the basis of fruit characters (Vashishtha, 2001). There are many cultivars of ber suitable for different parts of the country. For Indian arid zone, the main selection criteria could aim at identification of genotype having i) high productivity, ii) large fruit size (over 35 g), iii) small seed kernal iv) sweet in taste v) high TSS (above 16^0 brix), low acidity vi) attractive fruit colour besides resistance to extremes of temperatures, pests and diseases. Some well-known cultivars and species from other countries have also been introduced and adapted in arid and semi arid parts of the country.

Among commercial cultivars Gola, Seb, Mundia, Umran, Illaichi, Goma kirti, Peondi, Banarasi kadaka, Sanour-6 and Selections are popular in the country (Pareek *et al*, 2002). Besides this, a number of cultivars are grown in different parts of the country.

Propagation

In India, ber is generally propagated by grafting the scion buds of the desired cultivars on to rootstocks during the summer. The Chinese jujube is propagated by whip or tongue grafting onto rootstocks during the dormancy period at the end of winter through early spring, or by budding during the active growth period in summer. Chinese jujube is multiplied by stem cuttings and root suckers also. The rootstock are raised by sowing seeds of bordi (*Ziziphus rotundifolia)* directly in to the field or in nursery beds or in polythene bags.

Bordi is found most suitable vigorous root stock in hot arid region in comparison to other *Ziziphus* sp. Variability in boradi populations have been observed in Western Rajasthan (Vishal Nath *et al,*1998;Vishal nath and Singh,2000). Seedlings of *Z. nummularia* (Cheema, 1929; Singh *et al.,* 1988; Anon., 1981; Singh and Bal, 1986; Vashishtha, 1987, 1997,Vishal nath *et al,*2000), *ber* (Singhrot *et al.,* 1970; Singh *et al.,* 1973 ; Bal *et al.,* 1997), *Z. xylopyrus* (Singh *et al.,* 1988; Singh and Bal, 1986) and *Z. joazeiro* (Bal *et al.,* 1997) have been used as rootstocks for *ber.* No graft incompatibility has been observed between the scions of the commercial *ber* cultivars and the rootstock seedlings of *ber* (*Z. mauritiana*), boradi (*Z. mauritiana* var. *rotundifolia*), *Z. nummularia, Z. oenoplia , Z. rugosa, Z. spina-christi* or *Z. xylopyrus* (Pareek, 1983). Budding on *Z. nummularia* results in an inverted-bottle graft union, since the rootstock remains thinner than the scion portion (Vashishtha, 1987,). Budding success on *Z. nummularia* is also lower (50 %) compared to 70 % on bordi rootstocks (Anon., 1981).

Scions of cultivar Umran showed maximum vigour, girth and height on dwarf and spreading growth on *ber* (Coimbatore), inverted bottleneck bud union and dwarf growth on *Z. nummularia* and the least vigour and girth of stock and scion on *Z. xylopyrus* (Singh *et al.,* 1988; Singh and Bal, 1986). Out of the 12 rootstocks including 5 accessions of *ber,* Chinese jujube , *Z. joazeiro, Z. xylopyrus,* and 4 accessions of *Z. nummularia, ber* proved to be the best in terms of productivity of the Umran scion (Bal *et al.,* 1997).

In Israel, over 95 % success has been obtained when budding onto *ber* and *Z. abyssinica* rootstocks compared to less than 35 % success on *Z. spina-christi* (Nerd and Mizrahi, 1998). The grafts on *Z. spina-christi* developed a bottleneck form. Seedlings of *Z. mucronata* as a rootstock for cultivar Seb showed delayed incompatibility in Zimbabwe.

The rootstock seedling plants of different species, although from variable progenies, influence the vigour, growth form and productivity of the scion and presumably also its adaptation to adverse climatic and soil conditions. The best choice, at present, seems to be to raise rootstocks from seeds collected from vigorous trees of *ber* (*Z. mauritiana*) and bordi (*Z. mauritiana* var. *rotundifolia*) or other suitable wild species such as *Z. abyssinica*. Use of vegetatively propagated or seedling plants from identified genotypes of the selected species should in due course lead to the development of standard rootstocks to derive the desired effects on the adaptability, productivity, vigour and form of the scion and longevity of the stionic combination under different ecological habitats. In Chinese jujube, seedlings of *Z. jujuba* var. *spinosa* and other wild species are used as rootstocks (Ming and Sun, 1986).

Viability of seed

Seed stones collected from fallen fruits of *ber* often have poor viability. 50 -70 % of the seed stones have non-viable seeds. The variability of seed in ber depends upon maturity,seed storage and other environmental factors. Such seed stones float in 15 to 18 % brine and can be discarded. Grice (1996) observed that germination in seeds collected from the soil surface after 6 months was reduced to 31 %, after 12 months to 20 % and to 7 % if the seeds remain buried in soil (at 2 cm depth) for 6 months. Storage condition has an expected significant effect on germination of seeds. Storage at reduced temperatures e.g. at 4.5 ± 0.5° C in perforated polythene bags, result in retention of viability for a reasonable period (Kulkarni *et al.*, 1992). The viability of seeds can be tested by the triphenyl tetrazolium test (Shanker and Rangnath, 1974).

Seed treatments

Ber seeds can germinate only after a period of after maturity. Seed treatment is required to break dormancy. One or two months after extraction, germination increases and one year old seeds germinate better (Kajal and Singhrot, 1986). Reddy and Murthy (1990) recorded that sowing freshly extracted seeds gave only 33 % germination which increased after 2 months, reaching 50 % after storage for 8 months. Seedling vigour was also the highest when the seeds were sown after 8 months of storage. Murthy *et al.* (1989) observed that ABA-like substances are present in the testa of the seed and that germination was reduced as a result of soaking the seeds in 500 ppm abscisic acid solution. Treatment with 5 ppm abscisic acid also inhibited germination of Chinese jujube seeds (Kim and Kim, 1983 a). Therefore, treatments such as removal of endocarp and water soaking of kernels (Mankar *et al.*, 1997) result in significant improvement in seed germination. Cracking or partial removal of the stony endocarp also result in faster and higher seed germination (Bhambhota and Singh, 1971; Prins and Maghembe, 1994). Removal of the endocarp gave 56 % germination compared to only 3 % in seeds with the endocarp intact (Grice, 1996). In another study, shelling of the stones not only hastened germination but also gave 81 % germination compared to only 38 % in the control (Anon., 1981).

The scarification of seed stones by soaking in concentrated sulphuric acid for 6 minutes was also helpful (Chattopadhyay and Dey, 1992), mechanical extraction of the kernels is still better (Murthy and Reddy, 1989 a). Germination commences within a week when the seed kernels extracted by breaking the stones are sown as compared to 3-4 weeks when sowing seeds with intact endocarps (Bhambhota and Singh, 1971; Pareek, 1978, Anon., 1986; Murthy and Reddy, 1989 a). The seedlings raised by sowing extracted kernels are also more vigorous and attain buddable thickness earlier than sowing of stones. Soaking of seeds in water also enhances percent seed germination in the nursery. Seeds treated with cow dung also promote germination percentage of arid fruits.

Seed germination

Usually seed stones are sown to raise seedlings. Each stone contains as many as 3 seeds embedded in the endocarp of the drupe. The quickness and extent of germination depends upon viability and after-ripening status of the seed, presence of endogenous inhibitors, weathering of the stony endocarp and environmental conditions such as the temperature, moisture, salinity and alkalinity of the growing medium.

In Chinese jujube, one group of cultivars gives 85 % seed germination (e.g. Ya-tszao) and the other gives 98 % germination, e.g. Nikitskii 84, 92 and 94. Similarly, some Chinese jujube seeds take 1 to 3 days to germinate while others take 7 days (Kim and Kim, 1984).

Use of Plant Growth Regulators

Treatment of seed kernels by soaking in 100 ppm (Shanmugavelu, 1966), 200 ppm GA (Hore and Sen, 1994) or 500 ppm GA (Krishnan and Kulasekaran, 1984) or in 0.1 % thiourea (Murthy and Reddy, 1989 b; 1990 a) enhanced the germination percentage. Kulkarni *et al.* (1992) obtained over 45 % germination with seeds soaked in 100 ppm GA for 12 hours, with better seedling vigour in terms of root and shoot growth. *Ber* seeds collected from the faeces of cattle and pigs gave 88 and 77 %

germination, respectively (Grice, 1996). The per cent seed germination can be enhanced with treatment of growth regulators like GA,Cytokinins (Prasad and Singh, 2004).

In Chinese jujube, treatment of seeds without endocarps or of seeds with acid scarified endocarps with 400 ppm gibberellic acid, interrupted dormancy and increased germination. Treatment with 1000-2000 ppm cycocel (chlormequat) reduced the span of germination in Chinese jujube (Casini and Salvadori, 1980) as well as in *ber* (Hore and Sen, 1994). Application of 10 ppm GA or benzyladenine induced germination in immature seeds of Chinese jujube (Kim and Kim, 1983 b). However, there is no treatment required for germination of *Z. obtusifolia* seeds (Vora,1989).

Environmental Factors

Temperature

To raise rootstock seedlings, seeds should be sown at an appropriate time, when the temperature conditions are favourable. In *ber,* maximum germination takes place at 30°C (Murthy and Reddy, 1990 b). Germination is reduced at 25°C and at higher than 35°C. For germination of Chinese jujube seeds, 25°C proved to be optimum (Kim and Kim, 1983a) and temperature. Accordingly, the best time of the year can be selected for seed sowing. For *ber,* the best period has been found to be spring through summer (April to July) in tropical and subtropical climates (Pareek, 1978 , Singhrot and Makhija, 1979 a; Anon., 1981; Kajal and Singhrot, 1986).

Growing Medium

Sand, silt or clay and organic manure in the growing medium gives the best results in terms of seed germination and top and root growth of the seedlings. A mixture composed of equal parts of sand, clay and manure has been used to raise seedlings (Pareek, 1978 ; Bisla *et al.*, 1984). Salinity in the growing medium delays or inhibits germination of *ber* seeds (Dhankar *et al.*, 1980, Dahiya *et al.*, 1981; Dahiya and Dhankar, 1982; Mehta, 1982; Gupta *et al.*, 1989). Increasing chloride and bicarbonate concentration from 2 to 10 dSm^{-1} in irrigation <SPAN> </SPAN>waters tended to decrease seed germination and seedling survival with significant reduction at 4 $dSm^{-1}$of chloride waters and 7.5 dSm^{-1} of bicarbonate waters (Gupta *et al.*, 1989). Dhankar *et al.* (1980) observed that germination declined to 50 % at 6 dSm^{-1} and failed completely at 12 dSm^{-1}. Increasing soil boron levels from 4 to 16 ppm has no appreciable effect on final emergence but in combination with salinity delays the germination (Dahiya *et al.*, 1981). Alkalinity at ESP 40 in the growing medium also delayed seed emergence in *ber* (Mehta, 1982).

Seeds of Chinese jujube sown in the open and covered with plastic sheeting not only emerged 20 days earlier than non-covered seeds but also gave 22 % higher emergence so that almost 50 % more rootstocks could be raised (Ivanova, 1976). Maintenance of pre-irrigation moisture of 80 % in the nursery soil has been helpful in the development of over 98 % seedlings suitable for budding in the first year which had 44 % survival rate of vigorously growing buds and gave a yield of 98 % first grade transplants .

Raising of Rootstocks

Rootstock seedlings can be raised directly in the well prepared field or in the nursery beds by sowing healthy and viable seeds of the selected *Ziziphus* species during the spring to monsoon period at 2 cm depth (Kajal and Singhrot, 1986; Chattopadhyay and Dey, 1992). Seedlings raised in the nursery may be used to prepare budded plants or can be planted in the field for budding *in situ*.

Seed sowing is done during the monsoon season at demarcated locations directly in the field. The seedlings become ready for *in situ* budding during the following spring and summer, i.e. 7-12 months after sowing. Since no transplanting is involved, the plants retain the deep tap roots and are thus more drought hardy and vigorous. In arid region, *in situ* budding method is found suitable for establishment of orchards (Vishal Nath *et al*, 2002).

Sowing is done at 30 x 30 cm spacing (Bakhshi and Singh, 1974) any time from spring through the monsoon period. The seedlings can either be budded in the nursery 90-100 days after sowing (Chattopadhyay and Dey, 1992) or can be transplanted in the field with 20 x 15 cm earth ball (Singh *et al.*, 1982) for *in situ* budding. In subtropical climates, if sowing is delayed until the monsoon period, the seedlings can be budded in the nursery or can be transplanted in the field only after 7 to 12 months, i.e. during the following spring through the rainy season. Thus the plants have to be looked after in the nursery for a long time. Weeds often pose a problem in the nursery and are usually removed manually. Nursery bed raised seedlings have to be carefully handled while lifting from the beds, packing, transporting and transplanting in the field to obtain good survival. Even then, mortality is higher than in plants raised in alkathene bags (Singhrot *et al.*, 1970) because of root injury. By treatment of roots with 250 ppm IBA, transplanting success can be improved to 78.8 %. To obtain better survival, Singhrot and Makhija (1979 a) shifted the seedlings within the nursery one month after sowing and then transplanted them in the field after 7 months, i.e. during the spring. Seedlings can also be transplanted bare-rooted (without earth ball) when these are not growing actively during the winter in subtropical climates. A very high survival of 95 % has been obtained by treating bare rooted seedlings with 250 ppm IBA. Seedbed raised seedlings are not considered suitable for planting under rainfed conditions owing to loss of their tap roots while lifting from the seedbed.

Seedlings grown in the nursery in 28 x 23 cm earthen pots (Singh, 1979) and polythene bags of 23 x 10 cm size (Bhambota and Singh, 1971) can be transplanted in the field at any suitable time and give fairly good survival rates. Sowing time can also be adjusted depending upon the stage of growth required for transplanting in the field. The success rate is higher when transplanted during the monsoon. Singh (1957) raised seedlings by sowing 8-10 stones in a 30 cm pot and then pricking out the young seedlings at the 4 leaf stage into separate pots before finally transplanting in the field. Gupta *et al.* (1991) developed a device known as '*jaltripti*' for establishment of *ber* plants under extremely arid conditions in northwest India. The device comprises of two earthenware pots of the same height but of different diameters joined at the base, keeping the base of the inner pot open and that of the outer pot closed. The outer pot can regularly be filled with water. The *ber* plant is grown in the inner pot in a manure-soil mixture. The device resulted in very good growth of the plants and saved 75 % irrigation water.

Growth of transplanted seedlings can be improved by four weekly applications of 1g/litre urea solution (Bisla *et al.*, 1984). GA treatment has been found to enhance shoot growth of seedlings (Shanmugavelu, 1966). Rootstock seedlings raised in pots and polythene bags can easily be transported to distant places and give higher transplanting success of over 80.5 % (Singhrot *et al.*, 1970). However, due to coiled roots, they have shallow root development and reduced growth vigour when transplanted in the field.

Sowing is done in 300 gauge polythene tubes of 25 x 10 cm size (Pareek, 1978) during the spring .After 90-100 days, seedlings can be transplanted in the field for *in situ* budding. Polytube raised seedlings develop straight and uncoiled roots and thus make nearly as vigorous growth as seedlings raised by sowing seeds directly in the field. Alternatively, the seedlings can be budded while in the polytubes about

100 days after sowing. The polythene tubes have to be protected from disintegration by exposure to solar radiation either by keeping them buried in sunken beds or by proper covering when kept above ground.

Budding

Bud Wood

Bud wood becomes available during the active growth period in the summer i.e. May to July. The bud sticks, with well swollen and recently matured buds (but still not open) are collected. Immature and undeveloped buds from the upper part of the new shoots are unsuitable. Similarly, over mature and inactive buds should not be used. Buds collected from a flowering shoot give very poor success. Buds should be collected from juvenile shoots. Such shoots can be induced to grow by severe pruning of the mother trees.

After collection, the bud wood is often stored for some period or has to be transported. During this period, considerable loss of viability may take place. Bud wood is observed to retain good viability when kept under ventilated conditions and wrapped in moist jute cloth. Singh (1988) found that bud wood wrapped in polythene sheeting and moist sphagnum moss gave better results than those stored in moist jute cloth.

Methods of budding

Budding can be carried out by different methods such as I or T (shield), ring and patch (Anon., 1981; Mawari and Singh, 1992 a ,Vishal Nath *et al*,1998, 2000), chip (Anon., 1981), flute (Mawari and Singh, 1992 a, b) and forkert (Jyotishi *et al.*, 1967); however, shield (Jyotishi *et al.*, 1967) and patch budding are the most commonly used methods. The ring method is not popular because of the necessity for the scion stick to be of equal diameter to the rootstock limits its use on a large scale. Using patch budding techniques, 75 to 85 % success has been observed (Anon., 1981).

Budding should be done as close to the ground level as practicable to minimise the area for emergence of sprouts from the rootstock portion. Over 93 % success has been obtained by budding at a height of 10 cm on seedlings with a stem diameter of 0.53-0.55 cm (Singh *et al.*, 1981; Chattopadhyay and Dey, 1992). Lopping and topping of the rootstock from 8 (Singhrot and Kajal, 1986) to 15 days (Singh *et al.*, 1984 c) after budding in the nursery beds has been found to ensure 90 to 100 % success. Topping has also been found to increase growth of the budlings

(Anon., 1981). Pareek *et al.* (1999) have recommended that just before *in situ* budding in the field, rootstock seedlings should be lopped at 30-45 cm height. When budding onto seedlings raised in polythene tubes, the rootstocks should be prepared by removing all the side shoots and leaves from the stem up to a height of 15 cm above ground level. The seedlings are then topped just before budding using either the shield or patch method. *In situ* budding was found suitable method for orchard establishment under hot arid conditions (Vishal Nath and Singh, 2000). Tomar *et al.* (2010) reported that in situ budding is highly effective with minimum expenditure when local bordi (*Z. rotundifolia*) are grown naturally in plenty. I- budding method was used on newly developed shoots at farmers field in Jodhpur District.

Time of budding

The best time for successful budding is during the active growth period. The active growth period is indicated by easy and clear separation of the bark from the wood in both scion and the rootstock. Kaundal *et al.* (1984) observed maximum bud take (80-87 %) at ambient temperature between 30 and 34°C irrespective of variations in atmospheric relative humidity (RH) from 45 to 73.5 %. Bud take declines below 36% RH and when temperatures drop to 18.5-20° C. Such conditions occur during the summer and monsoon period (June to September) in the arid and semi-arid subtropics of northwest India (Pareek, 1978; Anon., 1981; Kaundal *et al.*, 1984; Mawani and Singh, 1992 a, b). Maximum sprouting of 93.2 % has been recorded by budding when the maximum temperature was between 33 and 36.5° C and minimum temperature was around 26° C (Kaundal *et al.*, 1984). Some success in budding can be obtained even during the mild winter months of the tropics (Jyotishi *et al.*, 1967). However, July to September months are found suitable for *in situ* budding in the field under arid conditions (Vishal Nath *et al.*, 1998).

Transplanting of Plants

Budded plants prepared in the nursery beds are lifted with earth balls at about 9 to 12 months after budding for transplanting in the field. A number of plants may be lost due to damage while lifting, packaging, and transport to the field and transplanting. These operations are also cumbersome and incur high costs due to the large earth balls. Attempts have been made to reduce mortality by i) shifting the plants within the nursery 25 days after sprouting (6 leaf stage) to harden them before final transplanting at about 8 months after budding (Singhrot and Makhija, 1979), ii) shifting the budlings into polybags of 30 x 20 cm size along with treatment of 12 % Waxol (Kundi and Singhrot, 1990), iii) transplanting the defoliated budlings without earth ball during the winter in the subtropics (Sandhu and Dhillon, 1983; Sandhu *et al.*, 1983), and iv) cutting the tap roots along with treatment with 200 ppm IBA (Singh, 1988) and wrapping the roots with moist sphagnum moss or jute cloth during transport to the field. All these methods have been helpful to increase establishment success, but only under irrigated conditions, as they tend to reduce the drought-hardy character of *ber* trees owing to loss of tap roots.

Budded plants prepared in polytubes become ready for transplanting at about 30 days after budding. The polytubes with the budlings are removed from the nursery and kept in shade for a week. These can then easily be transported and transplanted to the field with over 90 % survival (Pareek, 1978) since there is little stress or root damage during lifting, transport and planting. The roots of the plants do not coil and, therefore, retain the drought hardy character and vigour almost similar to plants raised *in situ*. For long distance and overseas transport, the soil from the roots of polytube raised sapling is washed and the roots are covered with sphagnum moss moistened in nutrient solution containing 0.2 g potassium nitrate, 0.8 g calcium nitrate, 0.2 g magnesium sulphate, 0.2 % potassium hypophosphate and 1 ml of 0.5 % ferric tartrate along with 0.2 % Dithane Z-78 (Pareek and Vashishtha, 1980).

Grafting

It is not commonly used in ber under arid environment. *Ber* has been successfully propagated by wedge grafting in Israel (Nerd and Mizrahi, 1998). But this method may not become popular since budding is more economical and easier to adopt. Chinese jujube is propagated by whip and tongue grafting. This is done during the winter when the plants are dormant. Slanting 3-5 cm long cuts are made on both stock and scion shoots of equal diameter. The cut ends of the stock and scion are joined and tied. The grafts become ready after one year for transplanting. One year old, primary shoots of Chinese jujube gave the best results for *in situ* grafting during spring on seedling rootstocks planted in the field (Kim *et al.*, 1982). Cutting back the reproductive shoot above the third node in the middle of June increased shoot length. Softwood grafting in early spring also gave a success of 90 % (Ivanova, 1976).

Stooling and air layering

Attempts have been made to propagate *ber* by stooling (Singh *et al.*, 1973; Anon., 1986) and air layering (Misra and Jauhari, 1970; Ughreja and Chauhan, 1983). In stool beds, nearly 100 % rooting has been obtained within a month in juvenile shoots induced by heading back one month old *ber* plants, and application of either 8000 to 12000 ppm (Singh *et al.*, 1973 b) or 15000 ppm IBA (Ughreja and Chauhan, 1983) immediately after ringing at 10 cm height during the monsoon (early August in north-west India. The rooted stools gave 76 % establishment in the field (Singh *et al.*, 1973). In air layers also, nearly 100 % rooting has been obtained by treatment with 7500 ppm IBA along with 100 ppm boron (Misra and Jauhari, 1970) or 10000 ppm IBA (Chatterjee and Rao, 1978).

In Chinese jujube, cuttings taken from root suckers before emergence give 76.8 % survival (Dong and Song, 1988). Root sucker differentiation from mature trees is 1000 to 1500 per hectare if 15 cm of the soil surface layer is removed in the orchard during March-April or September-October. Chinese jujube suckers when treated before stooling with 2500 ppm IBA gave 100 % rooting in sand, 66.7 % in vermiculture + perlite and 55.6 % in soil medium but did not root at all when stooled without IBA treatment. Chinese jujube suckers can be transplanted during favorable seasons.

Cuttings

Attempts to propagate *ber* by cuttings have been unsuccessful even with the use of growth regulators (Misra and Jauhari, 1970) although root initiation has been observed on treatment with 200 to 400 ppm NAA (Jauhari, 1960). *Ber* contrasts from Chinese jujube in this respect. In Chinese jujube, 5-10 cm long softwood cuttings having 2 3 leaves are taken from the middle or lower part of the shoots during June when the soil temperature stays consistently above 17° C at 8.00 h (Shen *et al.*, 1992). Shen *et al.* (1992) obtained best results by a quick dip in 250-500 ppm IBA or NAA or in a mixture of 250 ppm IBA and 500 ppm NAA. Exposure to 2000 ppm CO_2 for one week immediately after taking the cuttings (Kossuth *et al.*, 1982) and treatments with the strains A4 and TR 105 of *Agrobacterium rhizogenes* (Hatta *et al.*, 1996) have also been observed to induce early rooting and to increase rooting percentage. *Z. nummularia*, can be propagated by root cuttings (Pareek, 1983).Chinese jujube (*Ziziphus jujuba*) can be propagated by cuttings but success percentage was low under hot arid conditions.

Micro-propagation

Micro-propagation offers the advantage of rapid multiplication of clones of elite genotypes in terms of time and space with no limitation of growth seasons. Clonal propagation using stem segments/nodal explants or axillary buds obtained from mature trees has been reported in some cultivars of *Ziziphus* species.

Goyal and Arya (1985) induced adventitious root formation within 20 days in shoot segments of ber (*Ziziphus mauritiana)* Gola and Seb cultivars, in half strength MS medium supplemented with 0.5 mg per litre of IBA and kinetin. Nearly 60 % of the rooted shoots could be successfully transplanted into pots and the total time taken from initial culturing to transplantation was 150 days. The protocol developed by Rathore *et al.* (1992) involves culturing nodal explants on MS medium containing 7.5 mg benzyladenine + 0.1 mg IAA per litre for shoot proliferation, then in White's liquid medium containing 25 mg IBA/litre to induce rooting and then in White's solid medium without growth regulators followed by hardening off before transplanting into pots. Kabir *et al.* (1994) multiplied shoots through axillary bud elongation on MS medium containing 0.1 mg/litre NAA or IAA and 2-mg/litre bezyladenine, which were rooted by transferring the shoots to MS medium containing 0.5-2 mg/litre IBA. Large-scale multiplication of shoots can be carried out by repeated subculturing of the nodal segments of *in vitro* grown shoots after every 6-8 weeks.

Mathur *et al.* (1995) grew stem explants from mature trees on MS medium containing 3800 mg/litre potassium nitrate, 2475 mg/litre ammonium nitrate, 11 mm-M benzyladenine and 0.5 mm-M IAA, which produced 15-20 shoots per inoculums on successive sub culturing. Rooting in the shoots could be induced by pre-treatment with 50 m-M IAA or NAA for 24 hours followed by their transfer to auxin-free White's medium and then in soil and vermiculite mixture. Fougat *et al* (1997) reported that ber cv. Gola can be multiplied through *in vitro* technique.

By culturing stem segments of Chinese jujube lines A 17, A 27 and A 80 in MS medium containing 1 mg BA per litre, the highest percentage of clustered buds (66.7) was differentiated and then maximum root growth (95.9 %) was induced in the medium containing 0.4 mg/litre 1BA (Yan *et al.,* 1990). Culture medium containing 2 mg BA + 0.4 mg IBA per litre gave 81.8 % differentiated buds which could be excised continuously for multiplication. Kim and Lee (1988) obtained the best shoot and root growth from axillary buds of cultivar Geumsung after eight weeks when 500 mg/litre activated charcoal was added to half strength MS medium with 0.5 mg/litre benzylaminopurine whereas 1000 mg/litre activated charcoal was required for cultivar Bokjo. Rooting and callus growth of Geumsung microshoots increased as IBA concentration was increased to 3 mg/litre. Rooting was better when treated with paclobutrazol rather than with IBA, NAA or Rooton.

For producing rooted plants from shoot tips and single bud segments, MS medium supplemented with IBA was more effective than with IAA or NAA (Cheong *et al.*, 1987). Kinetin and 2iP (0.5 mg/litre) and BA (1 mg/litre) increased plant height, number of nodes and plant weight. A combination of 1 mg/litre IBA and 1 mg/litre BA or 0.5-1.0 mg/litre kinetin also increased plant height and weight. Addition of adenosine sulphate at 40 mg/litre to the shoot tip culture and at 80 mg/litre to bud segment culture in MS medium + 1 mg/litre IBA + 1mg/litre BA further increased plant height and number of nodes. Charcoal, when added to the culture medium decreased root number and length.

Attempts have also been made to regenerate plantlets of *Ziziphus* cultivars via callus formation from explants such as nodal segments, leaf, cotyledon, root, and hypocotyl followed by organogenesis and somatic embryogenesis (Mitrofanova *et al.*, 1994, 1997). This may not, however, provide a practical technique for multiplication of true-to-type plants, and there is a risk of somaclonal variation occurring.

The micro-propagation of ber (*Ziziphus mauritiana* Lam.) was studied through tissue culture and results shows that it is a perfect time for cutting explants from healthy growing plants in March and April. The optimum media were screened out through studies on the condition of initiation, proliferation and rooting culture. Main achievements were shown as follows: MS medium supplemented with BA 0.8 mg/L and IBA 0.4 mg/L was suitable for ber initation. MS medium supplemented with BA 1.2 mg/L and IBA 0.5 mg/L was the best for shoots multiplication. But the medium favourable for rooting was 1/2MS basal medium supplemented with IBA 3.0 mg/L. The results show that eight generations are suitable for shoot multiplication. After eight generations rooting culture should be carried out (Jiuru *et al*, 2003).

The limited research has been done in ber on micro propagation thus, the work on micro propagation needs to be strengthened in ber.

Top working

Old trees of *ber* and other wild *Ziziphus* species can be made productive by top working, i.e. replacement of the top by budding or grafting with desired scion cultivar. The top growth of the old tree is first removed by lopping at 60-100 cm height (Pareek, 1983) or at 2 m height (Pareek, 1983), if high-headed trees are to be developed. Several new shoots emerge from the remaining trunk in 15 to 30 days time, depending upon its age. Of these shoots one vertical, well-spaced and vigorous shoot on each limb or 2 to 3 such shoots facing different directions on the single trunk are retained. These shoots are budded with a suitable scion cultivar. The date of heading back is adjusted so that suitable new shoots become available for budding at the selected time. In the arid and semi-arid subtropics of India, lopping can be carried out between spring and summer time (Singh *et al.*, 1973; Pareek, 1983) for better sprouting and growth.

Quality Standards of Planting Material

There are no proper quality standards of planting material of fruits. However, in order to maintain the quality of planting materials, Govt. of India has passed the fruit Nurseries regulation Acts in each state to function the registration under the acts of licensing and regulation of nurseries of fruit plants. As per standards, the plant should be true to the type, healthy, vigorous, disease/pests free, use of proper root stock/ scions and maintenance of proper stock/registers of nursery production. The plant should be of proper age and genuine to the quality. The nurseries acts should be imposed strictly in each state to maintain the quality standards of fruit plants. It is very essential for quality production of fruits.

References

Anonymous 1986. Annual Report. All India Coordinated Fruit Improvement Project, Coordination Cell III, ICAR, India. Technical Document No. 19: 88.

Anonymous 1998-99. Vital Horticulture Statistics, Directorate of Horticulture, Govt of Gujarat.

Anonymous, 1981. Project Report 1976-1980. All India Coordinated Fruit Improvement Project, Coordination Cell III, ICAR, India. Technical Document No. 1: 6-8, 35.

Baily, L.H. 1947 .The standard Cyclopedia of Horticulture, Macmillon & Co, New York.

Bakhshi, J. C. and Singh, P. 1974. The ber - a good choice for semi-arid and marginal soils. Indian Horticulture, 19: 27-30.

Bal, J. S., Singh, M. P. and Sandhu, A. S. 1997. Evaluation of rootstocks for ber (*Ziziphus mauritiana* Lamk.) cv. Umran. Journal of Research, Punjab Agricultural University, 34(1): 60-63.

Bankar, G. J. and Prasad, R. N. 1992. Relationship between stomatal distribution and growth of ber rootstocks. Indian Journal of Horticulture, 49(2): 169-171.

Bhambhota, J. R. and Singh, A. 1971. A quick method of raising ber rootstock. Punjab Horticultural Journal, 11: 251-253.

Bisla, S. S. Singhrot, R. S. and Chauhan, K. S. 1984. Effect of growing medium and urea application on seed germination and growth of ber (*Ziziphus mauritiana* Lamk.). Haryana Journal of Horticultural Science, 13(3/4): 118-122.

Casini, E. and Salvadori, S. 1980. The germination of seeds of jujube (*Ziziphus sativa*). 1. The effect of gibberellic acid on seeds with and without endocarp [in Italian]. Rivista di Agricultura Subropicale e Tropicale, 74(1/2): 39-47.

Chattopadhyay, P. K. and Dey, S. S. 1992. Note on standardization of some aspects of ber propagation. Indian Journal of Horticulture, 49(1): 47-49.

Cheema, G. S. 1929. The value of *Ziziphus rotundifolia* as a rootstock plant. Agricultural India, 24: 46-47.

Cheong, S. T., Kim, S. K., Peak, K. Y. and Ahn, H. K. 1987. *In vitro* rooting and branching responses of jujube shoots as affected by growth regulators [in Korean]. Journal of the Korean Society for Horticultural Sciences, 28(1): 53-60.

Dahiya, S. S. and Dhankhar, O. P. 1982. Studies on salt tolerance of ber. Haryana Journal of Horticultural Science, 11(1/2): 53-58.

Dahiya, S. S., Dhankar, O. P. and Khera, A. P. 1981. Studies on the effect of soil salinity levels on seed germination of ber (*Ziziphus rotundifolia*). Haryana Journal of Horticultural Science, 10: 20-23.

Dhankhar, O. P., Makhija M. and Singhrot, R. S. 1980. Effect of salinity levels on germination of seeds and growth of transplanted seedlings of ber (*Ziziphus rotundifolia*). In *Proceedings of the International Symposium of the Arid Zone Research Development*, Jodhpur, India, ed. Mann, H. S. pp. 351-356.

Dong, Y. X. and Song, E. T. 1988. Summary of propagation techniques of *Ziziphus jujuba* with its suckers. Shanxi Fruit Trees, No.2, 24-25.

Fougat,R.S.,Joginder,S.,Tashlim,A.Arya,M.D.,Godhani,P.R.,Singh,J. and Ahmed, T.1997.*In vitro* studies in ber (*Zizyphus* mauritiana Lamk. cv.Gola). Jour.Applied Hort.Navsari, 3:45-47.

Gopalan, C., Ramashastri, B.V. and S.C. Balasubrramanian 1985. Nutritive value of Indian Foods, National Institute of Nutrition, ICMR, Hyderabad.

Goyal, Y. and Arya, H. C. 1985. Tissue culture of desert trees. II. Clonal multiplication of *Ziziphus in vitro*. Journal of Plant Physiology, 119: 399-404.

Grice, A. C. 1996. Seed production, dispersal and germination in *Crytostegia grandiflora* and *Ziziphus maurititana*, two invasive shrubs in tropical woodlands of northern Australia. Australian Journal of Ecology, 21: 324-331.

Gupta, I. C., Sharma, B. D. and Singh, P. M. 1989. Effect of bicarbonate and chloride in irrigation water on germination and growth of ber and soil properties. Journal of the Indian Society of Soil Science, 37: 428-430.

Gupta, I. C., Singh, P. M., Yadav, N. D. and Sharma, B. D. 1991. Note on performance of ber during establishment under conventional and *Jaltripti* irrigation system. Indian Journal of Horticulture, 48(3): 222-223.

Hatta, M., Beyl, C. A., Garton, S. and Diner, A. M. 1996. Induction of roots on jujube softwood cuttings using *Agrobacterium rhizogenes*. Journal of Horticultural Science, 71(6): 881-886.

Hore, J. K. and Sen, S. K. 1994. Role of pre-sowing seed treatment on germination, seedling growth and longevity of ber (*Ziziphus mauritiana* Lamk.) seeds. Indian Journal of Agricultural Research, 28(4): 285-289.

Ivanova, R. B. 1976. Raising planting material of large-fruited jujube cultivars [in Russian]. *Trudy N II Sadovodstva Vinogradavstva i Vinodeliya R.R. Sherdera* No. 37, 107-116.

Jauhari, O. S. 1960. Some observations on vegetative propagation of *Ziziphus mauritiana* Lamk. (ber) from gootees and cuttings with the aid of growth regulators. Current Science, 29: 30-31.

Jawanda, J. S., Bal, J. S., Josan, J. S. and Mann, S. S. 1981. Ber cultivation in Punjab. Punjab Horticultural Journal, 21: 17-22.

Juiru, Xu, Li. Chunli and Sun Jianshe 2003. Micro-propagation of ber (*Ziziphus mauritiana* Lam.). Jour. Beijing Forestry Univ, 2003-03.

Jyotishi, R. P., Kashyap, R. and Maurya, K. N. 1967. Budding of ber, *Ziziphus jujuba* for top working can be done throughout the year. JNKVV Research Journal, 1: 91-92.

Kabir, A., Islam, R., Haque, A., Khalekuzzaman, M. and Joarder, O. I. 1994. Micropropagation of *Ziziphus mauritiana* from nodal segments of mature trees. Bulletin of Sericultural Research (Bangladesh), 5: 5-8.

Kaundal, G. S., Grewal, S. S. and Bal, J. S. 1984. Influence of meteorological factors on bud take efficiecy of ber (*Ziziphus mauritiana* Lamk.) cv. Umran. Journal of Research, Punjab Agricultural University, 21(3): 372-374.

Kim, D. S. and Lee, S. P. 1988. Study on practical micropropagation of jujube cultivars through axillary bud culture [in Korean]. Journal of the Korean Forestry Society, 77(4): 445-452.

Kim, W. S. and Kim, Y. S. 1984. Changes in carbohydrates, proteins, RNA and hydrolytic enzyme activity of jujube seeds during germination [in Korean]. Journal of the Korean Society for Horticultural Sciences, 25(2): 109-115.

Kim, Y. S. and Kim, W. S. 1983 a. Factors influencing seed germination of *Ziziphus jujuba* Mill. [in Korean]. Research Reports, Office of Rural Development, South Korea Horticulture, 25(10): 125-130.

Kim, Y. S. and Kim, W. S.1983 b. Studies on germination of *Ziziphus jujuba* Mill. seeds at different stages of growth and development [in Korean]. Research Reports, Office of Rural Development. South Korea, Horticulture, 25(10): 47-53.

Kim, Y. S., Kim, W. S. and Hong, K. Y. 1982 . Studies on the grafting of *Ziziphus jujuba* Miller [in Korean]. Research Reports, Office of Rural Development. South Korea Horticulture, 24(12): 117-122.

Kossuth, S. V., Biggs, R. H., Webb, P. G. and Portier, K. M. 1982. Rapid *propagation techniques for fruit crops.* Proceedings of the Florida State Horticultural Society, 94: 323-238.

Krishnan, B. M. and Kulasekaran, M. 1984. Studies on seed germination in wild ber (*Ziziphus rotundifolia*). South Indian Horticulture, 32(3): 153-154.

Kulkarni, A. R., Bhat, N. R., Supe, V. S. and Kale, P. N. 1992. Influence of storage conditions and pre-sowing treatments on germination and seedling growth in ber. Journal of the Maharashtra Agricultural Universities, 17(1): 132-133.

Kundi, S. and Singhrot, R. S. 1990. Propagation studies in ber (*Ziziphus mauritiana* Lamk.). X. Effect of Waxol dipping on the transplanting success of budlings. Haryana Journal of Horticultural Science, 19(3-4): 225-231.

Mankar, S. W., Dod, V. N. and Bharad, S. G. 1997. Effect of different methods of seed germination in ber. Crop Research (Hisar), 14(3): 437-439.

Mathur, N., Ramawat, K. G. and Nandwani, D. 1995. Rapid *in vitro* multiplication of jujube through mature stem explants. Plant Cell Tissue and Organ Culture, 43(1): 75-77.

Mawari, P. B. and Singh, S. P. 1992 a. Effect of method and time of budding success in ber (*Ziziphus mauritiana*) cv. Gola. Part II. Horticulture Journal, 5(2): 79-83.

Mawari, P. B. and Singh, S. P. 1992 b. Effect of method and time of budding success in ber (*Ziziphus mauritiana*) cv. Gola. Part I. Horticulture Journal, 5(1): 31-35.

Mehta, P. K. 1982. *Studies on the Effect of Soil Sodicity and Boron on Seed Germination, Growth and Mineral Composition of Young Ber* (*Ziziphus* spp.) *Plants*. Thesis Abstracts, pp: 47-48.

Ming, W. and Sun, Y. 1986. Fruit trees and vegetables for arid and semi-arid areas in northwest China. Journal of Arid Environments, 11: 3-16.

Misra, A. K. and Jauhari, O. S. 1970. Root induction in layers and stem cuttings of *Morus alba* L. and *Ziziphus mauritiana* Lamk. with special reference to plant growth regulators. Indian Journal of Horticulture, 27(3/4): 141-146.

Mitrofanova, I. V., Cheboter, A. A. and Mitrofanova, O. V. 1994. Capacity for *in vitro* morphogenesis in vegetative buds and embryos of *Ziziphus jujuba* Mill. as affected by material genotype and culture conditions. Russian Journal of Plant Physiology, 41(6): 826-831.

Mitrofanova, I. V., Mitrofanova, D. V. and Pandei, D. K. 1997. Somatic embryogenesis and plant regeneration in *Ziziphus jujuba* Mill. *in vitro*. Russian Journal of Plant Physiology. 44(1): 94-99.

Murthy, B. N. S. and Reddy, Y. N. 1989 a. Effect of different methods on seed germination in ber. Journal of the Maharashtra Agricultural Universities, 14(3): 296-298.

Murthy, B. N. S. and Reddy, Y. N. 1989 b. Improvement of seed germination in jujube (*Ziziphus mauritiana* Lamk.) with growth regulators and nitrogenous compounds. *Indian* Journal of Agricultural Science, 59(4): 269-270.

Murthy, B. N. S. and Reddy, Y. N. 1990 a. Evaluation of Umran and local type of ber (*Ziziphus mauritiana* Lamk.) for raising rootstocks. Indian Journal of Horticulture, 47(4): 389-391.

Murthy, B. N. S. and Reddy, Y. N. 1990 b. Temperature dependence of seed germination and seedlings growth in ber (*Ziziphus mauritiana* Lamk.) and their modification by pre-sowing treatments. Seed Science and Technology, 18(3): 621-627.

Murthy, B. N. S., Murti, G. S. R. and Reddy, Y. N. 1989. Effect of abscisic acid on seed germination and its endogenous levels in ber (*Ziziphus mauritiana* Lamk.) seeds. Indian Journal of Plant Physiology, 32(3): 275-277.

Nerd, A. and Mizrahi, Y. 1998. Ber experience in Israel. Abstr. *International Workshop on Ziziphus mauritiana*, Harare, Zimbabwe, 13-16 July, 1998.

Pareek, O. P. 1983a. The Ber. ICAR, New Delhi, India.

Pareek, O. P. and Vashishtha, B. B. 1980. Soilless ber budlings for safe transport. Indian Horticulture, 25: 11-12.

Pareek, O. P. and Vishal Nath. 1996. Ber. In *Coordinated Fruit Research in Indian Arid Zone - A two decades profile (1976-1995)*. National Research Centre for Arid Horticulture, Bikaner, India: 9-30.

Pareek, O. P., Vishal Nath. and Vashishtha, B. B. 1999. *In situ* technique to establish ber orchards in arid region. Extension Bulletin 8. National Research Centre for Arid Horticulture, Bikaner, India.

Pareek, O. P.; B. B. Vashishtha; R.S. Singh and Vishal Nath. 2002. Sushka kshetra mein-Ber ki kheti se Adhik Amdani, Extension Bulletin,(IInd edition), CIAH,Bikaner,pp16.

Pareek, O. P.1978. Quicker way to raise ber orchards. Indian Horticulture, 23, (1): 5-6.

Prasad,J and A.K.Singh 2004. Practical manual on propagation of fruits, N.D.U.A.&.T,. Kumar ganj, Faizabad.(UP)pp.50.

Prins, H. and Maghembe, J. A. 1994. Germination studies on seed of fruit trees indigenous to Malawi. *In.* Forest Ecology and Management. Elsevier: 111-125. Rajasthan, Jaipur, pp.171.

Rathore, T. S., Singh, R. P., Deora, N. S. and Shekhawat, N. S. 1992. Clonal propagation of *Ziziphus* species through tissue culture. Scientia Horticulturae, 51(1-2): 165-168.

Reddy, Y. N. and Murthy, B. N. S. 1990. Studies on germinability and seedling vigour at different intervals of seed storage in ber (*Ziziphus mauritiana* Lamk.). Indian Journal of Horticulture, 47(3): 314-317.

Sandhu, A. S. and Dhillon, B. S. 1983. Success in bare-rooted transplanting of ber at grower's fields. Punjab Horticultural Journal, 23(1/2): 57-58.

Sandhu, A. S., Dhillon, B. S. and Singh, Z. 1983. Studies on bare-rooted transplanting of ber budlings. Punjab Horticultural Journal, 23(1/2): 53-56.

Shanker, G. and Rangnath, A. S. 1974. A quick test for aonla, ber and guava seed viability. Current Research, 3(2): 18-19.

Shen, X. D., Gao, F. G., Chen, B. X. and Wang, Y. 1992. An experiment of propagation of *Ziziphus jujuba* by green wood cuttings [in Chinese]. Ningxia Journal of Agroforestry Science and Technology, No. 2: 32-34.

Singh, K. 1979. Vegetative Propagation of Ber. M.Sc. thesis, Haryana Agricultural University, Hisar, India.

Singh, K. K. 1957. The Ber in India. ICAR Farm Bulletin No. 18. New Delhi, India

Singh, K., Singhrot, R. S. and Chauhan, K. S. 1981 b. Effect of budding height and diameter of rootstock on budding success and budding growth of ber. Haryana Journal of Horticultural Science, 10(1/2): 69-71.

Singh, K., Singhrot, R. S. and Chauhan, K. S. 1982. Vegetative propagation of ber (*Ziziphus mauritiana* Lamk.). VI. Effect of various transplanting treatments on the growth of ber seedlings and success of budding. Haryana Jour. of Horticultural Science, 11: 5-12.

Singh, L. 1988. Propagational Studies in Ber (*Ziziphus mauritiana* Lamk.). Ph. D. Thesis, Haryana Agricultural University, Hisar, India.

Singh, M. P. and Bal, J. S. 1986. Performance of Umran ber (*Ziziphus mauritiana* Lamk.) on different rootstocks. Horticultural Science, 21: 797.

Singh, M. P., Bal, J. S. and Sandhu, A. S. 1988. Effect of different rootstocks on vigour, yield and fruit quality of ber (*Ziziphus mauritiana* Lamk.) cv. Umran. *National Symposium Dryland Horticulture*, Hyderabad, India, Abstr. p. 3.

Singh, P., Bajwa, M. S. and Singh, R. 1973 . Propagation of ber (*Ziziphus mauritiana* Lamk.). I. Effect of ringing and IBA on the stooling. Plant Science, 5: 137-139.

Singh, R.S. 2013. Ber (*Ziziphus mauritiana* Lamk,) *In.* Fruit Production in India (*ed.* by Dhillon WS), Narendra Publishing House, New Delhi, pp.127-137.

Singhrot, R. S. and Kajal, R. S. 1986. Studies on the vegetative propagation of ber (*Ziziphus mauritiana* Lamk.). I. Effect of after budding treatments on the budding success and budling growth. Haryana Journal of Horticultural Science, 15(3/4): 162-169.

Singhrot, R. S. and Makhija, M. 1979 . Vegetative propagation of ber (*Ziziphus mauritiana* Lamk.). II. The transplanting success of grafted ber plants at various stages of scion shoot growth. Haryana Jour. of Horticultural Science, 8(1/2): 4-8.

Singhrot, R. S., Bakhshi, J. C. and Singh, K. 1970. Vegetative propagation of ber (*Ziziphus mauritiana* Lamk.). 1. Relative performance of seedlings raised in field and in alkathene bags. Punjab Horticultural Journal, 10 (3/4): 181-186.

Tomar, A.S., Singh, R.P., Bhati, P.S., Badgujar,V.K. and Bhandari, M.C. 2010.Performance of *in situ* budding technology in ber at farmers field under rainfed conditions in Jodhpur District of Rajasthan, In National Conference on Production of Quality seed and Planting material, New Delhi,11-14th March,2010, p.27.

Ughreja, P. P. and Chauhan, K. S. 1983. Effect of plant growth regulators on rooting in air layers of ber (*Ziziphus mauritiana* Lamk.). Haryana Journal of Horticultural Science, 12(3/4): 160-164.

Vashishtha, B. B. 1987. Horticultural qualities. In *Bordi* (*Ziziphus nummularia*), (eds. Mann, H.S. and Saxena, S.K.), CAZRI, Jodhpur pp. 31-33.

Vashishtha, B. B. 1997 *Ziziphus* for drylands - a perennial crop solving perennial problems. Agroforestry Today, July-September, 1997: 10-12.

Vashishtha, B. B.2001. Ber Varieties: A Monograph, Agrobios (India), Jodhpur, 97p.

Vishal Nath and Singh, R.S. 2000. *In- situ* budding-Sushka kshetra mein Ber Bag sthapana ki Vidhi(in Hindi), Udam Patrika, 31:12-13.

Vishal Nath, Saroj P.L.; Singh, R.S. ; Bhargava, R. and Pareek, O.P. 2000. *In situ* establishment of ber orchard under hot arid ecosystem of Rajasthan. Indian Jour.Hort.57(1):21-26.

Vishal Nath, Singh, R.S. ; Bhargava, R. and Pareek, O.P. 1998. Patch and 'I' budding of ber (Z. *mauritiana* Lamk.) on boradi (Z. *mauritiana* var. *rotundifolia*) root stock, Curr. Agric.

Watt, G.1908. Commercial Products of India, pp1143-74.

Yan, R. L., Liu G. R., Zhang, L., Wang, Z. X. and Yang, E. Q. 1990. Effect of exogenous plant hormones on rapid multiplication of *Ziziphus jujuba* test tube plants. Journal of Plant Science, 7(4): 231-233.

Chapter-9

Pomegranate (*Punica granatum* L.)

R.S. Singh and B.D. Sharma
Central Institute for Arid Horticulture, Bikaner-334 006

Pomegranate (*Punica granatum* L.) also known as anar, dalim is an ancient fruit and cultivated in semi-arid to arid regions of the country. However, pomegranate plants are found throughout the country from temperate to tropical climate. It has enjoyed a reputation for its healthy dietetic and medicinal properties. Pomegranate has versatile adaptability, drought tolerance, high value fruits, steady and high yields of the crop (Vishalnath *et al.*, 2007). The pomegranate fruit has therapeutic value (Julie Jurenka, 2008); good keeping quality, high export potential and is mainly grown for table purpose, though several value added products are also available.

Pomegranate fruit has good national and international demand in the market. It is commercially grown in the states of Maharashtra, Gujarat, Tamil Nadu, Karnataka, Rajasthan and Andhra Pradesh. Maharashtra is leading state in the country for its cultivation and export of fresh fruits to Middle East, Germany, Holland, U.K. and The Netherlands. Pomegranate in its wild form exists in hilly slopes of Himachal Pradesh, Uttarakhand and Jammu & Kashmir. The wild pomegranate called as "*Daru*" is highly acidic. Under arid condition, the performance of introduced anardana type from H.P., has been evaluated. The fruit arils are dried called anardana is used as a condiment and acidulent in many Indian curries, chutneys (Singh e*t. al.*, 2011) and also used for various industries like tanning, dying and colouring. Amlidana, a hybrid variety of pomegranate is highly acidic (16.18%) and marketed as condiment in North India for use in culinary preparations (Jalikop *et al.*, 2002).

In India, the area of pomegranate cultivation is about 1.3 lakh ha with production of 11.0 lakh tonne. Pomegranate occupies an area of 754 ha with estimated production of 6000MT. In western parts of Rajasthan, it is under cultivation, where the summer season is long and dry, cool winters and irrigation facilities are available. It is a drought hardy fruit plant. Due to availability of proper soil moisture and mild temperature, flowering during August-September month is suitable under arid conditions. The tree requires hot and dry climate during fruit development and ripening. Pomegranate can be grown on a wide range of soils but is found to do well on light and medium type of soil. The soil should be well drained. Fruit quality and colour development in light soils is good but poor in heavy soils. Under hot arid environment, it has been observed that upper top portion of plants/ tender shoots are slightly affected by frost during winter season.

Nutritive Value and Uses

Pomegranate is a nutritious fruit and known for its antioxidants, which eliminate harmful free radicals in the body (Damania, 2005). Pomegranate has edible portion 68 per cent, moisture 78 per cent, protein 1.6 per cent, fat 0.1 per cent, fibre 5.1 per cent, carbohydrates 14.5 per cent and mineral matter 0.7 per cent, calcium 10 mg 100 mg^{-1}, magnesium 12 mg 100 g^{-1}, phosphorus 70 mg 100 g^{-1}, iron 0.3 mg 100 g^{-1}, oxalic acid 14 mg 100 g^{-1}, thiamine 0.6 mg 100 g^{-1}, riboflavin 0.10 mg 100 g^{-1}, nicotinic acid 0.30 mg 100 g^{-1}, vitamin C 14 mg 100 g^{-1} and calorific value 65 mg 100 g^{-1} (Anon., 1952). Pomegranate fruits vary in colour from pale yellow to red. The juicy pulp in the arils (seeds) varies from almost colourless to blood red. The seeds in some pomegranate varieties are very soft, while in others they are large and hard. Pomegranate fruits consist of about 60-67 per cent seeds and 33-40 per cent peel (Jagtap *et al.*, 1992). Whole pomegranate fruits contain 45-61 per cent juice, while arils yield 76-85 per cent juice (Patil and Sanghavi, 1980). The seed pulp in superior quality pomegranate fruits is thick, fleshy and very juicy. The pulp taste varies from sweet and aromatic to sour and insipid. The seed coat varies in hardness, the softer-seed varieties being known as seedless. Lack of lignifications of the testa is the main cause of seedlessness in pomegranate (Saxena *et al.*, 1987). Pomegranate juice is of great value in digestive disorders. It is an appetizer, a digestive food for patients suffering from colitis and mucus. Pomegranate fruit is used for making delicious juice, jelly. Now-a -days, pomegranate arils are being extracted by machines and frozen arils are packed for marketing.

Pomegranate fruits are the good sources of carbohydrates and minerals such as Ca, Fe and S and a moderate source of pectin. The vitamin and mineral contents of pomegranate are given in table 1. Pomegranate is a poor source of vitamin C (Trapaidze and Abuladze, 1989). Citric acid is the predominant organic acid in pomegranate, amounting to about 7-30 mEq/100 g fresh weight (Malhotra *et al.*, 1983, Ulrich, 1970 and Lee *et al.*, 1974). Other acids include malic, oxalic, succinic and tartaric acids. Glucose and fructose are the main sugars in pomegranate. Pomegranate fruit juice contains 5.46 per cent glucose, 6.14 per cent fructose and almost no sucrose. However, Siddappa and Bhatia (1954) reported about 8.9 per cent reducing sugars and 0.1 per cent sucrose.

Table 1 : Vitamin and mineral contents of pomegranate (edible portion).

Vitamin / Mineral	**Content (mg 100 g^{-1})**	
	Jagtap *et al.* (1992)	**Sood *et al.* (1982)***
Vitamins		
Thiamine	0.06	-
Riboflavin	0.1	-
Niacin	0.3	-
Vitamin C	16.0	5.3 – 7.7
Minerals		
Calcium	10.0	24 – 145
Phosphorus	70.0	33 – 44
Iron	0.30	0.62 – 0.69
Magnesium	12.0	-
Copper	0.17	-
Sodium	4.0	-
Potassium	17.1	-
Sulphur	-	25 – 28

* Variation in three varieties

The fruit juice easily ferments and may be used for the production of wine. The juice of wild pomegranates in Azerbaijan (USSR) is used in the manufacture of citric acid and sodium citrate for medicinal purpose. The bark of the stem and root contain a number of alkaloid belonging to the pyridine group. It is also employed in the therapeutics in dysentery and diarrhea. El-Shaaraway and Nahapetian (1983) reported that pomegranate seed contains about 15 per cent oil with a high refractive index, iodine value and very low melting point. The oil has a potential for industrial use. The seeds also contain 1.09 mg oestrone 100 g^{-1} of seeds and 0.036 mg coumestrol (a non-steroidal oestrogen) 100 g^{-1} of seeds (Moneam and El-Sharaky, 1988). Pomegranate seed oil is ideal in nourishing and improving skin elasticity (especially fine lines). High in anti-oxidants, it is extremely moisturizing. The pomegranate is native to ancient Persia, Iran, has been valued for centuries as a medicinal plant. Which is extensively cultivated in the Mediterranean countries like Spain, Morocco, Egypt, Iran, Afghanistan and Baluchistan. It is also grown to some extent in Mynamar, China, Japan, USA (California), USSR and India. It is also found growing in Bulgaria and Southern Italy. Today, we know that the secret of the pomegranate can be found in a single drop of its precious oil, pressed from the seeds found in the red arils of the fruit.

Pomegranate seed oil contains nutrients that can add moisture to the skin, making more elastic and resistant to wrinkles. It also has anti-inflammatory properties which can relieve the skin of irritations such as eczema and damage from sunburn. There are also recent studies which prove that pomegranate seed oil can stimulate regeneration of skin cells, which can make the skin look softer and younger. Pomegranate seed oil also contains conjugated fatty acids which is anti-inflammatory and can relieve muscle pains. It can also relieve skin rashes that may appear because of too much exposure to the sun during summer. The peel of fruits is used for making natural colour for dying cloths and in tooth powder. The bark both of the stem and roots of pomegranate tree is well known for its anthelmentic properties for destroying parasitic worms. Juice of pomegranate is used to prepare sodium citrate, citric acid which is used in medicines.

Botanical Description

The genus *Punica* has two species: (i) *Punica protopunica* and (ii) *Punica granatum* . *Punica protopunica* is found wild in the Socotra Island and *Punica granatum is* cultivated in tropical and subtropical regions. *Punica granatum* belongs to the family Punicaceae has been classified into two subspecies, chorocarpa and porphyrocarpa. Chlorocarpa is found in Transcum cacus region and porphyrocarpa in Central Asia.

Pomegranate is a bushy plant, which attains 3-5 m height with multiple branch system. The leaves are elliptic to lance late, dark green in colour measuring 7.5 cm long. The inflorescence in pomegranate is cymes. Male and hermaphrodite flowers are borne in one cluster. Flower bud of spring flush is borne on mature wood of one year old shoot whereas during July- August flush, these come on current season's growth. Orange red flowers are borne in clusters mainly in the axis towards the end of branchlets. Terminal flowering is also found in pomegranate. The calyx is tabular, persistent, five to seven lobbed, the petioles lance- shaped inserted between the calyx lobes. The ovary is embedded in the calyx tube and contains several compartments in two series, one above the other. It is diploes with the basic chromosome number n = 16 or 18. Time of dehiscence of anthers in pomegranate varies with variety and existing environmental conditions particularly temperature and humidity. Although, male and hermaphrodite flowers bear fertile pollens, the fertility rate in hermaphrodite flower is quite high. The stigma attains maturity one day prior to anthers and receptivity remains at peak up to two days of anthesis. Both self and cross-pollination recorded in pomegranate however, higher percentage of fruit set has been observed by hand pollination and open pollination (Bose and Mitra, 1984). Flowering takes place throughout the year in pomegranate in three flushes. To take fruiting of one flush according to climate conditions of the region, crop regulation or bahar treatment is essential. Fruit is juicy, sweet and used for fresh eaten.

To regulate flowering/fruiting, irrigation is withhold for about 2 months in advance of normal flowering season. To obtain higher fruit yield during a particular period, plants are given a resting period by which the natural tendency of the tree is alerter with artificial means. Flowering can be induced in June. Two main flush *i.e.* ambe (February -March) and mrig bahar (July -August) is common for fruiting, however, late flowering in mrig season is called hast bahar (August -September). For better fruiting, choose only one bahar according to growing place and regulation of flowering through artificial method like withholding irrigation, spray of chemicals, deblossoming, etc.

Genetic diversity has been observed in pomegranate because of orchards of seeds originated. Varietal improvement in pomegranate has been attempted both by selection of promising types from the indigenous ones and through controlled hybridization. Exotic introductions including Russian cultivars under the climatic conditions of Deccan plateau give sparse flowering and poor fruit set. The fruits are highly acidic, hard seeded but with arils of very attractive red colour. Pomegranate has the major problem of the fruit borer which could be tackled using the *Bt* genes and the systems for micro-propagation and somatic embryogenesis have been attempted and protocols are available. There is a wide amount of variability and lot of genetic resources are available, which can be characterized using biochemical and molecular markers, but limited work has been reported on these aspects. Fruit cracking is another problem in pomegranate cultivation in many areas.

Several types of pomegranates cultivated in India are distinguished by the shape of the fruit, colour and thickness of rind, colour of the seeds (aril) and the taste. On the basis of yield and physico-chemical characters, various cultivars have been recommended for the commercial cultivation in different states in India, *viz.*, Ganesh, Muskat, G-137, Mridula, Phule Arakta and Bhagwa in Maharashtra, Dholka in Gujarat, Jalore Seedless, Jodhpuri Red, Jodhpuri White and Seedless (Bedana) in Rajasthan, Bassein Seedless and Madhurigi in Karnataka, Chawla, Nabha, Country Large Red in Haryana. The description of some important cultivars of pomegranate is as given below.

i) **Ganesh** : Prolific yield, medium-sized fruits, soft seeds, pinkish flesh and sweet juice, agreeable taste.

ii) **Dholka** : Fruit large in size, rind greenish white, fleshy testa, pinkish white or whitish with sweet juice, seeds soft, juice acidic.

iii) **Kabul:** Fruit large in size, rind deep red mixed with pale yellow, thick, fleshy testa dark red with slightly bitter juice.

iv) **Kandhari** : Fruit large in size, rind deep red, fleshy testa blood-red or deep pink with sweet, slightly acidic juice, seeds hard.

v) **Jalore Seedless**: It is suitable for cultivation in arid and semi arid condition, but fruit cracking is major problem. Medium fruit yield, light red aril colour, sweet in taste and medium fruit yielder.

vi) **Mridula:** Fruits are dark red in colour, sweet in taste, good yielder, hybrid cultivar developed from MPKV, Rahuri, It is being exported because of its better fruit quality.

vii) **Jyoti** : Attractive yellowish red fruit colour, medium fruit size, red aril colour and soft seeds. It has a yield potential of over 18,000 kg fruits per ha (Anon., 1985).

viii) **G-137**: Fruit colour is yellowish-red, sweet in taste, good fruit yield and export potential.

ix) **Bhagawa** : 'Bhagawa' variety of pomegranate presently under commercial cultivation known by different names *viz.*, *Shendari, Ashtagandha, Mastani, Jai Maharashtra* and '*Red Diana*' is a heavy yielder and possesses desirable horticultural characters. This variety matures in 180-190 days with average yield of 30 kg fruits / tree. Bigger fruit size, sweet, bold and attractive arils, glossy, very attractive, saffron coloured thick skin makes it suitable for distant market (Waskar *et al.*, 2003).

x) **Phule Arakta** :Phule Arakta variety of pomegranate is a heavy yielder and possesses all desirable fruit characters. The fruits are bigger in size, sweet with soft seeds, blood red arils and posses' glossy, attractive dark red skin (Waskar *et al.*, 2003). It is also suitable for distant market.

Propagation

Pomegranate is propagated by seeds but it is not advisable due to variability in seedlings. It is largely propagated through vegetative methods *viz.* stem cuttings and layering (gooties) (Hayes,1953). Micro-propagation has been found successful method to multiply true to type pomegranate plants (Chauhan and Kanwar, 2012). Looking to problem of Bacterial blight, oily spot disease in Maharashtra, multiplication of disease free planting material through micro-propagation is very essential and useful. To tackle this aspect, work has initiated to raise tissue culture pomegranate plants on large scale.

Seed

It has been observed that pomegranate plantations in India have been raised from seedlings but the trees from seedlings show wide variation in characters. Low yields and poor quality of fruits are mainly due to seed propagation. In order to obtain fruits of uniformly good quality, it is essential to establish orchards with vegetatively propagated plants.

Vegetative propagation

Vegetative propagation of pomegranate (*Punica granatum* L.) through cutting is the most convenient and cheap method of obtaining true to the type, fully developed plants in considerably lesser time and in order to improve the rooting and reduce the mortality of rooted cuttings, the use of growth regulators and time cuttings has been standardized.

Cutting

The common method of vegetative propagation is by stem cuttings. It is propagated by both soft and hard wood cuttings. The cuttings are taken from fully mature wood, about one-year old and 20-25 cm long and the suckers which arise from the base of the stem are usually utilized for the purpose. Soft wood cuttings were used to multiply plants under mist conditions during January and February and success was achieved. Root formation is markedly improved by using growth substances. Phadnis (1974) reported higher percentage of rooting and large number of roots by treatment with IAA at 200 ppm. Basal and medium cuttings rooted to the extent of 75 per cent by soaking the base of cuttings in 50 ppm IBA for 18 hours and to 66 per cent by quick dip for 20 seconds in 5000 ppm IBA (Anon., 1985). Bose and Mandal (1972) obtained 85-100 per cent rooting in pomegranate using IBA at 3000 ppm under mist and the rooted cuttings also showed very high percentage of survival. Bicket and Bilderback (1982) reported that softwood cuttings of pomegranate when treated with 0.5 per cent potassium salt of IBA as quick dip and planted in peat-perlite media under intermittent mist gave high percentage of rooting. Pomegranate may also be propogated by air layering or gootee. Treatment of cuttings with 10000 ppm IBA in lanolin as carrier was about to improve rooting (Suryanarayan and Venkateswara Rao, 1982). Treatment of cuttings with 1000ppm IBA gave better rooting in pomegranate under semi arid conditions (Singh, 1994). Prasad and Bankar (2003) also reported that cuttings treated with 1000ppm IBA for 30 seconds gives 60-80% success in rooting in different season. Third week of February to second week of March is optimum time for planting of cuttings.

Maximum rooting, root number and root length is observed with IBA 500 ppm + Borax 1% both in semi hard and hard wood cuttings in pomegranate. Field survival of the cuttings treated with IBA 500 ppm + Borax 1%, IBA 300 ppm + Borax 2% and IBA 5000 ppm remained maximum. Hard wood cuttings respond better to the hormonal treatment as compared to semi hard-wood cuttings (Sharma *et al.*, 2009). For improving the rooting and growth potential of pomegranate stem cutting, the basal end of semi hard wood cuttings should be dipped for 24hrs. in IBA @100ppm before planting on 15th January (Navjot *et al.*, 2010). A decline in rooting and plant growth was recorded with increasing dose of IBA.

Air-layering

Pomegranate is largely propagated through air layers (Srivastava, 1961). Air layering is carried out in the month of May/June. In this method a 1-2 year old, healthy, vigorous mature shoot of 45-60 cm in length and pencil thickness is selected. A circular strip of bark about 3cm wide just below a bud is completely removed from the selected shoot .Rooting hormones like IBA & NAA 50mg each in lanoline paste are applied over this portion. Moist sphagnum moss is packed around this portion and tied with polyethylene sheet to prevent the loss of moisture. Application of such hormones promotes early rooting. Light brown roots are visible through the polythene wrap in the month of July-August. The rooted shoot is slowly detached by giving 2-3 successive cuts over a period of week before finally detached from the parent plant. The polythene sheet is removed before planting them in pots. They are planted in pots and kept in nursery under shade. Top of the shoot is cut back to maintain a proper ratio of leaves: roots. The grafts can be transplanted in the field in the month of September/ October (Kalloo *et al.,* 2005).

Grafting method has also been tried but not very much successful for large scale multiplication.

Micro-propagation

Micro-propagation of a dwarf variety of pomegranate has been reported using stem portion (Omura *et al.*, 1987a), leaf explant (Omura *et al.*, 1987b), cotyledons hypocotyl, embryos (Jaidka and Mehra, 1986, Raj Bhansali, 1990). Tissue cultures of anther (Moriguchi *et al.*, 1987) and petals (Nataraj and Neelambika, 1996) for haploid and somatic embryogenesis have been reported. Browning of explants and media are biggest problems to initiate pomegranate culture from mature plants. The growth of cultures could be maintained successfully by using charcoal containing media and quick subculturing on fresh media.

The apical and nodal segments of pomegranate cultivars *viz.*, Jalore Seedless and Ganesh can be used for clonal multiplication by using 0.5 - 1.0 mg /l BAP and 0.1 mg/l NAA or without any auxin. Higher concentration of BAP (2-3 mg/l) produced multiple shoot formation from nodal bud as well as from apical tips. Elongated shoots initiated roots from the basal cut end on soaking in 200-500 mg/l IBA and then cultured on without auxin medium. In this way, a large number of plants could be regenerated, hardened and successfully grown in the fields. A rapid and efficient procedure was outlined for *in-vitro* clonal propagation of an elite cultivar of pomegranate (cv. Ganesh) using nodal stem segments of a mature tree. Bud break and axiallary shoot proliferation were induced in the nodal segments on MS medium supplemented with BA, Zeatin riboside (ZR) or TDZ, ZR being most effective. The greatest number of shoots were developed on a medium containing 2.0 mg/ l ZR. However, the shoots failed to elongate. This problem was overcome by transferring the shoot clumps to a medium with a lower level of ZR. TDZ was the least effective of all the cytokinins tested. Rooting was induced in 86 per cent of the regenerated shoots in a half-strength MS medium supplemented with 1.0 mg/l IBA. The plants were acclimatized and established in soil (Naik *et al.*, 1999).

According to Naik and Chand (2011) systematic biotechnological advances has been made in pomegranate. These encompass optimization and establishment of *in vitro* culture techniques including micro-propagation, somatic embryogenesis, synthetic seed production, plant regeneration via callus-mediated shoot organogenesis, adventitious shoot regeneration, anther culture, tetraploid induction and genetic transformation.

Micro-propagation could be a beneficial approach to mass produce cultivars with ideal characteristics, including insect and disease resistance. Research has demonstrated the most efficient way to micro-propagation is by enhancing axillary bud branching (Chauhan and Kanwar , 2012). However, more research is needed to refine micro-propagation methods.

Quality Standards of Planting Material

The plant should be true- to -type and free from diseases. Plant should not be very old, and vegetatively propagated plants of improved varieties should be used for planting in the field. To avoid spread of oily spot diseases, plants should be procured from reliable source or from registered nursery. At present, wilt disease is becoming a major problem in cultivation of pomegranate in other parts of country.

References

Anon. 1952. The Wealth of India - A Dictionary of Raw Material and Industrial Products. Vol III, CSIR, New Delhi.

Anon. 1985. Progress Report. 3rd National Workshop on Arid Zone Fruit Research, MPKV, Rahuri held during July 5-8., 1985.

Bicket, J. and Bilderback, T. E. 1982. Softwood cutting in pomegranate. Amer. Nurseryman, 156 : 85-86.

Bose, T. K. and Mandal, D. P. 1972. Studies on rooting in pomegranate. Indian Hort., 17 : 25-26, 30.

Bose, T. K. and Mitra, S.K. 1984. Pomegranate, Tropical and subtropical Commercial fruits Naya Prakash, Calcutta.

Chauhan, R.D. and Kanwar Kamlesh,2012. Biotechnological advantages in pomegranate *(Punica granatum* L.*), In Vitro* Cell Dev. Boil- Plant, 48: 579-594.

Damania, A.B., 2005. The Pomegranate : its origin, folklore, and efficacious medicinal properties. In. Proceedings of the International Conference on Agricultural Heritage of Asia.,Hyderabad. pp. 175-183.

El-Shaaraway, M. I. and Nahapetian, A. 1983. Oil content and iodine value of pomegranate seed. *Feete Scifen Anstrichitted*, King Faisal Uni. Saudi Arabia. 85 : 123-126.

Hayes, W.B. 1953. Pomegranate, Fruit growing in India. Kitabistan, Allahabad, p. 344.

Jagtap, D. B., Desai, U. T. and Kale, P. N. 1992. Chemical composition of some indigenous and exotic cultivars of pomegranate. Maharashtra J. Hort., 6 (1) : 10-12.

Jaidka, K. and Mehra, P. N. 1986. Can. J. Bot., 64 : 1644-1653.

Jalikop, S.H., Tiwari, R.B. and Sampath Kumar, P. 2002. Amlidana- A new pomegranate hybrid. Indian Horticulture, 47(2):22-23.

Julie Jurneka, M.T. 2008.Therapeutic applications of Pomegranate (*Punica granatum* L.): A review. Alt.Med.Rev.,13: 128-144.

Kalloo, G.; Pandey, S.K., Chakrabarti, S.K. and Kaushik,S.K.2005.Technologies for Production of Quality Seed and Planting Material of Horticultural Crops, Tech. Bulletin No.71, CPRI, Shimla,pp 91.

Lee, S. W.; Kim, K. S. and Kim, S. D. 1974. Studies on changes in composition of pomegranate fruit during maturation. 1. Changes in sugars, organic acids and the respiration rate. J. Korean Soc. Hort. Sci., 15 : 57.

Malhotra, N. K.; Khujuria, H. N. and Jawanda , J. S. 1983. Studies on physico-chemical characters of pomegranate cultivars II. Chemical characters. Punjab Hort. J., 23 : 158.

Moneam, N.M.A. and El-Sharaky, A. S. 1988. Oestrone and Coumestrol content of pomegranate seeds. J. Chromatography, (438) : 438-441.

Moriguchi, T.; Omura, M.; Matsurta, N. and Kozaki, I. 1987. Hort. Science, 22 : 947-948.

Naik, S. K.; Chand, P. K. and Pattnaik, S. 1999. *In vitro* propagation of pomegranate (*Punica granatum* cv. Ganesh) through axillary shoot proliferation from nodal segments of mature tree. Scientia Horti., 79 : 3-4.

Naik, S. K.; Chand, P.K., 2011. Tissue culture- mediated bio technological intervention in pomegranate- A review. Plant Cell Rep. 30 707-721

Nataraj, K. and Neelambika, G. K. 1996. Indian J. Exp. Biol., 34 : 719-721.

Navjot, Brar,J.S and Dalal, K.P.S. 2010.Role of time and growth regulators on rooting and growth on stem cutting of pomegranate. In. Abstracts of National Conf. on Production of quality seed and planting material held at New Delhi from 11-14March,2010 pp.147.

Omura, M., Matsuta, N.; Morighuchi, T. and Kozaki, I. 1987 b. Hort. Science, 22 : 133-134.

Omura, M., Matsuta, N.; Morighuchi, T., Kozaki, I. and Sanada 1987a. Bull. Fruit Tree Res. Stn., Japan, 14 : 17-44.

Patil, A. V. and Sanghavi, K. U. 1980. Performance of different varieties of pomegranate (*Punica granatum* L.) in dry regions of Western Maharashtra. Ann. Arid Zone, 19 : 485.

Phadnis, N. A. 1974 Effect of growth hormones on rooting of cuttings in pomegranate. Indian Hort. 19 : 8-13.

Prasad, R.N. and Bankar, G.J., 2003. Cultivating pomegranate in arid region scientifically. Indian Horticulture, 47:(4) 26-29.

Raj Bhansali, R. 1990. Somatic embryogenesis and regeneration of plantlets of pomegranate. Ann. Bot., 66 : 249-253.

Saxena, A. K.; Menah, J. K.; Berry, S. K. 1987. Pomegranate post-harvest technology : Chemistry and Processing. Indian Food Packer, 41 (7/8) : 43.

Sharma, Nirmal; Anand, R.; Dharminder Kumar, 2009. Standarization of (*Punica granatum* L,) propagation through cuttings. Biological forum - An Internation Journal 1(1): 75-80.

Singh, R. S. 1994. Effect of plant growth substances on rooting in pomegranate cuttings. Current Agric., 18: 87-89.

Singh, R. S., Sharma, B.D., Bhargava, R. and More, T.A. 2011. Introduction and evaluation of Anardana types pomegranate under hot sarid conditions. Acta Hort.,890: 239-242.

Sood, D. R.; Dhindsa, K. S. and Wagle, D. S. 1982. Studies on the nutritive value of pomegranate fruits. Haryana J. Hort. Sci., 11 (3-4) : 175-177.

Srivastava, R.P, 1961. Propagation of pomegranate (*Punica granatum* L.) by air layering, Sci. & Cult. 27: 350-351

Suryanarayan, V. and Venkateswara Rao K. 1982. Air layering in pomegranate. Natl. Semi. Plant Propagation, Calcutta. 70-71.

Trapaidze, T. G. and Abuladze, L. S. H. 1989. Pomegranate cultivars resistant to cracking. Subtropicheskie Kultury, (2) : 95.

Ulrich, R. 1970. Organic acids, In: Biochemistry of Fruits and Their Products (Hulme, A.C.E., Ed.), Academic Press, New York, p. 89.

Vishal Nath, R.S. Singh and Bikas Das 2007. Pomegranate In. Fruit Production Technology (ed. by Yadav, P.K.,), International Book .Dis..Cop, Lucknow, pp 307-322.

Waskar, D. P.; Kulkarni, S. R.; Navale, A. M. and Shete, M. B. 2003. Bhagawa and Phule Arakta. MPKV Publication No. 410. MPKV, Rahuri.

Chapter-10

Date palm (*Phoenix dactylifera* L.)

R.S. Singh
Central Institute for Arid Horticulture, Bikaner-334 006

Date palm (*Phoenix dactylifera* L.) known as *Khajoor, Kharek* is an important fruit tree of semi-arid and arid regions of the world. It grows well under poor desertic soils due to its hardy plant characteristics and deep root system. It is also one of the most suitable fruit plant, which can be easily grown under saline irrigation conditions. It is well known fact for date palm that its feet in the water and its head in sun. It requires dry hot climate for growth and development of fruits. In arid region, crop production is risky preposition, where date palm cultivation contributes in achieving food security, high nutritive value food, crop diversification, desertification control, income generation and foreign exchange earnings. It can be grown at such areas where adequate irrigation facilities are available besides other climatic requirements (Chandra *et al.*, 1992; Sharma and Singh, 2013).

Date palm has religious significance. It is believed that it is oldest food plant in the world. Egyptians too honored it. Later it was important to threes of the world's major religious groups, the Jews, Muslims and Christians. The Jews are also believed in the date palm. In almost all culture and religions of the world, dates are supposed to have special importance. For instance, in Muslims Holy book 'Quran' date palm has been referred to more than 20 times. The palm was also said to have been the building material for Mohammed's home. Likewise in many other religions, dates are important on the religious occasions like Christians tradition and it is believed by some that the 'tree of life' in the Bible. The fruits are used during holy month of *Ramzan* by Muslims. As Saudi Arabia is principal producer of date palm in the world, the palm tree is incorporated in the national emblem of Saudi Arabia to represent vitality and growth. In many other countries mentioned on coins, currency and as symbols.

Date palm requires specific climate for fruiting and fruit quality hence, in India it is grown in the states of Rajasthan, Punjab, Haryana and Gujarat. However, commercial date palm growing countries of the world are Iraq, Saudi Arabia, Algeria, Iran, Egypt, Morocco, Tunisia, Pakistan and Sudan besides many other Near East and North African countries. Date palm has played a vital role in the history of mankind by providing food and shelter to millions of people. Pakistan is the fourth largest date producing country, contributing 11% to the world production (Zaid *et al.*, 2007). In India, date palm production is not meeting our domestic demand and maximum dates are imported from date-growing countries of the world. However, India also exports small quantity of dates to neighboring countries like Nepal, Myanmar, Bhutan, Maldives and Sri Lanka.

The extremely dry hot areas comprising Jaisalmer, Barmer, Bikaner and Jodhpur in Rajasthan, Bhatinda, Fazlika, Abohar in Punjab, Sirsa, Hisar, Mahendergarh, Bhiwani in Haryana and Kachchh region in Gujarat are the potential areas for date palm cultivation in India. At present, the area of cultivation under this crop is limited in our country and now it is being increased by planting of imported tissue culture plants.

In Rajasthan, earlier area of date palm cultivation was about 50- 60 hectare, now it has increased by planting of tissue culture plants in districts of Jaisalmer, Barmer, Jodhpur and Bikaner. At Bhojka (Jaisalmer), 14,125 plants of cultivars *viz.*, Barhee, Khadrawy, Khuneizi, Khalas, Zamali, Saggai and male palms have been planted in August, 2009 whereas at Khara Farm, Bikaner total of 5,268 plants have planted of same cultivars during August 2010. Rajasthan Horticulture Development Society, Directorate of Horticulture has initiated a programme for plantation of date palm in the state.

In Kachchh region, the maximum date palms are natural and as grove in depressions, on riverbank, *nala* and boundary of the farms (Singh *et al.*, 2004). An estimated area of 8, 973 ha. with production 53,836 tonnes was during the year 2000– 01, now which has been increased to 12,493 ha with annual fruit production of 85,352 tonnes during 2004-05 in the Kachchh region (Bhuj, Anjar, Khedoi, Mundra, Mandavi, etc.) as reported by Muralidharan *et al.* (2008). During the year 2009-10, the area of date palm cultivation in Gujarat state has increased to 16,688 hectare with production of 123490 tonnes fruits. The tissue culture plants of cultivar Barhee imported from UK through UAE have been planted in about 500 acre areas on farmers' field in Kachchh district and started fruiting after 3-4 years of planting. In these plantations, suckers production is not allowed by the growers for early fruiting.

Further, the area under tissue culture plantations has been increased in Kachchh region with the support of Gujarat state under Government financially assisted a programme. A private organization M/s Kutch Crop Services Ltd., Gajod c/o Excel Crop Care Ltd., Mundra has supplied date palm tissue culture plants either imported cv. Barhee or produced local elite type at Tissue Culture Laboratory, Mundra. Few hectare plantations of date palm has also done near Didigul, Tamil Nadu state by the farmers.

Looking to its future prospects, new date palm plantation has been done in arid region of Rajasthan and Gujarat. There is a large chunk of potential area for the development of date palm groves in arid region of north- west India. The total estimated area suitable for date palm cultivation is about 5.5 million ha. This includes low-lying stagnant and saline lands, natural wastelands and *Khadins* (natural water harvesting areas), canal command areas and well-irrigated lands. This potential area can be brought under date palm cultivation where other crops may not be that remunerative.

At present, India is importing date fruits in the form of processed products like *chhuhara,* pind, etc. (2, 98,423 tonne) from Gulf countries which involves about 9.90 million $ foreign exchange. Besides this, India also exports dates to neighboring countries, amounting 1, 792, 000 dollars (FAO Stat).

Nutritive Value and Uses

Date palm fruit is nutritious and favorite fruit throughout the world. It has medicinal value. Fruits of date cultivars viz. Halawy, Khalas, Khuneizi, Chipchap, Braim and Barhee are eaten as fresh (*doka* or Khalal stage). Besides fresh consumption, several value added products dry date (*Chhuhara*), pind, beverages, jam, RTS, paste, chutney, pickle, etc are prepared from *doka* stage fruits. It is known to produce more well mineralized, highly flavored and nutritious food per hectare than any other fruit of even temperate zone. The mineral content of dates amounts to about 2 per cent of cured fruits on weight basis. It is a good source of iron and potassium and fair source of calcium but a poor source of phosphorus (Nixon, 1966). Chlorine, copper, magnesium and sulfur are also present in moderate quantities. The dates contain high amount of vitamins (Vitamin A = 220 I.U.; Vitamin B1 (Thiamine) = 0.35 mg, Vitamin B2 (Riboflavin) = 0.38mg, all being vital for the normal metabolic processes of the body. Protein content varies from 1.75 to 2.75 per cent of the flesh on wet weight basis. The flesh of the date contains 0.31 – 1.9 per cent fat and 2.52 – 7.42 per cent wax on wet weight basis. The date fruit is slightly acidic. High acidity is often associated with poor quality.

This fruit can supplement the dietary needs of the desert people where very less nutritious food is available. Dates are cured as dry dates (*Chhuhara*) and soft dates (*Pind Khajoor*), which has great demand and market in the country. In fact, it was the only staple food available to the desert people in the Near East and North African region till oil was discovered there. Even today, it forms a prominent component of daily diet of people in these countries. The per capita consumption of dates in Saudi Arabia, Iraq, Libya and Egypt is 28.9, 17.3, 28.0 and 10.8 kg/ year, respectively. Date fruits provide abundant quantities of sugar, iron, potassium, calcium and nicotinic acid. Small amounts of protein, copper, magnesium, chlorine, and sulphur, Vitamins A, B_1 and B_2 are also present in date pulp. The flesh of dates with a moisture content of 20 per cent contains 60-65 per cent sugar, about 2.5 per cent fiber, 2 per cent protein and some what less than 2 per cent fat, mineral matter and pectic substances, provides about 3,550 calories per kg (Nixon and Carpenter, 1978). Fresh 100g date fruit contains 59.2 g moisture, 1.2 g protein, 0.4 g fat, 1.7 g minerals, 3.7g fiber, 33.8 g carbohydrates, 22.0 mg calcium, 38.0 mg phosphorus while 100 g dry dates (*chhuhara*) provides 2.5 g protein, 2.1g minerals, 3.9g fibers, 75.8g carbohydrates, 120 mg calcium, 50 mg phosphorus, 7.30 mg iron, 0.02 mg riboflavin, 0.9 mg Niacin, 3.0 mg vitamin C (Gopalan *et al.*, 1985).

The flesh of date fruits contains about 80 per cent sugar on dry weight basis which being easily digestive provides a ready source of energy to the human body. Such fruits provide about 3,150 calories per kg. The sugars are mostly of the invert type (glucose and fructose). However, some dry and semi-dry date varieties contain relatively higher content of cane sugar (sucrose). The walls of the cells in dates are primarily made of cellulose which together with other non-soluble solids constitutes about

85 per cent of the dry matter in the very small green stage. However, with the increase of sugars, cellulose content decreases. Fully ripe Zahidi dates contained from 4.5 to 10.1 per cent of crude fiber, whereas a fully cured date contained not more than 2.0 per cent of crude fiber of cellulose. Similarly, starch content decreases with ripening of fruits.

Due to presence of tannins, many dates in green stage are astringent in taste. However, a few varieties/cultivars such as 'Halawy' and 'Barhee' are not astringent when green. These are palatable and can be eaten fresh (*doka stage*). As the colour changes from yellow to red, tannin is deposited as insoluble grain and thus the astringency disappears.

Date-palm leaves are used for preparation of paper, baskets, mats and brooms. The tree trunk provides timber for making furniture. The date palm tree also furnishes shelter, clothing, fuel, building material, fiber, starch, sugar, oil, wax and many other products. Wine is also prepared from dates. The date palm trees are very valuable as a cover crop for ground crops in hot and dry regions. However, all plant parts of date are useful since history of cultivation (Chao and Krueger, 2007) in several ways to make handicrafts items. The shade also allowed other plants to grow in the heat of the desert. In Iraq, two tiers of crops are grown below the date palms *viz.* fruit trees like olive, almond, pomegranate, peach, orange and lemon and below them vegetables, berries and flowers. Growing of citrus plants under date palm tree's shade has been found successful under Abohar conditions, Punjab. Further, cultivation of turmeric and ginger was also found suitable in inter space of date palm plantations at Abohar, Punjab.

Due to better keeping quality of dry and soft dates and their increasing demand it does not face any problem of perishability or glut in the market. The fruits have better keeping quality under normal conditions and it can be stored for long period under refrigeration. Besides, doka dates, *chhuhara* (dry dates) and *Pind khajoor* (soft dates) are in high demand for raw eating (Vishal Nath and Singh, 2003). Fruits are also used for making squash, pickles, chutney, beverage and syrup. Its delicious fruits at doka (khalal) stage are eaten as fresh and also used for processing to different value added products (Singh *et al.*, 2004). Caramel colour has been produced from dates which can be used in food and pharmaceuticals industries (Zaid *et al.*, 2007). Moreover, all plant parts of date palm are useful since its history of cultivation (Chao and Krueger, 2007) in several ways to make handicraft items. Its seed kernel is used in the preparation of cattle feed. In Gulf countries, several value added products are made from date palm plant parts as well as fruits. Date biscuits were prepared by adding of 10 and 20 per cent dried date pulp in wheat flour, ghee, milk, sugar in local bakery. The biscuits were tested and it was found suitable on organoleptic score basis (Singh *et al.*, 2009).

Delicious drinks, RTS can also be made from *doka* fruits of local red and yellow type's berries of date palm in Kachchh region of Gujarat. Many date products are exported to other countries where date cultivation is not being done.

Botanical Description

Datepalm (*Phoenix* sp.) belongs to family Arecaceae (Palmae) and consist twelve species, out of which some are grown for edible fruits and others as ornamental purposes. *P. dactylifera* has 36 chromosomes (n=18 2n=36).The two wild species *P.sylvestris* (L.) Roxb and *P. canariensis* are most widely available; both are crossable to *P. dactylifera.* Besides date palm, five species viz. *P. atlantica* Chev., *P. reclinata* Jacq., *P. pusilla Gaertn. (syn.P. farinifera* Roxb.), *P. humilis* Royle, *P. acaulis* Roxb. also bear edible fruits (Chandra *et al.*, 1992; Zaid, 1999). The Latin name of the date palm, *Phoenix dactylifera* is derived from a Phoenician name "*Phoenix*", which means date palm and *dactylifera* derived from a Greek word "daktulos" means a finger.

In India, generally two species of *Phoenix* are found. Out of which, *P. sylvestris* called "desi khajoor" is found through out the country but it produces inferior quality of fruits. The fruit contains large stone with less pulp content. From its crown parts, a liquid is extracted known as "*neera*" or *taddi*, which is used as fresh drink and also for preparation of jaggary. The tree of desi khajoor (*P. sylvestris*) is found in Bundelkhand region of the country and it is suitable tree component for agro forestry system (Tewari *et al.*, 2001). However, P. *dactylifera* is commercially cultivated through out the world for fresh fruits and value added products.

Twelve species of the genus "*Phoenix*", along with their geographical distribution, were first listed by Chevalier (1952):

Species	Common Name	Distribution
Phoenix dactylifera L.	Date Palm	Mediterranean Countries, Africa and part of Asia; introduced in North America and Australia
P. atlantica A. Chev.		Africa and Canary Inlands
P.canariensis	Canary Palm	Occidental Canary Islands and Cape Verde Chabeaud.
P. reclinata Jacq.	Dwarf Palm	Tropical Africa (Senegal and Uganda) and Yemen (Asia)
P. sylvestris Roxb.	Wild Date Palm or taddi Palm	India and Pakistan
P. humilis Royle.	-	India, Burma, and China
P. hanceana Naudin.		Meridional China and Thailand
P. robelinic O'Brien.		Sri Lanka, Tonkin, Annam, Laos and Thailand
P. farinifera Roxb.	Pigmy Palm	India, Sri Lanka and Annam
P. rupicola T. Anders.	Rocky Date Palm	India
P.acaulis Roxb. ex.Benth.	Dwarf Palm	Bangladesh and India
P. paludosa Roxb.	Hental or Juliana Palm	Bangladesh, Tenasherim, Andaman, Nicobar Island and Thailand

Besides date palm, five of the above species bear edible fruit (*P. atlantica, P reclinata, P. humilis*, and *P. acaulis*). Most of the *Phoenix* species are well known as ornamentals, the highly valued is *P. canariensis* commonly called the Canary Island Palm. *P.sylvestris* Roxb. is widely used in India as a source of sugar. *P.dactylifera* is distinguished from the above two species by several characteristics which could be summarized as follows:

- production of offshoots;
- its tall, columnar and relatively thick trunk. If the crown of fronds is included, the date palm could reach a height of over 20-25 m
- its dark green leaves,

Close relationship among the 12 species is illustrated by the ease of hybridization and cross-pollination (Munier, 1973). Several natural hybrids were hence obtained: *P. dactylifera* X. *P. sylvestris* (India); *P. dactylifera* X *P. canariensis* (Morocco, Algeria and Israel); *P. dactylifera* X *P. reclinata* (Senegal). *Phoenix dactylifera* has 36 chromosomes (n=18; 2n=36) (Beal, 1937), but polyploidy cases were reported by Al- Salih and Al- Jarrah (1987) with Iraqi date varieties (2n=64). The same authors reported differences between varieties; Sayer as an early variety (2n = 32) and Khasab, a late variety (2n=36). Furthermore, in both varieties, aneuploidy and euploidy were observed: (Sayer: 32, 34, 36 and 64 and Khasab: 32 and 36).

Root system: Being a monocot plant, date palm has no tap root system. Its root system is fasciculate and roots are fibrous, similar to a maize plant. Secondary roots appear on the primary root which develop directly from the seed. These secondary roots produce lateral roots (tertiary roots and so on) of the same type with approximately the same diameter throughout their length.

All date palm roots present pneumatics, which are respiratory organs. Root are found as far as 25 m from the palm and deeper than 6 m, but 85 percent of the roots are distributed in the zone of 2 m depth and 2 m on lateral sides in a deep loamy soil (Munier, 1973). It is worth mentioning that date roots can withstand wet soil for many months, but if such conditions spread over longer periods, it becomes harmful to the health of the roots and to fruit production. It is clear that the date palm root system is divided into four zones (Oihabi, 1991):

- *Zone I, called respiratory zone*: It is localized at the palm base's surrounding area with no more than 25 cm depth and a lateral distribution of a maximum of 0.5 m away from the stipe. Found in this zone are mainly roots of primary and secondary nature. Most of these roots have a negative geotropism and play a respiratory role.

- *Zone II, called nutritional zone*: It is a large zone and contains the highest proportion of primary and secondary roots. It could contain 1000 roots per m^2 and more than 1.60 gm of roots/100 gm soil (Oihabi, 1991). They develop between 0.90 and 1.50 m depth and could laterally be found outside of the projection of the tree's canopy. In the case of Deglet Nour variety, lateral roots were found up to 10.5 m from the trunk. Recently planted offshoots develop their roots at zone II then at zone III. At one year old, they could reach 1 m while 3 m depth is easily reached at the second year.

- *Zone III, called absorbing zone*: The importance of this zone is dependant on the type of culture and on the depth of underground water. It is usually found at a depth of 1.5 to 1.8 m. mostly primary roots with a decreasing density from top to bottom are found here per m^2.

- *Zone IV*: The largest portion of this zone is dependant on underground water. At a shallower depth, it becomes difficult to distinguish between Zone III and Zone IV as both types of roots are found here. While when the underground water is deep, roots of this zone could reach a greater depth. They usually are presented as vessels with a positive geotropism.

The root type and distribution illustrates the role of the date palm. Its root development and distribution depends on the soil's characteristics, type of culture, depth of the underground water and variety. The lack of roots in the top soil allows other cultures such as barley, Lucerne and vegetables to be inter-cropped. While, the high concentration and deep presence of primary roots allows the date palm to benefit from underground moisture and consequently, unlike most fruit palms, resist water stress and drought conditions.

Stem: The stem is erect, tall (30-40 m) bearing crown of leaves. The date palm trunk, also called stem or stipe is vertical, cylindrical and columnar of the same girth all the way up. The girth does not increase once the canopy of fronds has fully developed. It is brown in colour, lignified and without any ramification. Its average circumference is about 1 to 1.10 m. The trunk is composed of tough, fibrous vascular bundles cemented together in a matrix of cellular tissue which is much lignified near the outer part of the trunk. Being a monocotyledon, date palm does not have a cambium layer. The trunk is covered for several years with the bases of the old dry fronds, making it rough, but with age these bases weather and the trunk becomes smoother with visible cicatrices of these bases. Vertical growth of date palm in ensured by its terminal bud, called phyllophor, and its height could reach 20 meters.

Horizontal or lateral growth is ensured by an extra fascicular cambium which soon disappears, and which results in a constant and uniform trunk width during the palm's entire life. However, the terminal bud could experience an abnormal growth caused by a nutritional deficiency, which leads to shrinkage of the trunk. This stage is mainly caused by drought conditions. In Kachchh region, branching in date palm is observed and locally called *modo* (Singh *et al.*, 2004). However, sometimes date palms show a branching phenomenon which was studied by Zaid (1987) and found to be attributed to several different causes as follows:

- Branching in date palm is a result of either dichotomy, axillary bud development, polyembryony or attack by a disease.
- Branched date palms are fertile and can produce as much fruit as a single headed palm.

- There is a need of an analysis of the vascular system of branched date palm by cinematographic techniques. This anatomical study is necessary to show the continuity of growth from the single to the divided state of the shoot.
- It is necessary to study *in vitro* the regenerating capacity of divided portions of the apical meristem and axillary buds of these specimens in the hope of establishing a rapid mass propagation technique for date palm.

Leaves: Depending on variety, age of a palm and environmental conditions, leaves of a date palm are 3 to 6 m long (4 m average) and have a normal life of 3 to 7 years. The greatest width of the frond midrib attains 0.5 m, but elsewhere it is only half this size and rapidly narrows from the base upwards. The frond midrib or petiole is relatively triangular in cross section with two lateral angles and one dorsal. It is bare of spines for a short distance but full of spines on both sides thereafter. Intermediate zones have spine-like leaflets, also called leaflet like spines. At the tip of the leaf, there may be a single terminal leaflet or two leaflets forming a V. Leaf structure is variety and environment dependant, but usually the whole length of a frond has the following proportions:

- the distance from the fiber at the base of the frond to the base of the spine-leaflets is about 28% of the whole frond;
- the non spiny-leaflets occupy about 4%;
- the leaflets occupy about 62 %; and
- the terminal leaflets occupy about 6%.

All these characteristics coupled with others are used as a taxonomical index to differentiate between varieties. Unlike other fruit trees, dead or old leaves are not shed and do not drop on their own, but they are removed under cultivation. An adult date palm has approximately 100 to 125 green leaves with an annual formation of 10 to 26 new leaves. The functional value of the leaf to the palm declines with age and no two leaves are the same age. Furthermore, leaves which are four years old are only about 65 percent as efficient in photosynthesis per unit area, compared to leaves of one year old (Nixon and Wedding, 1956). Under good cultural conditions a leaf can support the production of 1 to 1.5 kg of dates. Leaflets or pinnae, are 120 to 240 per frond, entirely lanceolate, folded longitudinally and obliquely attached to the petiole. Their length ranges from 15 to over 100 cm and in width from 1 to 6.3 cm. Their arrangement depends on variety and could be in groups of 1,

2, 3, 4 or 5 pinnae. Depending on their position in the tree's canopy, leaves could be divided into 3 categories:

- on the outside, leaves are green and photosynthically active;
- at the centre, fast growing green leaves;
- on the inside, at the palm's heart, juvenile leaves, not yet photosynthetic with a white colour.

On an average, there are 40% of juvenile leaves, 10% fast growing leaves and 50% photosynthetic leaves. Leaves are grouped in 13 nearly vertical columns, spiraling slightly to the left on some palms and to the right on others. The grower must only count the number of leaves in one of these columns and multiply it by 13. According to Nixon and Carpenter (1978) and in order to allow for uneven pruning at the base, counts could be made on opposite sides and divided by two. This technique will allow the grower to calculate the total number of leaves on the palm. A ratio of 8 leaves per fruit bunch will indicate how many bunches to leave on that palm. At the base of each leaf, there is an axillary bud which could yield an inflorescence at the palm's top level or an offshoot at its base. Three distinct development phases have been reported:

- Juvenile phase which is sterile and leads the palm to produce more inflorescence buds than vegetative ones, which will abort very soon.
- Second phase called vegetative, where vegetative and flowering buds are produced in equal numbers; however, vegetative buds are the ones which develop.
- Third phase, usually after the palm is more than 10 years old, where most of the buds produced are flowering ones.

Spines: As described by Dawson (1982), the base of the frond is a sheath encircling the palm. This sheath consists of white connective tissue ramified by vascular bundles. As the frond grows upwards, the connective tissue largely disappears leaving the dried, and then brown, vascular bundles as a band of tough, rough fiber attached to the lateral edges of the lower part of the midribs of the fronds and unsheathing the trunk. Varieties differ in the height to which the fiber grows up the central column of unopened fronds and in the texture of the fiber and also somewhat in colour.

Spines, also called thorns, vary from a few centimeter to 24 cm in length and from a few millimeter to 1 cm in thickness. They are differentially arranged on the two outer edges of the fronds while their number varies from 10 to about 60. Spines can be single, in groups of two, or in groups of three.

Reproductive organs: Date palm is a dioecious species with male and female flowers being produced in clusters on separate palms. These flowering clusters are produced with axils of leaves of the previous year's growth. In rare cases both pistillate and staminate flowers are produced on the same spike while the presence of hermaphrodite flowers in the inflorescence has also been reported. The unisexual flowers are pistillate (female) and staminate (male) in character; they are borne in a big cluster (inflorescence) called spadix or spike, which consists of a central stem called rachis and several strands or spikelets (usually 50 – 150 lateral branches).

Inflorescences/flowers: The inflorescence, also called flower cluster, in its early stages is enclosed in a hard covering/envelope known as spathe which splits open as the flowers mature exposing the entire inflorescence for pollination purposes. The spathe protects the delicate flowers from being shriveled up by the intense heat until these are mature and ready to perform their function. The spathe at the beginning is greenish in colour, becoming brown when near splitting – splitting is longitudinal. The male spathes are shorter and wider than the female ones. Each spikelet carries a large number of tiny flowers as many as 8,000 to 10,000 in female and more in male inflorescence (Chandler, 1958). The annual number of spathes born by a palm varies from none to about 25 in females and to even more in males, but the average is a dozen for females and more for males.

The male inflorescence is crowded at the end of the rachis, while branches of the inflorescence of the female cluster are less densely crowded at the end of the rachis. These characteristics allow the recognition of the inflorescence's sex before its opening. The male flower is sweet-scented and normally has six stamens, surrounded by a waxy scale-like petals and sepals (3 each). Each stamen is composed of two little yellowish pollen sacs.

The female flower has diameter of about 3 to 4 mm and has rudimentary stamens and three carpels closely pressed together and the ovary is superior (hypogynous). The three sepals and three petals are united together so that only tips diverge. On opening the female flowers show more yellow colour while the male ones show white colour dust, produced on shaking. The pollen sacs usually open within an hour or two after the bursting of the spathe. Pollination is done manually after opening of spathes.

Only one ovule per flower is fertilized, leading to the development of one carpel which in turn gives a fruit called a date; the other ovules aborted. The aborted carpels persist as two brown spots in the calyx of ripe fruits.

Fruit: Depending on the cultivars, climatic conditions and management practices *viz.*, fertigation, pollination, fruit thinning etc. employed, fruit characters vary greatly. The date fruit is a single, oblong, one-seeded berry, with a terminal stigma, a fleshy pericarp and a membranous endocarp. The colour of the berry also varied from green to yellow and red at Khalal *(doka)* stage.

Seed: The stone characters vary according to variety, climate and growing sites. The stone weight could range from 0.5 – 4.0 g, length from 12.0 to 36.0 mm and breadth from 6.0 to 13.0 mm. The seed is usually oblong, ventrally grooved, with a small embryo, and with a hard endosperm made of a cellulose deposit on the inside of the cell walls.

Date varieties have been developed by thousands of years of selection of seedlings. More than 1000 varieties of date palm are found all around the world. In India, about 50 cultivars have been introduced for evaluation at different research stations. The varieties have been classified according to fruit maturity (early/medium/late) and uses. The descriptions of some important varieties are given below:

Halawy: An early maturing, regular bearing high yielding and suitable variety to grow in large areas for fresh fruit and processing. Its fruits are yellow and sweet in taste at doka stage with an average weight 9.6g; size 3.56 x 2.10 cm and TSS 31.0%. It produces about 70- 80 kg doka fruits/tree per year.

Barhee: Its origin is Basrah, Iraq. It is a medium to late maturing cultivar and good for fresh consumption. In this variety, sucker production is low. The fruit at khalal stage (doka) has excellent flavour, yellow and attractive fruit colour and sweet in taste. The stone is very small in size with high pulp: stone ratio. It has high market potential in the world's market. The yield potential of this variety is 80-120 kg per plant which goes up to 200 kg per palm.

Medjool: The origin of this variety is Morocco from where it was introduced to the other countries of the world. It is No. 1 variety in the world and very high yielding cultivar due to its bigger attractive colour berries (size 4.5 x 3.2 cm). It is also known as Mejhool. It also matures late and it is highly suitable for dry date productions. Its doka fruits are astringent in taste. The colour of berry is yellow-orange at doka stage. It has high commercial value and most important variety for export purposes. Average yield is 80-100 kg/palm.

Khalas: It is a medium maturing type cultivar. Its berry is big in size at doka stage and sweet in taste. The yellow colour doka fruit is suitable for fresh eating and processing. This variety is medium to heavy yielder.

Khuneizi: It is an early to mid season maturing variety. The colour of berry is red and suitable for fresh consumption at *doka* stage. The *doka* fruits are very sweet in taste. The yield potential of this variety is 60-80 kg/tree.

Zahidi: It is a late maturing variety and suitable for pind preparation. Its doka fruits are astringent in taste and solid with hard skin. The size of berry is small, attractive yellow colour but it is a heavy yielder (125 kg *doka* fruits/tree). The bunch of this variety is compact with 18 to 24 berries per strand.

Shamran: It is a mid season to late maturing and high yielding variety. The average fruit weight is 6.5 g and fruit size of 3.50 x 2.18 cm. The colour of berry is light yellowish red and astringent at doka stage. This variety is grown commercially. It is suitable for processing in to dry dates and other value added products. It is a prolific bearer as fully-grown tree of Shamran has an average yield potential of 100 kg doka fruits per tree.

Sewi: The trees are tall and erect with moderate canopy. The fruits are yellowish green in colour at *doka* stage and oblong in shape. The average fruit weight is 6.1 g and fruit size of 2.9 cm x 1.90 cm. It has pulp thickness of 0.6cm having 40.0 per cent TSS at full *doka* stage. The fruits are sweet at full *doka* stage and therefore suitable for both raw eating as well as soft dates. It is late in ripening as fruits of this variety reach full *doka* stage by the first fortnight of August at Bikaner. The fully-grown tree of Sewi has an average yield potential of 50 kg *doka* fruits per year.

Chipchap: This variety has been introduced in India from Iraq during 1997 and suitable for fresh as well as for processing purpose. Its fruits are yellow, bigger in size and sweet in taste at doka stage because of less astringency (Singh *et al., 2005*). It is medium in yield and maturity period is also less in arid region. It is suitable for dry date preparation.

Zagloul: The trees are vigorous in growth having erect growing habit. The fruits are red in colour and large in size. The fruit bunches are bigger in size. The average fruit weight is 9.5g and fruit size is 3.90 cm x 2.20cm. It has pulp thickness of 0.64 cm having 28.0 per cent of TSS at *doka* stage. The fruits are astringent at *doka* stage. The *dang* and *Pind* fruits are relatively less sweet but have pleasant taste. It is suitable for soft dates and for preparation of fruit products such as date juice and preserved products. It is medium in ripening as fruits of this variety reach full *doka* stage by the beginning of second fortnight of July at Bikaner. It is prolific bearer and average yield potential is 150 kg *doka* fruits per year.

Khadrawy: The plants of this variety are dwarf in growth. The fruits at *doka* stage are greenish yellow in colour and oblong in shape with broad apex. The average fruit weight is 7.7 g and fruit size of 3.10 cm x 2.10 cm. It has pulp thickness of 0.66 cm having 30.0 per cent TSS at *doka* stage. The fruits are highly astringent at full *doka* stage and are badly spoiled due to rains at the time of ripening. This variety is suitable for preparation of both soft dates as well as dry dates. It is medium in ripening as the fruits of this variety reach full *doka* stage by the end of July. Average yield potential is 40-50 kg *doka* fruits per/palm.

All the commercial date cultivars have developed through selection of chance seedlings based on local needs. A rich genetic diversity is available in costal belt of Kachchh region of Gujarat in India which should be exploited (Singh *et al.*, 2004). In other parts of country, seedlings are less in population. At CIAH, Bikaner, 60 genotypes/ cultivars have been conserved in National field gene bank collected from different available sources. Further, SDAU, DRS, Mundra, (Gujarat); CAZRI, Jodhpur, SKRAU, Bikaner, PAU, RFRS, Abohar, (Punjab); Central State Farm, Jetsar, Sri Ganganagar, has also maintained and evaluated the date palm germplasm.

Propagation

Date palm is propagated by seed and suckers of mother palm. There are three techniques to propagate date palm *viz.*, seed, offshoot/sucker propagation (traditional methods), and the recently developed tissue culture techniques. Date palm is propagated both by seeds as well as vegetative methods but it is commonly multiplied through suckers (rooted) and also by tissue culture technique.

Seed propagation

In date palm variability and sex ratio is the main problem by seed propagation and 50:50 ratio of male and female plant has been observed in seedlings. The germination of seeds in date palm have been observed about 80-90 per cent when seed sowing take place in the month of July-August. Moreover, growth in seedlings is very slow at initial stages. It has been observed that seed soaking enhanced germination and seedling growth by pre-treatments of seeds with cow dung for 24 hours in *Phoenix* species (Singh and Bhargava, 2009). Seedlings varied considerably with respect to vegetative and reproductive characteristics (Al-Ghamdi *et al.*, 1988). However, date palm propagation through seedlings can provide an ideal material as some times seedlings palms are outstanding in performance (Chandra *et al.*, 1992). In Kachchh region, the seedlings are used for planting on farm boundary due to lack of planting material (Singh *et al.*, 2004).

Suckers

Presently, date palm is mainly multiplied through suckers / offshoots. The offshoots bear fruits 2-3 years earlier than seedlings. But production of suckers is a limiting factor because a palm tree produces about 10-20 suckers in whole life. Out of them, some are aerial offshoots and have no roots and fail to establish after detachment from mother palm. It is a main constraint in increasing area under date palm cultivation. Survival of offshoots is about 40-50% under field conditions. The appropriate size and rooted suckers are suitable for better establishment and growth of plant. Work on initiation of roots in suckers and better survival is also very important for increasing the area of cultivation in date palm. Sodagar *et al.* (1996) observed that the best time of planting is April – May for higher survival and rapid growth of plant in Kachchh region. At the time of removal of suckers from mother palm, leaves should be pruned and tie up.

In date palm, basal as well as aerial offshoot suckers are produced. Basal offshoot generally has well developed root system compared to poorly developed roots in aerial suckers. After 4-6 years of plantation, suckering starts and 2-3 suckers can be removed every year from each mother palm tree for a period of 8-10 years as per observations made at Abohar (Fig. 1).

Fig. 1 : Mounding for rooting in arial suckers in date palm.

The offshoots should be well rooted and of proper weight for better survival in the field. Well rooted offshoot of medium weight (range 8-12 kg) is better in survival performance in the field compared to poorly rooted small sized or bigger sized offshoots (Singh *et al.*, 2003). The offshoots weighing between 8 and 12 kg have been reported to give maximum rooting (Mohammed, 1978). High offshoots, which emerge from stem, perform poorly because of weak root development. Such offshoots of cultivar Medjool even after dipping in IBA (1000 or 3000 ppm) and planting in nursery failed to strike roots under arid irrigated conditions at Abohar, Punjab. However, in an experiment when such offshoots were covered at the base with moist moss grass enclosed in polythene and attached to mother palm treated similarly with wooden shavings, there was 44.9, 7.5 and 22.0 per cent increase in average root length, root thickness and root weight, respectively (Vij *et al.*, 1977). The treatments of suckers with organic such as cow dung solution, cattle urine etc also enhanced the survival of plants.

Tissue culture techniques: Date palm is multiplied through micro-propagation, which is also called *in vitro* propagation. By both organogenesis and embryogenesis plantlets have been raised in Gulf countries. In our country, date palm tissue culture work is limited. The micro propagation technique has many advantages in date palm cultivation like:

- Production of genetically uniform plants which is a limiting factor at present situation.
- Propagation of healthy selected female cultivars.
- No seasonal effect on plants because they can be multiplied under controlled conditions in the laboratory throughout the year.
- Clones to be propagated from elite cultivars already in existence, or from the F1 hybrids or previous selections, and seed-only originated palms.
- Ensure an easy and fast exchange of plant material.
- Economically reliable when large production is required.
- Large scale multiplication.

In a situation such as this, recourse to tissue culture technique for production of tissue culture raised plants has been adopted. The physiological performance of these plants is at par with those of sucker plants (Bhargava *et al.*, 2003). Thus, it seems apparent that this may provide a method for mass multiplication of date palm mother plants. In addition to this, supply of genuine plant material to farmers is another responsibility of scientists. Therefore, there is a need to standardize the biochemical markers for identification of superior plantlets produced from tissue culture.

Tissue culture is the only tool to produce a large number of plantlets in date palm to meet out the requirement. In India, groves of seedling date palm are found on the coastal belt in Kachchh region of Gujarat. In Kachchh area, some promising selections have also been made (Pareek and Sodagar, 1986). The promising selections should be multiplied on large scale, as some selections are better than exotic varieties.

At Central Institute for Arid Horticulture, Bikaner, exotic as well as indigenous cultivars/genotype of date palm is under evaluation. In order to develop phyto-chemical markers for delimitation of date palm cultivars, a study was undertaken using five cultivars. The foliar flavonoids were extracted and separated on TLC plates coated with cellulose. The spots thus developed were compared and subjected to cluster analysis using computer software. The results demonstrated that date palm cultivars can be identified using either specific spot or spot combinations. The results of the cluster analysis demonstrated that the selected five cultivars can be grouped into three distinct groups on the basis of their genetic distances. The group A was represented by cultivars having closer genetic distance with Sewi and the group C was produced with the group A and B within these three groups the cultivars. Khuneizi and Dayari have the closest genetic distance (42%).

The tissue culture work also explains the background to the cloning methods applied to the date palm and explores the wide range of results obtained with embryo culture, meristematic tissue (shoot tips and buds) and highly differentiated somatic tissue (leaf, stem, inflorescence and root sections).

Embryo culture: Embryo culture involves excising an embryo-aseptically from the seed and planting it in a sterile nutrient medium. Embryo culture is suggested to have several potential applications in plant research. It is used to save embryos that fail to develop naturally in the fruit or seed, or grow out embryos from interspecific hybridization where defective endosperms are common. Embryos culture may also be used to reduce lengthy dormancy periods due to physical and/or chemical inhibitors present in the fruit or seed. Excised embryos cultured *in vitro*, free from these inhibitors, usually germinate immediately. Isolated embryos were also chosen as explants material in metabolic studies. The culture of isolated embryo segments may be useful to study the development of the primary meristems, organogenesis and the interactions between different organs. The culture of embryo outside the seed was first performed with crucifers. It has since become a routine procedure.

With regard to date and other palms, callus initiation and embryoid induction was first observed by Rabechault (1962) working with oil palm embryos. Reuveni (1979) reported that callus and roots developed from the date palm embryo cotyledonary sheath tissue in media containing naphthalene acetic acid (NAA). This callus continued to proliferate and to differentiate roots when subcultured if a piece of the cotyledonary sheath was present. Ammar and Benbadis (1977) established organogenic callus from date palm cotyledonary sheath of zygotic embryo germinated *in vitro*.

Reynolds and Murashige (1979) cultured embryo explants of *Chamaedores costaricana* Oerst, *Howeia forsteriana* Becc., and *Phoenix dactylifera* L. *in vitro.* Green date palm fruits, harvested two to three months after pollination were planted in a medium enriched with 2, 4-dichlorophenoxy acetic acid (2, 4-D), and a creamy-coloured grainy callus was subsequently developed. Transfer of this callus to an auxin-free medium resulted in the development of numerous asexual embryos. Mature zygotic embryos cultured in nutrient media containing charcoal with high auxin levels, 10 and 100 mg/1NAA, also produced nodular callus. Repeated culture resulted in the formation of plantlets. Tisserat and DeMason (1980), described plantlet formation from date palm tissue cultures. The morphological development of asexual embryos from callus closely paralleled excised zygotic embryo germination *in vitro.*

Zaid and Tisserat (1984) performed a survey study to determine excised embryo callus production. In the Arecaceae, embryo excised from mature seeds of 38 species were cultured on modified Murashige and Skoog (MS) medium containing 3g/L^{-1} activated charcoal; with 100 mg L^{-1}, 2,4-D and 3 mg L^{-1} N6- (2 – isopentyl) adenine (2-iP). Embryo cultures from 18 of these species produced prolific callus after repeated reculturing for six months.

Culture of meristematic tissues: When comparing shoot-tips and lateral buds *in vitro* versus culturing other explants sources, the following advantages become apparent:

- Shoot-tips and lateral buds are protected by bud scales and leaves, and are usually easier to surface-sterilize than root or stem explants.
- By culturing shoot-tips or buds, an entire shoot is already present, thus only root induction is required to produce a whole plant.
- The cells of the shoot-tips and buds are more uniformly diploid than those derived from less meristematic regions. Presumably, plantlets derived from naturally meristematic regions are likely to be clonal and generate faster than other explant sources.

A distinction is made between bud and apical meristem cultures. Lateral bud culture involves the growth of an entire rudimentary vegetative shoot. Apical meristem culture, ideally involves only the excision and growth of apical dome of the shoot usually less than 0.1 mm in diameter and 0.25 mm in length, sometimes with, though preferably without, a few leaf primordial attached. In contrast to culturing herbaceous angiosperm shoot apices, few woody angiosperm shoot-tips have been established *in vitro.*

***In vitro* culture of buds:** Most buds of date palm were reported to die within the first 30-50 days after planting *in vitro*. Only the largest and most distinctly differentiated buds grew. These buds exhibited leaf expansion and produced additional leaves. Tisserat (1979) and Zaid (1999) also investigated the conditions for bud development, and found that in nutrient medium shoot-tips and lateral buds grew equally well on the same medium. Callus cultures have been initiated from axillary buds of 2 to 4 year old date palm offshoots. It was found that subcultured lateral buds callus on nutrient media devoid of charcoal and supplemented with 0.1 mg/1 NAA, produced adventitious plantlets. Tisserat and DeMason (1980) found that on a medium devoid of 2,4-D and 2-iP, sectioned buds callus consisted of two distinct types of tissues; a loose friable tissue and compact aggregates. The friable portion of the callus was composed of large non-meristematic cells and disorganized clumps which were highly vacuolated and ranged in diameter from 20-40 μm. This tissue was not involved in embryo formation and was generally found surrounding the aggregate clumps which consisted of densely cytoplasmic cells containing few vacuoles and usually were 8-20 μm in diameter. The formation of vascular bundles within the asexual plantlet at the 8 weeks old stage corresponded to that found in zygotic seedling.

Starting from the bottom of young leaves, soft tissues shoot tips or axillary buds of date palm offshoots and using MS half strength or Beauchesne medium supplemented by various auxins at a low concentration, buds were obtained after six months of *in vitro* culture.

Early rooting of date palm tissues reduce bud multiplication and is, occasionally, responsible for the inhibition of bud initiation. In order to solve this problem, the bottom of young leaves of date palm offshoots were cultured on eight different nutrient media with different levels of growth regulators. High level of auxins, especially NAA, allowed root initiation. These roots showed a rapid growth after sub-culturing on a medium containing a lower level of auxins. Furthermore, organogenesis was inhibited on media with al low concentration of auxins.

Vitrification phenomenon of date palm tissues is a handicap for the successful *in vitro* multiplication of some date varieties and selected clones. In order to overcome such problem, four culture media with different ammonium / total nitrogen ratio were tested, and bottom young leaves from offshoots of cultivar Aguellid were used. It was found that ammonium plays an important role in vitrification (46 to 53% of cultures); while this phenomenon is reduced to 14 – 19% in media with low levels of ammonium nitrate.

Culture of Highly Differentiated Date Somatic Tissues *in vitro*

(i) ***Leaf cultures:*** Callus developed from a seedling date palm leaf, and gave rise to roots several months later. Primordial leaves survived in culture and expanded, especially in the presence of light. The addition of plant growth regulators at concentrations of 0.1 mg/1 and above was injurious to cultured leaves. Eeuwens and Blake (1977) working with date palm leaf found development of root initials to be enhanced by the presence of a low level of gas and auxins, and by a reduction in either the level of minerals or sucrose. *Phoenix* leaf petiole explants have initiated roots within 6 weeks when subcultured onto a medium with high levels of auxin. Root initiation was not prevented by the presence of high cytokinin or low sucrose levels, but occurred more frequently in media containing high sucrose and reduced cytokinin levels. Zaid (1999), working with date palm leaf explants from adult trees, offshoots, seedlings and asexual plantlets, found that only subcultured leaf callus from seedling and asexual plantlets produced roots.

(ii) ***Stem culture***: *Phoenix* stem explants reportedly enlarged considerably in size during the first few weeks of culture (Tisserat, 1979). Repeated culture to fresh media resulted in the formation of non-friable nodular callus. Plantlets were developed from this callus.

(iii) ***Inflorescence culture:*** Inflorescences of several species have been cultured *in vitro* (Nitsh, 1963). Since 1973, several workers attempted to culture palm inflorescences. Explants of female and male oil palm inflorescences were cultured on a variety of media and usually developed somewhat normally, but callus was not obtained. A high auxin level was speculated to be necessary to disrupt normal development. This has subsequently been confirmed in date palm.

Date palm ovules, carpel tissue, parthenogenetic endosperm, and the fruit stalk blackened within 24 hours after culturing on nutrient media, and subsequently died. Also cultures of date palm floral bud reproductive tissues and especially male anthers, usually turned brown and died after a few weeks in culture.

Vestigial female date carpals on surviving male flowers enlarged and became quite prominent. White friable callus usually initiated from the floral bud strand. In some cases, root and embryoid were initiated from explants of *Cocos* inflorescences rachillae and from date palm. Roots have not been initiated on inflorescence rachis explants which lack leaf or meristem tissue. Date palm inflorescence culture was also largely investigated by Drira (1981). Morphogenetic responses were found dependent on the origin and physiological stage of the explants.

(iv) ***Root culture***: Schroeder (1970) was the first to investigate root cultures in palms *in vitro*. Oil palm root and root primordial failed to develop. It was observed that date palm root pieces in turn developed secondary rootlets but did not produce shoots. Eeuwens (1978) found that isolated roots excised from cultured explants of date and coconut palms continue growth and produce laterals when subcultured on liquid static media. Callus was also reported to form at the root tip region of young date palm seedlings. This callus had produced leaves and shoots. Other investigators reported no growth for cultured date palm roots. Usually, severe browning and death of root explants occurred within the first few weeks of culture. However, Zaid and Tisserat (1984) obtained some callus from seedlings and asexual plantlets roots when callus failed to exhibit any morphogenic response.

(v) ***Browning of tissues and media***: During the course of *in vitro* growth and development, plant tissues not only deplete the nutrients that are furnished in the medium, but also release substances that can accumulate in the cultures. These substances, such as phenols, may have profound physiological effects on the cultured tissues. Date palm tissue cultures, like those of many other plants, have been commonly observed to release discolouring substances into the medium which inhibit their own growth. For date, injury through cutting of tissue is accompanied by secretion of the substances into the medium. The intact organ, as exemplified by embryos or whole leaves on tips do not brown and thus grow well in culture. Browning of tissue and the adjacent medium is assumed to be due to the oxidation of polyphenols and formation of guinones which arc toxic to the tissues.

To minimize browning, Murashige (1974) has suggested the pre-soaking of explants in ascorbic and citric acid solutions and adding them to the culture medium. Zaid and Tisserat (1983) soaked their date palm explants in an anti-oxidant solution (150 mg/1 citric acid and 100 mg/1 ascorbic acid) prior to the surface sterilization treatments. Addition of a combination of adsorbents including citrate, adenine and glutamine, retarded browning in date palm explants.

Addition of other adsorbents to nutrient media such as di-hyroxynaphtalene, dimethylsulfoxide, were ineffective against browning in date palm explants. The browning could be eliminated by a nutritionally balanced medium. Excision of browning explant parts during culture was also advocated to prevent this problem.

The use of charcoal is preferred over cysteine and other adsorbents because the latter are often toxic to the plant tissues at higher concentrations. Addition of 3 % charcoal has caused substantial root and shoot growth of date embryos. It has reported that than addition of charcoal to a culture medium drastically alters the properties of the medium. Hence, growth regulator substances are tested at high levels (e.g. 10 and 100 mg/1) with charcoal included in the nutrient media to obtain beneficial effects on tissues.

(vi) ***Cryopreservation of shoot tips***: Studies on the cryopreservation of date palm for germplasm collections were initiated by Towill *et al.* (1989). Shoot-tips were excised from 2 month-old seedlings derived from the cultivar "Medjool", precultured for 2 days and then cooled to liquid nitrogen (LN) temperatures using procedures described for potato and mint species. Viability of treated shoot-tips was assessed by growth *in vitro*. Dimethylsulfoxide (DMSO) in concentrations up to 10% was not toxic, although growth was slower than untreated shoot-tips. Several combinations of DMSO and sucrose were effective in obtaining survival after LN exposure. In most cases, the LN-treated shoot tips developed directly into a shoot without callus formation.

Production of Tissue Culture Plants

Many tissue culture laboratories in the world have tried to propagate date palm by tissue culture techniques. The tissue culture laboratory in Al -Ain University, UAE, Egypt and UK, are also producing plants on large scale of different cultivars. However, in India, there are many tissue culture laboratories engaged in tissue culture work; however, commercial production of date palm plants is yet to be done. M/s Excel Crop Care Ltd. is doing work on date palm tissue culture in Mundra, Gujarat to multiply elite type local date palm and selling @ 3000 - 3500/- per plant. The commercial laboratory of the "Domaine Agricole El Bassatine" (Morocco), which since its start had produced > 500,000 plants, is reserving all its production for national use. No significant sale outside Morocco has been implemented because all the production is destined to rehabilitate the Moroccan Date plantations destroyed by the Bayoud disease.

Acclimatization and hardening-off of tissue culture plants: Although *in vitro* mass plant propagation has become commercially feasible, many problems hinder its application to economically important crops. One of the major obstacles concerning the practical application of plant tissue culture to mass propagation has been the difficulty of successful transfer of plantlets from *in vitro* conditions to a soil medium. Losses from 50 to 90 % of *in vitro* propagated plantlets of many species have been encountered at the time of transfer to soil (Zaid and Hughes, 1989). This is unfortunate because the ultimate success of plant tissue culture as a commercial means of plant propagation depends on the ability to transfer plantlets out of culture, on a large scale, at low cost and with a high survival rate. It is appropriate at this level to differentiate between the acclimatization of date palm *in-vitro* plants at the laboratory's glasshouse and their hardening-off at the farmer's nursery. Acclimatization presents challenges at least equal to those posed by the initiation of cultures because it marks the end of artificial control and the beginning of autonomous plant growth. The culture of date tissue *in vitro* with almost 100 % relative humidity within the culture vessel can lead to various abnormalities in the plant structure (Zaid and Hughes, 1989). Plants of many species produced *in vitro* often show morphological, structural, and physiological and biochemical differences from those produced conventionally. It is worth mentioning that loss of viability is attributed to poor control of water loss from the date plants and their heterotrophic nature.

Stomatal development and frequency can be affected by water availability, light intensity, temperature, humidity and osmotic concentration of the culture medium. Even when gradual hardening has been used, poor survival and slow growth of date plantlets have commonly been reported. Such a low survival rate (that sometimes reaches below 50 %) is caused by several factors which are mainly young physiological stage of plantlets to transfer, inadequate root system, unsatisfactory irrigation schedule, and lack of technical care at the *in vitro* laboratory stage.

Several techniques have been used to acclimatize date plantlets and improve their survival during establishment under greenhouse conditions. The effectiveness of these methods depended upon ambient conditions, and most methods have involved environmental modifications. Three most important factors are to be taken into account by the manager of a date palm propagation laboratory in order to ensure a high survival rate and fast growing conditions of date palm tissue culture-derived plantlets.

Hardening-off: Plantlets received from a laboratory are usually about 35 to 45 cm long with 4 to 5 leaves among which are 0 to 2 pinnae leaves (called also permanent leaves). The plant must have a thick shoot system and the base must be similar to that of a large bulb sized. As stated above, the plant must have a well developed root system. Transportation of these plants must be realized in a proper manner and plants must preferably not be stacked on top of each other to avoid stem breakage and/or leaf damage. Transport must preferably also be in one stage and if plants/truck are to stay over somewhere, it must be in a shaded area; watering should not be neglected if transport takes up several days. It is recommended that, upon reception of this material by the date grower, plants are transferred to larger bags (7 to 10 liters capacity) with an adequate substrate, usually sand (soil), vermiculite and gravel at a ratio of 1:1:1, respectively. Transplanting should be done properly with no disturbance to the root system. Original substrate around the roots should stay intact. Plants are then left in the nursery for approximately 8 to 12 months depending on surrounding conditions and care given, till most of them reach the 4 pinnae leaf stage.

The nursery size and type are related to the number of plants to be hardened-off. An average size of 150 m^2 will be adequate for 1,000 plants. A ultra-violet resistant shade net of 80 % is recommended during the first 6 months. During the summer time, the top of the nursery should have a double layer of the shade net for insulation purposes. The nursery should be well located (close to several trees to benefit from their shade) but also in a protected area to avoid sand storms and severe wind. A water tap should be installed inside the unit for easy irrigation and the unit must be enclosed to avoid animals getting in and eating the plants. Irrigation is an important factor and must be implemented once a week in winter time and at least twice a week during summer. Water should never be sprayed on top of the plant; soil is to be mounted around the base of the plant so water can not get into its heart. Control of diseases and pests is also recommended and the use of Benlate (or any other large spectrum fungicide) has proven to be highly efficient. Foliar spray of Benlate is to be applied every 3 to 4 weeks.

Performance of Tissue Culture Plants

Tissue culture plants of eight varieties procured from M/s A.V. Thomas Company, Cochin, Kerala, are also under field evaluation and showed good survival, vegetative growth under hot arid conditions. The tissue culture plants of cultivars viz., Dhamas, Nagal and Khuneizi have started flowering and fruiting after five years of planting. Moreover, it has been observed that the number and size of bunches produced on tissue culture plants are comparatively less and small in size than that of offshoots plantations. The better vegetative growth has been observed in tissue culture plants of cv. Barhee planted during 2009-10 under hot arid conditions. Further, evaluation of tissue culture plants is in progress.

A private company M/s Atul Comp. Ltd., has started planting of tissue culture plants of date palm in western Rajasthan imported from UAE and it is under initial stage of establishment. However, fruiting has started in few plants after three years of planting. Plants are maintained under drip irrigation/fertigation. Suckers are not allowed to grow with mother palm. In Gujarat state, a tissue culture plant of date palm cv. Barhee has been planted and the vegetative growth performance with offshoots under field condition is superior. It has been observed that the fruiting starts in tissue culture plants after 3-4 years of planting. The tissue culture plants were maintained and growing well under drip irrigation system on farmer's field in Kachchh region. It was observed that the emerging suckers have not allowed in mother palms for good vegetative growth and early flowering/fruiting.

Quality Standards of Planting Material

Because of dioecious nature, in general, planting of seedling is not advisable. The offshoot of appropriate size and weight (8 -12kg) should be used for planting. Suckers having proper root system performed better survival and establishment under field conditions. Tissue culture plants arc comparatively better in comparison to suckers planting for high survival rate and establishment. Availability of sucker is also a limiting factor in date palm for large scale plantation. Tissue culture raised plant should be well hardened, having 4-5 well developed leaves, 1-1.5 feet height and free from diseases. Very young sapling of primary leaves should not used for planting.

References

Al-Salih, A.A. and A. Al-Jarrah 1987. Chromosome number of a date palm male: Cultivar Ghannami Akhdar. Date Palm Journal, 5 (2): 128-137.

Al-Ghamdi, A.S., Hassan,A.L. and Jahjah, M. 1988. Evaluation of eight date plam seedling male and their effect on fruit characteristics of three female cultivars. Arab Gulf J.Sci. Res., 6: 175-189.

Ammar, S. and Benbadis, A. 1977. Multiplication vegetative du palmier dattier (*Phoenix dactylifera* L.) par la culture de tissus de jeunes plantes issues de semis. C.R. Acad. Sci. Paris. Series D. 284: 1789-1792.

Beal, J. M. 1937. Cytological studies in the genus *Phoenix*. Bot. Gaz., 99: 400-407.

Bhargava R., Singh R.S., Vashishtha, B.B. and Dhandar, D.G. 2003. Photosynthetic parameters in tissue culture and sucker plants of date palm cultivars. *In*. Abstract Paper presented in 2nd International Congress of Plant Physiology, New Delhi.

Chandler, W.H. 1958. Evergreen Orchards. Lea and Fabiger, Philadelphia, USA.

Chandra, A., Chandra, Anju and Gupta, I.C. 1992 .Date palm Research in Thar Desert. Scientific Publishers, Jodhpur p.116.

Chao, C.T. and Krueger, R. 2007. The date palm (*Phoenix dactylifera*); an overview of biology, varieties and genetics. Hort Science, 42: 1077-82.

Drira, N. 1981. Multiplication vegetative et micropropogation du palmier dattier (*Phoenix dactylifera* L.). Faculte des Sciences de Tunis. 138.

Dawson, V.H.W. 1982. Date production and protection with special reference to North Africa and the Near East. FAO Technical Bulletin No. 35. pp 294.

Chevalier, A. 1952. Recherche sur les *Phoenix africains*: R.B.A., Mai-Juin, 1952.

Eeuwens, C.J. 1978. Effects of organic nutrients and hormones on growth and development of tissue explants from coconut (*Cocos nucifera* L,) and date (*Phoenix dactylifera* L.) palms culture in vitro. Plant. Physiology, 42: 173-178.

Eeuwens, C.J. and Blake, J. 1977. Culture of coconut and date palm tissues with a view to vegetative propagation. Acta Horticulture 78: 277-286.

Gopalan, C.; Ramasastri, B.V. and Balasubramanium, S.C.1985. Nutritive Value of Indian Foods, National Institute of Nutrition, ICMR, Hyderabad, p87..

Mohammed, S. 1978. Problems in date palm propagation. Indian Hort. 23(3): 15-18.

Munier, P. 1973. Le Palmier-dattier-Techniques agricoles et productions tropicales: Maison Neuve et Larose, Paris., 17pp.

Muralidharan, C.M., Tikka, S.B.S. and Piyus Verma 2008, Date palm cultivation in Kachchh, Tech. Bull.No. 02/2008 S.D. A.U. Mundra, Kachchh, 36p.

Murashige, T. (1974. Plant propagation through tissue culture. Biologist 21: 87-93.

Nitsh, J.P. 1963 . The *in vitro* culture of flowers an organ culture. Symposium, 198-214.

Oihabi, A 1991. Effect of vescular arbuscular Mycorrhizae on Bayoud disease and date palm nutrition. Ph. D. thesis at the University of Marrakech: 199pp.

Pareek, O.P. and Sodagar, N.N. 1986. Date palm grooves in Kachchh. Indian Hort., 31:21-27.

Rabechault, H. 1962. Researches sur la culture *in vitro* des embryons de palmier a huile (*E. guineneensis* J.) 1. Effects de l ' acide B-indoleacetic. Oleag. 17: 757-764.

Reuveni, O. (1979). Embryogenesis and plantlets of date palm (*P. dactylifera* L.) derived from callus tissues. Plant Physiology, 63: 138

Reynolds, J.F. and Murashige, T. 1979. Asexual embryogenesis in callus cultures of palms. *In vitro*. 15:383-387.

Schroeder, C.A. 1970. Tissue culture of date shoots and seedlings. Date Growers Institute. 47: 25-27.

Singh, R.S., Samadia, D.K and Vashishatha, B.B. 2004. Genetic diversity in date palm (*Phoenix dactyliferra* L.). Indian Horticulture, 49 (3): cover page II & 24.

Singh, R.S., Bhargava, R. and Dhandar, D.G. 2003. Status and prospects of Date palm propagation, Current Agric., 27(1-2): 9-16.

Singh, R.S., Bhargava, R. and Dhandar, D.G. 2005. Performance of Date palm introductions under hot arid conditions. Indian Jour. Plant Genetic Res., 18 (1): 65-66.

Singh, R.S. and Bhargava, R. 2009. Effect of Seed treatments on Germination and Growth behavior of *Phoenix* species under hot arid conditions. Journal of Tropical Forestry, 25: (1) 42-48.

Singh, R.S., Bhargava, R., Singh S. K. and More, T. A. 2009. Post harvest utilization of date palm fruits. In Book of Abstracts "International Conference on Nurturing Arid Zones for People and the Environment: Issues and Agenda for 21st Century", November 24-28, 2009, CAZRI, Jodhpur, India.p310-311.

Sharma S.K and Singh, R. S. 2013. Date palm *In*. Fruit production Technology (ed. Dhillon, WS), Narendra Publishing House, New Delhi. pp.213-225.

Sodagar, N.N.; Murlidharan, C.M. and Ramdeoputra, M.V. 1996. .Date palm propagation problems and prospects with special reference to Kachchh. Jour. Applied Hort., 2:102-05.

Tewari, R.K. Solanki,K.R.,and Bajpaie, C.K. 2001. Krishi vaniki ke liye upyukt desi khajoor. *Phal-Phool,* Jan.-March issue,37-38.

Tisserat, B.H. (1979). Tissue culture of date palm. Jour. Hered. 70: 221-222.

Tisserat, B.H. and DeMason, D. 1980. A histological study of the development of adventive embryos in organ cultures of *Phoenix dactylfera* L. Annals of Botany, 45: 465-472.

Towill, L.E., Jaid. A.and Hughes, H.G. 1989. Cryopreservation of date palm shoot tips, Abstract 1989, ASHS Meeting p 091..

Vishal Nath, and Singh, R.S. 2003. Boost up date cultivation. Intensive Agric., 41:3-7.

Vij, V.K.; Kalra, S.K. and Bajwa, M.S. (1977). Studies on rooting of high offshoots in date palm. Punjab Hort. Jour., 17 (3&4): 135-138.

Zaid, A. (1987). Branching phenomenon in date palm. Date Palm Journal, 5: 48-58.

Zaid, A. (1999) Date palm Cultivation, FAO, Rome. 238p.

Zaid, A. and Hughes, H. 1989. *In vitro* acclimatization of date palm (*Phoenix dactylifera* L.) plantlets. Third Intern. Conf. IPBNET p. 43.

Zaid, A and Tisserat, B.H. 1984. Survey of morphogenetic potential of excised palms embryos *in vitro*. Crop Research 24: 1-9.

Zaid, A.; Hegarty, V.; Kaabi, H.H.S. 2007. Proceedings of the III International Date Palm Conference. Acta Horticulture, 736, p. 586.

Chapter-11

Aonla (*Emblica officinalis* Gaertn.)

A.K. Singh and R.S. Singh
Central Horticultural Experiment Station, Godhra, Gujarat

Indian gooseberry or *aonla* (*Emblica officinalis* Gaertn.), is an important fruit crop of 21st century and it has been regarded as '*Amritphal*' in ancient literature, and has been identified as an ideal plant for various kinds of wastelands *viz,* moisture stress, eroded, ravines, upland, riverbed and the areas with undulated topography (Korwar *et al.*, 2006) as it possesses some specific characters like intensive and deeper root system, summer dormancy of zygote, reduced leaf area, synchronization of fruit growth and development with moisture availability period and selective absorption of cations and anions which enable it to grow in fragile agro-climatic conditions. India is principal aonla growing country and Uttar Pradesh is the leading state and Pratap garh district, U.P. is well known for aonla cultivation throughout the country. Aonla is a deciduous tree, small to medium in size and distributed in tropical and sub-tropical forests in India. Looking to its high demand and value added products, now, it is being commercially cultivated in different parts of country. It can tolerate alkalinity and salinity and can be grown in wide range of soils.

In India, aonla is grown in about 50,000 ha area with annual production of 1.50 lakh tonnes with productivity of 3.03 tonne /ha (Mehta and Singh, 2003). Under South Indian conditions, aonla is flourishing well with very good fruits yield even twice in a year. However, under North Indian conditions, fruits available from November to January month. Outside India, aonla trees are also reported to be found in the natural forest of Cuba, USA, Pakistan, Sri Lanka,

Malaysia, China, Java and West Indies. The area under aonla cultivation is increasing rapidly in arid and semi-arid regions of Gujarat, Maharashtra and Rajasthan, Haryana, Andhra Pradesh, Karnataka, Tamil Nadu, Punjab and Himanchal Pradesh. However, the productivity is relatively low owing to moisture stress and susceptibility of crop by frost/low temperature in hot arid regions. It has been observed highly susceptible plants to frost than other horticultural crops (Singh *et al.*, 2012). It is mentioned in many ancient book and literatures. Aonla tree has aesthetic, nutritional and medicinal values. It can be planted in kitchen garden also by managing its plant canopy.

Nutritive Value and Uses

Aonla fruit is the rich source of vitamin C next only to Barbados cherry. It has become important fruit crop of arid and semi-arid region of the country owing to its hardiness, high productivity, suitability for growing under varied agro-climatic conditions and its utilization in cosmetic, pharmaceutical and processing industry, which attracts the growers for its cultivation under rainfed condition. The demand of organic produce is more in the market (Pathak *et al.*, 2005). The fruit contains chemical substances known as leucoanthocyanins (polyphenols) that retard the oxidation of vitamin C. The fruit is also rich in pectin and minerals such as iron calcium and phosphorus. Aonla fruit is very well known for its medicinal properties as mentioned in literature of both *Ayurvedic* and *Unani* system of medicines. It is acidic, cooling, diuretic and laxative. Although the fresh aonla has no table value, juice of the fresh fruit is popularly consumed by masses. After storing in brine for few days, it is fondly consumed. Dried fruit is useful in chronic dysentery, diarrhoea, diabetes, and dyspepsia, cough, anemia and jaundice. The excellent nutritive value of the fruits has great potentiality for processing into various products like *Murabba*, candy, shred, jam, pickles, herbal jam, *chawanprash, trifala ,mouth freshner, sweets,* etc. Fruits are also used in shampoos, oil and hair dyes preparations (Pathak *et al.*,1993; Singh *et al.*, 2008).

Botanical Description

Indian gooseberry or aonla (*Emblica officinalis* Gaertn.) comprises about 350 (Hooker, 1973) to 500 species (Bailey, 1971) mostly shrubs, few herbs and trees.there are other species which are also used for pickling etc. as *Phylanthus acidus, Emblica fischeri,and Phylanthus longiflorus.* Aonla is much-branched tree and ranges 9-12 m height.

Seedling trees may sometimes grow beyond 20 meters. Aonla tree is observed to be deciduous under north Indian conditions. Amla or Aonla is a member of the family Euphorbiaceae. However it is considered as evergreen in the tropics (Bajpai, 1957). The tree bark is glossy and cracks irregularly. Aonla trees bear two types of shoots on the basis of growth characteristics. These have been categorized as long or indeterminate and short or determinate (Bajpai, 1968; Ram, 1971). The indeterminate shoots are longer and continue to put growth in the season. These shoots neither fall from the tree nor bear flowers irrespective of period of emergence. While on the other hand, determinate shoots appear on the nodes of indeterminate shoots and their number in each node may vary from 3-5 in different cultivars, These shoots bear small sized (10-13mm length, 2-3mm width) leaves arranged so closely that apparently the shoot appears to be pinnately compound leaf.

Inflorescence is recemose type. Flowers are minute, unisexual with short pedicel. Male flower first appear in clusters; perianth 6, yellowish green to deep pink in colour with valuate aestivation. Androecium consists of 3 stamens, each profusely branched, filament attachment is basified or innate, number of segments varies from 5 to 7, but commonly 6 are observed. Ovary hypogynous, Carpel 3-4, 3 chambered, placentation axile, two ovules per locule, margin straight to crescent shaped, ovarian chamber shallow to deep.

Flowering: In North India, aonla tree begins to shed its determinate shoots from February onwards resulting the indeterminate shoots devoid of foliage by the end of February to mid March. Flowering under rainfed condition of western India starts from month of February. The new determinate shoots, which emerge at nodes from the margins of the scars formed after abscission of determinate shoots in the previous season, begins to appear by the end of February and continue to appear till the first week of April. Blossom buds appear on the newly developed determinate shoots but shoots that emerge after mid April do not bear flower and remain vegetative. During the year 2012, the flowering time was noted to be 20-25 days late as compared with previous years in different genotypes of aonla. Among various genotypes of aonla, it was studied that initial fruit set (%) through open pollination was recorded to the tune of 20-50% at 15 days after anthesis. During floral study, it was found that average number of male flowers varied from 378-511, while female flowers varied from 3-15. Sex-ratio per determinate shoot of different aonla genotypes was also calculated. A preliminary trial on the study of self-incompatibility indicated that in selfing, there is no fruit set in comparison with various crosses made and open pollination.

After 15-20 days after crossing, the highest initial fruit set was recorded 55-88%, which was reduced to 10-15% at final fruit set (fruit retention).

Pollination and fruit set : Aonla is a cross- pollinated plant. Wind, honeybees and gravity all play important role in effective pollination (Singh *et al.*, 1998). Common visiting hours of honeybees are late evening and mornings (Bajpai, 1957). Bajpai (1968) reported that there is no self-incompatibility in aonla and cause of poor fruit set may be due to high percentage of male flowers. However, Pathak *et al.* (1994) have observed self-incompatibility in initial studies in flowering and fruit set in Indian gooseberry. They also noted considerable variations with respect to sex ratio which varied from 28:70:1 to 355:49:1 in different cultivars. Number of female flowers varied from 2 to 8. NA-6 had the lowest sex ratio followed by NA-7, Kanchan, Chakaiya while NA-9, Banarasi and Krishna showed higher sex ratio. Pollen viability with acetocarmin showed 82.32 to 96.40 per cent, while only 16. 71 to 36. 90 pollen germinated in the sucrose solution (Pathak, *et al.*, 1993).

The productivity of cultivar depends upon the percentage of female flowers on the branches. The phenomenon of flowering requires cross pollination, thus, fruit set depends on pollinizer (Singh *et al.*, 2001). Occurrence of self- incompatibility is also reported in some cultivar which needs for mixed planting of cultivars to have better fruit set and yield.

The work on improvement in aonla has been done through selection only at various places in the country. The earlier varieties of aonla such as Banarasi, Francis and Chakaiya have some limitations. Banarasi is a shy bearer, Francis is affected by necrosis and Chakaiya bears small sized fibrous fruits. Pathak *et al.* (1993) made some varietal improvement in aonla at NDUAT, Faizabad. Some varieties have also been developed by GAU, Gujarat,CISH, Lucknow, CHES, Godhra. Concerted efforts have been made to develop aonla varieties for enhancing the production. These varieties are not having only high yielding potential but also possess many other horticultural traits.

Propagation

Traditionally, aonla is propagated by seeds, however, inherent limitations associated with the seedling progenies; seed propagation is limited for the raising of rootstocks only (Singh, 1998). True to type planting materials can be produced through vegetative methods. It can be propagated through cuttings, budding and grafting (Wagh *et al.*, 2001) and also through micro-propagation (Mishra *et al.*,1999).

Seeds

Seed propagation is the most common method of propagating aonla. The aonla seeds germinate in 15-25 days after sowing. The seedlings become ready for transplanting in spring or next Monsoon. If the seeds are sown too deep, seedling emergence is delayed and there may be chance of rotting due to poor aeration. Seeds may also be sown in polythene bags, as it facilitates an easy handling of rootstocks and grafted plants. The orchard raised by seedlings is not true to type and exhibits variability.

Raising of Seedlings in Nursery Beds

For raising of seedlings in the nursery, beds are thoroughly cleaned and ploughed. For better germination, higher survival and establishment, well rotten FYM should be mixed with the soil before sowing of seedlings. Fresh seeds can directly be sown because of no dormancy. Polyethylene tubes can also be used to raise the seedlings. Mixture of FYM, sand and soil (1:1:1), should be mixed before filling in the polythene tubes. Seeds are sown in the polythene tubes or bags during rainy season for better germination and survival. The foliar sprays of plant growth regulators *i.e.* gibberellic acid (GA_3) and IBA (both at 250, 500, 750 and 1000 ppm), and potassium nitrate (250, 500, 750 and 1000 ppm) improves seedling vigour by means of improved growth of stem and roots.

Raising Root Stocks in Polythene Bags

Raising of rootstock in nursery beds and lifting of budded plants with earth ball in highly sandy soils is practically not feasible. Transportation of plants for a long distance may also cause high mortality particularly under hot semi-arid and arid environment. To reduce the time for raising root stock and to avoid damage during handling and transportation, polythene tubes and polythene bags may be used on commercial scale. Generally, polythene bags (25cm x 10cm) are used for raising the rootstocks. Small holes are made in the bottom and sides of polythene bags for drainage and aeration and filled with porous rooting medium or pot mixture for raising rootstocks. Generally, 1-2 seeds are sown in each poly bags and then placed in trench bed, so that it can be irrigated easily. Sometimes coiling of root may become problem, hence root pruner is also used for trimming of roots. Polythene bags of the size of 25 cm x10 cm, 25cm x 15 cm and 30 cm x 15cm as per the requirement are being used for raising the rootstocks in India. About

6-12 months old seedling of uniform size having stem of pencil thickness are used as rootstock for budding and grafting. Plants raised in the polythene bags can easily be transported to distant places with higher success after planting (Saroj *et al.*, 2000).

In-situ Establishment of Rootstocks

For getting maximum survival and success of budding under rainfed conditions of hot semi-arid ecosystem, sowing of seeds should be done directly in the field during monsoon (June-July) at demarcated location, and the seedlings become ready for *in-situ* budding during June, 10-11 months after sowing. If there is long gap between two spell of rain, plants should be irrigated as and when required. Proper irrigation is required to save the plants in the field during summer. Since no planting is involved in this case, the plants retain the deep tap roots and thus become more hardy and vigorous after *in-situ* budding.

Vegetative methods

The vegetative propagation techniques are recommended for commercial orcharding in aonla. Among the vegetative propagation techniques, patch budding and softwood grafting are most successful for methods of multiplication of aonla under rainfed conditions. Now-a-days, patch budding has been adopted commercially for multiplication of aonla.

Selection of bud wood

Bud wood becomes available during the active growth period in summer season. The bud sticks (2-3 months old), with well swollen and recently matured buds (but still not open) are collected. Immature and undeveloped buds from the upper part of the new shoots are not suitable. Similarly, over mature and inactive buds should not be used. The active growth period is indicated by easy separation of the bark from the wood of scion sticks. After collection, the bud wood is often stored for some period or takes sometime in transportation. During this period, considerable loss of survivability may take place. Bud woods retain good survival when kept under ventilated shade and wrapped in moist jute cloth.

Patch budding

In this method, a healthy bud is selected from the axils of leaf. Leaf blade is removed with the help of a sharp knife leaving petiole intact.

The upper cut is given about 1-1.5 cm above bud which goes downwards up to 1.0-1.5 cm below the bud without wood portion and then lower cut is given about 1.0 cm below the bud. The similar rectangle incision is made on the rootstock by placing the bud on the root stocks to mark the exact size of the bud on them and after removing the bark of root stock, the bud is placed at the juncture. The bud is pressed by hand to remove open space if any and tied tightly except the place of bud with white polythene strip (200 gauge thickness and 2 cm wide). In case, the cuts on rootstock are wider, at least one side bark of scion and stock must be matched properly. The rootstock is cut about 10 cm above the bud to facilitate bud to sprout. After union, the top of the rootstock is cut a little above the bud union and polythene strips are removed carefully. One year old, 0.8-1.20 cm thick rootstocks budded in June showed better success (Pathak *et al.,* 1991). Under rainfed conditions, patch budding was found successful when performed in the month of June before onset of monsoon under Gujarat conditions and recorded 98.21 per cent bud take (success) (Table 1). Patch budding in aonla was found suitable under Maharashtra conditions (Wagh *et al.,* 2001). For successful multiplication of aonla plants through budding in June month has also been reported by Pandey and Prasad (1980).

Table 1 : Standards for patch budding in aonla under semi-arid region.

Type of rootstock	Straight & active growth stage
Raising rootstock	In polythene bag
Size of polythene bag	25 x 15 cm
Age of rootstock	6 to 12 month old
Diameter of rootstock	0.80 – 1.20 cm
Age of scion shoots	2- 3 month old
Diameter of scion	0.70 – 1.00 cm
Size of patch	1.5-2 cm x 1cm
Budding height	20-25 cm above the ground
Bud union	Smooth
Plant height	50-60 cm
Root type/ Architecture	Well developed root system free from coiling
Foliage	Healthy and green foliage
Disease/Pest incidence	Plant should be free from insect, pests and diseases.

Soft wood grafting

With the onset of the monsoon season in the middle of the June, seedlings grown in polythene bags are planted in the field after first shower of monsoon. The polythene bags are carefully removed and seedlings are planted in the centre of the pit. The tap root system should not be disturbed while lifting the plants from nursery. After planting, the soil around the plants is pressed firmly to avoid formation of air pockets. Thereafter, basin is prepared around the plant for irrigation or collection of rain water. After planting, seedlings are irrigated thoroughly and thereafter, as and when required. Seeds may also be sown directly in the field if irrigation water is available for summer irrigation, otherwise the seeds should be sown directly in the field during rainy season under rainfed condition (Table 2).

In- situ soft wood grafting through cleft method is done in the month of June under semi-arid environment of western India. The growth below the graft union is removed regularly to encourage the sprouting and subsequent growth of the scion shoots. The bud sprouts within 15 to 20 days of grafting with proper establishment of bud's vascular connection with rootstock. The polythene strips are carefully removed after completion of the union. The plants are given support with the help of stakes to protect them from stormy winds (Amin, 1978). High temperature and relative humidity during June- July has helped in early sprouting and better graft success, because of fast establishment of vascular connection with rootstock.

Table 2 : Standards for soft wood grafting in aonla under hot semi-arid conditions.

Method of propagation	Soft wood/Wedge grafting
Type of rootstock	Straight and active growth stage
Raising rootstock	In polyethylene bag
Size of polyethylene bag	25 x 15cm
Age of rootstock	8 to 11 months
Diameter of rootstock	0.8 – 1.2 cm
Age of scion shoots	2- 4 months old
Diameter of scion	0.8 – 1.00 cm
Length of scion sticks	12 - 15 cm
No. of buds on the scion stick	5-8 buds
Grafting height	20-25 cm above the soil
Root type/ architecture	Well developed root system without coiling
Plant height	45-60 cm
Foliage	Healthy and green foliage with vigorous growth
Disease/pest incidence	Plants should be free from insect, pests and diseases

Precautions: Shifting of poly bags in nursery is required at one and half months intervals to discourage tap root. Grafts are to be handled carefully during lifting, packaging and transportation to avoid damage to the union or scion portion.

Budded/grafted plants

Plants in nursery beds are lifted with large earth balls 9 to 12 months after budding for transplanting in the field. Some mortality may occur due to damage occurred during lifting, packaging, transporting and planting. These operations are also cumbersome and incur high costs due to the large size of the earth balls. Budlings prepared in polythene bags become ready for transplanting after 50-60 days of budding or grafting. The budlings in bags are removed from the nursery and kept in the shade for a week and these can easily be transported and planted in the field with maximum survival. The roots of the plants do not coil and therefore, retain the drought hardy character and vigour almost similar to the plants raised *in-situ.*

Effect of bud wood storage on graft success

Scion shoots covered with moss grass and wrapped with news paper followed by moist jute cloth can be used for grafting up to the 3th day from the date of detachment from the mother plant successfully under hot semi-arid ecosystem of western India.

Micro-propagation

Looking to the huge demand of quality planting material, importance of micro-propaglion can not be ignored.In case of aonla, limited tissue culture work has been done so far. The efforts on micro-propagation in aonla has been made by Mishra *et al.* (1999) for mass multiplication but this work is basic in nature and none of them has reached the stage of commercial production. The tissue culture protocol have yet to be developed and made available for large scale multiplication and propgation of selected cultivars. *In vivo*-micrografting consist of raising of root stock of star goose berry (*Phyllanthus acidus* Skeeb.) for two and half months and grafting by the micro bud of a desired scion in a split of decapitated rootstock. In this technique, a micropipette tips are capped over the grafted portion for better graft union which takes place in 2-3 weeks (15-21 days). Micrografting done during January-February resulted in 90% sucess. This seems to be easy and more economical method of vegetative propagation in aonla (Kadam, *et al.*, 2010).

Top working

Singh (1953) reported that inferior trees can be headed back to a height of 120cm from ground level during March and the stumps can be shield budded in early June with scion of improved cultivars. It is found suitable in case of rejuvenation of old aonla orchards. An encouraging result has been found in top working carried out on 15years old seedling plants in Forest Department, West Bengal. Among three cultivars, the scion of cv. Balwant gave highest percentage of bud take (90%) at one month after grafting followed by Krishna (80%) and Chakaiya (60%) (Ghosh and Dutta Ray, 2003).

Care of Nursery Plants

Aonla plants at nursery stage are likely to be damaged by scorching sun and hot desiccating wind under hot semi-arid ecosystem of western India. During summer, the irrigation should be given at 5-7 day intervals depending upon agro-climatic condition. The beds/ polybags should be kept free from weeds by regular weeding/hoeing. In western India, plants should be irrigated in the evening hours and should be protected from Agro- shade net (50%) to protect the plants from direct sunlight and wind. Weeding should be done properly so that plant could not compete of nutrient during active growth period under nursery conditions. Before transplanting in the field, shifting of plants is necessary for hardening process.

Quality Standards of Planting Material

Healthy and disease free planting material of improved cultivars should be prepared for plantation. The rootstock should be of known plants. Scion should also be taken from mother orchards maintained for nursery plants. Plants should of proper age and height for transplanting in the field.

References

Amin, R. S. 1978. Soft-wood grafting-a new technique for hard wood plants. Current Sci., 47: 468-469.

Bajpai, P. N. 1957. Blossom biology and fruit set in *Phyllanthus emblica* L. Indian J. Hort., 14:99-102.

Bajpai, P. N. 1968. Studies on flowering and fruit development in aonla. Hort. Adv., 7: 38- 67.

Ghosh, S.N. and Dutta Ray, S.K. 2003.Standardizationof some agro-techniques for cultivation of amla in red laterite zone of West Bengal. *In.* Amla in India, Aonla Growers Association of India, Salem,pp.104-11.

Hooker, J. D. 1973. The Flora of British India Vol. V1[st] Indian Reprint, Periodical Experts Delhi.

Kadam,J.H.,Karale,A.R,Desai,U.T.and Athani,S.I.2003. In vivo-micrografting-A new technique of vegetative propagation of amla. In. Amla in India, Aonla Growers Association of India, Salem, TN, pp.63-64.

Korwar, G. R., Pratibha, G., Ravi, V. and Palanikumar, D. 2006. Influence of organic and inorganic on growth, yield of aonla (*Emblica officinalis* Gaertn) and soil quality in semi-arid tropics. Indian Journal of Agricultural Sciences, 76 (8): 457-61.

Mehta, S.S and Singh,H.P. 2003. Amla in India, Aonla Growers Association of India Salem-636 016,(T.N)pp.266.

Mishra,M.; Saxena,R.P. and Srivastav,A. K.1999.Studies on micro- propagation of aonla. Prog.Hort.,31: 116-122.

Pandey,I..C. and Prasad,R.S.1980. Propagation of aonla by budding.Prog.Hort.,11:27-30.

Pathak, R. K. Ram R. A. and Sukla, S. K. 2005.Cultivating *aonla* organically. Indian Hort. 50:4-9.

Pathak, R. K., Srivastava, R. K., Dwivedi, R. and Singh, H. K. 1993. Aonla Descriptor. Department of Horticulture, N. D. University of Agriculture and Technology, Faizabad, Uttar Pradesh.

Pathak, R. K.,Ojha, C. M., Dwivedi, R and Hari Om 1991. Studies on the effect of method and duration of budding in aonla. Indian J. Hort., 47:207-12.

Ram, S. 1982. Induction of off-season flowering in Aonla trees. Prog. Hort., 14 (2-3):180-83.

Reddy, C.S. and Janki Bal 1979. A study on the pollination in *Emblica officinalis* Gaertn. New Bot. 4: 155-160.

Saroj, P. L., Nath, Vishal and Vashistha, B. B. 2000. Effect of polycontainers on germination, seedling vigour, root characters and budding success in aonla. Indian J. Hort, 57(4):300-304.

Singh, H. K., Srivastava, A. K. and Dwivedi, R. 1998. Pollination and fruit set behavior of aonla. Indian J. Agric. Sciences, 68(4):204-205.

Singh, H. K., Srivastava, A. K. and Dwivedi, R. 2001. Effect of pollinizer on fruit set and fruit quality of NA-7 aonla. Indian J. Agric. Sci., 71(1):65-66.

Singh, I. S. 1998. Aonla Production Technology -A Bulletin, Deptt. of Hort., NDAUT, Faizabad.

Singh, L.B. 1953. Top working of Aonla (*Phyllanthus emblica*).Indian J.Hort.,10:59-61.

Singh, R.S.; R. Bhargava; Garima Pal and S.K. Sharma. 2012. Effect of frost on fruit plants and its management in hot arid region. Crop Improvement, 39 (spl. Issue).39: 723-724.

Singh, S., Singh, A. K. and Joshi, H. K. 2008. Post harvest management of aonla (*Emblica officinalis* Gaertn) and Ber, Technical Bull. CIAH, Bikaner.

Wagh, A.P.,Choudhary,M.H.,Jadhav, B.J. and Joshi,P.S.2001.Studies on propagation in aonla. Jour. Soils & Crop,11(10:74-77.

Chapter-12

Bael (*Aegle marmelos* Correa.)

D. Pandey, S.K. Shukla and G. Pandey
Central Institute for Sub-tropical Horticulture, Lucknow- 227 107

Bael (*Aegle marmelos* Correa.) also known as *Shri Phal, Baelpatra,* Bengal quince, Beel, Bel is a nutritious and medicinal fruit plant, which is most suitable for water scarce areas of the country. It is indigenous tree to India and known since ancient time. Bael fruit is mentioned in ancient literature viz., *Vedas, Ramayana, Upavan Vinod, Brihat Samhita* (Singh and Roy, 1984). Methods of ripening of the bael fruit are also given in Buddhist and Jain literature 800 BC – 325 BC. During the *Ramayana* period, it has been mentioned that bael trees were found growing in the *Chitrakut hills* and *Panchvati*. The bael is grown throughout the country and main bael growing states are Uttar Pradesh, Bihar, Madhya Pradesh, West Bengal, Odisha and Rajasthan. It is found growing in neighboring countries like Nepal, Sri Lanka, Pakistan, Bangladesh, Myanmar, Thailand and most of the Southeast Asian countries. Besides systematic orchard, bael can also be grown in parks, temple gardens, roadsides etc. The bael plant is sacred to the Hindus and is of common occurrence in temples of 'Lord Shiva' because of its mythological significance.

The data on area and production is not available since it is underutilized fruit crop. However, in recent years, some organizations have started the survey, collection, evaluation and establishment of germplasm blocks at their experimental farms. Some progressive farmers have also started planting bael varieties either in the form of small orchards or as boundary plantation of other fruits. As an estimate, about 27 ha area is under bael plantation with about 150 tones of production of improved varieties. Moreover, the fruits available mostly in the markets are harvested from the trees either naturally growing in the forest areas or from seedling plantations. Thus, the total production per year is estimated to be about one thousand ton.

The bael tree grows successfully and produces higher yield in sub-tropical climate where summer is hot and winter is mild. It has wide range of adaptability to adverse soil and climate. Further, the seedling *bael* tree is reported to be cold hardy is found to be grown up to an elevation of 1200 m/sl and is not damaged by temperature as low as -7^0 C but under arid conditions of Bikaner (Western part of Rajasthan), the leaves and tender twigs of improved budded *bael* cultivars were affected at a temperature $<5^0$C. However, plants tolerate high temperature and drought by shedding leaves during summer.

Nutritive Value and Uses

The bael is one of the most nutritious fruit and rich in riboflavin, vitamin A, carbohydrate etc. as given in Table 1 (Gopalan *et al.*, 1985). Various chemical constituents viz., alkaloids, coumarins and steroids have been isolated and identified from different parts of bael tree. Marmelosin is probably the therapeutically active factor of bael fruit and is known as the panacea of the stomach ailments. The aroma component of bael fruits was studied by Tokitoma *et al.* (1982). All the part of tree, whether stem, bark, root, leaves or fruit at different maturity stages has some use or the other. The medicinal properties of bael have been dealt in Charak Samhita . It is also used in the preparation of several Ayurvedic medicines since ancient times (Rai and Dwivedi, 1992). The pulp of fresh ripe fruits is eaten. They have laxative properties and good for heart and brain. Fruit pulp is used for preparing *sherbet* and syrup. The *sherbet* and marmalade prepared from fruit pulp is prescribed for diarrhea and dysentery. It is said that the ripe fruit is a tonic, an astringent and laxative, whereas unripe fruit is an astringent and digestive. The pulp of fruits is used for the preparation of several value added products such as jam, squash, powder, nectar, toffee etc.

Bael fruits have mucilaginous substance. Gum is also obtained from stem and seed locules. The use of bael gum has been reported as early as in 1961 by Haskar and Kendurkar. The gum is used to prepare adhesive, waterproofing and oil emulsion coating. The wood is used for making small agricultural implements. The wood pulp is used for manufacturing wrapping paper. The dried twigs are used as firewood. The leaves are used to worship 'Lord Shiva' according to Hindu mythology. The leaves are also used as a fodder for feeding of sheeps and goats. The leaves contain 0.6 per cent of essential oil, mostly composed of limonene. The use of bael leaf powder as a protectant against storage pests of paddy has been reported by Prakash *et al.* (1983). Antifungal activities of essential oil of bael has been reported by Jain (1977). Banerjee and Maiti (1980) reported that bael seeds contained protein (62 % water insoluble-60 % and water soluble-2 %), oil (32 %), carbohydrate (3%) and ash (3%). The antimicrobial and anthelmintic properties as well as chemical properties and medicinal importance of bael seeds have been also studied by Singh *et al.* (1983) and Pal *et al.* (1993) respectively.

Table 1 : Nutritive value of bael fruit

Constituents	**Value (per 100 g. fruit)**
Water	61.5 g
Protein	1.80 g
Fat	0.30 g
Fibre	2.90 g
Minerals	1.70 g
Carbohydrates	31.80 g
Mucilage	12.7 - 19.0%
Carotene	55.00 mg
Calcium	5.00 mg
Iron	0.60 mg
Thiamine	0.13 mg
Riboflavin	0.03 mg
Niacin	1.1 mg
Acidity	0.25-0.36%
Vitamin A	186 IU
Vitamin C	8-18 mg

Botanical Description

Bael (Aegle marmelos Correa.) belongs to the family Rutaceae. The generic name is of Greek origin and the species marmelos is of Portuguese origin. The chromosome number is x=9 and 2n=36. The genus *Aegle* consists of 2 or 3 species.

The tree is a medium to tall, deciduous, slow growing and 6-8 m in height with whitish /grayish stem. The bark is furrowed and corky yellowish-brown in colour. Its leaves are green aromatic, trifoliate; branches have spines, and trunk is strong and stout. Misra *et al.* (1999)

have reported great variations in leaf characters, development pattern and shoot growth of 8 selected genotypes of bael under *Tarai* conditions of Uttarakhand. The bisexual flowers are nearly 2 cm wide and borne in clusters and greenish white in colour. The calyx is shallow with 5 short, broad teeth, pubescent outside. There are 5 petals, which are oblong oval, blunt, thick, pale greenish white, dotted with glands. Stamens are numerous, sometimes coherent in bundle. The ovary is oblong ovoid, slightly tapering, the axis being wide; cells are numerous 8-20, small and arranged in a circle, with numerous ovule in each cell. Fruit is hard-shelled berry, globose, pericarp smooth and round to oblong in shape, greyish yellow in colour.

The seedling trees require 7-8 years to bear fruit while budded plants start bearing fruits at the age of 4 years of planting. Bael tree flowers in May-June, fruit setting takes place by the end of May and continues up to July. Fruiting occurs on one-year-old shoots. Srivastava (1982) defined the phytophases of bael and depicted by phenobiogram. Development of bael fruit follows a single sigmoid curve and has three distinct periods of its growth. The initial growth is slow for about one month followed by a rapid increase for four month and then more or less stationary phases until the fruits are harvested. Clonal variation in inflorescence and floral morphology was also studied by Jaiswal and Misra (1996). The studies on variation in fruit set, retention and yield of 10 eight-year-old budded trees of bael (*A. marmelos*) clones indicated that Pant Sujata followed by Pant Shivani set maximum fruits while Pant Bael-10 retained maximum fruits. Yield efficiency in terms of number of fruits per unit volume of tree was found maximum in Pant Bael-10 while it was minimum with Pant Bael-3. Pant Shivani produced maximum weight of fruits per unit volume followed by Pant Sujata. Pant Bael-10 and Pant Bael-1 produced higher number of fruits per tree while Pant Shivani produced highest yield in terms of fruit weight per tree followed by Pant Sujata (Jaisawal and Misra, 1998). The developmental physiology of bael (*Aegle marmelos* Correa) fruit variety Kalyani Selection-1was also studied by Bhadra *et al.* (2000). Variation in fruit set and fruit retention was also observed in different germplasm under eastern Uttar Pradesh conditions (Anon., 2001). Fruit drop and cracking in bael can be minimized by the application of growth regulators (Saini *et al.*, 2004).

Bael is one of the fruits that remain intact on the tree for longest time. The tree flowers in May-June and fruits become ready for harvest by April-May. The fruit is categorized as climacteric fruit (Roy and Singh, 1980). In general, ripening of fruit is judged by the turning colour of the fruit skull, flesh colour, TSS and aroma of the fruit. The changes in composition of bael fruits at various stages of fruit growth and development have been described in details by Singh and Roy (1984), which can be used as reliable indices for judging the maturity of bael fruits. Roy and Singh (1981) reported that ripening of bael fruit is accelerated by combination of high temperature and exogenous application of ethylene but ethylene induced ripening is not effective at lower temperature. Preliminary studies on the availability of ripe *bael* fruit in off-season were also attempted by Roy *et al.* (1972). Ripe fruits can be made available 2-3 months ahead of schedule by giving pre-harvest treatment of 1000-1500 ppm ethereal and storage of fruits at 30 ^{0}C after harvesting them in January. It takes 18-24 days for the fruits to be artificially ripened. The composition of fruits, whether ripened artificially or naturally, does not vary much; only the accumulation of sugars is slightly less in artificially ripened fruits as compared to natural ones.

Seeds are numerous, arranged in seed cavity/locules. Mucilage is present in seed locules. The testa is white with wooly hairs. The embryo has large cotyledons (Reuther *et al.*, 1967 and Roy and Singh, 1979).

Bael plant is found growing naturally in Uttar Pradesh, Bihar, Jharkhand, Madhya Pradesh, Odisha, West Bengal and Chattisgarh with large genetic variability, which should be exploited. In Uttar, Pradesh, Deoria, Basti, Gonda, Faizabad, Sultanpur, Jaunpur, Pratapgarh, Mirzapur, Allahabad, Lucknow, Etawah, Agra etc. are the districts where large number of promising genotypes are either growing naturally or planted near the houses. Rai and Dwivedi (1992) reported vivid account of bael genetic diversity available in India. Apart from the tree morphological characters, wide variability exists in fruit size and shape, bearing habit, flesh colour, texture, fibre content, sugar content, mucilage content, etc. Some types have more number of seeds, gum locules and thick pericarp (Mishra *et al.*, 2000). However, some selections have been made at NDUAT, Faizabad and GBPUAT, Pantnagar, which are gaining popularity for commercial cultivation. Besides this, some germplasm were also collected and maintained in the field repository at CAZRI, Jodhpur; NDUAT, Faizabad, GBPUAT, Pantnagar and TNAU, Arrupukottai.

The limited efforts have been made for crop improvement in bael so far. The available promising types are through the clonal selection only. Some basic works on floral biology, fruit set, fruit drop and fruit retention and pollen viability have been done on seven years old plants of NB-4, NB-5, NB-7 and NB-9 at NDUAT, Faizabad. It was observed that that the flower bud emergence in bael cultivars started from third week of May and continued up to last week of May, it was earliest in NB-9 followed by NB-4, NB-7 and latest in NB-5. The flower bud took 15-22 days from flower bud emergence to anthesis. The maximum flowers open between 6-7 AM followed by 5-6 AM and 7-8 AM in all the bael types under study. The maximum flowers dehisced between 6.30-7.30 AM. The stigma receptivity was highest with NB-9 closely followed by NB-5 and lowest with NB-7. The stigma receptivity was maximum on the day of anthesis. The pollen viability was highest in NB-7 followed by NB-5, NB-4 and lowest in NB-9. The fruit retention was highest in NB-9 closely followed NB-5 and lowest in NB-7 (Srivastava and Singh, 2000). Variation in physico-chemical parameters in bael fruits have been observed by Ram and Singh (2003).

The clustering pattern of genotypes was random and not parallel to geographic distribution. The characters viz. ascorbic acid content, fruit weight, fruit length, number and weight of seed per fruit, fibre content, petiole length contributed maximum in genetic divergence. These characters may be used in selecting diverse parents in hybridization and intercrossing between clusters to develop superior clones of bael with most desirable tarits (Deepti Rai, *et al.*, 2002).

Among local types, Ayodhya, Kagzi, Etawah, Gonda and Mirzapuri bael are popular in Uttar Pradesh due to their good yield potential and quality fruits. In recent past, some promising varieties of bael have been developed through clonal selection at G.B. Pant University of Agriculture and Technology, Pantnagar. Pant Bael-3 and Pant Bael-1 produced vigorous growth and tallest trees while Pant Bael-9 (Pant Aparna) and Pant Bael-12 (Pant Sujata) produced dwarf trees. Pant Bael-10 and Pant Bael-1 produced higher number of fruits per tree while Pant Bael-5 (Pant Shivani)) followed by Pant bael –12 (Pant Sujata) produced highest yield in terms of fruit weight per tree. In general, Pant Shivani, Pant Aparna and Pant Sujata were found better for physico-chemical characters of fruits as compared to other clones. From Narendra Deva University of Agriculture and Technology Faizabad (U.P.), the cultivars NB-5 and NB-9 are performing well under arid conditions of western Rajasthan. The vegetative growth parameters and physico-chemical characters of some bael genotypes at NDUAT, Faizabad (Anon., 2000) were recorded and given in Table 2. Dhara road is a prolific bearing another bael suitable for dry areas (Singh and Vishal Nath, 1999). Tree size is very tall and number of fruits/plant is higher than the other local cultivars. Recently, CISH-B-1 and CISH-B-2 selection of bael have been developed at Central Institute for Sub-tropical Horticulture, Lucknow (Pandey *et al.*, 2001).

Table 2 : Evaluation of bael genotypes

Genotypes	NB 4	NB 5	NB 7	NB 9
Plant height (m)	3.80	4.12	4.38	4.30
Av. Plant Spread (m^2)	5.30	6.26	6.00	5.20
Fruit Wt. (kg)	0.75	1.10	2.30	1.40
Fruit length (cm)	22.75	27.27	27.50	26.50
Fruit Width (cm)	27.20	31.00	29.00	23.00
No. of seeds/fruit	60.00	61.00	72.00	84.00
TSS (%)	32.00	34.00	38.00	35.00
Acidity (%)	0.15	0.10	0.26	0.23
Ascorbic Acid (mg/100 g of pulp)	19.02	21.40	25.10	24.00
Yield (fruits/tree)	30.00	46.00	19.00	41.00

Propagation

Bael is propagated both sexually and asexually. On large scale, it is multiplied through budding and grafting methods. Budding is most common method of nursery men. Now a days, it s multiplied through tissue culture.

Seed

Earlier, bael was propagated by seeds but now seeds are used for raising of rootstocks. Plants of seedling origin exhibit a lot of variability in plant and fruit characters. For seedlings, the seeds are sown in nursery bed. The seeds have no dormancy and they germinate within 10-15 days after sowing. The young seedlings should be protected from frost. The orchard raised by seedlings is not true to type and exhibits variability. Three days after extraction the seeds were treated in several ways, then dried back to their original weight and stored in alkathene bags at room temperature for up to 132 days. Some seeds were sown 12 days after treatment and others at 132 days. Germination of seeds sown 12 days after treatment was >88 per cent, except in seeds treated in boiling water for 25 minutes (3.3%). After 132 days of storage germination was highest in seeds treated with 1per cent KH_2PO_4 for 24 h (91.66%) or coated with Paste-1 (Agromin at 0.75 g 100 g^{-1} seeds + 100 p.p.m. GA + 0.75 g KH_2PO_4 100 g^{-1} + 0.75 g KNO_3 100 g^{-1}) which gave 90 per cent germination (Hore and Sen, 1985).

The effect of sodicity on seed germination and seedling growth was studied by Pandey *et al.* (1988). A pot experiment was conducted to study the effect of artificially maintained sodicity levels (18.5, 29.0, 38.0 and 49.5 ESP (Exchangeable Sodium Percentage)) on germination of seeds and growth parameters of bael seedlings (Shukla and Singh, 1996). Delayed and poor seed germination and reduced plant growth were observed in response to increased sodicity. However, satisfactory performance of bael with respect to seed germination and plant growth was observed in sodic soils up to 29.0 ESP without application of any chemical amendments. The media requirement for seed germination and seedling establishment has been suggested by Chattopadhyay and Mahanta (1989). Seed germination and growth of bael seedlings was affected by soil sodicity (Shukla and Singh,1996).

The effect of plant biological regulators, *i.e.* gibberellic acid (GA) and IBA (both at 250, 500, 750 and 1000 ppm), and potassium nitrate (250, 500, 750 and 1000 ppm) on the seedling quality of Bael (*Aegle marmelos*) was studied. Seedling growth, in terms of height, number of branches, leaves, secondary roots, and dry weight of leaves, stem and top, was found maximum with foliar sprays of GA at 1000 ppm, while the number of fibrous roots per seedling, length of tap root and dry weight of roots were found maximum with foliar sprays of IBA (1000 ppm). The shoot root ratio was found maximum with GA (1000 ppm) and minimum with IBA (1000 ppm). The correlation among most of the morphological traits was found positive and significant, but the correlation of the number of branches per seedling with dry weight of leaves; fresh weight of top with dry weight of leaves and stem; and dry weight of top with fresh weight of top, dry weight of leaves and stem, were more significant (Misra and Jaiswal, 2001).

Vegetative propagation

Vegetatively propagated plants come in to bearing earlier than the seedling trees. Therefore, vegetatively multiplied plants should be used for planting in the field.

Suckers: Some root suckers which, arise from the roots of the bael trees, can be separated during monsoon and transplanted into field. But, the success percentage is very low. It is not recommended for arid conditions.

Cutting: Ray and Chaterjee (1996) reported that Growth regulator and etiolation treatments were significantly effective in inducing roots in ringed stem cuttings of *A. marmelos*. An invigoration treatment (the production of water shoots following removal of large branches) accompanied by growth regulator and etiolation treatments significantly increased root quality of cuttings. The highest rooting rates of 75-80% were achieved using 5000 ppm IBA, etiolation and invigoration.

Budding: Patch budding in the month of June or July proved to be the best. Singh (1954) reported that budding in *Aegle marmelos* in June with bud wood from one month old scion gave 80 per cent success. Moti *et al.* (1976) also suggested patch budding in the month of June for propagation of bael. The effect of four scion genotypes (Pant Aparna, Pant Shivani, Pant Sujata, and Pant Urvashi) on patch budding of seedling rootstocks of bael was studied at G.B. Pant University of Agriculture and Technology, Pantnagar. Pant Aparna gave higher bud sprouting, lower unsprouted green and unsurvived buds, higher final survival, sprout length, number of leaves per budding and maximum sprout diameter while Pant Urvashi performed poorly in most of these characters. Pant Shivani had minimum sprout diameter while Pant Sujata had higher percentage of unsprouted green buds (Mishra and Jaiswal, 2001). Singh *et al.* (1976) propagated bael by patch budding, T-budding at monthly intervals from July-October and from march-June. Bud take was 100 per cent with patch budding in June or July. Further the effect of genotype, method and time of budding was studied in *bael*. The results showed that more sprout length was found in Pant Aprana while leaf area was greater in Pant Shivani. Per cent survival and number of leaves per sprout did not differ significantly in genotypes. In all characteristics, patch budding excelled over 'T' and inverted 'T' budding. Among interaction, higher percentage of survival was recorded with patch budding performed on 30th June in Pant Aparna and patch budding on 8th July in Pant shivani (Rai *et al.*, 1999).

Patch budding is ideal method with 90 per cent success rate and June-July is the best time for it (Chadha, 2001). Under arid conditions of Bikaner (Rajasthan), more than 90 per cent success was obtained through patch budding in July in polybag raised rootstocks.

Kumar *et al.* (1994) studied the effect of time and method of budding on bud sprout and survival in bael cultivar Narendra Bael-5 under Faizabad conditions (Table 3).

Table 3 : Effect of time and method of budding on bud sprout and survival in bael

Time of budding	Bud sprouted (%)	Bud survival (%)
Mid of May	92.22	88.88
Mid of June	88.88	85.55
Mid of July	76.66	73.33
Mid of August	51.11	47.77
Mid of September	31.66	28.33
Method of budding		
Modified ring	73.33	70.00
Patch	77.33	74.34
Shield	53.66	50.00

Tripathi and Kumar (2004) conducted studies on the effect of method and time of budding in bael cv. Gonad Selection at CCS HAU, Hisar. The highest bud take was observed when budding was done during the last week of July (70 %), followed by second week of August (63.33%). Bud take was highest with patch budding (57.50%), followed by modified ring budding (49.16%) and shield budding (41.66%).

The anatomical studies of bud union indicate that callus formation took place after two week of budding where callus cells were fewer in number and not much differentiated. After third week of budding, intermingling of callus tissues between stock and scion occurs. At four week stage, the clear union was observed between stock and scion (Kumar and Pathak, 2004).

Grafting: Maiti *et al.* (1999) reported that bael can be propagated successfully by grafting and among different grafting methods tried, whip grafting had given good response (70 % success and best shoot growth). It can also be grafted on the *Aegle fraeglegabonensis, Aeglopsis chevalier* and *A.paniculata.* However, it is not commonly used for raising of *bael* plants.

Work at CISH, Lucknow has resulted in standardization of soft wood grafting under polyhouse/ shade net house conditions. In this technique one year old bael seedling raised in polybags or raised in nursery and shifted in polybags after attaining pencil thickness of stem are used as rootstock. Slightly mature scion shoots cut into pieces are used for grafting. This technique has given almost cent percent success in bael. It reduces the time period for preparing new plants considerably as compared to budding.

Air-layering: Air layering is very successful in bael (90% rooting and 77% survival of rooted layers) provided that mother trees are given invigoration treatment by heading back to few of the thick branches during April. Air layers are prepared in the second week of August by bark ringing and application of IBA at 10,000 pm in lanolin paste.

Micro-propagation: Arumugam and Rao (1996) reported that cotyledonary node explants, excised from 15-day-old seedlings of *A. marmelos* were placed MS medium supplemented with BAP [benzyladenine], IBA, IAA or NAA. BAP induced the best production of multiple shoots and subsequent plant regeneration. The highest number of shoots (75.2/explant) was observed on MS medium supplemented with BAP at 3 mg $litre^{-1}$. The number of shoots was further enhanced by (i) using nodal explants of in vitro regenerated shoots as microcuttings, and (ii) repeated subculture of the original explants on the same medium after excising the shoots. Several shoots were obtained from a single explant within 5 months. Regenerated shoots produced roots in 30% of cultures when transferred to a medium containing IBA at 4 mg $litre^{-1}$. Plantlets were transferred to soil, acclimatized and transplanted to the field.

Rapid clonal multiplication of *Aegle marmelos* was achieved by enhanced axillary bud proliferation in young single-node segments of a 25-year-old tree cultured in MS nutrient medium. Bud break was dependent on cytokinin supply, but the synergistic combination of 2.5 mg 6-benzylaminopurine (BAP) [benzyladenine] and 1.0 mg IAA $litre^{-1}$ induced the formation of 12.1 shoots of up to 5.2 cm in length in 48 per cent of the explants after 7 weeks of culture. Explants of in vitro-grown shoots (node, whole leaf, shoot tip and internode) were subcultured in the presence of 0.05-2.5 mg BAP $litre^{-1}$ to produce 11.3, 18.4, 5.3 and 3.2 shoots and shoot buds at a rate of 100, 70, 95 and 40 per cent, respectively, in 7 weeks. Different shoot nodes and leaves were equally regenerative and adventitious organogenesis in the latter was confined to cut petiolar ends. Nodal explants responded most favourably at low BAP (0.05-0.1 mg/litre) and produced uniform (3.8-5.3 cm) shoots facilitating their simultaneous harvest for rooting. Repeated subculturing through five cycles of nodes and leaves of shoot cultures enabled continuous production of healthy callus-free shoots without any sign of decline. Shoot cuttings (3.0-5.2 cm) were best rooted in half-strength MS medium with 0.5 mg IAA/litre (70%) or 10.0 mg IBA/litre (90%). Some 85 per cent of the rooted plants were established in polybags after hardening (Ajithkumar and Seeni, 1998).

Callus of *A. marmelos* was initiated from stem explants on MS medium supplemented with different concentrations of kinetin, 2,4-D and NAA. Meristemoids developed in the callus when subcultured on medium supplemented with 1 mg kinetin + 5 mg NAA litre^{-1}. In the presence of BA, alone or in combination with NAA, these calluses showed shoot development. Multiple shoot induction from nodal explants was achieved on MS medium augmented with different concentrations of BA, kinetin and NAA. The shoot buds that developed from nodal explants were most numerous in medium supplemented with kinetin and NAA. Rhizogenesis of shoots was achieved in the presence of IAA (Varghese *et al.*, 1993).

A protocol for organogenesis from nucellar explants excised from fertilized ovules of immature fruits was developed. Adventitious buds were initiated on MS medium containing various combinations of benzyladenine (BA), NAA, IAA and gibberellic acid. Medium containing 4.4 µM BA and 2.7 µM NAA produced the greatest number of adventitious buds per explant. Shoots were elongated by transferring explants with shoot buds to medium with a low concentration of BA (0.44 µM). Rooting of in vitro-regenerated shoots was obtained in half-strength MS medium with 4.9 µM IBA. This is the first report of plant regeneration from nucellar explants of *A. marmelos* (Hossain *et al.*, 1993).

Further, the protocol for organogenesis from leaf explants of in-vitro grown seedlings of *A. marmelos*, an adventitious buds were initiated on MS medium containing various concentrations and combinations of BA and IAA. The maximum frequency of shoot organogenesis (61.2%) and the highest number of shoots per explant (38.4) were obtained when 1.5 mg BA and 0.5 mg IAA/litre were used. Twenty-day-old seedlings proved to be the optimum source for explants. Shoots were elongated on the same basal medium without auxin and supplemented with 0.5 mg BA litre^{-1}. Elongated shoots were rooted on half strength MS medium supplemented with 0.1 mg IBA litre^{-1} (Islam *et al.*, 1993). Bhargava *et al.* (2008) reported that surface sterilization with mercuric chloride was better with respect to production of aseptic culture and reducing time of bud break. The Further, increase in duration of exposure to either of sterilizing agents decreased the number and length of shoot, which may be because of toxic effect of these chemicals.

Rootstocks

In general, seeds extracted from local inferior fruit types are being used for raising of rootstocks. No work on standardization of bael rootstocks has been done so far, which needs attention to assess vegetative compatibility and vigour, fruiting, fruit quality, pest and disease reactions, usefulness under degraded sites etc. of component plant.

Top working

Inferior, old and uneconomical bael plants can be converted into superior and productive phase by adopting top working/ frame working technique . A large number of trees seedling in origin with small and poor quality fruits are existing in Vindhyan and Aravali hills, which can be rejuvenated by this technique(Jauhari and Singh,1971).

Quality Standards of Planting Material

The planting material should be of improved cultivars. Proper root stock/seedling should be used for vegetative propagation. Seedling should be of single stem and free from diseases and insects. The earth ball is required for shifting of plants in the nursery. Well hardened plants of appropriate height and age should be used for planting. Always procure planting material from recognized source/nursery.

References

Ajith kumar, D. and Seeni, S. 1998. Rapid clonal multiplication through in vitro axillary shoot proliferation of *Aegle marmelos* (L.) Corr., a medicinal tree. Plant Cell Reports, 17(5): 422-426

Anon., 2000. Annual Report, AICRP on AZF, CIAH, Bikaner.

Anon., 2001. Annual Report, AICRP on AZF, CIAH, Bikaner.

Arumugam, S. and Rao, M.V. 1996. In-vitro production of plantlets from cotyledonary node cultures of *Aegle marmelos* (L.) Correa. Advances in Plant Sciences, 9(2): 181-186.

Banerjee, N. and Maiti, M. 1980. Some physico-chemical studies on a protein rich fruit seed (*Aegle marmelos* Correa). J. Indian Chem. Soc., 57(9): 945-46.

Bhadra, S., Chakraborti, K. and Sen, S.K. 2000. Developmental physiology of bael (*Aegle marmelos* Correa) fruit var. Kalyani Selection-1. Environment and Ecology, 18(2): 470-471.

Bhargava, R., More, T.A., Singh,R. S.; Sharma, B. D., Raj Kumar and Sarita Chauhan 2008. Development of contamination free culture in bael (*Aegle marmelos* Correa.). Indian Jour. Arid Hort.3(1):64-65.

Chadha, K. L. 2001. Bael. In: A Handbook of Horticulture, ICAR, New Delhi.

Chattopadhyay, R.K. and Mahanta, S.K. 1989. Media requirement for seed germination and seedling establishment. Indian Hort., 36(3): 27.

Deepti, Rai., Misra, K.K. and Singh, V.P. 2002. Analysis of genetic divergence in bael (*Aegle marmelos* Correa) germplasm. Prog. Hort., 34(1): 35-38.

Gopalan, C. B. M., Ramashastri, B. V., and Balasubramanium, S. C. 1985. Nutritive value of Indian foods, ICMR, Hyderabad.

Haskar, C.N. and Kendurkar, S.G. 1961. Uses of bael gum. Print India, 11(1): 135.

Hore, J.K. and Sen, S.K 1985. Effect of seed treatments on longevity of bael (*Aegle marmelos* Correa.) seeds. Haryana J. Hort. Sciences, 14 (3-4): 204-210

Hossain, M., Karim, M.R., Islam, R. and Joarder,O.I. 1993. Plant regeneration from nucellar tissues of Aegle marmelos through organogenesis, Plant Cell, Tissue and Organ Culture, 34: (2): 199-203.

Islam, R., Hossain, M., Joarder, O.I.and Karim,M.R. 1993. Adventitious shoot formation on excised leaf explants of in-vitro grown seedlings of *Aegle marmelos* Corr., J. Hort. Science, 68(4): 495-498.

Jain, N.K. 1977. Antifungal activities of *Aegle marmelos* Correa. Indian J. Microbiology, 17(1): 51.

Jaisawal, H.R. and Misra, K.K. 1996. Clonal variation in inflorescence and floral morphology of bael (*Aegle marmelos* Correa). Prog. Hort., 28(3-4): 119-123.

Jaisawal, H.R. and Misra, K.K. 1998. Studies on fruit set, retention and yield of bael (Aegle marmelos Correa) clones under Tarai conditions. Prog. Hort., 30(3-4): 164-167.

Kumar, D. , Pathak, R. K. and Wahid Ali,1994. Studies on effect of duration and method of budding in bael. Indian Jour.Hort.,51(2):150-153.

Kumar, D. and Pathak, R. K. 2004. Anatomical studies on callusing and bud union in bael. Indian Jour.Hort.,61:173-174.

Maiti, C.S., Nath, A. and Sen, S.K. 1999. Studies on the propagation of bael (*Aegle marmelos* Correa.) by different grafting methods in West Bengal, J. Applied Hort., 1(2): 131-132

Misra, K. K., Singh, R. and Jaiswal, H.R. 2000. Performance of *bael* (*Aegle marmelos*) genotypes under foothills regions of Uttar Pradesh. Indian J.Agric. Sci., 70(10): 682-683.

Misra, K.K and Jaisawal, H.R.2001. Effect of plant bioregulators and potassium nitrate on seedling quality of bael (*Aegle marmelos* Correa). Advances in Hort. & Forestry, 8: 67-74.

Misra, K.K and Jaisawal, H.R.2001. Effect of scion genotypes on patch budding in bael (*Aegle marmelos* Correa). Advances in Hort. & Forestry, 8: 61-66

Misra, K.K., Sigh, R. and Jaisawal, H.R.1999. Studies on leaf characters, development pattern and shoot growth in bael genotypes. Prog. Hort., 31(3-4): 144-150.

Moti., Dhar, L. and Chaturvedi, O.P. 1976. Propagation of some tropical and subtropical fruits by budding. Punjab Hort. J., 16(1-2): 33-38.

Pal, P.K., Bhattacharya, D.K. and Ghosh, S. (1993). Chemical examination of seeds of *Aegle marmelos* (Bael fruits) and its medicinal importance. J. Oil Technologists Association India, 25(3): 55-59.

Pandey, D., Pathak, R. K. and Pramanik, P. K. 2001. Shri Phal Bael. Udyan Rashmi (In Hindi), 2(2):13-18.

Pandey, G. Pathak, R.K. and Om, H. 1985. Effect of sodicity on mineral composition, clorophyll content and injury symptoms in bael. Prog. Hort., 17(1): 5-8.

Pandey, G. Pathak, R.K. and Om, H. 1988. Effect of sodocity on seed germination and seedling growth in bael (Aegle marmelos). Indian J. Agril. Sciences, 58(7)577-578.

Prakash, A., Pasalu, I.C. and Mathur, K.C. 1983. Use of bael (*Aegle marmelos*) leaf powder as protectant against storage pests of Paddy. Agril. Res. J. Kerala, 21(2): 79-80.

Rai, M. and Dwivedi, R. 1992. Bael (*Aegle marmelos* Correa) diversity in India. *In:* Underutilized fruits of medicinal value, IBPGR Newsletter, 10:13-14.

Ram, D. and Singh, I.S. 2003. Physico-chemical studies on bael (*Aegle marmelos* Correa) fruits. Prog. Hort., 35(2): 199-201.

Ray, D.P. and Chaterjee, B.K. 1996. Effect of different concentrations of growth regulators, etiolation and vigoration treatments on the rooting of stem cutting of bael (*Aegle marmelos* Correa.). Orissa J. Hort., 24(1-2): 36-41.

Reuther, W., Webber, H.J. and Bachlor, L.D. 1967. The *Citrus* industry, Vol. I, Pub. Division of Agriculture Sciences, University of California. Pp. 407-409.

Roy, S.K. and Singh, R.N. 1979. Preliminary studies on storage of *bael* fruits (*Aegle marmelos*). Prog. Hort., 11:3

Roy, S.K. and Singh, R.N. 1981. Studies on ripening of bael fruits (*Aegle marmelos*). Punjab Hort. J., 21(122): 74-82.

Roy, S.K., Pandey, R.M. and Singh, R.N. 1972. Preliminary studies on the availability of *bael* fruit (*Aegle marmelos*) in off-season. Indian J. Hort., 29(3-4): 344-347.

Saini, R.S., Surender Singh and Deswal R.P.S. 2004. Effect of micronutrients, plant growth regulators and soil amendments on fruit drop, cracking, yield and quality of bael (*Aegle marmelos* Correa) under rainfed conditions. Indian J. Hort., 61(2): 175-176

Shukla, S.K. and Singh, G.N. 1996 . Seed germination and growth of bael (*Aegle marmelos* Correa.) seedlings as influenced by soil sodicity, J.Tropical Forestry, 12(4): 242-245.

Singh, K.V., Bhatt, S.K. and Sthapak, J.K. 1983. Antimicrobial and anthelmintic properties of the seeds of *Aegle marmelos*. Fitoterapia, 54(6): 261-264.

Singh, L. B. 1954. Propagating bael (*Aegle marmelos*) vegetatively. Sci. & Cult., 19: 485-486.

Singh, R. N. and Roy, S. K. 1984. *The Bael*. I.C.A.R., New Delhi, p.25.

Singh, R. S. and Vishal Nath 1999. *Anupayogi Bhoomi mein – Upayogi Bael Vriksh Lagayein*. Unnat Krishi, 38 (3):3-8.

Singh, U. R., Pandey, I. C., Upadhaya, M. P. and Prasad, R. S. 1976. Propagation of bael (*Agele marmelos*) by budding. Punjab. Hort. J., 16:57-59.

Srivastava, K.K. and Singh, H.K.2000. Floral biology of bael (*Aegle marmelos*) cultivars. Indian J. Agriculture Sciences, 70(11): 797-798.

Srivastava, R.K. 1982. A note on phenological observations on *Aegle marmelos* Correa. Indian Forester, 108(80: 593-595.

Tripathi,A. and Kumar, R. 2004.Studies on the effect of method and time of budding in bael (*Aegle marmelos*). Haryan Jour. Hort. Sci.,33:195-198.

Varghese,S.K., Inamdar, J.A., Kiran Kalia., Subramanian,R.B., Nataraj,M. and Kalia,K.1993. Micro-propagation of *Aegle marmelos* (L.) Correa. Phytomorphology, 43: 1-2, 87-92

Chapter-13

Karonda (*Carissa carandas* L.)

K.K. Misra and Subhash Chandra Singh
G.B. Pant University of Agriculture & Technology, Pant Nagar-263 145

Karonda (*Carissa carandas* L.) is found growing throughout India, mainly in the semi-arid and arid regions. It is widely cultivated in the home gardens, farmer's fields and orchards as hedge plant. Karonda is commonly used as a condiment or additive to Indian pickles and spices. Karonda is one of the few fruits indigenous to India while 30 species of genus *Carissa* have been reported; many species are found growing wildly in India while other species came from Malaysia and South Africa (Hayes, 1957). It is cultivated throughout India in tropical and sub-tropical areas for fruits (Rai and Gupta, 1996). In India *Carissa* are found growing most widely in plains and Siwalik hills and grow wild in Deccan Peninsula, Maharashtra along the western ghats, parts of Gujarat, Punjab, lower ranges of J & K, U.P., Uttarakhand, Chhattisgarh M.P. and Aravalli hills, Mount Abu (Rajasthan). It also cultivated in other countries like Bangladesh, South Africa, U.S.A., Denmark, Ghana, Israel, Nepal and Pakistan. It is also grows in the Sri Lanka lowland rain forests. It can be grown in any type of soil however; it is most suitable for planting in marginal lands (Sharma and Badiyala, 2002). It is a very hardy, drought-tolerant plant that thrives well in a wide range of soils. However, it is susceptible to very low temperature/frost under hot arid conditions (Singh *et al.*, 2012).

Nutritive Value and Uses

It is used as bio-fencing around farm because of presence of spines and bushy type. The fruit is a rich source of iron, so it sometimes used in treatment of anaemia. The karonda fruit is a rich source of iron and contains a fair amount of Vitamin C and therefore is an antiscorbutic. Mature fruit is harvested for pickle and vegetable preparation. Karonda is good appetizer. It contains pectin and accordingly is a useful ingredient in jelly, jam, syrup and chutney which have great demand in International market (Banik *et al.*, 2012). Ripe fruits exude white latex when severed from the branch. The roots of the plant are heavily branched, making it valuable for stabilizing eroding slopes. In Rajasthan, karonda fruits are commonly cooked with green chillies to make a tasty dishes.

Botanical Description (Hooker, 1882)

A genus of densely branched, spinous, erect shrubs belonging to the family Apocynaceae. There are some 30 species of which 4 are of Indian origin including *C. carandas*. The systematic botany of Carissa as follows :

Habit	:	Large shrubs or small tree with spine densely branched.
Leaves	:	Oblong to broadly ovate, opposite small coriaceous, apex obtuse and round at base.
Flowers	:	Terminal and axillary pendunclel, 3 chotomous cymes, white or pink colour.
Calyx	:	5-partile, ciliate sepals, glandular within or not, segment acute.
Corolla	:	Tube cylindrical, throat necked, glabrous, lobes overlapping to the right.
Stamens	:	Inserted at the top of the tube.
Anthers	:	Lanceolate, cells rounded at the base.
Disc	:	Absent.
Ovary	:	2-celled, bilocular.
Style	:	Filiform.
Stigma	:	Fusiform or columnar, minutely bifid.
Ovules	:	1-4 each cell, rarely more.
Fruits	:	Berry, ellipsoid or globose, 2 (or by abortion 1) celled.
Seeds	:	Usually 2, pellately attached to septum, albumen fleshy and cotyledons ovule.

Time and Duration of Flowering

The flowering started in karonda from 3rd to 4th week of March and continued for about 7 weeks and plants are in full bloom during April.

Flower Bud Development

Karonda takes 19 to 21 days for the full development of floral buds.

Anthesis : Anthesis takes about 2 to 3 hours to open the buds completely. The maximum number of flowers open between 8 to 10 A.M. Chaturvedi and Saxena (1963) reported that anthesis in Karonda takes place in morning.

Dehiscence : Dehiscence of anther takes place before anthesis between 12 noon to 4 P.M. The dehiscence taken place in the longitudinal fashion and complete within an hour.

Pollen grain : It appear as fine whitish mass. It is round in shape. The average diameter of fertile pollen ranges between 25 to 37.5 whereas sterile pollen grains are smaller in size.

Stigma receptivity : The maximum receptivity of stigma is found two days before anthesis and two days after anthesis. The maximum receptivity is, however, found on the day of anthesis.

Cultivated Species of *Carissa*

(a) *Carissa carandas* L. (syn. *Carissa congesta* Wt.) : It is also known as Karonda and Christ's Thorn. It is evergreen shrub or small tree and attain height of 3 to 6 m when allow to grow. The species is thorny bush commonly found in degraded areas and ideal for use in hedges. Sturrock (1948) divides Carissa carandas into two botanical varieties amara, the Karonda and dulcis, the perunkila. Fruit is nearly 3 cm. in diameter and has a pleasant flavour.

(b) *Carissa grandiflora* DC (Natal plum) : The fruits are dark red, with peppery skin and a few small seeds fruits are available throughout the year. The bush is spiny with bifurcated thorn having leaves of deep green colour.

(c) *Carissa bispinosa* Desf. (*C. arduina* Lamk.) : This species posses strong spines often 5 cm. long. The leaves are ovate, subcordate glabrous and entire. The fruits are of the size of cherries and closely resemble to Natal plum.

(*d*) *Carissa edulis* Vahl. : It is straggling shrub. The leaves are persistent ovate and acuminate. Fruits are oval in shape, red in colour but turn back at maturity.

(*e*) *Carissa ovata* : It is native of Australia. The shrub is more open than other species. The fruits are small and edible and are used in jam making.

(*f*) *Carissa spindrum* DC : This species bears small fruits. It is native of India. This is the hardiest species of *Carissa* and is known to thrive in poor and rocky soil.

(*g*) *Carissa macrophylla* Wall. Cat. : Shrubby, erect, quite glabrous except puberulous cymes, leaves 5-10 cm., ovate or elliptic ovate acute or acuminate. It is found in Deccan Peninsula and Konkan region of Maharashtra (Hooker, 1882).

On the basis of fruit colour, the cultivars of Karonda can be classified as follows :

(i) Green fruited, (ii) White fruited. There is a quite resemblance in the shape and size of their fruits. However, there is a tremendous scope for improvement using selection force and vegetative method of propagation. Some clones were selected in Maharashtra. These are K1, K2, K3 (Joshi *et al*., 1986). Characteristics of these clones are as follows:

Characteristics	**Clones**		
	K1	**K2**	**K3**
Berry size	Medium 1.93 × 1.83 cm	Small to medium 1.81 × 1.75 cm	Big 1.91 × 1.99 cm
Shape	Oval	Roundish	Roundish
Skin colour	Shining dark Bluish black	Dull bluish black	Shining dark Bluish black
Pulp colour	Red	Pink	Dark red
Fruit weight	3.65	3.28	4.33
T.S.S. (Brix)	19.20	19.10	22.20
Total sugar (%)	11.45	10.20	12.80
Acidity (%)	0.358	0.41	0.256
Ascorbic acid (mg/100 g)	1.84	1.74	1.68

At Rahuri in Maharashtra, some promising clones as No. 3, 12, 13 and 16, have been identified (Karale *et al.* 1989). At Pant nagar three clones have been selected and named as Pant Manhar, Pant Sudarshan and Pant Suvarna. The detail is as follows :

(a) **Pant Manhar :** Plant medium size, dense bushes, fruit size 2.13 × 1.69 cm., colour dark pink blush on white background, fruit weight 3.49 g, seeds 3.94/fruit, flesh 88.97 percent, dry weight 12.77 percent, T.S.S. 3.92 percent, total titratable acidity 1.82 percent and yield 27 kg/plant.

(b) **Pant Sudarshan :** Plant medium size bush, fruit size 2.16 × 1.69 cm, colour pink blush on white background. On ripening fruits become dark brown. Average fruit weight 3.46 g, seeds 4.68/fruit, flesh 88.47 percent, dry weight 11.83 percent, T.S.S. 3.45 percent, total titratable acidity 1.89 percent, yield 29 kg/plant.

(c) **Pant Suvarna :** Plants are upright growing and sparse, fruit size 2.25 × 1.69 cm., colour dark brown blush on green background, average fruit wight 3.62 g, seeds 3.89/fruit, flesh 88.27 percent, dry weight 12.39 percent, T.S.S. 3.83, total titratable acidity 2.30 percent, yield 22 kg/plant, on ripening fruit colour changes into dark brown.

At CHES, Godhra, Gujarat, a large number of collections have been made and out of which a promising genotype (CHES K-1) has identified of high yielding and red fruit colour type.

Propagation

Carissa species are mainly propagated by seeds but the seedlings are slow in growth, variable and are ordinarily not ready to plant until about 2 years old. The other drawback with seeds is that they are recalcitrant which hampers the availability of planting material throughout the year. The other methods of propagation such as hard wood stem cuttings in open nursery, hard/semi- hardwood stem cutting under mist, softwood cuttings with the use of auxins under mist, air layering, stooling and grafting have been tried in the past but with variable success. The above methods are cumbersome, time consuming, labour expensive, weather dependent and produce limited number of plants per unit time to meet the growing demand for planting material of superior types.

As *Carissa* is planted a closer distance and requires large number of plants per unit area, some alternative method is required to produce large number of uniform plants. Hence in order to remove this major constraint in increasing area under *Carissa* cultivation and to protect the valuable germplasm from being eroded, micro propagation can be used.

All the species of Karonda (*Carissa* spp.) are multiplied through seeds. Vegetative propagation such as cutting, budding, air layering, stooling and grafting in *Carissa carandas, Carissa grandiflora* and *Carissa bispinosa* have been studied.

Seed propagation

The most common method of propagation of Karonda is by seeds. Fresh seeds are sown in the nursery in the month of August and September for raising the seedlings since seeds usually lose their viability within 4-5 weeks. The germination of seed starts 40-45 days after sowing and is completed within 60 days. Besides, few species, Carissa do not set any fruit (Bandopadhyay *et al.*, 1982). However, for hybridization and vegetative propagation, seedlings have to be raised and quick and enhanced germination to be secured. Banker (1987) obtained early and enhanced germination (68%) with increased vigour of seedlings when seeds were soaked in 100 ppm GA_3 for 24 hrs. Seedlings are raised in polythene bags during rainy season. Seedlings become ready in a year's for transplanting.

Gibberellic acid has been found to hasten the growth of young seedling of many plant species. Misra and Jaiswal (1998) concluded that for better and early growth of Karonda seedlings, GA_3 (1000 ppm) may be sprayed six times at monthly intervals. Improvement of germination of the seeds after longer storage period by treatment with fungicide, Dithane M-45 (Mancozeb) had been an observation put forward by Singh and Tiwari (1998).

Misra and Jaiswal (1997) observed that the survival of bare root seedlings of Karonda decreased with increasing period of storage independent of season. Survival in Karonda was 86 percent after 7 days of storage. Although seedling height and number of new leaves per seedling decreased with increasing period of storage. The survival of seedling was found significantly higher in rainy season planting followed by spring and autumn season planting independent of storage period. The increase in plant height and number of new leaves per plant three month after replanting were found higher in spring season planting followed by rainy and autumn season planting.

Manish Prakash (2003) found that maximum height of plant, fresh and dry weight of shoot per plant were with 500 ppm concentration of gibberellic acid under protected condition while shoot : root ratio was maximum with 0.5 percent concentration of urea under open condition. Further it was concluded that the foliar application of 0.4 percent $ZnSO_4$

under protected condition can be recommended for best and rapid vegetative growth of Karonda seeding.

Vegetative methods

Vegetative propagation is most important for getting true-to-type plants. Karonda can be multiplied through hard wood cuttings, stooling and air layering.

Cutting

Karonda can be propagated by hardwood stem cutting successfully, though cuttings are difficult to root. They may also be planted under polythene tent which gives higher higher rooting. Where mist facilities are available, soft and semi-hardwood cutting can be planted for successful rooting. Verma *et al.* (1971) observed significant rooting with the aid of 100 ppm + 4 percent sugar and 100 ppm IBA+ 5 percent sugar in *Carissa carandas.* Cutting from young plants and hedge sources proved to be best for rooting and shoot growth (Mahata and Singh, 1960).

Divekar (1984) reported more rooting with 1000 ppm IBA treatment as quick dip. No rooting was obtained without growth regulators treatments. If the cuttings are treated with 500 ppm of IBA before planting, significant increase in rooting can be obtained (Bandopadhay *et al.*, 1982). Aqueous solution of IAA successfully rooted cuttings in Natal plum (Cooper and Knowlton, 1940).

Significant rooting, number of primary and secondary roots and longer length per cutting in softwood cuttings of *C. grandiflora* and *C. carandas* were reported under mist in polyhouse condition irrespective of auxin treatment. IBA 8000 ppm was found better for rooting in both the species (Tyagi *et al.*, 1999). Sugar 4 per cent + 100 ppm IAA gave the maximum percentage of the success, maximum number of roots, more number of lateral roots and maximum length of root per cutting in karonda (Verma *et al.*, 1971). Shanker and Shanker (1974) observed maximum length of roots (24.9 cm) in semi hard wood cuttings of Natal plum with IBA, 200 ppm. The higher concentration of NAA (400 ppm) gave cent per cent success with hard wood cuttings of *Carissa grandiflora* and *Carissa spinarum* in August and October. The best results (50%) was recorded with semi hard wood cutting of *Carissa carandas* in the month of August. NAA 200 ppm gave cent per cent rooting with hard wood stem cuttings of *Carissa spinarum* in August.

Air layering

Air layering is quite successful in Karonda using growth regulators. It is done in the beginning of monsoon. Rooted layers are obtained after 5.0 month of layering. Jauhari and Nigam (1958) found high percentage of success with air layering when layers were treated with mixture of 7500 ppm each of IA and NAA. A field study was carried out at Pantnagar (Misra and Singh, 1990) on 17 year old bushes of clone white pink. Highest rooting and survival were obtained with 5000 ppm IBA. IBA was more effective than NAA in promoting rooting in air layers. Air layering with the help of IBA and NAA is also proved effective in Natal plum (Misra and Jaiswal, 1993b). The best season for air layering is beginning of monsoon i.e. June-July though it can be extended further upto September and air layers should be treated with 4000 ppm IBA to get promising results (Divekar, 1984).

Carissa spinarum can be propagated with greater success and early rooting by air layering with the aid of seradex (Srivastava *et al.*, 1962).

Stooling

Karonda is most suitable plant for stool layering as it produces many side shoots in the nursery during early period of its growth (Misra and Jaiswal,1993a). This method can be utilized commercially for vegetative propagation of superior clones of Karonda. Lambourne (1935) has reported that mound layering was satisfactory in Karonda. Misra and Jaiswal (1993b) reported that 10000 ppm IBA was most effective for rooting (93.33%) in stool layers which significantly increased the length of longest roots. Karonda (*Carissa carandas* L.) produces many lateral shoots in nursery during early periods of its growth and is most suitable for stooling. IBA 1000 ppm proved most effective and gave an average of 93.33 percent rooting, number of roots/layer (18.2) and survival (83.3%) in stool layering (Misra and Jaiswal, 1993).

Natal plum (*Carissa grandiflora*) can be commercially propagated by stool layering using Beta IBA, 7500 ppm in lanolin paste (Misra and Jaiswal, 1992).

Grafting

Carissa carandas can be successfully grafted on *C. carandas, C. grandiflora* and *C. bispinosa* rootstock and *C. grandiflora* can be successfully grafted on *C. grandiflora* rootstock. Some success was obtained with *C. grandiflora* on *C. carandas* rootstock and with *C. bispinosa* on *C. grandiflora* and *C. carandas* rootstocks (Shanker, 1972). Natal plum is more productive when grafted on Karonda (Hayes, 1957). In *Carissa carandas*, grafting is also done to a limited extent by inarching (Singh *et al.*, 1967).

Budding

Propagation of Karonda by shield budding is economic and convenient. Promising types of Karonda and Natal plum could be budded on one year old rootstock of *C. carandas* (Banker *et al.*, 1994). The best time for budding is July to August in low rainfall areas.

Micro-propagation

Micro-propagation is a worldwide applied tool for propagation in the horticultural industry. Clonal propagation through tissue culture facilitates rapid multiplication, easy storage and safe exchange of germplasm. Due to limitation of recalcitrant seeds in *Carissa* species a well planned programming of in vitro production is necessary. Therefore, micro-propagation of elite clone of *Carissa* species may prove as a corner stone for the production of large number of disease free and true to type plants in a short period which will reduce the cost of individual plant and make available elite planting material in bulk. Hence, there is a urgent need to undertake a systematic work to standardize a protocol for successful micro-propagation of *Carissa* species.

Rai (2001) found that both *Carissa carandas* and *Carissa grandiflora* can be successfully propagated by *in vitro* shoot tip culture. The longer shoot tip (1.5 cm) collected during spring season gave better shoot proliferation and growth. Maximum shoot growth in *Carissa carandas* and *Carissa grandiflora* were obtained with 3.0 mg lit^{-1} +0.2 mg lit^{-1} IBA, respectively. Rooting was enhanced with 0.8 mg lit^{-1} IBA +0.2 mg lit^{-1} NAA and 0.6 mg lit^{-1} +0.2 mg lit^{-1} NAA in *Carissa carandas* and *Carissa grandiflora*, respectively. Shoot tip from field grown mature plants of *Carissa carandas* cv. Sudershan were cultured (Rai and Misra, 2005).

Quality Standards of Planting Material

In recent years, the importance of sanitation during propagation and growing has become widely recognized as an essential part of propagation. During propagation losses of young seedlings, rooted cuttings, tissue cultured rooted plants and grafted nursery plants to various pathogens and insect pests can sometime be devastating especially under warm humid conditions found in propagation houses.

Ideally sanitation strategies should be considered even in the construction phase of propagating structure. Maintenance of seed cultivars requires that genetic variability be controlled during production. Genetically pure seed has the following attributes : (a) trueness-to-name, (b) trueness to type and (c) freedom from pathogen, mixture of other crop seeds and weeds.

One of the keys to successful vegetative propagation is the proper source of planting material. Plants have names and a successful nursery industry requires that plant being produced and sold be correctly labeled either by scientific name, cultivar name or both. The process starts by identifying the source of plant material at the time of collection and is continued by maintaining record during propagation.

Fingerprint methods to distinguish among cultivars of specific crops are laboratory biochemical tests which produce characteristics pattern of lines or dots on a gel which can be correlated to specific cultivar. The propagation source must not only be the correct cultivar but the characteristic of the progeny plant must confirm to an accepted norm for that cultivar when grown in its ultimate location i.e. true-to-type. Viruses and related organisms such as phytoplasma, viroids and fastidious bacteria have plagued clonally propagated cultivars that are why concept of pathogen tested source trees used for propagation was developed. During clonal or vegetative propagation more emphasis must be given for the correct cultivar, maintaining trueness to type and producing pathogen free material. Selection, maintenance, and distribution programme must be ensure for the hightest quality material to consumers.

References

Bandopadhayay, D.P.: Nath, N.: Pandey, H.S. and Yadav, L.P. 1982. Effect of growth regulators in propagation in Carissa spp. by stem cutting. Indian Agriculturist, 26 (1) : 57-63

Bankar, G.J. 1987, A note on influence of gibberellic acid on seed germination and vigour of seeding in Karonda (*Carissa carandas* L.). Prog. Hort., 19 (1-2) : 90-92.

Bankar, G.J.; Verma, S.K. and Prasad, R.N. 1994. Fruit for the arid region : Krounda. Indian Hort., 39 (1) : 46-47.

Banik, B.C.,Ghosh, S.N. and Singh, S.R. 2012.Resrach and development in Karonda (*Carissa carandas*)-A semi wild fruit in India, Acta Horticulture,948:61-66.

Chaturvedi, R.B. and Saxena, G.K. 1963. Preliminary studies on anthesis, dehiscence and pollen grain viability in Karonda (*Carissa carandas* L.). Allahabad Farmer, 27 (4) : 19-24.

Cooper, W.C. and Knowlton, K.R. 1940. The effect of synthetic growth substances on rooting of subtropical plants. Proc. Amer. Soc. Hort. Sci., 37 ; 1093-8.

Divekar, P.B. 1984. Effect of growth regulators on air layering of karonda (*Carissa carandas* L.) M.Sc. Thesis submitted to Konkan Krishi Vidyapeeth, Dapoli, 75 p.

Hayes, W.B. 1957. The Carissa. Fruit growing in India. Kitabistan, Allahabad, pp. 416-417.

Hooker, J.D. 1882. The flora of British India, L. Reene and Co. Ltd., The Oast House, Book No. Asford, Kent, Vol. III pp. 630-32.

Jauhari, O.S. and Mehra, R.C. 1959. Propagating *Carissa carandas* (Karonda) vegetatively. Trop. Agriculturist, 115 : 29-35.

Jauhari, O.S. and Nigam, V.N. 1958. A note on propagation of karaunda by air layering. Allahabad Farmer, 32 (5) :174.

Joshi, G.D.; Prabhudesai, V.G. and Salvi, M.J. 1986. Physicochemical characteristics of Karonda (*Carissa carandas* L.) fruits. Maharashtra J. Hort., 3(1) : 39-44

Karale, A.R.; Keskar, B.G.; Sheta, M.B.; Dhawale, B.C.; Kale, P.N. and Chaudhari, K.G. 1989. Seedling selections of Karonda (*Carissa carandas* L.) Maharashtra J. Hort., 4 : 125-29

Lambourne, J. 1935. The etiolation shoot method of fruit propagation. Malay. Agric. J. 23 : 514-27

Mahata, K.C. and Singh, S.M. 1960. Influence of source and time of planting on the performance of hardwood cuttings of *Carissa carandas* L. Indian J. Hort., 17 : 31-37.

Manish Prakash 2003. Effect of foliar applications of gibberellic acid, $ZnSO_4$ and urea on seedling quality of Karonda (*Carissa carandas* L.). M.Sc. Thesis submitted to G.B.P.U.A & T., Pantnagar, 69 p.

Misra, K.K. and Jaiswal, H.R. 1992. Effect of beta indole butyric acid on rooting and survival of stool layers of Natal plum (*Carissa grandiflora* DC). Ann. Agric. Res., 13 (4) : 414-15.

Misra, K.K. and Jaiswal, H.R. 1993. A note on the effect of indole butric acid on rooting and survival of stool layers of Karonda (*Carissa carandas* L.). Haryana J. Hort. Sci., 22 (3) : 209-210.

Misra, K.K. and Jaiswal, H.R. 1993 a. Effect of growth regulators on rooting and survival of stool layers of Natal plum (*Carissa grandiflora* DC). Ann. Agric. Res., 14(2) : 232-34.

Misra, K.K. and Jaiswal, H.R. 1993 b. Effect of growth regulators on rooting of stool layers of Karonda (*Carissa carandas* L.). Indian J. Forestry, 16(2) : 181-2.

Misra, K.K. and Jaiswal, H.R. 1997. Effect of planting season and storage period on survival and growth of bare root seeding of Bael (*Aegle marmelos* Correa.) and Karonda (*Carissa carandas* L.). J. Non-Timber Forest Product, 4(3-4) : 170-172.

Misra, K.K. and Jaiswal, H.R. 1998. Effect of foliar spray of gibberellic acid on growth of Karonda (*Carissa carandas* L.) seedlings. Indian J. Forestry, 21(1) : 70-71.

Misra, K.K. and Singh, R. 1990. Effect of growth regulators on rooting and survival of air layers of Karonda (*Carissa carandas* L.). Ann. Agric. Res., 11(2) : 208-210.

Rai, M. and Gupta, P.N. 1996. Distribution and diversity of indigenous tropical fruits. NBPGR, New Delhi.

Rai, R. 2001. Standardization of micropropagation technique in Carissa spp. Ph.D. Thesis submitted to G.B.P.U.A&T., Pantnagar, 121 p.

Rai, R. and Misra, K. K. 2005. Micro-propagation of karonda (*Carissa carandas*) through shoot multiplication.Sci.Hort.,103(2):227-232.

Shanker, G. 1972. A note on the cultivation and uses of *Carissa bispinosa* Desf. Allahabad Farmer, 48(2) : 161-2.

Shanker, V. and Shanker, G. 1974. Studies on propagation of four Carissa spp. by hard and semi-hardwood cuttings with and without growth regulators under mist. Allahabad Farmaer, 48 (4) : 539-540

Sharma, Shashi K.and S.D. Badiyala 2002. "Karonda, fruit plant for marginal lands", India Tribune, June 1, 2002.

Singh, R.S.; R. Bhargava; Garima Pal and S.K. Sharma. 2012. Effect of frost on fruit plants and its management in hot arid region. Crop Improvement, 39 (spl. Issue).39: 723-724

Singh, J. and Tiwari, R.K.S. 1998. Germination studies in Karonda (*Carissa carandas* Linn.). Advances in Plant Sciences, 11(1) : 313-315.

Singh, S.: Krishnamurhty, S. and Katyal, S.L. 1967. Fruit Culture in India. Indian Council of Agricultural Research, New Delhi.

Srivastava, V.K. 1964. Floral biology of Karonda (*Carissa carandas* L.), Indian J. Hort. 26 : 151-5.

Sturrock, D. 1948. The Karonda as a commercial fruit. Proc. Fla. Hort. Soc., pp. 289-91.

Tyagi, S.; K.K. and Jaiswal, H.R. 1999. Effect of auxins on rooting of softwood stem cuttings of Carissa spp. under mist. Scientific Horticulture, 6 ; 37-43.

Verma, A.N.; Srivastava, D.C.; Pujari, M.M.; Misra, H.R. and Sharma, R.K. 1971. Rooting in Karonda (*Carissa canrandas* L.), mulberry (*Morus alba* L.) and phalsa (*Grewia asiatica*) cuttings as affected by sugar alone and with plant growth regulators. Punjab Hort. J., 11 (1-2) : 57-60.

Chapter-14

Fig (*Ficus carica* L.)

R. S. Singh
Central Institute for Arid Horticulture, Bikaner-334 006

Fig (*Ficus carica* Linn.) also called 'Anjeer' is a nutritive and delicious fruit. Its tree is native to western Asia. It is rich in calories (269), protein, and calcium (higher than milk), iron and fibre content. The small deciduous plants of fig are grown in different parts of the country. Fig has been cultivated since ancient period (4000 BC.) and well known through out the world. It is originated in Middle East and naturalized in many parts of the world. Fig tree plantations cover 426,244 ha. area in the world with an overall yield 1070676million tones. The largest fig growing countries are Turkey, Egypt, Iran, Greece and Algeria (Pasqual and Ferreira, 2007). It is also grown in some parts of Australia, California, New- Zealand and Argentina. Turkey is main fig producing country in the world with its 30 % production share amounting to 280,000 tonne followed by Egypt 170,000 tonne of produce. The United States, Turkey, Greece and Spain are the primary producing nations of dry fig. In Turkey, 65% fig trees are in the Western Aegean Region (Aksoy *et al.*, 2001). The common fig plants are cultivated in sub-tropical to dry regions of the country. In India, Fig is considered as minor fruit crop and commercial cultivation is confined in Pune, Solapur, Rahuri, Aurangabad, Sasawad areas in Maharashtra; Bellary, Chitradurga, Srirangapatnam in Karnataka, Lucknow, Saharanpur (U.P.), Coimbatore (Tamil Nadu), Gujarat and Andhra Pradesh (Pareek and Nath, 1996). At present, fig cultivation is being done in about 3,000 ha area in Maharashtra state and adjoining areas. In dry land area, the cultivation of fig is possible since it requires dry climate and steady irrigation. In arid parts of Rajasthan, few growers have planted fig but the size of fruit is small and the performance is yet to be ascertained. The fig fruit has great potential for domestic consumption, processing and export.

The fig can be grown on a wide range of soils; light sand, rich loam, heavy clay or limestone, provided there is sufficient depth and good drainage. Sandy soil that is medium-dry and contains a good deal of lime is preferred when the crop is intended for drying. Highly acid soils are unsuitable. The pH of soil should be between 6.0 to 7.0. The tree is fairly tolerant to salt and drought conditions. It can tolerate a fairly high level of sulphate or chloride salt.

There is a rich genetic resource of Fig available in the state of Maharashtra and Karnataka and other sub-tropical dry parts of the country. In arid region, small fruits type figs are found. Vegetative growth of plant has been found better under dry condition. Fruit drop has also been observed. Conservation and evaluation of fig crop is being carried out at MPKV, Rahuri, Jadhavwadi, I.I.H.R, Bangaluru, ARS, Anantpur, MAU,Parbhani. The elite type collections have been made from these areas and are under evaluation. Daulatabad, Dinkar, Yercaud, Maishram, Mediterrnean figs from Chalisgaon, M.S. and other selections have been made from available gene pool and are being evaluated at different centers for growth and production.

Nutritive Value and Uses

Its fruits are used fresh as well as for drying purpose. It is highly nutritive fruit with high levels of calcium and iron. and ripened fruits are used for table purposes.The total sugar content in fresh fig is 16% and dried 52%. The chemical composition and flavor varies with the cultivars. Commercially, figs are peeled by immersion for 1 minute in boiling 1.0l water or a boiling solution of sodium bicarbonate. In warm, humid climates, figs are generally eaten fresh and raw without peeling, and they are often served with cream and sugar. Peeled or unpeeled, the fruits may be merely stewed or cooked in various ways, as in pie, pudding, cake, bread or other bakery products, or added to ice cream mix. Home owners preserve the whole fruits in sugar syrup or prepare them as jam, marmalade, or paste. Fig paste (with added wheat and corn flour, whey, syrup, oils and other ingredients) forms the filling for the well-known bakery product, "Fig Newton". The fruits are sometimes candied whole commercially. Dried fruits are very delicious. In Europe; western Asia, northern Africa and California, commercial canning and drying of figs are industries of great importance.

The latex of the unripe fruits and of any part of the tree may be severely irritating to the skin if not removed promptly. It is an occupational hazard not only to fig harvesters and packers but also to workers in food industries, and to those who employ the latex to treat skin diseases.

Seed oil: Dried seeds contain 30% of a fixed oil containing the fatty acids: oleic, 18.99%; linoleic, 33.72%; linolenic, 32.95%; palmitic, 5.23%; stearic, 2.1 8%; arachidic, 1.05%. It is edible oil and can be used as a lubricant.

Leaves: Fig leaves are used for fodder in India. They are plucked after the fruit harvest. Analyses show: moisture, 67.6%; protein, 4.3%; fat, 1.7%; crude fiber, 4.7%; ash, 5.3%; N-free extract, 16.4%; pentosans, 3.6%; carotene on a dry weight basis, 0.002%. Also present are bergaptene, stigmasterol, sitosterol, and tyrosine. In southern France, there is some use of fig leaves as a source of perfume material called "fig-leaf absolute"—a dark-green to brownish-green, semi-solid mass or thick liquid of herbaceous-woody-mossy odor, employed in creating woodland scents.

Latex: The latex contains caoutchouc (2.4%), resin, albumin, cerin, sugar and malic acid, rennin, proteolytic enzymes, diastase, esterase, lipase, catalase, and peroxidase. It is collected at its peak of activity in early morning, dried and powdered for use in coagulating milk to make cheese and junket. From it can be isolated the protein-digesting enzyme *ficin* which is used for tenderizing meat, rendering fat, and clarifying beverages. In tropical America, the latex is often used for washing dishes, pots and pans.

Medicinal uses: The latex is widely applied on warts, skin ulcers and sores, and taken as a purgative and vermifuge, but with considerable risk. The leaf decoction is taken as a remedy for diabetes and calcifications in the kidneys and liver. Fresh and dried figs have long been appreciated for their laxative action. Fig fruits are used as medicine after preserving in vinegar for treatment of liver and spleen enlargement in children. It is also effective in chronic cough and piles (Chandra *et al.*, 1994). Fig fruits valued for its laxative properties. In traditional medicine, the roots are used in treatment of leucoderma and ringworms and its fruits which are sweet have antipyretic, purgative aphrodisiac properties and have shown to be useful in inflammations and paralysis. Locally the leaves are used in the treatment of jaundice (Patil and Patil, 2011)

The drying of fig is done in Poona, India, and there is currently interest in solar-drying in Guatemala. Fig is one of the first fruits to be preserved by drying. Apart from canning and drying, it is processed in to paste and jelly. Usually, the fruits are allowed to fully ripen and partially dehydrate on the tree, then are exposed to sulphur fumes for about a half hour, placed out in the sun and turned daily to achieve uniform drying, and pressed flat during the 5- to 7 days process. 'Black Mission' and 'Kadota' figs are suitable for freezing whole in syrup, or sliced and layered with sugar. Dried figs have been roasted and grind as a coffee substitute. In Mediterranean countries, low-grade figs are converted into alcohol. An alcoholic extract of dried figs has been used as a flavoring for liquors and tobacco.

Botanical Description

Fig (*Ficus carica* Linn.) belongs to the family Moraceae. There are several species found under *Ficus* genus. The fig has been grouped into four types based on the sex of flowers and pollination. Tree is small to medium much branched having broad leaves. Large sized figs belong to common fig group. However, more than 700 varieties of fig are available in the world. The varieties are divided into green and yellow figs and dark-colored figs. Latex is present in plants part which used for different purposes. Flowering and fruiting takes place during March and harvesting of fruit is done May to June. Fruit drop and cracking is common problem in this crop. Fruit development in fig is done without pollination. In arid region frost/low temperature is affected to the vegetative growth of plants. However, plants recovers during spring after employing cultural practices. The descriptions of some of the important varieties are as follows.

Green and Yellow Fig

Alma- A small to medium-sized golden brown, pear-shaped (pyriform) fig with amber pulp, sweet in taste and delicate flavor. Bred by Texas A&M Univ. and released in 1975. Well adapted in the Southeast. Highly resistant to fruit rot diseases. Wood is very hard.

Blanche - An old, reliable variety usually called Lemon in the South. A medium to large greenish yellow fig with white pulp and many seeds. Top-shaped (turbinate) without a neck. Sweet and delicate flavor with a nutty texture from the soft seeds. Well adapted in the South. Fairly hard plant type.

Brunswick - A medium to large fig with bronzy yellow skin and rich flavor. Oblique- turbinate. Well adapted in the Southwest and drier areas of the South. The fruit is ruined by excessive rain since it has open eyes. Fairly hardy

Flanders - A greenish yellow medium-sized fig with violet stripes and amber pulp. Bred and released by Ira Condit in 1975. Pyriform with a long, slender neck. Fine flavor. Plants are vigorous but not particularly hardy. Good on the West Coast.

Green Ischia- A small grass-green fig with dark strawberry pulp. Oblate to spherical shape. Small, fairly well-closed eye. Good quality but not as good as Brown Turkey. Matures late. Fairly hardy.

Gulbun - It is also known as Jewel. A large light green to pale yellow fig with a translucent pulp tinged with pink. Good flavor. Seems moderately hardy. Bred by Ira Condit.

Ischia - A small- to medium-sized yellow fig with fair flavor. Oblate to spherical with or without a short neck. Well adapted in coastal California. Its quality is poor in South part of the country.

Jurupa - A very large green fig with pink pulp and a medium-sized closed eye. Pyriform. A Condit hybrid selected by Julius Enderud. Good flavor. Under trial in the South. Somewhat hardy but tends to leaf out early—a negative point in areas with late frost.

Kadota- A greenish white small- to medium-sized fig. Pyriform. Vigorous. Delicious fresh or dried. It is a rich, sweet, all-purpose fig and the most common canned fig. Well adapted in the Southwest and drier areas of the South. Fairly hardy.

King - A medium-sized greenish yellow fig with strawberry pulp. Pyriform to oblique in shape. It ripens between late June and August. A San Pedro type. It sometimes sets main-crop figs without pollination. Sweet and rich. Well adapted in the Northwest and cooler areas of the South. Fairly hardy.

Dark-Coloured Fig

Beall - A medium to large purplish black fig with amber pulp. Brebas are pyriform with prominent necks; main-crop figs are oblate to pyriform with short, thick necks. Very good flavour. Well adapted in California and hardy in the South. A chance seedling was found in California in the 1920.

Brown Turkey - A small- to medium-sized light brown to violet fig with strawberry pulp. Turbinate to oblique, mostly without a neck. Small eye has a reddish color from very early stage (unlike Celeste),cold hardy. It fruits on new growth if winter killed. Often bears two crops a year. Its fruit are very sweet but not rich in nutrition. The tree is prolific. The main crop, beginning in mid-July, is large; the early crop is small. The cultivar is well adapted to warm climates. It is grown on all the islands of Hawaii. Moreover, this variety is used as rootstock in India (Anon, 2009).

Black Ischia- A small purplish black fig with strawberry pulp. Turbinate. Fairly sweet and rich in flavors. Well adapted in coastal

California but not very productive in the South. Not particularly hardy. Has been replaced by Celeste. It is grown in near by Lucknow, Saharanpur in U.P. Fruits become black after ripening.

Conadria - A medium to large yellow-green fig with light strawberry pulp and rich flavor. Bred by Ira Condit and released in 1957. Pyriform in shape. Well adapted in California and the Southeast. Hardy with good rebound from freezes. Its performance has been found better under Indian conditions.

Excel - A medium-sized yellow fig with amber pulp. Also bred by Ira Condit and released in 1975. Oblate to spherical in shape. Well adapted in California. Early trials in the Southeast are very promising. Seems to be very hardy. Superb flavor.

Deanna- It is an exotic variety. Medium size yellowish, bigger fruit size and weight (55 g) and TSS 22.6 %. It is being commercially grown in Maharashtra and Karnataka states.

The exotic varieties viz. Conadria, Deanna and Excel are performing better under Bangalore conditions and these are good for drying, canning and table purposes. The performance evaluation showed that Fig cultivars Deanna, Excel and Conadria are suitable for dry areas. Excel and Conadria develop small canopy and are suitable for high density planting. Poona fig and Dinkar cultivars are being grown in drier parts of Maharashtra (Anon. 2009-10).

Indian Cultivars

Poona fig- Fruits are bell-shaped, of medium size, weighing about 20-30 g. thin-skinned; light purple with red flesh, of sweet, good flavor. It is most popular variety and it is commercially cultivated near by Poona. It is suitable for fresh consumption and drying purpose.

Dinkar- It is an improvement over Daulatabad type for yield and fruit quality. This variety has been released by Marathawada Krishi Vidyapeeth, Parbhani and it is gaining commercial cultivation in Maharashtra state. It is medium type variety in yield. It produces about 200-300 fruits/ tree. The performance of some cultivars of fig are presented in Table-1.

Table 1 : Performance of different variety of Fig under Rahuri conditions (MS).

Varieties	Av. weight of fruit (g)	No. of fruits /tree	Weight (kg/tree)	TSS (%)
Poona fig	40.86	1382.66	44.69	19.58
Dinkar	36.59	1156.33	37.50	19.11
Excel	26.61	797.66	21.33	19.89
Conadria	43.47	930.00	37.43	20.69
Deanna	46.91	824.66	37.19	21.12
Chalisgaon	34.90	541.66	11.56	19.41
Maishram	30.47	406.66	12.94	17.61
CD at 5%	7.43	49.13	5.43	NS

Propagation

Fig trees have been raised from seed; even seed extracted from commercial dried fruits can germinate and produce full-fledged plants. Ground- or air layering can be done satisfactorily, and rapid mass multiplication by tissue culture has been achieved in Greece. For commercial production, the plant is commonly propagated by cuttings of mature wood (Pareek and Sharma, 1991). Planting must be done within 24 hours but, first, the upper, slanting end of the cutting should be treated with a fungicide to protect it from disease, and the lower, flat, end with a root-promoting hormone. Trees of unsatisfactory varieties can be top worked by shield- or patch-budding, or cleft- or bark-grafting. Figs are easy to propagate because they root very easily. The most common method is to root leafless cuttings taken in late winter or early spring.

Seed propagation

In fig, seed propagation is used only for improvement and it is not common method. The seed germination of two introductions of *Ficus septica* was about 60 per cent. Seed sowing was carried out in a warm greenhouse. Seeds are used for breeding/hybridization purpose. Female trees may also produce apomictic seeds i.e. without pollination and subsequent fertilization.

Vegetative propagation

Air layering

Gooties or air layering is very popular in South India and is widely practiced in fig. This crop is easily propagated through air layering. For layering one year old two cm thick branches are preferred. Air layers are done during early spring; they will root well by mid summer (Hartmann *et al.*, 1997). Plants are later transferred to pots then planted in the main field.

Budding/ grafting

Budding or grafting also has been used to top work a tree to a better variety. Several *Ficus carica* cultivars were successfully budded or grafted onto *F. glomerata, F. cocculifolia, F. Graphalocarpa* and *F. palmata;* no incompatibility was noted after a year or more. Under Bangalore conditions, the plants of fig varieties are propagated on Brown Turkey root stock which is found suitable root stock..

Chip budding was best for the economical production of large numbers of container-grown trees. Ring budding and side grafting can also be performed for multiplication of propagules. Figs are grafted on a nematode resistant rootstock such as *Ficus cocculifolia* and *F. glomerata* but these rootstocks are cold sensitive, unlike cuttings, and the roots must be protected from the cold in the winter. Results obtained at Kerala Agricultural University, Vellanikkara indicated that ring budding and patch budding were found successful in *Ficus* species Particularly rare cultivars may be propagated on rootstocks, or older trees, top worked by whip, cleft or crown grafting, or chip or patch budding.

Cuttings

Fig is generally propagated through stem cuttings. Leafless, fully mature, dormant hardwood cuttings of 1.5 to 2 cm thickness and 25-30 cm length are taken from parent plant. Cuttings root quickly and plants so raised may produce fruits before they are one year old. One year old layered plants also produce the first crop in the first year itself and a fair crop can be obtained from third year onwards. Select foot-long pieces of dormant wood, less than 1-inch diameter, with two -year-old wood at base. One-year twigs with a heel of two-year branch at the base may also be used. Dip the cuttings in a rooting hormone and allow them to callus one week in a moist place at 50-60° F. Summer cuttings may also be made, but they do best if defoliated and winterized in refrigeration for 2-3 weeks before potting. Leafy shoots require a mist bed. Antunes *et al.* (1996) suggested the treatment of cuttings with 100 ppm IBA and then planting them in sand: soil (1:1) mixture without stratification for better rooting and shoot development.

The cuttings of Fig (EC 538098) Brazilian varieties were introduced from Niger at CIAH, Bikaner during the year 2004, but did not sprouted and dried after some time due to small size cuttings and low temperature during the month of January. Therefore, appropriate size of cuttings should be standardized for successful establishment.

Planting is generally done in the season from June to January. Cuttings are placed in bundles with basal end upwards in callusing pits for about 1.5 to 2 months before being planted. Single bud cuttings from the zone between the third and sixth nodes take minimum number of days for rooting. The development of root and leaf was best on 30-35cm long cuttings. Cuttings having 20cm of more length resulted in 30 per cent rooting. Rooting was faster and more prolific from the basal nodes. Rooting was best (up to 90 per cent) in cuttings from 3-year-old wood, followed by up to 85 per cent rooting in 2-year and up to 55 per cent in 1-year-old wood. Cuttings prepared in autumn and planted in spring rooted better than those prepared and planted in spring; autumn-planted cuttings gave poor results. An application of N at 30 kg + P at 22 kg + well rotten FYM at 10 t/ha in the nursery increased the production of standard quality transplants by up to 34%. Fig trees are readily propagated by dormant hardwood cuttings; fig propagation using hardwood stem cuttings under field conditions is, therefore, of great importance, as this produces plants producing vigorous and potentially sellable plants. Rooting was examined in hard wood cutting by applying various concentrations of IBA to the cuttings obtained in different periods. One-year-old hardwoods cutting of three-fig culture were planted in perlite. Figs are easy to propagate as they root very easily. The most common method is to root leafless cuttings taken in late winter or early spring. Take cuttings which are 3 to 6 inches long and pencil to finger thick. The best cuttings will have some of last year's wood on them. Using 4" deep plastic pots, pack a half sheet of newspaper tightly into the bottom of the pot. Put a little mix in the bottom, stand 1 to 4 cuttings upright in each pot and fill the pot with the mix. Water the pots thoroughly and stand them in a very bright, but not in sunny day. The water should apply only when they are very dry. After a few days, it will be time to pot up the new plants. Sometimes there will be 4 or 5 leaves and few if any roots. The vigorous growth takes place after sometimes of establishment.

Earlier experiments were conducted at the University of California to determine the extent and degree of root and shoot regeneration in tip cuttings of fig and their response to various concentrations of IBA. The treatments resulted in injury to 3 of the cultivars and increased regeneration in the fourth. Treatment of cuttings with Indole butyric acid (IBA) at 2,000-10,000 ppm resulted in better rooting. Treatment with IBA and NAA showed higher percentage of rooting and larger root number, compared to IAA, and the effective growth substance also promoted shoot growth. Treatment with NAA and IBA at 20 ppm proved very effective in improving rooting in cuttings of fig in a medium containing soil and 10 kg FYM/sqm.

Plants raised in containers develop coiled roots. For success in dry lands, plants must have root architecture with a strong tendency to penetrate deep into the soil. The plant density mainly depend upon the plant type, soil fertility status and management practices while planting system to be adopted in dry lands depends largely upon the topography of the land, fruit species and soil type. In the plains, planting is generally done in square or rectangular system. The plant spacing depends on variety and soil type. It is generally planted at a distance of 5- 6 meter. However, Excel and Conadria can be planted at 2.5-3.0 m spacing. Planting of fig trees should be done during dormant stage. Spring is the best time. In warm areas, bare-root trees can be set out in fall or early winter, but where late spring frosts are common, it is best to set them out in spring after the danger of hard winter freezes has passed. In arid region, the best time of planting is onset of monsoon. Container-grown plants should always be planted in the spring. For best growth, fig trees need full sunlight and freedom from competing trees and shrubs.

Micro-propagation

Though micro-propagation of fig is possible to obtain pathogen free plantlets and this is one of basic requirement for successful commercial orchard. This technique has advantages to produce large scale planting material when ever needed. Pasqual and Ferreira (2007) has also tried micro-propagation in fig tree. Mother plants must be in good phytosanitary and nutritive conditions to facilitate the decontamination of explants for micro-propagation .Kumar *et al.* (1998) used apical meristem from mature tree for *in vitro* regeneration in Fig (*Ficus carica* L. *cultivar Gular*).

An *in vitro* protocol for *Ficus carica* cv. 'Roxo de Valinhos' was optimized by Chrystiane Borges Fra´ Guas *et al.* (2004). Nodal explants containing two buds were excised from field-grown mature plants, and transferred to different proliferation media consisting of combinations of distinct concentrations of activated charcoal with benzyladenine (BA), kinetin with gibberellic acid (GA_3), and WPM (woody plant medium) with kinetin. The regular strength of WPM in combination with

0.5 mg/l kinetin was the best condition for shoot proliferation of *Ficus carica* 'Roxo de Valinhos' plants. The addition of activated charcoal in the medium completely inhibited shoots proliferation. The inclusion of BA in the medium induced excessive callus formation as well as small and vitrified shoots, while GA_3 induced excessive elongation associated with vitrification, chlorosis, and tip-burned shoots.

Quality Standards of Planting Material

Looking to the potential of crop, the new introductions should be made from iso-climatic region of world. Identification of drought hardy rootstock for arid conditions is essential. Efforts should be made for production of Quality planting material by registered nurseries in the country. Development of agro-techniques for different agro-climatic zones is required in this crop. Promotion of Fig cultivation with disease free planting material should be advocated. Disease resistant, high yielding, better shelf-life varieties should be developed.

References

Aksoy,U., CAN, Hepoksay, H.Z., and Sahin, N. 2001. Fig cultivation. TARP, Turkey Agricultural Research, Izmir, Turkey.

Anonymous 2009-10. Biennial Report, AICRP on AZF, CIAH, Bikaner, 202p.

Antunes, L.E.C., Chalfun, N.N.J., Pasqual, M., Dutra, L.F. and Cavalcante-Alves, J.M. 2003. Factors affecting on rooting of figs (*Ficus carica* L.) cuttings. Acta Hort., 605:141-146

Chandra, A., Chandra, Anju and Gupta, I.C. 1994. Arid Fruit Research, Scientific Publishers , Jodhpur pp.250-54.

Chrystiane Borges Fra´ Guas, Moacir pasqual*, Leonardo Ferreira Dutra, and Jairo osvaldo Cazetta 2004. Micro- propagation of Fig (*Ficus carica* L.) 'Roxo de Valinhos' Plants, In Vitro Cell. Dev. Biol.—Plant,40:471–474,

Hartmann, H. T., Kester, D. E., Davies, F. T. and Goneve R. L., 1997. Plant Propagation- Principle and Practices. Published by Prentice Hall of India Pvt. Ltd., New Delhi, pp. 635.

Kumar ,V.; Radha, A, and Chita,S.K.1998. *In vitro* plant regeneration of fig (*Ficus carica* L. cv. Gular) using apical buds from mature tree.Plant Cell Report,17:717-720.

Pareek, O.P. and Sharma, S. 1991. Fruit trees for arid and semi-arid lands. Indian Farming., 41:25-30.

Pareek, O.P. and Vishal Nath 1996. Coordinated Fruit Research in Indian Arid Zone- A Two-Decade Profile. NRC for Arid- Horticulture, Bikaner, pp75-79..

Pasqual, M and Ferreira, E.A. 2007. Micro-propagation of Fig tree (*Ficus carica* L.) In. Protocol for Micro-propagation of Woody trees and Fruits (ed. SM.Jain and H. Hoggman), pp.409-416.

Patil, V.V. and Patil, V.R. 2011. *Ficus carica* Linn. An Overview. Research Journal of Medicinal Plants, 5:246-253.

Chapter-15

Manila Tamarind (*Pithecellobium dulce* Roxb.)

R. S. Singh and S.K. Maheshwari
Central Institute for Arid Horticulture, Bikaner-334 006

Manila tamarind (*Pithecellobium dulce* (Roxb.) Benth.), also known as 'Jangle jalebi', 'Madras thorn', Sweet inga and 'Monkey pod' has a potential to become an important tree of semi- arid and arid regions since it is a drought hardy plant and tolerates harsh environmental conditions as well as salinity. In general, it is planted on the boundary of orchard, road and field sides as protective hedge plant. It is a medium size, much branched tree, suitable for silvi-pastoral system (Gupta *et al.*, 1994). It is drought resistant species but susceptible to frost under arid conditions, coming up well in areas of low rainfall due to its extensive root system. Names like *dulce* (sweet) and "Manila tamarind" reflect the wide use of the pods as food. The most valuable and commonly used part of Manila tamarind tree is the pulp, which is white or red, very sweet and has a pleasant flavour. The large aril represents 50 per cent of the fruit (Seeds 25%, husk 25%). The major constraint in commercialization of this tree species is lack of improved genotypes, agro-techniques for cultivation in different ago-climatic conditions and limited commercial value.

Nutritive Value and Uses

It produces tamarind like nutritious pods which is used for fresh consumption and making delicious drinks (Chundawat, 1990). It is valued primarily for fuel wood, fodder for sheep and goats (Kundu, *et al.*, 1983) and ornamental hedge tree. Flowers are an important nectar source for honeybees. Its seed oil is edible and used in soap industry.

The tree is also very popular as an ornamental and is used in topiary (plant sculpturing). Trees with variegated leaflets are available as ornamental in Hawaii. The ripe fruit is used as a snack and is very popular among the Mexican farmers, especially children who love eating Manila fruits straight from the tree in the form of natural popcorn. In some places it is also used in atoles. In India the seeds are used fresh or in curries. The fruit is eaten raw or for making sweet sour drinks.

The leaves are browsed by goats, sheep, cattle and horses and can withstand heavy browsing. From the viewpoint of standing cattle pressure and for silvipastoral system (Chouhan, *et al.*,1992), *P. dulce* has been reported to be promising tree species (Gupta *et al.*, 1994 and Foroughbakhch *et al.*, 1988). The leaves are reported to contain tannin content of 6.96mg/g (Malarvizhi, *et al.*, 1999). The dry leaves contain 4.91 per cent N, 0.78 per cent P_2O_5, 1.04 per cent Ca and 2.67 per cent K_2O (Anon, 1976, (CSIR, 1948-1976). According to Kundu *et al.* (1983), the leaf composition in DM, with digestibility in parenthesis was crude protein 20.3 (71), ether extract 7.5 (40), crude fiber 19.8 (30), nitrogen free extract 43.0 (72), ash 9.3, calcium 2.2 and phosphorus 0.3%; DM was 34.5 (58), Total digestible nutrient was 58.4. digestible crude protein 14.4 per cent and daily balances were N, 3.93; Ca, 0.64 and P, 0.43.

Flowers are an important nectar source for honeybees. The honey obtained is of high quality. The seeds contain greenish oil (20%), which after refining and bleaching can be used for food or in making soap. The press cake rich in protein (30%) may be used as stock feed. The bark is very high in tannins of the catechol type (up to 37%) and is used in tanning industry in India and Philippines. It has potential source for natural dyes and produces yellow dye.

Manila tamarind is a nutritious fruit with a variety of uses. The properties of this species have been extensively studied, particularly with reference to the components of whole fruit, fruit pulp and seed. The average composition of which is given in Table 1 & 2.

The wood is strong, durable wet soft and flexible. Its wood is used as timber and for many other purposes including construction. The wood is used for making boxes, crates and wagon wheels in Philippines. The wood although used as a fuel is not of high quality due to smokiness and low calorific values (5500 K cal/kg). Tree seems promising for the cultivation of lac insect, which could be harvested and sold as stick lack for the production of varnishes and lacquers.

Table 1 : Average composition (whole fruit) of Manila tamarind per 100g fruit

Constituents	Aril+Husk+Seed (58% refuse)
Moisture	32.7%
Protein	1.3g
Fat	0.2%
Total carbohydrate	7.6%
Fiber	0.5%
Ash	0.2%
Calcium	5.0mg
Phosphorus	18.0 - 42.0mg
Iron	0.2- 0.5mg
Sodium	8.0 - 19.0mg
Potassium	93- 222.0mg
Â carotene	5.0 me
Thiamine	0.10 mg
Riboflavin	0.2 mg
Niacin	0.2 mg
Ascorbic acid	56 mg

(***Source:*** Leung *et al.* (1972))

Table 2 : Composition of manila tamarind seed

Constituents	Value mg/100g
Water	13.5g
Protein	17.7g
Fat	17.1g
Starch	41.4g
Fiber	7.8g
Ash	2.6g

Botanical Description

Manila tamarind *(Pithecellobium dulce* (Roxb.) Benth,) a member of the dicotyledonous, family leguminosae and sub family Mimosideae is one of the 100-200 species in this genus. It is a diploid species with a chromosome number of X=13 and 2n=26. The generic name refers to the curly pod, that mimics an ape's earing (*pithekos ellobium*), and the species name "*dulce*" refers to the sweet pod. The genus includes several other important species such as *P. arboreum, P. unguiscati, P. flexicaule, P jiringa, P. paviflorum* and *P. avaremotenum. P. dulce* is known by a variety of common name/vernacular names.

It is a spiny tree commonly 10-15m high, but ranges from 5-20 m. The tree habit is spreading and has a short bole of about 1m thick with irregular branches, greyish bark becoming rough, furrowed and then peeling. Leaves are parppinate with one single pair of leaflet per pinna, leaflets oblong 2.0-4.0 cm long 1.0-1.5 cm wide. There are small thorns 2.0-15.0 mm long axillary pairs inserted on each side of the leave pedicel. Thorn less individuals do occur. Leaves are deciduous but the new leaves appear while the old ones are being shed giving the tree an evergreen appearance. Flowers are disposed in small spherical glomerules with heads 1 cm in diameter, forming short axillary panicles of 5-30 cm in length. Flowers are white greenish slightly fragrant with often a sweet smell 1.0-1.5 mm in diameter, with a hairy corolla, 50 thin stamens united in a tube at the base. Flowering begins in 3-4 years and is seasonal. The fruits or legumes are pinkish-red to greenish brown rather thin 10-15cm long and 1-2.0 cm wide. The pods are irregular, sickle shaped or circinate and are spiraled in a set of 1-3 whorls and strangled between the seeds (lomentaceous). There are about 10 seeds per pod, which are black and shiny about 1cm in diameter hanging in the pods by a red funicle. The seed is surrounded by an edible whitish pulpy aril (Galang, 1955). The pulp is red or white, sweet and rather dry and mealy. The pod with an inner sweetish white edible pulp (the funicles) is sweet and has chestnut like flavour (Stone, 1970).

The tree is believed to have been originated from tropical and subtropical America and is native along coasts from California through Mexico to South America but is now found throughout the tropics. *Pithecellobium dulce* followed the Spanish gallion route through the Pacific and Asia to Africa. In Fiji "cultivated often as a street tree", has now naturalized in dry zone along roadsides. It is now common and naturalized in India, Tropical Africa, especially along the coasts. It is notably weed in the Caribbean Islands (including Cuba, Jamaica, Puerto Rico and St Croix) and in Florida and Hawaii, USA.

In most countries, Manila tamarind is a subsistence tree mostly meeting local demands. Although grown throughout the tropics and sub tropics, its cultivation is sporadic and the exact acreage under this crop is not known. The major growing sites are in the American and Asian continents. The tree has been successfully planted in the South Sahelian and North Sudanian ecozones but not over large acreage.

The genetic variation of *P.dulce* has been based on phenotypic observation such as fruit characteristics, length of pod, shape of pod (curved, semi curved, straight), pod weight, seed number, pod colour, sweetness of the pulp, pod bearing ability and tannin content etc. Good genetic diversity of Manila tamarind is found in Southern Florida, tropical Africa, America and tropical Asia while in India it is growing in Periyakulam, Virudhnagar, Coimbatore, Tamil Nadu; Bastar, Chhindwara, district of Chattisgarh, parts of Madhya Pradesh, Godhra, Panchmahal district of Gujarat, Udaipur and Alwar districts of Rajasthan and Delhi. The variations have also been observed against tolerance to drought, wind, poor soils, high and low pH and grazing. Phenological diversity also exists and tree to tree variation is common in flowering and in maturing fruits, which may reflect either genetic environmental interaction or genotype x environmental interaction or both. Three collections made from Jodhpur, Periyakkulum and Godhra were under evaluation at CIAH and were found susceptible to frost.

There is a wealth of *P. dulce* germplasm across the regions where it is naturally grown however, no systematic germplasm collection and evaluation has been attempted to date in spite of its future potential. Most countries have not considered it as priority species for conservation. The tree is also not considered as an endangered or vulnerable species and hence not received priority in its conservation. Although Manila tamarind docs not contribute substantially to the economy in national and international market, it may be of major importance to the local populations in rural areas and worthy of conservation.

Genetic variability within a species is the basis for any systematic tree improvement programme. Germplasm conservation and improvement programme should focus on determining the important traits and the improvement carried through the use of variation in natural population. Since large variability in germplasm exist in *P.dulce*, germplasm collections have to be made for each region extending from tropical and subtropical America (South west USA, North west Mexico, Central and Southern America) tropical Asia (India) and Africa (Burundi). These collections should be subjected to systematic and intensive evaluation to select desirable types.

Propagation

Manila tamarind is propagated by both seeds and vegetative methods. It is mainly multiplied through seeds but germination is poor and irregular. Manila tamarind seeds remain viable for months and will germinate in a week after sowing.

Seed propagation

Manila tamarind (*P. dulce*) is traditionally grown from seed (Hayes,1953). Under natural condition the seeds are dispersed by birds. These seeds germinate and result in natural re-generation. When propagated by seed, unblemished mature ripe pods are collected during June-July; cleaned and dried seeds are used for sowing. Seed viability is longer under dry cool storage and in refrigerator at 3-5°C (Kanak Saha *et al.*, 1999). Germination in *P. dulce* is epigeal and seed germination begins within 1-2 days following sowing. The radicle emerges from one end of the seed and descends rapidly. Humus/decomposed coir dust (1:2) and decomposed sawdust is recommended for potting and raising seeds of *P.dulce* (Rimando, 1981). It has been reported that germination of seed is poor due to hard seed coat and dormancy in Manila tamarind. To enhance germination percentage and vigour of plants, pre-treatment of seed has been tried (Parotta, 1991). In case of seed germination, occurrence of abnormal seedlings has reported by Mamta and Jamaluddin (1986) in *P.dulce*. Seed viability is long under dry cool storage. For seeds to germinate, although nicking may improve and hasten the process. Germination occurs quickly, normally in 1-2 days. Application of *Rhizobium inoculum* to seeds is suggested prior to sowing. It has reported that direct seed sowing is successful. Germination starts from 7-8 days and is completed within 2-3 weeks.

Owing to the hard seed coat, the seed germination in this species is poses problems and hence, the plant stands are highly irregular. In order to overcome this, an attempt was made to pre-treat seeds with BA, KNO_3, GA_3, PEG and water to speed up germination, increased germination percentage and seedling growth under hot arid conditions at CIAH, Bikaner (Table 3). Seeds of PKM manila tamarind-1 were used for the study. Among the various treatments, seeds treated with 50 ppm BA and GA_3 exhibited maximum germination percentage, seedling growth and biomass production (Singh *et al.*, 2011).

Table 3 : Effect of pre- treatment of seeds on germination in *P. dulce* under arid conditions.

Treatments	Concentration	Duration of seed soaking	Seed germination (%) days after sowing				
			7	14	21	28	35
BA	50 ppm	12 hrs.	10.0	30.0	33.3	56.6	60.0
KNO_3	50 ppm	12 hrs.	6.6	36.7	36.7	36.7	36.7
GA_3	50 ppm	12 hrs.	6.6	26.7	46.6	60.0	63.3
PEG	2 M Pascal	12 hrs.	6.6	26.7	30.0	33.3	33.3
Soaking in water	-	24 hrs.	0.0	13.0	13.0	13.0	13.0
Control	-	-	3.3	40.0	43.3	46.6	50.0

Vegetative propagation

Manila tamarind can be propagated vegetative methods by branch or stem cuttings. Coppice shoot collected from a selected tree of *P. dulce* in Kadiam village of East Godavari District, Andhra Pradesh were studied for rooting response using different growth regulators. Shoots 14-15 cm long taken from coppiced shoots were sterilized with 5 per cent formalin and treated with IBA, IAA, NAA by a quick (5s) dip before being planted in a vermiculite and kept under intermittent misting in the green house with 70-80 per cent humidity. Rooting percentage was best in 2500 ppm IBA and least with 500 ppm IAA and IBA. Root length was maximum in cuttings with 2500 ppm IAA, IBA and NAA both in individual and combined hormone treatments (Venkaiah *et al.*, 1994).

Although *P. dulce* plays an important role in the diet of people living in tribal tracts and dry areas particularly during famine and food shortage, very little effort has been made by the research workers to improve the tree as a crop plant. There is often a lack of awareness by the local people on the need to plant, protect and manage the plants. The research work carried out by different research organizations is scattered. There is therefore, a great need for a clear and prioritized research for the development of *P. dulce* as a "*Tree for the Future*". It is an underexploited fruit tree of dry region. Development of improved variety is needed for quality production. Techniques should be standardized for large scale production of quality planting material.

References

Anon, 1976. The Wealth of India, CSIR (Council of Scientific and Industrial Research). 11 Vols. New Delhi.

Chouhan,G.S.; Mathur,A.N.; Bhandari, M.M.C.; Jat,R.K and Tauro,P.; and Narwal, S.S. 1992. Allelopathic effect of some tree species on associated grasses under silvipastoral system. *Proc. First National Symp. Allelopathy in agroecosystems (Agriculture &Forestry)*, Feb. 12-14, 1992 held at CCS Haryana Agric. Univ., Hisar-125 004,

Chundawat, B.S. 1990. Arid Fruit culture. Oxford, IBH, Publisher, New Delhi pp 158-163.

Foroughbakhch, P.R. and Penaloza, W.R. 1988. Introduction of 10 tree species to the matorral of northeastern Mexico. Results after two growing seasons in the Linares region, Nuevo Leon, Mexico. *Reporte-Cientifico,-Facultad-de-Silvicultura-y-Manejo-de-Recursos-Renovables,-Universidad-Autonoma-de-Nuevo-Leon.* No. 8, pp33.

Galang, F.G. 1955. Fruits and nuts grown in the Philippines. Malabon, Rizal AIA, Printing Press.

Gupta,B.N.; Singh,A.K.; Bhowmik,A.K. and Banerjee, S.K. 1994. Suitability of different tree species for coal mine overburdens. Annals of Forestry. 2 (1): 85-87.

Hayes,W.B.1953. Fruit Growing in India , Kitabistan, Allahabad.

Kanak, S. 1999. Effect of storage on seed viability in some leguminous taxa. Indian Forester, 125 (6): 609-612.

Kundu, H.; Panda,N.C. and Sahu, B.K. 1983. Leaves of inga dulcis (*Manila tamarind; Pithecellobium dulce*) as a fodder for goats. Indian Journal of Animal-Sciences, 53 (6): 669-671.

Malarvizhi, P. and Khan, A.K.F. 1999. Tannin content in different fodder crops. Madras Agricultural Journal, 86 (7-9) :510-511.

Mamta, P. and Jamaluddin 1986. Abnormal seedlings in *Pithecellobium dulce* (Roxb.) Benth. Journal of Tropical Forestry, 2(2): 172-173.

Parrotta, J.A. 1991. *Pithecellobium dulce* (Roxb.) Benth, Guamuchil, Madras thorn. SO-ITF-SM-40. New Orleans: USDA Forest Services, Southern Forest Experiment Station. 5 p.

Rimando, R.M. 1981. Growth and development of some fuelwood species in different potting media. Sylvatrop, 6(3): 91-100.

Singh, R. S.; Bhargava R. and Garima Pal 2011. Effect of seed treatment on germination and growth behavior of Manila tamarind (*Pithecelobium dulce*) under hot arid conditions. Jour. Trop. Forestry, 27: (4).

Stone, B. 1970. The flora of Guam.Micronesica,6: 302-303..

Venkaiah, K.; Bramaramba, B.; Praveena, P.L.R.J. and Kumar, D.V. 1994. Rooting response of coppice shoots of Pithecellbium dulce Benth. Indian Journal of Forestry, 17 (4): 367-368.

Chapter-16

Custard Apple (*Annona squamosa* L.)

R. S. Singh and A. K. Singh
Central Institute for Arid Horticulture, Bikaner-334006

Custard apple (*Annona squamosa* L.), also known as sitaphal, sharifa, sugar apple is a nutritious, edible fruit of family Annonaceae. It is a drought hardy, deciduous fruit tree and most suitable for dry land horticulture. Annonaceous fruits originated in tropical America and are widely distributed in tropics and sub tropics. Among Annona fruits, custard apple is the most favourite in our country. Its plants come up unattended in parts of Andhra Pradesh, Assam, Bihar, Karnataka, Maharashtra, Madhya Pradesh, Odisha, Rajasthan and Tamil Nadu. Due to hardy nature and escape from animal's damage, this plant species is grown in different tropical and sub-tropical regions of the country. Of late, custard apple has gained commercial significance and exclusive orchards are emerging in Maharashtra, Andhra Pradesh, Gujarat and Rajasthan. A wide range of genetic diversity is available in Anonaceous fruits in many parts of country. It is naturally grown in Aravali belt especially in Udaipur, Sirohi district of Rajasthan and its potential should be exploited for improvement (Chundawat, 1990). The plant of this species is shrub or small size tree and very important in both horticulture and in ethno-botanical use. However, the maximum area of its cultivation is in Maharashtra and Karnataka state. The area are being expanded because of its prospects in semi arid and dry parts of the country.

Nutritive Value and Uses

Custard apple is mostly consumed as table fruit and used for making value added products like jam , jelly, ice-cream. Further, the immature fruit, seeds, leaves and roots are of considerable medicinal value both in Ayurvedic and Unani systems of medicines. Its fruit are good tonic enrich the blood, increase the muscular strength, burning sensation and relieve in vomiting (Ayurveda). Ripe fruit is mild laxative and anthihelmentic. The fruits contain carbohydrates, fat, vitamins (Riboflavin, Thiamine, Niacin, Ascorbic acid) and minerals viz. calcium, iron, phosphorus, etc. (Table 1). The seeds are abortifacient and roots are drastic purgative .Seed oil (30 %) is used in soap and paint industry while seed cake (4% N) as cattle feed and as manure. It is used as insecticide with Neem oil.

Table 1 : Nutritive value of Annona fruits (g/100g edible portion)

Constituents	Sweet soup	Cherimola	Atemoya	Sour soup
Moisture	75.97	68.71	78.7	80.1
Protein	1.89	1.54	1.4	0.69
Fat	0.57	0.13	0.6	0.39
Carbohydrate	20.82	28.95	15.8	18.23
Fiber	1.41	-	2.5	0.95
Ash	0.75	0.67	0.5	0.58
Energy (kc)	360	460	310	247
Calcium (mg)	17.0	9.0	17.0	9.0
Magnesium(mg)	22.0	-	32.0	22.0
Phosphorus(mg)	54.0	24	-	29.0
Potassium(mg)	142	-	250	320.0
Sodium(mg)	2.0	-	4.0	22.0
Iron (mg)	0.3	0.25	0.3	0.82
Ascorbic acid (mg)	35.9	12.20	43.0	16.4
Thiamine	0.10	0.11	0.05	0.07
Riboflavin	0.06	0.11	0.08	0.12
Niacin	0.89	1.0	0.8	1.52

Annona Fruits and their Fruit Characters

Besides custard apple, other annona fruits are cultivated on a limited scale. Bullock's heart (*Annona reticulata*) is more commonly found in south India than in north India. It is usually associated with gardens and compounds and not commercial orchards. Cherimoya is mostly restricted to Assam and hills of south India. Atemoya and sour soup are cultivated in some gardens as miscellaneous fruits. Atemoya, cherimoya and *Annona diversifolia i.e.*Ilama which is not popular in our country, also provided excellent opportunities for a large-scale exploitation in India.

Custard Apple (*Annona squamosa*) also known as Sweet soup, Sugar apple, Sitaphal, Sharifa- The plants are woody, semi deciduous. Fruit weight is from 250 to 300g; globular; green skin; sweet (20% sugar); non acidic; pulp creamy white; distinct segment, 60-80 seeds /fruit.

Bullocks Heart; West Indian custard apple (*Annona reticulata*)**,** Ramphal- Plant is semi-deciduous reaching 6-7m height; fruit large (350-400g); Heart shaped; yellowish red; Smooth rind with hexagonal markings, Pulp pale, gritty, highly flavored, 12.5% sugar and few seeds (30-40). It is cultivated in very limited area in Maharashtra. However, scattered tree are found in forests.

Cherimola & Cherimoya (*Annona cherimola***),** Local name- Hanuman phal -Semi deciduous, tree reaching to a height of 8m.Fruits wt. 250-300g, pale green when ripe, sub-globose, Pine like aroma 18% sugar, segment fused, 10-15 seeds /fruit.

Atemoaya *(Annona atemoya* **)** **(***A. squamosa* x *A. cherimoya*), local name Lakshaman Phal –Semi-deciduous, large spreading tree, 5m ht. Fruit wt. 500g globular green; white smooth pulp is very juicy with excellent sugar acid blend large segments 10-15 seeds/ fruit.

Sour soup & prickly custard apple (*Annona muricala*) **–** Its trees are ever green , 6-8 m in height, fruits 1.5-3 kg heart shaped; dark green; fleshy pines, pulp is white; fibrous, juicy with mango like flavor; 11-14 % sugar.

Botanical Description

Custard apple (*A. squamosa* L.) is a medium, deciduous tree of genus Annona and family Annonaceae are collectively known as Annonaceous fruits. Genus Annona has 120 species, 6 of them having pomological significance. Annona fruits are formed by fusion of pistil and receptacle into a large flashy aggregate fruit. Annonaceous fruits have morphological affinity for each other but each type is unique in its taste, flavour, pulp color and texture. Among the edible annonas, cherimoya (*Annona cherimola.*), sugar apple (*A. squamosa*) and the hybrid between the two, atemoya, are most popular. The other less important edible *Annona* species are *A. reticulata* L. (Bullock's Heart), *A. diversifolia* (Ilama) and *A. muricata* (sour sp). The edible annonas species differ for fruit traits and are amenable for genetic manipulation, as they cross with each other easily. Each species can benefit from the other edible annonas for one or the other specific fruit and/or plant traits and hybridizing them should generate useful recombinants as illustrated by the popular *atemoya*.

Among the annonaceous fruits, sugar apple, locally called *sitaphal* or *sharifa* is by far the most relished and widely consumed fruit of India. Most of the annona varieties in cultivation were often developed by clonal or seedling selection by mainly exploiting the intra species variability. Sugar apple plants come in wild abundantly in the vast arid tracts of the country. There are few varieties like Balanagar, Red Sitaphal, Local Sitaphal, British Guinea, Mammoth and Washington. Presence of several big seeds and poor shelf life are the major constraints limiting cultivation of *sitaphal* fruit on a commercial scale.

Anonas is a tree growing to a height of 6-8 meters or less. Leaves are shiny, oblong to oblong-lanceolate, up to 20 cm. long, 2 to 5 cm. wide, pointed at the tips, with petioles 1 to 1.5 cm. long. Flowers are greenish-yellow, fragrant, 2 to 2.5 cm. long, occurring in two or three together on lateral peduncles. Outer petals appear as in the inner one is very small, narrowly oblong, or wanting. Fruit is large, heart-shaped, brownish-yellow, about 8 cm. or more in diameter, with pentagonal areoles outside. Skin is thin, covering a cream-colored, fair-flavored, juicy, and sweet pulp. Fruits are available from September to November month.

Flowering and Fruit Set

The time of flowering and fruit development in different varieties varies considerably. Custard apple flowers from March to July and sets fruit in June- July and takes 4 months for fruit development where as Cherimoya flowers from May to June and July to September .It takes 6 months for fruit development. On the other hand Bullocks heart flowers from August to October and takes 8 months for fruit development. Custard apple is cross pollinated crop, a wide variation in fruit size as well as colour of pulp are available. It is aggregate fruit and sweet in taste.

Low marketable yield is a major problem in various annonaceous types. The reasons for low fruit set are dichogamy, lack of pollinizers and production of flowers without attractive color, poor pollen germination, environmental and tree condition. Close planting may increase the pollination. Venkatraman, (1979) and Thakor and Singh (1965) reported that stigma remains receptive from one day before anthesis to about 2-3 hrs after anthesis but it decreased abruptly so by the time of dehiscence stigma turned almost unreceptive and observed protogynous dichogamy. Similar trend was reported by Limaye (1966) and Rajput (1985). Ahmad (1935) reported that the best time for artificial pollination was from 5 am to 8 am and also reported that drier and hotter period adversely affected the fruit set. Similar result was reported by Karale *et al* (1989). Rajput (1985) reported that there are no insect pollinizers for Annona because the flowers are unattractive. Pollen germination was found to be poor ranging from 11.5 % to 20.00 %. Work done at CHES, Vejalpur on germplasm evaluation of eight varieties of Annona revealed that maximum No of flowers per plant were produced in Washington (4133.60 no.) however fruit set and retention was the least (0.70%and 0.19% respectively). In cultivar Balanagar though least no of flowers were produced (350.68) fruit set and retention was found to be highest, 26.29 % and 14.66% respectively, (Hiwale, 2002).

Annona breeding was initiated at Indian Institute of Horticultural Research, Bangalore, several intra- and inter-specific hybrids were produced and evaluated, which resulted in isolation of variety 'Arka Sahan' from the cross atemoya (cv. Island Gem) x Sugar Apple (cv. Mammoth). This exemplifies the use of allied species in breeding of arid fruits, if the traits of interest are not detected in the same species. Hybridization work is also being done in Custard apple at MPKV, Rahuri and Hybrid No. -6, Hybrid No.-13 and Hybrid No.-22 have been found promising. Hybrid No.- 13 is bigger fruit size and less seed percentage (7.02%) and TSS 26% with higher yield (12.3kg/plant) than other hybrids.

Hand pollination is a simple and fast technique that results not only in good fruit set but also produces big size and attractive uniform shape fruits with no loss in edible attributes. Thus fruit yield can be significantly increased by artificial pollination. Fruits developing from hand pollination command premium price in the market making this technology a commercially viable one. It should be affected in the morning before 9.30 hrs. This process comprised of: 1. collect pollen of custard apple (*sitaphal*) in a plastic cup from the previous day opened flowers, 2 place the cup in the shirt pocket and dip a clean dry paintbrush (No. 2 or 3) into the cup, and 3 using the pollen coated brush evenly smear the pollen on the stigma of 'Arka Sahan' flowers. The performance of Arka Sahan variety is better in respect of production of quality fruits (Anon, 2011).

'**Arka Sahan**' variety is remarkable for sweetness (>32° B TSS; 22.8% total sugars), scarce small seeds (9/100 g fruit weight) and slow ripening (6-7 days). The pulp is snow-white (76%), mealy and juicy with a mild pleasant aroma. As in other annonas, in this hybrid also few flowers of (about 1-2%) develop into fruits owing to male and female structures maturing at different time, besides limited insect and wind pollination. To supplement this, simple artificial hand pollination was worked out by testing different pollen source.

APK (Ca)-1: It is a variety of custard apple developed through selection and released by RRS, T.N.A.U., Aruppukottai for cultivation under rainfed conditions. It has good bearing habit. The average fruit weight is 182.2g, pulp weight 11.13g per fruit and TSS 27.9^0 brix.

Propagation

Custard apple is propagated by both through seeds and vegetative methods (Sammddar and Yadav, 1970). For multiplication of custard apple plants, stooling in plants was also reported by Rathore (1976) but success percentage was low.

Seed propagation

Seeds propagation is most common method (Hayes, 1953) of the annona. The seeds viability lasts for 3-4 years. Seed germination in nature is low being about 30-40%. However, fresh seeds germinate better when soaked for 24 hrs and resulted in highest germination of 71 percent (Ratan *et al.* 1993). Hard seed coat can be softened either by soaking the seeds in water for 2-3 days of keeping them under running water for 50-70 hours. The scarification of seeds enhances germination. Treating seeds with GA_3 at 5000 ppm increased germination (Smet *et al.*1999). Seeds are sown 2cm deep either in nursery beds or in pots under partial shade. Regular watering is necessary to maintain good soil moisture. Seeds are slow to germinate and take 3 weeks. Nevertheless it may extend to as long as 8-10 weeks. When seedlings are 10-12 cm tall, they are transferred to pots or plastic containing sand and peat or equal parts of garden soil, sand and decomposed farmyard manure. The 30cm tall seedlings become ready for transplanting in the field.

Seed originated plants are not true-to-type, lack of precocity and vigour, whereas grafting or budding helps largely to overcome to these drawbacks. *A squamosa, A. reticulata, A. cherimola and A. atemoya* are grafted or budded on their own species and each other. *A. muricata* can be grafted on *A. reticulata* and *A. glabra.* However, *A. reticulata,* which promotes vigour and shows good graft congeniality, is commonly employed as a rootstock for most of the annona (Khan and Rao, 1953).

Vegetative propagation

Custard apple is multiplied through vegetative methods like budding, soft wood grafting and also by tissue culture technique. Improved variety, true to type plants can be multiplied through vegetative methods (Satyanarayana, 1962). Budding and soft wood grafting is commonly used for plant multiplication.

Budding

In general, 18 months old or 30cm tall plants having pencil thickness are selected for budding or grafting. Scion of well-matured wood from which the leaves have dropped at the end of the dormant phase is used to graft either by veneer or cleft technique. Shield or T-budding carried out in spring (February-March) is suitable for south Indian conditions. Patch and chip bedding are other methods. Large buds, about 4 cm lengths are collected from 1-year-old wood after the leaf drop give good success. Since, the propagation by cutting and air-layers gives poor results, they are not widely practiced.

Soft Wood Grafting

Propagation studies were undertaken to find out best method viz. bud take percentage, soft wood grafting success, rooting of cutting and air layering in annona (*Annona squamosa* L.) cv. Balanagar. It was recorded that soft wood grafting was the most successful method for commercial propagation of Annona. Maximum graft take of 58.37 percent was observed in cv. Balanagar fallowed by budding (33.54%) (Table 2). Per cent success in propagation was least in air layering (8.68%). Soft wood grafted plants showed maximum increase in the growth parameters compared to budded plants six months after grafting (Hiwale *et al.*, 2010). Gholap *et al.*, (2000) also reported maximum success has achieved in softwood grafting fallowed by budding.

Table 2 : Effect of method of propagation on per cent success in Custard apple cv. Balanagar

Methods of propagation	Per cent success
Softwood grafting	51.74
Patch budding	33.54
Cutting	13.62
Air layering	8.68
SEm+	14.09
CD 5%	58.57

In another experiment on comparative performance of time (January to June) and method (Soft wood grafting and budding) of propagation revealed that soft wood grafting in the month of March gave the maximum success of 59.37 percent, compared to 41.20 percent in budding in the same month, when leaf fall starts. Least success was recorded in the month of June (9.57 and 7.0 %)in both the methods (Table 3). Growth of the plant one month after softwood grafting / Budding revealed maximum growth in softwood grafted plants. The vegetative propagation of Annona by softwood grafting method in the month of March was found to be the best (Hiwale *et al.*, 2010) under semi-arid conditions.

Table 3 : Effect of time of Grafting/Budding on per cent success in Custard apple cv. Balanagar.

Month	Success (%)	
	Softwood Grafting	**Budding**
January	28.45	17.4
February	41.72	33.07
March	59.37	41.20
April	30.27	27.17
May	12.03	9.17
June	9.57	7.02
SEm+	0.988	7.90
C.D. at 5%	2.96	23.83

Micro-propagation

In case of *Annona*, the commercial production of plants through tissue culture is not so successful. However, multiple shoot production from leaf explants of seedlings and root initiation from shoots are successful. Limited work has been carried out on micro-propagation in custard apple. Multiple shoot formation was induced from excised leaf explants of *Annona squamosa* Linn. (Custard apple) seedlings on a Murashige and Skoog basal medium containing benzyl aminopurine (BAP) and kinetin. Various auxins in combination with the above medium produced callusing of the explants. In an investigation of environmental factors affecting shoot induction it was seen that the maximum number of shoots was obtained using the leaf base with petiole at a temperature of 27°C and a light intensity of 1000 lux. Roots were initiated erratically when individual shoots were treated with an auxin and then transferred to an auxin free medium. The process of the development of adventitious buds in leaf culture was analysed histologically by Nair *et al* (1984). For axillary shoot proliferation nodal explants from field grown mature plants showed the best results on MMS medium containing 0.2 mg/l BA (6- benzyl adenine) + 0.1 mg/l NAA (α- naphthalene acetic acid). The highest degree of axillary shoot proliferation from nodal explants was found 80% with 6.0± 0.42 shoots per culture. Root forming performance of IBA (Indole-3-butyric acid) was found to be the best among three auxins (NAA, IAA/ Indole-3-acetic acid, IBA) tested. The highest percentage of root formation was obtained with 2.00± 0.21 roots per cutting on the medium containing 0.1 mg/l IBA. At the high concentration (0.1 - 0.2 mg/l) of all auxins callus formed from the cut margin of the micro cuttings. The regenerated shoots with well developed roots were gradually acclimatized and successfully transferred to the soil under natural conditions (Muhammad *et al*, 2002).

There is no quality standards fix for custard apple plants however; plants should be healthy, disease free of appropriate age and improved varieties for planting.

References

Anonymous, 2011. Improvement of Arid Zone Fruits Biennial Report, Group Workers Meet, AICRP on Arid Zone Fruits, CIAH, Bikaner pp. . 87-96.

Ahmad, M.S. 1935 Pollination and fertilization in *Annona squamosa* under dry climatic condition in Egypt. Hort. Abstract 4:76

Chundawat,B.S.1990.Arid Fruit Culture, Oxford IBH Publication, New Delhi.

Gholap, S.V., DOD, V.N., Bharad, S, G. and Wankar, A.M. 2000. Crop research (Hisar), 20:158-159.

Hayes, W.B. 1953. Fruit Growing in India, 3rd ed. Kitabishtan Allahabad, India

Hiwale, S.S. 2002. Annual Report, Central Institute for Arid Horticulture, Bikaner pp,12-13.

Hiwale, S.S., More, T.A .and Bagle, B.G. 2010. Vegetative propagation in Annona. (*Annona squamosa* L) In. National conference on Production of quality seeds and planting material –Health management in horticultural Crops at New Delhi from 11 - 13 March 2010,pp.26.

Karale, A.R. Kesar. B.G. Dhawale, B.C. and Kale P.N. 1989. The influence of tree condition at the time of hand pollination as a factor on fruit set in Annona species. Maharashtra J Hort. 2:114-116.

Khan, K.F. and Rao I.K.S. 1952. Indian J. Hort.10:140-144.

Limaye, S.P. 1966. Hand pollination studies in Atemoya. Agril. College Mag., Parbhani. 2 (1) 41-42.

Muhammad, N. Amin , Nahar,Kamrun,Firoj, Ahmed and Ahmed Shehreer, 2002.Micro-propagation of *Anona squamosa* Linn. Using explants (Shoot tip and nodes) of field grown mature plants, Pak. J.Bio. Sci.,5(4):394-397.

Nair, S.;Gupta, P.K., Shurgurkar, M.V.and Mascarenhas, A.F. 1984. *In-vitro* organogenesis from leaf explants of *Annona squamosa* Linn. Plant Cell Tissue Organ culture, 3 : 29-40.

Rajput, C.B.S 1985. Custard apple, Fruits of India: Tropical and Sub-tropical. Naya Prakash Publication, Calcutta pp: 479-488.

Rathore, D.S. 1976. Preliminary studies on the stooling of custard apple (*Annona squamosa* L.). Indian J. Hort. 33: 244-45.

Ratan, P.B., Reddy, S.E., and Reddy, Y.N., 1993. South Indian Hort., 41: 171-173.

Samaddar, H.N. and Yadav, P.S. 1970. A note on vegetative propagation of custard apple, South Indian Hort. 18 (1&2): 47-49.

Satyanarayana Swamy, G. 1962. Vegetative Propagation in Custard apple and Atemoya. *Fruit nursery practices in India*, ICAR, New Delhi, pp. 134-39.

Smet, S.de.,Damme,P.Van,Sheldman,X and Romero.1999. Acta Horticulture, 497:269-288.

Thakor, D.R. and Singh, R.N. 1965. Studies on floral biology of Annona. Indian J. Hort. 22(3-4): 238-53

Venkataratnam, L. 1979. Floral morphology and blossom biology on some Annona. Indian J. Agric Science, 29 (4): 68-76.

Chapter-17

Tamarind (*Tamarindus indica* L.)

J. Suresh, D. Sarla Devi, N. Kumar and S. Anbu
Horticulture College and Research Institute, Tamil Nadu Agricultural University, Periyakulam – 625 604

Tamarind (*Tamarindus indica* L.) is an important tree for social forestry, agro-forestry, wasteland development and dry land horticulture. The tree assumes great significance due to its various uses and capacity to withstand adverse agro-climatic conditions. The pulp of the mature ripe fruit has considerable export value in many parts of the world. The fruits are used for making processed products.

It is a common tree throughout the warmer parts of India, except where frost occures *i.e.* sub-Himalayan tracts and North-West Punjab. Rough estimates indicate that about 80,000 metric tonnes of tamarind pulp per annum is produced in India, bulk of which comes from the states of Tamil Nadu, Andhra Pradesh, Maharashtra, Karnataka, Kerala, Orissa, Madhya Pradesh and to a limited extent from Bihar. India is perhaps the main producer and consumer of this fruit in the world. In Maharashtra, the area under tamarind crop is around 13,186 hectare.

Nutritive Value and Uses

It is the fruit with low water content and high levels of proteins, carbohydrates and minerals. The ripe fruit pulp is composed of 20.9 g moisture, 3.1 g protein, 67.4 g carbohydrate, 5.6 g fibre, 0.1 g fat, 2.9 g total minerals and 230-283 k cal/ 100 g energy value per 100 g edible portion (Anon., 1976). According to Chauhan *et al.* (1991) the mineral composition of fruit pulp (on dry weight basis) of common (Brown) and Red type is given in table 1.

Table 1 : Mineral composition of two types of tamarind.

	Per cent						ppm						
	P	K	Ca	Mg	Na	S	Fe	B	Cu	Zn	Al	Sr	Ba
Brown	0.13	0.81	0.20	0.10	0.12	0.02	54	10	11	18	21	34	3
Red	0.19	0.82	0.52	0.24	Low	0.03	48	13	14	25	32	47	4

The tamarind is considered to be the sweetest and the sourest of all fruits due to exceptionally high sugar and acid content in the ripe stage. The total sugar content of tamarind fruit is about 61 per cent of fresh weight (Hasan *et al.* , 1978) stated that ripe fruit pulp is fairly rich in sugar content. Total sugars are 18.2 per cent while free sugars are 17.6 per cent of the pulp. The free sugars constitute of D-glucose (47.7 %), D-mannose (24.5 %), D-maltose (20.4 %) and D-arabinose (2.8 %). Amine sugars were absent.

It is the most acidic of all fruits with a total acidity range varying from 12.2 to 23.8 per cent as tartaric acid (Ulrich, 1970). The unripe (green) fruit is a better source of free tartaric acid than the ripe one (Iyengar, 1938). In ripe pulp, almost half of the tartaric acid is in combined form mainly as potassium bitartarate (cream of tartar) and to a lesser extent as calcium tartarate. The ascorbic acid content of tamarind is extremely low, varying from 2 to 20 mg/100 g (Lefevre, 1971). The fruit also contains about 2 per cent of other acids chiefly malic acid and minor amounts of lactic, oxalic, succinic and citric acids (Kalyankar *et al.*, 1952; Singh, 1973). The pH of the pulp is 3.12 - 3.22 . Of the total nitrogen in the pulp, 55 per cent is present as non-protein nitrogen and of this 70 per cent is free amino-nitrogen. The chief amino acids identified are proline serine and pipecolinic acid (Lakshminarayan Rao *et al.*, 1954). Other free amino acids of tamarind pulp are b alanine, Phenylalanine and leucine (Hasan *et al.*, 1978). Tamarind pulp is also unique in its high mineral and vitamin content (Anon., 1976 and Anon., 1979 a) as given in table 2.

Table 2 : Minerals and vitamin contents of ripe tamarind pulp

Minerals (mg/100 g fresh weight)		**Vitamins (values/100 g fresh edible portion)**	
Calcium	170	B carotene (μg)	60
Phosphorus	110	Riboflavin (mg)	0.07
Fe	17	Niacin (mg)	0.7
Mg	41	Thiamin (mg)	0.44
Cu	0.20	Vitamin C (μg)	3
Mn	0.55		

Tamarind pulp: The sweet acidic pulp is used in the Indian medicine as a refrigerant, carminative, antiscorbutic and laxative and is also commonly prescribed in febrile diseases and bilious disorders. In India, the pulp is used mainly for culinary purposes; while in other countries, it is processed into nectar, fruit punch, juice, sauce, concentrates, glaced and crystalized fruit. It is the fruit base in the preparation of jams, jellies, ice-creams, sherbets, beverages including carbonated drinks. Pulp contains 1.27 per cent pectin.

Pulp powder: Pulp powder is reported to be rich in calcium (74 to 143 mg%), potassium (23.8 to 27.7 mg %) and copper (1.2 to 3.4 ppm) (Manjunath *et al.*, 1991).

Seed: Tamarind seeds (locally termed as *chinchoka*) contain 63 per cent starch, 16 per cent protein and 5.5 per cent oil. The nutritive value of kernel protein comprises favourably well with cereal protein. Products from the seeds have many industrial applications. Tamarind kernel powder (TKP) is extensively used for starching of cotton yarns, jute fabrics and woolen. Seed powder also finds use in jelly, fortified bread, biscuits and in the confectionery industry for manufacturing chocolates.

Juice concentrate: The industrial method used in India for the manufacture of tamarind juice concentrate is based on the extraction of all soluble from the fruit pulp with boiling water. The juice concentrate is prepared by clarification of the fruit extract, concentration of the clear extract under vacuum to 65 68 per cent total soluble solids and filled hot into sterile grass containers (Nagaraja *et al.*, 1975). This product sets to a jam like consistency upon cooling. The recommended tamarind concentrate specifications were 12 per cent minimum acidity as tartaric acid, acid-insoluble ash not to exceed 0.8 per cent, water insoluble solids not to exceed 3 per cent, 65 per cent minimum TSS and NaCl content not to exceed 1 per cent. Jaleel *et al.* (1980) used pectic enzyme concentrate (PEC) for de pectinizing the pulp. Even at PEC doses of 1 per cent almost maximum extraction is reached. From 5 kg of tamarind pulp, 3.46 kg of free flowing concentrate was obtained which had the following characteristics: ^{0}Brix 75.0, acidity 18.1 per cent, pH 2.2.

Pulp powder: For the production of dehydrated "tamarind powder" with a free flowing nature, tamarind pulp is manually cleaned, deseeded and fibre is removed. The pulp is processed under standardized conditions of temperature, humidity and mill settings in order to obtain an acceptable and hygenic product in powder form possessing good reconstitutional characteristics. The product can be packed in HDPE pouches of desired unit packings.

Beverage: Benero *et al.* (1972) prepared a tamarind drink containing 9 to 12 per cent pulp and adjusted to 21.5 °brix soluble solids level with sugar. The ranges in composition were 20.5 to 21.4 °brix, 0.35 to 0.44 per cent acidity (as tartaric acid), 19.29 to 20.86 per cent total sugars and a pH ranging from 2.80to 2.96. The drinks were pasteurized at 85°C, canned in plain tin containers, cooled, dried and stored at 29.4°C. After 1 year of storage, the drinks were found acceptable by a taste panel. Utilization of tamarind fruit wine for the preparation of vermouth, an ascorbic fruit beverage has been reported (Lingappa *et al.*, 1993).

Tamarind kernel powder: The powder commercially known as TKP finds extensive use as a sizing material in the textile industry. When the powder is boiled, gruel is obtained which on drying in thin layers produces strong smooth continuous and elastic films - properties useful for sizing. According to available data, approximate 20000 tonnes powder is annually produced. The TKP is liable to deteriorate on long storage particularly in most areas. It should therefore be stored in a dry place in moisture proof containers and suitably fumigated at regular intervals. The powder may be mixed with 0.5 per cent of sodium bisulphite before packing to prevent enzymic deterioration.

Leaves are regarded as a good fodder with 12-15 per cent proteins (Laurie, 1945). Leaves are a good source of calcium. Reddish yellow dye is commercially extracted from leaves. Leaves can be tried in green manuring of soil. The bark is given in diarrhoea in lotions and poultices. Bark extracts reduce mosaic virus infection. Timber is invariably used for making agricultural implements, constructional work and furniture. The testa contains colouring agents and water soluble tannins to the extent of 20 to 40 per cent. The use of testa in dyeing and leather tanning industry has been reported. The utilization of testa as a raw material for preparation of plywood and paper adhesive has also been suggested.

The powder obtained by pulverizing deshelled tamarind seed is used in the manufacture of industrial gums and adhesives. Seed powder if added to soil at 0.125 to 1 per cent improves soil structure, water holding capacity and availability of N (Varade and Badhe, 1969). The amber (dark brown) coloured fatty acid (6 to 8 %) obtained by solvent extraction of deshelled tamarind seed powder is reported to be useful in preparation of paints, varnishes, soap making for burning lamps and also for edible purposes after refining. Oil is also used to give shiny finish to toys. The fatty oil naturally present in kernel powder imparts softness to yarn during weaving and knitting.

Botanical Description

The genus *Tamarindus* is a monotypic one containing the only species *indica,* belongs to sub-family Caesalpiniaceae of the family Leguminoseae. Cytogenetical studies revealed that the chromosome number of *T. indica* is 2n = 24 (Anitha Karun, 1985). There are several types of tamarind. They can be broadly grouped under three types based on fruit size and shape, colour of pulp, taste etc. as follows:

On the basis of fruit size and shape

Bailey (1947) recognized two types of tamarind.

(i) *East Indian type* having long pods with 6-12 seeds.

(ii) *West Indian type* having shorter pods containing 1 to 4 seeds.

Tamarinds in the America are of the shorter type. Paulos (1975) recognized a tamarind type known as "Valakatchi" which bears long and rectangular pods as against some other types, which produce short and cylindrical pods. Shivanandam (1980) and Karale (1998) reported that tamarind pods vary considerably in size and shape. Many are sickle shaped while some have straight long pods (16 to 22 cm) while others were with short pods. Seeds also exhibit a wide range of variation in shape size, colour and the ornamentation of the seed coat.

On the basis of pulp colour

There are two distinct phenotypes of tamarind based on mainly their pulp colour.

(i) *The yellow or brown pulp type* turning dark brown on storage. It is harvested after full maturity.

(ii) *The reddish pulp type* is locally known as "Raktichinch". The term red type covers the various shades of pink-pulped fruits. The red type is sweeter than brown (common) type because it has lower content of free acids and is generally harvested when fruits are immature and green. It is mostly preferred for making preserves. A high yielding Red type (Yogeshwari) has been released by Marathwada Agricultural University, Parbhani, and Maharashtra.

On the basis of organoleptic taste

The cultivated types could be broadly classified into two groups namely sweet and sour types.

(i) **Sweet type**: The ripe fruits of this type have sweeter pulp coupled with less acidity and fruits are mainly used for dessert purpose (Karale, 1998). Makham Waan, Secthong breed, Manila sweet are few cultivars found mostly in South East Asia (Thailand). A sweet type *viz.*, No. 263 has been located by MAU, Parbhani.

(ii) **Sour type**: It is highly acidic in taste and pulp is commercially marketed. The tamarind variety 'Pratisthan' released by Fruit Research Station, Aurangabad is a sour type and is reported to constitute of 61 per cent pulp, 12 per cent seed and 27 per cent shell (Anonymous, 1984). The sour type selections 'Urigam' and 'Cumbum' (good yielder) are popular in Tamil Nadu. A high yielding sour type PKM-1 (Periyakulam-1) has been released during 1992 from Horticulture Research Station, Periyakulam, Tamil Nadu (Kumar *et al.*, 1993). This cultivar is preferred for its early bearing habit and claimed to be suitable for high density orcharding (160 plants/acre against 40 plants/acre under conventional planting).

Tamarind is highly heterozygous in nature exhibiting large genetic diversity in bearing, pulp colour, pod shape and size and fruit quality i.e. sweetness, acidity, flavour, taste etc., in tamarind growing areas. Such innumerable types among seedling trees offer good scope for improvement of this fruit crop by seedling selection method. It involves identifying a genotype of good productivity coupled with good fruit quality.

Flowering: In tamarind tree, blossom bud differentiation begins with the resumption of vegetative growth just after leaf fall and flowers appear from April to July in southern, northern and western India. In the eastern parts of the country, new leaves appear in May and flowering continues upto September. The flowers are borne in small terminal drooping racemes on the current season's growth. The time taken for complete development of flower bud from its visible initiation to anthesis varies from 16 to 20 days (Bajpai *et al.* 1968;). Nicodemus *et al.* (1997) reported that a single inflorescence bloomed in a period of 12 to 19 days and the number of flowers opening per day per inflorescence was either zero or one and rarely two. A visibly differentiated bud was observed to take approximate 15 days to develop into a flower. The length of inflorescence and number of flowers/inflorescence varied from 7 to 14.2 cm and 20 to 41 respectively in sweet pulp type while it varied from 4 to 17 cm and 10 to 36 respectively in Red pulp type tamarind.

The anthesis start in the morning hours as early as 4.30 A.M. and continues upto 8.30 A.M. with a peak of anthesis at 6.30 A.M. and after 8.30 A.M. there was no anthesis. The anther dehiscence commence after anthesis i.e. at 10.30 A.M. and continues upto 11.30 A.M. only. The stigma is receptive a day prior to anthesis and two days after anthesis. The presence of clesitogamy has been recorded (Anitha Karun, 1985).

The shape of the pollen is spherical with an average diameter 1.5μ. The pollen fertility was 66.7 per cent in 8 per cent sucrose solution containing 1 % agar (Prasad, 1963). Acetocarmine stainability was 96.54 per cent (Thimmaraju *et al.* 1977).

Maximum pollen germination of 60 per cent occurred in 15 per cent sucrose solution whereas in control with distilled water, the pollen grains did not germinate (Anitha Karun, 1985). Nicodemus *et al.* (1997) reported that pollen fertility was high (87.2 %) and *in vitro* pollen germination was 90-95 per cent when the basal medium was supplemented with 10^6 M of kinetin. Pollen grains were viable upto 12 hrs at room temperature under ambient temperature and upto 30 days under cold storage (4^0C).

Pollination: Tamarind is a protogynous and highly cross pollinated crop. Pollination is reported to be carried out by flies, bees and red ants. Since the pollen production is small in quantity, the importance of insects over wind as pollinating agents is emphasized. Nagarajan *et al.* (1996) reported that flowers showed strong entomophilous adaptation, open pollination fruit set was between 1 and 2 per cent. Controlled pollinations indicate that tamarind is a predominantly out crossing species with extremely low level of selfing, apomixis was absent. Long term pollen storage was possible, variation in pollen size was observed and pollen viability was very low. Fruits showed colour dimorphism.

Fruit set

The fruit set is the most important factor, which determines the pods yield. Though the flowering is profuse in tamarind, the fruit set under natural conditions is low ranging from 3 to 5 per cent, which can be increased to about 50-55 per cent under artificial cross-pollination. Pollen limitation, floral abnormalities (Karale *et al.,* 1997a) are some of the reasons for low fruit set under open pollinated condition. The higher fruit set in controlled cross-pollinated flowers as compared to selfing indicated the existence of partial self-incompatibility in tamarind (Usha, 1987). The fruit set was found to be increased when the number of pollinators (honey bees) visiting the tree was high. Ant species *Oecophylla smaragdina* Fab. and *Camponotus compresses* were also found in large populations collecting pollen (Nicodemus *et al.*, 1997).

Fruit Growth and Development

After completion of pollination and fertilization, a series of physical and physiological changes take place in the fruit that results in the final differentiation of peel and pulp in the pericarp which starts from the 10th week after fruit set. According to the morphological changes taking place in the fruit from anthesis until harvest, three distinct stages (growth, maturation and ripening) were distinguished. Usha (1990) reported that tamarind fruit took about 252-280 days from fruit set to harvest showing three distinct stages i.e. pre-maturation, maturation and ripening.

During the pre-maturation (growth) period of 22 weeks, which corresponds to 63 % of the total period where maximum cellular activity took place. The colour of the fruit remains green while the colour of the pulp was greenish white.The second phase, maturation, lasted 20 % of the total growth period. During this stage of growth, the colour of the peel was changed from green to brown due to lignification and formation of stone cells. Maximum increase in fruit weight and fruit size was noticed during this stage.

The last phase, ripening, started about 29 weeks after anthesis and lasted 17 per cent of total growth. The significant changes during ripening were natural separation of the peel from the pulp, steep reduction in moisture content of the fruit which cause large increase (40 %) in the percentage of total sugars giving this sour fruit as sweet in taste and two fold increase in acid content due to effect of concentration.

Propagation

Tamarind is traditionally propagated through seeds which exhibits considerable genetic variation as it is a cross pollinated crop. Hence, the seedlings do not produce true to type tress. Further, prolonged juvenile phase has also been observed in plantations due to predominant seed propagation (Karale *et al.*, 1997). The vegetative propagation of superior genotypes is necessary for shortening juvenile phase, production of trees uniform growth habit and quality yields.

Seed propagation

Tamarind is normally grown from seed. Seeds should be collected from high yielding, pest and diseases free trees with well - formed, fully ripe pods, although they may not come true due to out crossing. Ripe pods should be collected from the trees in March – April by shaking the branches or by hand – picking. The pods should be dried in the sun and the seeds removed from the pulp by hand – kneading or washing in water. Washed seeds are then dried in the shade and stored in well – ventilated gunny bags (or) paper bags, in a cool place (Gunasena and Huges, 2000).

The seeds of tamarind are ex-albuminous and consist of an outer hard brown testa. The behaviour of tamarind seed is orthodox (Riley, 1981 quoted by Hong *et al.*, 1996). The seed retain its viability for several months if kept dry. Under field conditions seed life can be extended for up to or more than a year if the seeds are well dried, mixed with sand and kept in air tight containers. (Albrecht 1993 quoted by Hong *et al.*, 1996) reported that seeds could be stored for several years in air tight packs at 10 °C and 7-15% moisture content.

The seeds normally begin to germinate within 5-10 days following sowing and may take a month to complete. As tamarind seeds are hardy in nature due to the presence of hard seed coat. By which, delay and poor germination can occur. The overage tamarind seeds begin to germinate in about 10 days after sowing. Germination percentage was more in heavy or bigger size seeds compare to that of light seeds (Masano, 1994 and Parameswari *et al.*, 2002).

The germination percentage of fresh or well as stored seeds vary from 30 -70%. Generally no pre treatment is found necessary for germinating the seeds (Oza, 1978). However it has reported that the acid scarified seeds perform better than the non-scarified seeds. Studies carried by the Institute of Forest Genetics and tree breeding, Coimbatore (1993) revealed 59% germination in seeds that were subjected to scarification for 5 min in conc. H_2SO_1 and 96% germination with 10 min in H_2SO_4. Sankanarayanan *et al.* (1994) proved that seeds soaked in cow urine (82.8%) and cow dung solution increased germination (72.6%) rates considerably. Duration of acid scarification varies with the size of the seeds. Bigger sized seeds require 20 min. Of acid scarification with commercial H_2SO_4 @200 ml/kg of seeds, while the duration was reduced to 15 and 10 min. With medium and smaller sized seeds, respectively (Parameswari *et al.*, 2001). Besides scarification, the influence of certain plant based extract has also been reported. The scarified and surface dried seeds, pelleted with Arappu (*Albizzia amara*) and pungam (*Pongamia pinnata*) leaf mixture (1:1) at the rate of 200-300 g/kg of seeds using 10% maida @ 200 – 300 ml/kg as adhesive registered the highest germination (93%) and seedling growth at nursery (Parameswari *et al.*, 2001).

The scarified seeds treated with chlorine or iodine based halogen mixture @ 4g/kg of seeds maintained their viability and germination potential up to 9 months of storage under ambient condition, but none of the treatment were effective at 12 months of storage period due to insect infestation. Mechanical scarification has helped to improve the germination. Masano (1994) reported that slicing off the end of the seed was the best, with 92% germination compared to the control (82%). Slicing the seed also hastened germination. Germination began at seven and was completed at nine days. Sowing immediately after extraction or cleaning could enhance seed germination. Scraping the seed coat by rubbing on a rough surface also speed up the rate of germination (Megehembe, 1995).

Tamarind seeds collected from high rainfall zone produced elite seedlings and rootstocks for vegetative propagation (Parameswari *etal.*, 2002). Seeds collected from various sources are often contaminated with fungi after collection and deteriorate during storage. Sometimes seeds become contaminated with pathogens while still in the pods on the tree. These seed borne pathogens can also affect germination and subsequent growth.

Seed dressing with carbendazim or copper oxychloride (3g/ kg.seed) have been recommended for control of seed borne fungi (Purohit *et al.*, 1996). Sand, well rotten FYM, cow dung manure and their combination proved suitable media for higher germination, better growth and vigour of seedlings (Chattopadhayay and Mohanta, 1998).

The best medium for seed germination is sand or soil mixed with cow dung. Chatopadhyay and Mohanta (1998) reported that using cow dung and sand in the propagation medium could encourage seed germination. However, a normal nursery potting mixture containing three parts of topsoil, one part of sand and one part of compost, can be successfully used for germinating tamarind seed.

Seeds can be germinated in nursery beds, seed boxes, pots or plastic bags. When grown in nursery beds, the recommended spacing is 20 to 25 cm in both directions. The best germination results if 1.5 cm of loose sandy loam or a mixture of loam and sand covers the seeds. If the seeds are planted too deep, seedling emergence is delayed and there may be some rotting due to poor aeration. Seeds may be sown in polyethylene nursery bags provided the bags are deep enough to accommodate the tap root without causing undue distortion. Bag raised seedlings should be planted before they get root bound. If there is a need to delay seed sowing, well-dried seeds packed in a polythene bag, can be stored in a refrigerator for about a month (Webster, 1924). Seedlings grow rapidly in the early stages and produce a long taproot, which may reach 30 cm or more within two months of germination (Troup, 1921). Seedlings should attain a height of at least 80 cm before being transplanted to the field at the beginning of the rainy season. Under normal conditions seedlings are large enough to be planted in the field within a year.

Vegetative propagation

Vegetative propagation is preferable to seed propagation, as the latter does not produce true-to-type progenies. Vegetative propagation has the advantage of producing true-to-type progeny, which can be taken from selected, superior, mother trees. Tamarind can be propagated vegetatively by many methods. Proven methods include stem cutting, shield and patch budding, air layering and grafting (inarching).

Cuttings

The easiest and cheapest method of propagating tamarind is by stem cutting. Mascarentas *et al.* (1987) reported that vegetative propagation through rooting of cuttings was unsuccessful for tamarind. A few other reports however claim efficient rooting response from stem cuttings. Soft wood cutting of shoots bearing new flushes of fully turgid leaves are collected in the morning. They are dipped in IBS 1000 ppm for 10 seconds and in 50% isopropyl alcohol, planted in polythene bags containing media and placed in mist chamber with 70-80% humidity. IBA increased the rooting of cutting to over 94% compared to control, which recorded only 25%. The time taken for rooting was early (10-15 days) in IBA treated cuttings compared to control (40 -50 days). Softwood cuttings of 15 – 20 cm long taken 1-2 year old branches can also be rooted well (Swaminath *et al.*, 1990).

No rooting was observed in hardwood cuttings. The etiolated semi hardwood cuttings treated with 1000 ppm IBA kept under mist chamber with 75 – 80 RH registered the highest rooting percentage (19.00) followed by soft wood cuttings treated with 1000 ppm IBA (11.25). Similarly etiolated IBA treated semi hard wood cuttings planted in June recorded the highest percentage of rooting (20-60) as well as survival percentage (52-80) followed by September and October planted cuttings (Navaneetha *et al.*, 1990).

Layering

Three methods of layering adopted in tamarind are (i) ground layering (ii) air-layering or 'gootee' and (iii) mound layering.

In ground layering, a flexible branch is bent down and pegged to the ground, and the point of contact covered with soil. A small cut is made on the lower side of the stem where it touches the ground to impede sap circulation and encourage rooting. After 3-6 weeks, roots form at the point of contact with the soil and new plant can be obtained by severing the branch above the place of rooting.

In air layering, a branch is selected and the bark is removed to a length of 2-5 cm. This area is covered with moist soil mixture or a coir fibre dust, and kept in position by wrapping with polythene film. It is kept moist for 2- 3 months. When large quantities of roots are formed, the branch with its roots is severed from the parent tree and potted up for latter planting. The girdling of etiolated shoots reportedly stimulates air – layered roots to be produced within 10 weeks, and the process can be further accelerated through the application of IBA (Troup, 1921). The effect of rooting hormone has been well documented in air layering of tamarind.

Bhore (1986) reported that air layers made in brown pulped (sour) type gave 60 percent success during September by applying 4000 ppm IBA in lanolin followed by IBA 5000 ppm (54 % success). In another experiment with red pulped type, air layers with IBA 4000 ppm gave 50 per cent success followed by IBA 5000 ppm (40% success). Navaneetha *et al.*(1991) reported that air layers made in July with IBA 1000 ppm resulted in higher percentage of rooting and survival (24.38 %), followed by 5000 ppm of IBA in the same month. Gowda (1983) observed that girdling and etiolation and treatment with IBA or IBA + NAA increased the number and length of roots in air layers and reduced the time required for rooting from 10-6 weeks. Ganpathi (1989) reported that rooting percentage in air layer of tamarind increased (86.67%) with IBA at 2500 ppm. Patil *et al.* (1997) obtained 75 per cent success by treatment with 200 ppm IBA during June- July followed by August layering (60%) with 1000 ppm IBA + NAA treatment. Lokhande (1978) cautioned that air layered plants have comparatively weaker root system than seedling plant and the field success of air layered plants need utmost care.

Mound layering

The technique of mound layering consists of cutting down the selected tree to the ground level. Within two or three months, many buds start sprouting from the ground level. These are filled with sand and soil up to 15 – 20 cm and after 30- 40 days the shoots which turn from green to brown are cinctured and treated with root inducing hormone like IBA and again covered with sand. The cinctured shoots produce roots in 30 – 40 days. The number of layers that can be obtained by mound layering is low. In mound layering studies, highest percentage of rooting was recorded in 2000 ppm IBA treated shoots (80 %) followed by 2500 ppm IBA (70%) (Nache Gowda,1997).

Budding

The technique involved in budding is selection of a matured shoot from previous year's growth and the buds should be prominent in the leaf axils. The tamarind seedling used as rootstock is prepared by removing the side growth upto 20 – 25 cm from the base. The budding may be done either by shield or 'T' or by the ring and patch methods. Pathak *et al.* (1991) obtained 96% take in patch budding and 94% success in modified ring budding on a nine month old rootstock. Patch budding is a suitable method for large scale multiplication of tamarind. While working under Rahuri conditions on tamarind Karale (2001) reported maximum success of 63 per cent through patch budding in mid July. Budding done in mid August required considerably longer time for sprouting.

Grafting

Compared to other propagation methods, grafting is laborious. Different grafting techniques have been tried in tamarind with different degree of success. Vegetative propagartion by whip grafting, approach grafting, veneer grafting and softwood grafting are reliable methods, although they are expensive and time consuming.

Purushotham and Narasimharao (1990) reported 49 per cent success in **veneer** grafting during 2nd fortnight of December. For veneer grafting, rootstocks about 6 months old and of uniform size were selected. They were defoliated before grafting. Grafting was done 8 cm high on the rootstock. In green wood cleft grafting, maximum graft success was recorded during May (80%) followed by April (76.7%) and June (70%) and least success (26.7%) was observed in February month of grafting (Nache Gowda, 1997).

In recent years more efficient technique *viz.*, softwood grafting has been standardized and it is commercially employed in Maharashtra. This is by far the best method of propagation of tamarind since it is easier, more economical, gives a higher degree of success and is ideal for *establishing in situ* orchards. For softwood grafting, the rootstock seedlings are defoliated and their tops cut off at 15 cm high immediately before grafting. A vertical downward cut is made in the centre of the cut stem to about 4 cm depth and the scion sticks are cut into wedge shapes, inserted into the stock and wrapped using 2 cm wide 200 gauge polyethene ribbon. Purushotham and Narasimharao (1990) reported that percentage of graft take in softwood grafting was better (68%) and number of sprouts were also comparatively higher (3.22) than veneer grafting. Softwood grafting on 8-10 months old seedling stock (22 to 32 cm height and 0.3 - 0.4 cm diameter) in the month of March - April gave over 75 per cent success (Kulwal *et al.*, 1997). Softwood grafting was shown to be the best grafting method in terms of successful union and survival rates (Purushotham and Narasimharao, 1990).

Softwood grafting in August – September was found suitable under shade which gave up to 60% success (Naik *et al.*, 1997). Nachegouda and Baswanneppa (1997) hinted that out of all the various vegetative propagation methods, approach grafting appeared to be easy, very ideal and successful for propagation of tamarind in view of various advantages in cost effectiveness, easy and higher percentage of establishment and rapid multiplication of trees. In approach grafting, 74% graft success was observed compared to 25% using ordinary grafting (Swaminath and Ravindran, 1989). Attempts were made by Bhore (1986) to propagate tamarind by inarch grafting during November on 2-year-old seedling stock (having stem thickness of about 0.7 to 0.8 cm) and obtained 95% success. At present, approach grafting is one of the commercial methods of vegetative propagation in tamarind in South India.

Top working

Old, inferior seedling trees can be converted into productive ones of a desired superior type by dehorning the main limbs of tree, allowing new soft shoots to produce and using soft wood grafting (Karale, 1998). However, this is seldom practiced commercially.

Micro-propagation

Tissue culture based techniques like micro propagation, are used to complement conventional propagation methods, as these can be helpful in the multiplication and establishment of selected trees on a large scale. Though the practice has been gaining momentum for many forest tree species, only very few reports are available on *in vitro* clonal propagation of tamarind.

Prabakaran *et al.* (2000) reported that among the five different genotypes of tamarind tested for their suitability for multiple shoot induction, the genotype Urigam exhibited the highest percentage of bud break (93.63), with more number of shoots per bud (1.40) and the shoot length was also the highest (3.34 cm) followed by PKM 1, Pollachi-2, Asanoor-H-1and Salem0144. They also found that the explants collected (auxiliary buds) during the month of June showed the least contamination (7.32%) and the highest bud break percentage (84.14).

However, during the month of November, the contamination percentage was the highest (77) and the bud break percentage was also the lowest (45-39).

Micro-propagation of tamarind from hypocotyls and shoot tip of seedlings was reported by Mascarenhas *et al.* (1987). Shoot multiplication was facilitated on MS medium (Murashige & Skoog, 1962) and rooting was accomplished on white's medium (White, 1963) supplemented with cytokinins and auxins. Kopp and Nataraja (1990) described in vitro shoot multiplication methods from seedlings of tamarind. Splittstoesser and Yasseen (1991) observed multiple shoot regeneration from cotyledon nodes when Thidiazuron was used instead of BAP. Jaiswal and Gulati (1991) reported high frequency regeneration from excised cotyledon on MS medium containing BAP, Rooting was achieved on MS with IAA. They further reported multiple shoot regeneration from shoot tips and nodal explants with BAP. Similarly multiple shoot generation and rooting of microshoots was observed in tamarind seedling explants by Rao *et al.* (1997). MS containing 0.5 mg/I BA, 0.5 mg/I KN and 0.1 mg/l IIA was sufficient for promoting 2-3 multiple shoots as documented in their reports. Rhizogenesis, was facilitated on MS containing 0.2 mg/l each of IAA, IBA and NAA. Ganga and Balakrishnamoorthy (1997 b) worked out the optimum in vitro culture conditions for high frequency plant regeneration from seeding stm bits of a tamarind genotype "Urigam".

The maximum response to shoot bud differentiation, the earliest response and the maximum number of multiple shoots per culture were recorded from stem bits cultured in MS supplemented with 2.5 mg/l BA, and 0.5 mg/l GA. Induction of multiple shoots from auxiliary buds (Balakrishnamoorthy and Ganga 1997 a, 1997b) and in vitro rhizogenesis of these micro shoots of tamarind genotype urigam (Ganga and Balakrishna moorthy 1997 a) was also reported. The maximum response to auxiliary bud proliferation was on MS plus BA and GA. MS at half strength plus 0.5 mg/1 each of IA and IBA recorded the maximum rhizogenic response. It was also reported that tissue cultured raised plants showed better growth in height, branching habit, and spread of branches, early flowering and fruiting cycle (ICFRE, 1993 ; Mishra, 1997). Tissue culture raised plants started flowering at an average height of 3.7 m as against 8 m average height of flowering in case of seed raised plants. Hence, they have recommended planting of more number of tissue culture raised plants per unit area, with about 156 plants /ha at 9m × 9m spacing which could substitute the present planting practice of 12.5m × 12.5m spacing. A protocol for *in vitro* regeneration of plants via adventitious bud formation from the mature embryo axis of tamarind was standardized by Mehta *et al.* (2000). Explants consisting of the longitudinal section of the embryo axis with attached cotyledon were cultured in MS medium with various combinations and concentrations of NAA, BAP and sucrose. Induction of adventitious shoot buds was achieved on the cut surface of the axis when cultured in medium containing 2.69 mm NAA, 44.39 mm BAP and 4 per cent sucrose. A medium consisting of 0.91 mm Zeatin, 2.22 mm BAP, 0.41 mm calcium pantothenate and 0.40 mm biotin supported the differentiation of buds to form elongated shoots. The shoots developed roots in a half strength MS medium with 2 per cent sucrose following a 72 hr treatment in the dark with auxin mixture.

Quality Standards of Planting Material

Looking to the potential of export oriented crop, the planting material should be of improved varieties, insects, & disease free. For vegetative propagation proper known root stock should be used. The seedling plants should be healthy having single stem for budding or grafting. For large scale plantation, genuine and true- to -type plants should be used. The plants should be vigorous type and good foliage.

References

Anitha Karun. 1985. Cytomorphology and reproductive biology of tamarind. M.Sc. (Agri.) Thesis. GKVK, Bangalore.

Anon. 1976. The Wealth of India. A dictionary of Indian raw materials and industrial products. Raw maeterials, Vol. XI, CSIR, Pub. New Delhi, Pp. 114-122.

Anon., 1979. Res. Review sub-committee Report, Deptt. of Hort., M.P.K.V., Rahuri.

Bailey, 1947. Tamarind *In:* The Standard Encyclopedia of Horticulture. MacMillan, New York Inc. 3: 3306-3307.

Balakrishnamoorthy, G and M. Ganga. 1997 a. In vitro Induction of multiple shoots from Auxillary buds of the tamrind genotype, 'Urigam'. Proc. Nat. Sym. On Tamarindus indica L., Tirupathi (A.P.) organised by forest Dept. Of A.P., 27- 28 June, 1997, p. 110- 112.

Balakrishnamoorthy, G and M. Ganga. 1997 b. Influnce of season of explant collection on induction of multiple shoots from auxillary buds of the Tamarind genotype, 'Urigam". Proc. Nat. Sym. On Tamarindus indica L., Tirupathi (A.P.) organised by Forest Dept of A.P. 27-28 June, 1997, p. 98- 100.

Bhore, D.P. 1986. Inarch grafting in tamarind. Maharashtra J. Horti., 3: 55- 58.

Chattopadhyay, P.K. and S.K. Mohanta. 1998. Influence of the propagation medium on the germination and development of tamarind seedling. South Indian Horti., 36(6) : 324.

Chauhan, K.S., Pundir, J.P.S. and Shyam Singh. 1991. Studies on the mineral composition of certain fruits. Haryana J. Hort. Sci., 20: 210-213.

Ganpathi, T.1989. Effect of growth regulators on rooting of air layers in tamarind. Mysore J. Agric. Sci., 21:435.

Gowda, N. 1983. Studies on vegetative propagation of tamarind by air layering. M.Sc. (Agri.) thesis submitted to the University of Agricultural Science, Bangalore.

Gowda, N.V. (1997). Propagation of Tamarind by layering and green wood cleft grafting. Abstr. Proc. Nat. Symp. on Tamarindus indica L. 27-28 June. Tirupathi, p. 15.

Gunasena, H.P.M and A. Hughes. 2000. Tamarind. International Centre for underutilized crops, UK, p. 76- 82.

Hasan, S.K., Salam, A. and Wahid, M.A. 1978. A reinvestigation of Tamarind fruit. Pakistan Jour. of Sci., 30 (1-6) : 71-73.

Hong, T. D., S. Linnington and R.H. Ellis. 1996. Seed Storage Behaviour: A Compendium. Handbook for Gene banks No. 4 International Plant Genetic Resources Institute, Rome.

ICFRE. 1993. Tamarind (*Tamarindus indica* L.) Technical bulletin, Forest Research Institute, Dehradun, India, pp.16.

Iyengar, A.V.V. 1938. Isolation of tartaric acid in tamarind. Curr. Sci., 6-7 : 610-611.

Jaiswal, P.K. and A. Gulati. 1991. *In vitro* high frequency plant regeneration of tree legume, *Tamarind indica* (L.). Plant Cell Reports., 10 (11) : 569 – 573.

Jalee1, S.A., Narayana, K., Sreekantiah, K.R. and Ramanathan, P.K. 1980. Use of fungal pectic enzyme in the production of Tamarind concentrate. Indian Fd. Packer. 34(1/2) : 73.

Kalyankar,G.D.,Krishaswamy, P.R. and Sreenivasaya, M. 1952. Papayrographic characterization and estimation of organic acids in plants. Curr. Sci., 21: 210-222.

Karale, A.R. 1998. Status Report on "Improvement of Tamarind by selection" presented at Research Workers Group meeting of AICRP on Arid Zone Fruits held at CCS, HAU, Hisar, 14-16 Oct.

Karale, A.R. 2001. Final Report of Ad-hoc project on Tamarind submitted to ICAR, New Delhi. (Period 1.1.1998 to 31.12.2000).

Karale, A.R., Kaulgud, S.N. and More, T.A. 1997a. Status of Tamarind in Maharashtra. *Abstr. Proc. Nat. Symp.* on *Tamarindus india* L. 27-28 June, Tirupathi. p. 29.

Karale, A.R., S.N. Kavlgud and T.A. More 1997b. Studies of Tamarind in Maharashtra. Proc. Nat. sym. on *Tamarind indica* L, Tirupathi (A.P.) organized by Forest Dept. of A.P., 27-28 June, 1997, pp. 10-12.

Kopp, M.S. and K. Natarajan. 1990. In itro plantlet regeneration from shoot tip culture of *Tamarind indica* L. Indian J. Forestry, 13 (1) : 30-33.

Kulwall, V., S. Taydeg, M. H. Chaudari and R.M. Verma. 1997. Soft wood grafting- an easy method of propagation of tamarind. Proc. Nat. sym. On Tamarindus indica L., Tirupathi (A.P.) Organised by Forest Dept. of A.P., 27-28 June,1997.

Lakshiminarayan Rao, M.V., Subramanian, N. and Srinivasan, N. 1954. Free amino acids in tamarind pulp. J. Sci. Ind. Res., (India). 13B : 337-378.

Lefevre, J.C. 1971. Review of literature on the Tamarind. Fruits, 26: 687.

Lokhande, R.V. 1978. Air layering in Tamarind . M.Sc.(Agri.). Thesis submitted to Marathwada Agri. University, Parbhani.

Magehembe, J.A. 1995. Achievements in the establishment if indigenous fruit trees of the Nimbno woodlands of South Africa. SADC-ICRAF agro forestry project, Zomba, Malawi. Eds. J.A. Magehembe, Y. Nitupanyama, and P.W. Chirwa: 39-49.

Masaano, M.1994. The effect of pretreatment on the germination process of tamarind (*Tamarindus indica* L) Seeds. Bullotia penelition Hulan, 563: 33-41. In:Ed. Gunasena, H.P.M and a. Huges. 2000. Tamarind. International Centre for Underutilized Crops, UK, p.76 – 82.

Mascarenhas, A., S. Nair, V.M. Kulkarni, O.C. Agarwal, S.S. Khushpee and V.J. Mehta. 1987. In Cell and tissue culture in forestry, Vol. 3. J.M. Bonga and D.J. Durzan (Eds.), Martinus Nijhoff, Dordrecht, pp. 316 - 325.

Mehta, V.J., Krishnamurthy, K.V. and Hazra, S. 2000. Regeneration of plants via-adventitigus bud formation from mature zygotic embryo axis of Tamarind (*Tamarind indica* L.). Curr. Science, 78(10) : 1231-1234.

Mishra, R.N. 1997. *Tamarindus indica* L., An overview of tree improvement. Proc. Nat. sym. on *Tamarindus indica* L., Tirupathi (A.P) organized by Forest Dept. of A.P., 27-28 June, 1997, pp. 51 – 58.

Murashige, T. and F. Skoog. 1962. A revised medium for rapid growth and bioassays with tobacco tissue culture. Physiologia Plantarum, 15 : 473 – 497.

Nagaraja, K.V., Manjunath, M.N. and Nalini, M.L. 1975. Chemical composition of tamarind juice concentrate. India Fd. Packer, 29: 17-20.

Naik, K.R., Sulikeri, G.S., Patil, V.S. and Basavaraj, N. 1997. Soft wood grafting studies in Tamarind. Abstr. Proc. National Symp. on Tamarindus indica L. 27-28 June, Tirupathi, p.10.

Navaneetha, N., Palaniswamy, K.P., Khader, Md. Abdul and Kumar, N. 1990. Studies on rooting of cuttings of Tamarind. South Indian Hort., 38(4) : 220-224.

Navaneetha, N., Palaniswamy, K.P., Khader, Md. Abdul and Kumar, N. 1991. Studies on rooting tamarind air layers. South Indian Hort., 39(2) : 102-105.

Purushatam, K., and Narasimha Rao S.B.S., 1990. Propagation of Tamarind by Veneer and Soft wood grafting. South Indian Hort., 38(4) : 225.

Singh, P.P. 1973. Oxalic acid content of Indian foods. Qual. Plant Mater Veg., 22 : 335-347.

Thimmaraju, K.R., Reddy, M.A.N., Swamy, N. and Sulladmath, U.V. 1977. Studies on the floral biology of Tamarind (*Tamrindus indica* L.), Mysore. J. Agric. Sci.,11 : 293-296.

Usha, K. 1987. Studies on pollination biology of tamarind in relation to its productivity. Mysore J. Agril. Sci., 21 : 107.

Venkataratnam, L. 1979. Floral morphology and blossom biology on some Annona. Indian J Agric Science 29 (4): 68-76.

Nache Gowda, V. 1997. Propagation of tamarind by layering and cleft grafting. Proc. Nat. Syn. on *Tamarindus indica* L., Tirupathi (A.P) organised by Forest Dept. of A.P, 27-28 June, 1997, p. 89- 93.

Oza, G.M. 1978. Seed shapes in *Tamarindus indica* L. Indian Forester, 104 (5) 331-332.

Parameswari. K.P. Srimathi, K. Malarkodi and G. Sasthri. 2000. Influence of biocides on the storability of scarified seeds of tamarind. Prog. Hort., 32 (1) : 32- 37.

Parameswari. K.P. Srimathi, K. Malarkodi and G. Sasthri. 2001. Seed pelletization with botanicals for elite seedling production in tamarind. Madras Agric. J., 88 (1-3): 161- 163.

Parameswari. K.P. Srimathi, K. Malarkodi and G. Sasthri. 2002. Standardization of seed size and polypot for elite seedling production in tamarind. Madras Agric. J., (1 - 3): 48 – 51.

Parameswari. K.P. Srimathi and K. Malarkodi . 2001. Influence of seed size and duration of acid scarification on seed germination of tamarind. Madras Agric. J., 88 (1- 3):56 – 60.

Patil. V.S. ; M.B. Madalageri, G.S. Sulikeri and K.R. Naik. 1997. Influence of season and growth regulators on success of air layers in tamarind. Abstr. Proc. National symposium on tamarind, 27 – 28, June, Tirupathi, p. 9.

Prabakaran. G.N. Chezhiyan and G. Jansi Rani. 2000. Influence of season genotype on in vitro culture of tamarind. South Indian Hort., 51 (1 – 6): 76 – 82.

Parameswari. K, P. Srimathi and K. Vanangamudi. 2002. Prestorage treatment to prolong the storability of scarified tamarind seeds. Seed Tech News, 32 (1): 183.

Parameswari. K, P. Srimathi, K. Vanangamudi. 2002. Seed source variation on seed quality of tamarind. Seed Tech News, 32 (1): 183.

Purohit. M, M. Jamaluddin and G.P. Mishra. 1996. Testing seeds of some tropical species for germination and mycoflora. Indian Forester, 112: 492- 495.

Rao. J.S, K. Mary Mathew, K. Pradip Kumar, Salykutty Joseph, R. Laxmanan, K.J. Madhusoodanan. and S.N. Potty. 1997. Tissue culture studies on tree spices. Proc. Nat, Symposium on Biotechnology of spices, Medicinal and Aromatic Crops (BIOSAP), Calicut, 24 – 25th April, 1996, pp 39 – 44.

Sankaranarayanan. R,; R.M. Vijayakumar and P. Rangasamy. 1994. Cow urine is ideal seed for germination in tamarind. Indian Hort., 38(4): 15.

Swaminath. M. H, D.S. Ravindran and J. Mumtaz.1990. Propagation of *Tamarindus indica* through stem cuttings. My Forest, 26 (2): 207 – 208.

Troup, R.S.1921. Effect of Girdling and growth regulators on rooting of air layers. The Silviculture of Indian trees Vol. II. Leguminosae (Caesalpinieae) to Verbenaceae. 7. Tamarindus indica L: 263 - 363 p. Oxford, Clarendon Press.

Webster, P.J. 1924. Preparation and Packing of Seeds and Scions of trees and shrubs in the tropics. Philippine Agriculture Review, 17(1): 46-51.

Chapter-18

Wood Apple (*Feronia limonia* L.)

S.S. Hiwale
Central Horticultural Experiment Station, Godhra- 389 340

Wood apple (*Feronia limonia* (L.) Swingle) is also called kainth, elephant apple, monkey fruit, curd fruit, kath bel and other dialectal names in India. It is drought hardy, medium to tall sized, under utilized fruit tree. In Malaya, it is gelinggi or belinggi, in Thailand, Ma-khwit, in Cambodia, Kromsang and in Laos, Ma-fit. In French, it is pomm-d-elephant, pomm de bois, or citron des mois (Morton and Miami, 1987). It is one of the most highly valued fruit plant grown in drier region of the country. Its plants are deciduous, glabrous, tree with thorny branches. Limited work has been done on its post harvest utilization.

The wood apple is native to India and common in the wild form in dry plains of India and Sri Lanka. It is also found growing throughout South East Asia, in Northern Malaya and on Penang Island. In India, the fruit was traditionally a "Poor man's food" until processing techniques were developed in the mid-1950's. It occurs, wild or cultivated, up to an elevation of 1500 ft, in Western Himalayas, but more common in the Deccan; Thane and Chandrapur districts of Maharashtra. It is also reported to occur in parts of Hazaribagh, Palamau and Chota Nagpur in Jharkhand, in forest of Vidhyan hills of Uttar Pradesh and Chattishgarh. It is often cultivated on borders of fields and as a roadside tree near villages and sometimes planted in the orchards. It is grown in the states of Odisha, Karnataka, M.P., A. P. and Tamil Nadu.

There are no regular plantations however stray plants along the border of fields, roads, railway lines and banks of the river are the common places where the plants are found. Its plants are rarely grown in hot arid parts of Rajasthan, however, young plants are found susceptible to low temperature/frost in hot arid region.

Nutritive Value and Uses

Its fruit is nutritive and having medicinal value. Fruits are used for preparation of jelly and delicious chutney. The nutritional composition of edible portion (55-58 %) is given in table 1.

Table 1 : Nutritional composition of edible portion (100 g^{-1} basis)

Parameters	Fruit pulp	Seeds
Moisture (%)	64.20	4.00
Protein (%)	7.10	26.18
Fat (%)	3.70	27.10
Carbohydrates (%)	18.10	35.49
Ash (%)	5.00	5.03
Calcium (%)	13.00	1.58
Phosphorus (%)	110.00	1.43
Iron (%)	0.60	0.03
Tannins (%)	1.03	0.08
Mineral matter (g)	1.90	-
Fibre (g)	5.00	-
Riboflavin (μg/100g)	0.17	-
Vitamin C (mg/100g)	3.00	-
Carotene (μg)	6.10	-
Thiamin (μg)	0.04	-
Niacin (μg)	0.80	-
Energy (k/cal)	134.00	-

(*Source :* Gopalan *et al.* (1985); Morton and Miami (1987))

***Food value*:** The pulp constitutes 55-56 percent of the whole fruit. The pectin content of the pulp is 3-5 per cent (16% yield on dry weight basis). The seed contains bland, non-bitter oil with high unsaturated fatty acids. The fruits are so hard that rind be cracked with a hammer. The scooped-out pulp, though sticky, is eaten raw with or without sugar, or is blended with coconut milk and palm-sugar syrup and drunk as a beverage, or frozen as an ice cream. It is also used in chutneys and for making jelly and jam. A stiff jelly can be made from the wood apple, but the flavor is somewhat harsh, so it is seldom used alone, but more frequently mixed with such fruits as the guava. Anon. (1956) however, reported that the jelly having agreeable flavor. He also reported that syrup and chutney are other possible products. There are two types of wood apple, one with fruit larger and sweeter than the other and states that the ripe fruit pulp contains 2.3 per cent acid and 7.25 per cent sugars. Diluting pulp with water, passing it through a pulper to remove seeds and fiber, further diluting, makes nectar straining, and pasteurizing. A clear juice for blending with other fruit juice has been obtained by clarifying the nectar. Pulp sweetened with syrup of canes or palm sugar has been canned and sterilized. The pulp can be freeze-dried.

The consumption of its fruit increases appetite; facilitates digestion in human beings. Useful in renal diseases; has some antiseptic properties. It is used as root stock for grafting of citrus species.

Medicinal value: The fruit is much used in India as a liver and cardiac tonic, and when unripe, as a means of halting diarrhea and dysentery and effective treatment for hiccough, sore throat and diseases of the gums. The pulp is poultice on to bites and stings of venomous insects as is the powdered rind (Kirtikar and Basu, 1935).Juice of young leaves is mixed with milk and sugar candy and given as a remedy for biliousness and intestinal troubles of children. The powdered gum, mixed with honey, is given to overcome dysentery and diarrhoea in children. Oil derived from the crushed leaves is applied on itch and the leaf decoction is given to children as an aid to digestion. Leaves, bark, roots and fruit pulp are used against snakebite. The leaves are aromatic, carminative and astringent.

***Gum*:** The trunk and branches exude gum. The exudation is profuse after the rainy season. It is considered to be a good substitute for gum arabic and commercial samples of the latter are utilized as substitutes for, or adulterant of, gum arabic and is also used in making artists watercolors, ink, dyes and varnishes. It consists of 35.5 per cent arabinose and xylose, 42.7 per cent D-galactose, and traces of rhamnose and glucuronic acid.

Wood : The wood is yellowish gray to grayish white, hard and heavy (sp.gr.-0.83). It is somewhat refractory to season. It is durable both under cover and in exposed situations. It is used for house building, naves of wheels and oil crushers. It can be used as pattern wood for shoe lasts, penholders and similar articles. It is also used for agricultural implements and ornamental carving (Anon., 1956)

***Pectin*:** The pectin has potential for multiple uses in India, but it is reddish and requires purification. Fruit pulp contains 2.66 per cent pectin on fruit weight basis (Roy and Majumdar, 1988)

***Rind*:** The fruit shell is fashioned into snuffboxes and other small containers.

Botanical description

Wood apple belongs to family Rutaceae and chromosome no. is 2n=18. The slow-growing tree is erect, with a few up ward-reaching branches bending outward near the summit, where they are subdivided into slender branches drooping at the tips. The tree grows up to an elevation of 450 m in Western Himalayas. It is a small to moderate sized, deciduous, glabrous, tree with thorny branches, growing to a height of 10 meter and 0.6 - 1.6 meter in girth, trunk is short cylindrical with symmetrical crown of foliage (Troup, 1921). The bark is ridged, fissured and scaly and there are sharp spines: 2-5 cm long on some of the zigzag twigs. ; leaves pinnate, 8-10 cm long. The deciduous petiole and rachis, flat often narrowly winged, glabrous, entire, alternate. Leaves, dark-green, leathery, often minutely toothed, blunt or notched at the apex, are dotted with oil glands and slightly lemon-scented when crushed. The leaflets 3-9, opposite, sessile or short petiolate, obovate with crenate tip and cuneate base and glabrous. Flowers are dull-red or greenish; 1.25 cm wide and are borne in small, loose, terminal or lateral panicles. They are usually bisexual. The calyx is very small with 5-6 lobes, deciduous, petals 5-6 elliptic-oblong, spreading or bent downward. The size of pollen grains ranged from 28.4 to 36.1 μ. Pollination is done by insect and unpollinated flowers failed to set fruits. The fruit is round to oval, 5-12.5 cm wide, with a hard, woody, grayish-white, scurfy rind about 6mm thick. The pulp is brown, mealy, odorous, resinous, astringent, acid or sweetish with numerous small white seeds scattered through it.

Warm season appears conducive for the initiation of floral buds. The tree sheds its leaves and branches are bare for a short period during January and flowering starts in February-March. Flowering is on mainly on new shoots; emergence of panicles commences in the middle of February and continuous upto 3rd weeks of May. Opening of flowers starts in the 2nd week of March. The flowers are mainly staminate and hermaphrodite. Ovary, style, stigma in both hermaphrodite and male flower but rudimentary in the latter. Average fruit weight is 350g with rind thickness of 0.3-0.6 cm. Acidity 1.04 to 4.50 per cent and total sugar 4.08-4.47 per cent.

The systematic work on improvement of this crop has not been attempted so far except selections. The plants growing so far or of seed origin and found to have lot of variability which can be used for making selection of superior types. There are two forms one with large, sweetish fruits and the other with small, acidic fruit. Some selections have been made at UAS, Dharwad and the plants of superior genotypes have prepared through budding.

Seven germplasm lines collected from different places were evaluated at Central Horticultural Experiment Station, Vejalpur (Gujarat). The seed germination ranged from 15.56 to 62.40 per cent. Survival percentage ranged from 80.00 to 94.00 per cent. It is observed that percent germination was higher in Acc. No.-2 followed by Acc. No.-4 and per cent survival was higher in Acc. No.-5 followed Acc. No.-6.

There was significant variation in various parameters like fruit length, fruit diameter, pulp weight, skin weight, TSS, number of seeds and seed weight. Fruit weight ranged from 187 to 350.53 g, fruit length 71.50 mm to 95.06 mm, fruit diameter 69.84 mm to 86.00, pulp weight 83.23 to 155.69 g, skin weight 82.93 to 164.48 g, TSS ranged from 10.67 ^{O}B to 14.33 ^{O}B, number of seed 239.67 to 505.00 and seed weight 3.00g to 15.00g were recorded(Anon., 2003). Thus indicated that there is lot of variability in the species, which needs to be collected and evaluated for identifying high yielding cultivars for commercial cultivation (Table 2).

Table 2 : Evaluation of wood apple germplasm.

Acc. No.	Germi-nation (%)	Survi -val (%)	Fruit Wt. (g)	Fruit Length (mm)	Fruit Dia. (mm)	Pulp Wt. (g)	Skin Wt. (g)	TSS °Brix	No.of Seed	Seed Wt. (g)
1	22.88	87.00	289.83	81.90	80.33	106.72	125.86	12.67	409.67	9.00
2	62.40	90.00	257.16	74.73	82.33	83.23	116.84	12.33	262.00	7.00
3	15.56	80.00	217.83	73.06	79.53	92.47	87.72	14.33	505.00	15.00
4	33.47	83.00	215.16	75.13	77.33	92.03	82.93	11.33	224.00	13.07
5	25.16	94.00	350.53	86.20	84.16	155.69	164.48	14.33	451.33	12.67
6	21.14	91.00	328.67	95.06	86.00	148.15	158.42	10.67	239.67	3.99
7	–	–	187.00	71.50	69.84	86.63	86.30	12.50	247.00	7.33
CD (5%)	–	–	34.13	3.87	7.90	18.45	16.05	1.79	28.45	1.53

At CIAH, Bikaner, five germplasm have collected from Godhra, Aruppukottai (T.N.) and Udaipur and are being evaluated for growth and flowering / fruiting. It was observed that all genotypes were susceptible to frost and slow growth at initial stage under hot arid conditions.

Propagation

Wood apple is commonly propagated by seeds, although root cuttings or layers are said to be successful. Multiplication may also be done by budding onto self-seedlings to induce dwarfing and precariousness. Recently work done at CHES, Godhra showed that there is scope for softwood grafting for large scale multiplication of the plant.

Seed

It is propagated through seeds. Further, seedlings are used as root stock. In wood apple germination of seed is easy. Seedling growth is also good under nursery conditions. July is best time for sowing of seed in arid region to raise saplings.

Vegetative methods

Wood apple plants are propagated by vegetative methods *viz.* budding and soft wood grafting. Budding and soft wood grafting techniques have been standardized for large scale plant multiplication.

Budding

Wood apple can be successfully multiplied through budding methods. Angadi *et al.*, (2009) conducted experiment on the budding of wood apple (*Feronia limonia*) under different environments using differently aged rootstocks at UAS, Dharwad, Karnataka during 2007-08. The experimental results revealed that among the two environments tested, poly mist house condition produced better results compared to open condition. Among the differently aged rootstocks, nine-month-old rootstock resulted in better budding parameters. Maximum number of leaves (3.4), leaf area (5.9 cm^2) and sprout length (2.7 cm) were recorded in plants budded on eight-month-old rootstock. A minimum number of days (23.4) was required for sprout initiation under nine-month-old rootstock while eight-month-old stock produced the highest percent bud takes (92.8) and survival percentage (90). The interaction effect of age of rootstock and environments was significant. The scions budded on nine-month-old rootstock under poly mist condition took a minimum number of days (20.6) for sprouting while eight-month-old stocks produced highest bud take (99.3%) and survival percentage (93.3). The maximum number of leaves (3.9) was observed on eight-month-old rootstock under poly mist conditions which was on par with seven-and six-month-old stocks (3.1 and 3.1, respectively). A lower number of leaves and leaf area were registered with the scions budded on five-month-old stock in open condition (1.8 and 2.6 cm^2, respectively). A significantly higher leaf area (6.3 cm^2) was registered on eight-month-old rootstock which was on par with seven-month-old rootstock (6.0 cm^2).

Local seedling - Wood apple

Soft wood grafted seedling- Wood apple

Variability in fruit- Wood apple

Soft wood grafting

To reduce the gestation period of 12- 15 years required for fruiting in wood apple, a successful method of soft wood grafting has been standardized. Plants raised from seeds of same species (one year old) were defoliated ten days prior to grafting. These scions were grafted (on the one-year-old) local rootstocks seedling. Data presented in table 3 showed that per cent success was highest when grafting was performed in the month of January and February (48.88 and 53.33%, respectively) and lowest in December and April (30.24 % and 34.56 %, respectively) under semi –arid conditions at Godhra ,Gujarat.

Table 3 : Effect of time on success (%) of soft wood grafting in wood apple.

Months	Success (%)
Dec, 2003	30.24
Jan, 2004	48.88
Feb, 2004	53.33
March, 2004	44.28
April, 2004	34.56

Alloli *et al.* (2010) observed significant influence of different season and genotypes in wood apple. It was observed that wood apple. genotypes KWS-41 and KWS-19 recorded maximum sprout height above graft union, during the month from August to October in both shade house and poly house conditions.

Raghvendra *et al.* (2011) reported that multiplication of wood apple plants by softwood grafting among different age of rootstock under poly mist house condition found to be better than open condition at Dharwad. In ten months old rootstock (A6) less (23.8) days were required for sprouting with maximum number of leaves (13.7). Eight months old rootstock (A4) has produced highest (97.1) per cent success and maximum (96.2) survival percentage.

Quality Standards of Planting Material

Vegetatively multiplied wood apple plants should be transplanted under field conditions. The disease free rootstock seedling raised in polythene covers were selected for soft wood grafting. Plants should not be old for planting in the field. The joints of bud union should be appropriate. Root system should not coiled and damage.

References

Alloli,T.B., Athani, S.I, Sappandi, S. and Naik,K.R. 2010 .Influence of season and genotypes on sprout height of grafts under protected condition in wood apple. ***In.*** National Conference on production of Planting Quality Seed and Planting material, held at New Delhi,11-14March,2010,pp28.

Angadi, S.G., V.N. Raghavendra, T.B., Allolli, C.K. Venugopal, A.N., Mokashi, U.V. Mummigatti, 2009. Propagation of Wood apple (*Feronia limonia* L.) by Budding, *In.* IInd International Symposium on Pomegranate and Minor including Mediterranean- Fruits , held at UAS, Dharwad, pp.82.

Anon. 1956. The Wealth of India. A dictionary of Indian raw materials and industrial products, Vol IV, Pub. CSIR . New Delhi Pp. 18-20

Anonymous, 2003. Annual Report Central Institute for Arid Horticulture, Bikaner, Rajshthan, India.

Gopalan, C., Ramashastri, B.V. and Balsubramanium, S.C. 1985. Nutritive value of Indian foods. National Institute of Nutrition, ICMR, Hydrabad, India.

Kirtikar, K.R. and Basu, B.D. 1935. Indian Medicinal Plants, Vol. II, Pp. 496-497.

Morton J.F. and Miami, F.L. 1987. Fruits of warm climate, Wood apple, Pp. 190-191.

Raghvendra, V. Naik ; . Angadi, S. G;. Allolli, T. B, Venugopal, C. K. and Mummigatti, U. V. 2011. Studies on soft wood grafting in wood apple (*Feronia limonia* L.) , Karnataka J. Agric. Sci.,24 (3) : 371 - 374.

Roy, P. and Majumdar, B.C. 1988. Sci. and Cult., 55: 110-111.

Troup, R.S. 1921. In the Silviculture of Indian trees, vol.III, Govt. of India, Publication, pp.101-103.

Chapter-19

Phalsa (*Grewia subinaequalis* DC.)

D. B. Singh
Central Institute of Temperate Horticulture, Srinagar - 190 007 (J&K)

Phalsa (*Grewia subinaequalis* syn. *G asiatica*) is a minor fruit crop. It is the most hardy fruit, drought resistant and requires little care. This fruit crop is mentioned in vedic literature having medicinal properties. It can be grown almost in all parts of north India except at higher elevations. It cannot be grown in temperate climates with severe winters. Phalsa being very vigorous in growth can be an ideal plant for plugging gullies and ravines and for contours to protect bunds. It can also be grown as a hedge along the field boundary. It is also grown as filler tree(Chundawat,1990). It is capable of growing under marginal, waste lands and water scarcity conditions. The tiny fruits have to be picked from bush several times during the fruiting season. Fruits are used for fresh eaten and having short shelf life is suitable for disposal in local market.

The fruit is known by different names in different Indian languages, such as dhamana in Bengali, phalsa in Punjabi. It bears delicious fruits. In India, different parts of the plant are used for curing variety of ailments. Phalsa is commercially cultivated in the states of Punjab, Haryana, Rajasthan, Uttar Pradesh and M.P. Besides these states, it is also cultivated on limited scale in the states of Maharashtra, Gujurat, Andhara Pradesh, Bihar and West Bengal.

Nutritive Value and Uses

The fruits are nutritious and rich in vitamin and minerals. The edible portion of the fruit varies from 69 to 93% (Roy *et al.*, 1995). Ripe fruits are sub acidic and good source of vitamin A and C. They are fair source of phosphorous and iron. Fruits contain 50-60 % juice; 10-11 % sugar and 2.0-2.5 % acid (Aykroyd, 1963). Among the organic acids, citric acid is predominant with traces of malic acid (Khurdiya, 1979). The vitamin c content ranges between 27.39 and 39.5 mg/100g (Singh and Acharaya, 1999). Madan and Khurdiya (1987) reported that the seed coat and the kernel contain 2.0 and 23.7 % oil respectively, which is rich in linoleic acid. The pigment responsible for colour of phalsa fruit, is anthocyanin viz, delphinidin-3-glucoside and cynidin-3-glucoside (Khurdiya, 1979). The phalsa seed contains 7.17 % crude fat with a typical odour (Sharma, 1996).

The proximate composition of fruit is as follows (per 100g): Moisture (80.8%), Protein (1.3g), Carbohydrates (14.7g), Fiber (1.2g), Fat (0.9g), Total mineral (1.1g), Â carotene (419µg), Niacin (22.0mg), Vitamin C (22.0mg), Ca (129mg), P (39mg), Fe (3.1mg), K (351mg) and Na (0 4.0mg).

The fruits are excellent for making juice and squash. However, it is mostly used as fresh fruit and has cooling effect. The ripe fruits are used for making refresh drink in summer having cooling effect. Several beverages, viz., nectar, concentrate, squash, and crushed phalsa packed in glass bottles remain acceptable for up to 8 months in cool storage (Roy *et al.*, 1995). The fruits are highly perishable and need to be processed immediately after harvesting (Salunkhe and Desai, 1984: Roy *et al.*, 1995). The fresh leaves are valued as animal fodder. The root bark is applied in treating rheumatism.

The plants are grown on the boundary of fields as hedge. The shoots procured from pruning can be utilized either for making baskets or supporting vegetable crops. Pruning of one fully grown plant is sufficient for making one large or two small baskets. Pruning is done every year during December month for better fruiting. Phalsa baskets are fairly strong and can be used for transplanting fruits and vegetables. Ropes can also be made from the bark after treating, which yields fiber.

Botanical Description

Phalsa (*Grewia subinaequalis*) belongs to the family Tiliaceae. It is a small bush with many berries like fruits. In general, two types of phalsa plants are found. The plants are bush, small and other is erect tall type. The leaves are broad, round, 5-18cm. long with 1-1.15cm petiole. The flowers are hermaphrodite and fruits are drupe and are sweet to sour in taste. The flowers are produced in cymes as individual flower has diameter of 2cm, yellow in colour with 5 large sepals. The fruits are globular, smooth, deep reddish brown 5-12 mm in diameter, edible purple to black, drupe. In the dwarf genotype, flowers borne on all nodes, in the leaf axils of current season growth but in tall genotype 3 different flowering patterns have observed. Flowering takes place during March-April and fruits are ready for harvesting by the end of May and beginning of June. Ripening of fruits is uneven and irregular which require 3-4 picking. Hard seed is available in side the fruit.

Propagation

Phalsa is commonly propagated by seed as well as vegetative methods.

Seed propagation

Phalsa is commercially propagated by seeds. Seedling plants are fairly true to type. Freshly extracted seeds should be used for raising seedlings. Seeds loose their viability under ordinary storage after 90-100 days. The viability can be retained for a period of 6 months under cool storage. Seeds require 15-25 days for germination and seedlings get ready for planting in the field by 3-4 months. Germination percentage was highest (76%) and the germination period shortest (16 days) with seeds sown immediately after extraction (Singh and Jindal, 1979: Gauri Shankar 1985). The seed are sown during July - August in the nursery bed at the space of 25-30 cm. The seedlings are ready for transplanting during February-March or July - August. The best period for transplanting of phalsa is considered to be during February month.

Vegetative propagation

Cuttings

The mature cuttings when treated with plant growth harmones give some success. Cuttings during July – August treated with 100 ppm IBA for 54 hours has resulted in better success of the cuttings. (Singh *et al.*, 1961). Jauhari (1960) reported that treating phalsa cuttings with 100 ppm indole butyric acid (IBA) for 24 hours yielded 60 % success where as, Singh *et al.* (1961) recorded as high as 68.8 % success from hard wood cuttings treated with 100 ppm IBA. Both wood type and planting date influence rooting of phalsa. Singh and Kumar (1974) obtained 72.5 % rooting in cuttings when shoots were ringed 15 days before planting and treated with 2,500 ppm IBA. A report from Rajasthan Agric. College, Udaipur claims success with hard wood cutting, treated with 250 ppm ethrel for 5 minutes followed by 3000 ppm IBA for two minutes (Anon., 1981).

The best rooting was obtained following treatment of hardwood cuttings with IBA at 200 ppm. The semi hardwood cuttings did not root well as hardwood cuttings (Kathrotia and Singh, 1995). However, Srivastava (1996) reported that treatment of hardwood cutting with IBA (2000 and 3000 ppm) significantly increased the root length, number of roots/cutting and percentage survival.

The effect of IBA or NAA (100,200 or 300 ppm) on the rooting of semi hard woodcuttings of *G. asiatica* was investigated by Kathrotia and Singh (1995). The best rooting (82.5%) was obtained following treatment of hard woodcuttings with IBA at 200 ppm.

Hardwood cuttings when treated with 0,1000 , 2000, or 3000 ppm. IBA 0 or 5% sugar and 0 or 0.5% captan fungicide then planted in a mixture of sandy soil and FYM and placed in partial shade for 90 days. The longest root (3.83 cm) and highest percentage survival (36.27%) were obtained with 2000 ppm IBA and greatest root length of root (3.66 cm) and number of roots per cutting (2.95) were obtained with 3000 ppm IBA with no significant differences between these two treatments. Captan treatment significantly increased survival (from 19.5% without treatment to 26.81%) but had no significant effects on rooting. Sugar had no significant effects on rooting and survival.

It has been reported that cuttings were treated with combination of 3, 4 or 5 % cane sugar and 100 ppm IAA or IBA. Most effective treatment was 5% sugar + IBA for *Grewia asiatica*. Treatments with auxins (IAA, IBA, and NAA) improve rooting of difficult to root hard wood cuttings of Phalsa. Anatomical study of rooting in stem cutting of phalsa has been reported by Yadav and Rajput(1969).

Srivastava *et al.* (1996) obtained best results in terms of growth and survival of *G subinaequalis* cuttings was obtained with treatment with 2000 ppm IBA (the other concentrations tested were 100 and 300 ppm). Captan at 0.5% improved survival compared with untreated controls, but sugar treatment (5%) had no significant effect.

Jadhav (2007) conducted an experiment at UAS, Dharwad on propagation of phalsa through cuttings. The study pointed out that among different growth regulators formulations tried IBA 200 ppm formation helped in better induction of rooting 70.00 per cent as against 28.33 per cent in the control. Using NAA 200 ppm formulation was also found effective in promoting better shoot characters. However, use of higher concentration of growth regulators at 300 ppm was not promotive.

Among the different seasons tried, rooting of phalsa cuttings was best in the month of July (68.33% and 53.33%) in mist and open condition. The next best rooting was obtained in the month of August (66.67% and 50.67%) and June (53.33% and 37.00%) in mist and open conditions respectively. Whereas, October, November and December months were not suitable for phalsa cuttings. The studies revealed that the phalsa planting material can be raised with high success by pre-treatment of cuttings with IBA 200 ppm which are placed under mist conditions. However, among the seasons July month is suitable for rooting of phalsa cuttings under mist and open conditions.

Air layering

Air layering has been tried by some workers. Saxena (1963) reported that air layering yielded 50 percent success when the air layers were treated with a mixture of IBA, NAA, 2,4-D and boron at 10,000, 10000, 10000 and 100 ppm, respectively. The untreated air layers failed to root. Mohamed and Chauhan (1970) recorded 85% success when air layers were treated with a mixture of IBA and NAA at the rate of 1500 ppm each, but when the same mixture was used for ground layering, success was 50% only. Chatterjee and Rao (1980) recorded 80-90% success by treating the air layers with IBA or NAA at 5000 and 10000 ppm.

In phalsa the effect of different concentrations of IBA, NAA, Ethrel and phenolic compounds on rooting of phalsa through stool layering revealed that 10,000 ppm IBA showed pronounced effect on rooting, whereas, combined effect of IBA (5000 ppm) and Catechol (5000ppm) produced more number of roots and also increased the root length significantly (Table 1). During the process of regeneration initiated root primodia from secondary phloem and cork cambium as revealed by anatomical studies (Prasad *et al.*, 1989).

Table 1 : Effect of different treatments of Auxins, Ethrel and Phenolic Compounds on rooting and survival of phalsa shoots.

Treatments	Rooting (%)	Average number of root/shoot	Average length of root (cm)	Survival (%)
2500ppm IBA	52.50	21.25	11.03	40.47
5000ppm IBA	62.50	21.75	11.63	44.00
10,000ppm IBA	75.00	22.25	11.73	56.66
2500ppm IBA+2500ppm NAA	55.00	22.75	11.50	40.91
5000ppm IBA+5000ppm NAA	65.00	30.25	13.13	46.15
2500ppm IBA+2500ppm Catechol	60.00	30.00	24.63	50.00
5000ppm IBA+5000ppm Catechol	72.70	31.00	30.06	54.38
2500ppm IBA+2500ppm Ethrel	45.00	21.75	29.00	38.88
5000ppm IBA+5000ppm NAA	50.00	30.75	27.00	40.00
Control	30.00	19.00	10.00	29.17
S.Em±	4.40	0.44	1.42	0.32
C.D. at 5%	12.76	1.28	4.18	0.74

(*Source:* Prasad *et al.*, 1989).

Stooling

Sharma *et al.* (1987) obtained fairly high percentage of success in stooling of phalsa. The plants were headed back at 15 cm above the ground level in January. New shoots emerged out with the onset of spring. A ring of bark (2.0cm) was removed from the base of these shoots in August and the shoots were then treated with IBA and NAA in lanolin paste. The shoots were earthed up after 7 days. Rooting occurred in about 45 days. Maximum percentage (79%) of rooting and the longest roots were obtained with IBA at 7500 ppm followed by IBA 5000 ppm (72% rooting), while 62% rooting was noted with NAA at 10000 ppm.

Grafting

Amin (1987) reported a technique of soft wood grafting of phalsa with 100% success in which root stock were grown *in situ* for a year or more and pre defoliated (10 days before operation) scions were grafted.

Micro-propagation

Phalsa can be multiplied through tissue culture. In order to apply biotechnology through genetic transformation to enhance cold hardiness, a reliable and rapid micro-propagation system is needed. Thus, developing the most dependable micro-propagation protocols for phalsa was the primary goal of the research. Phalsa explants prepared from different tissues, including leaf, nodes, internode, and zygotic embryos, were collected from mature trees growing in the specialty plants house, cultured on MS medium supplemented with various cytokinins alone or along with auxins and incubated under a 10-hour photoperiod at ambient temperature. *In vitro* propagation of phalsa tissues through both organogenesis and somatic embryogenesis was achieved. Of these, single shoots were developed from nodal explants as a result of bud break on MS medium supplemented with BAP, kinetin, and zeatin separately. Somatic embryos were developed from the zygotic embryos when cultured on MS medium with 0.023 μM BA + 0.022 μM zeatin, for 2 weeks following a pulse treatment on NN medium supplemented with 5% sucrose, 0.11 μM BAP, 0.22 μM 2,4-D, and 29.20 μM L-glutamine. Somatic embryogenesis was also observed on modified basal medium supplemented with 13% sucrose, 58.40 μM L-glutamine, and 1.75 μM IAA. Enormous callusing was a major problem for in vitro studies with this species, irrespective of media composition. Further studies for multiple shoot development and higher frequency of SE induction are under way (Biswas *et al.*, 2005).

References

Amin, R.S. 1987. Current Science, 47:468-469.

Anonymous, 1981. Ann. Res. All India coordinated Fruit Improvement Project, pp30-

Aykroyd, W. R. 1963. The Nutritive Value of Indian Foods and the Planning of Satisfactory Diets (6th edition), ICMR, New Delhi, pp101-126.

Biswas, Bipul, Joshee, Nirmal.Yadav, A. and Yadav, A .K. 2005 *In vitro* Propagation and Somatic Embryogenesis in Phalsa (*Grewia asiatica*), Hort.Science,40.(4).

Chatterjee, B. K. and Rao, D. P 1980. Sci. and Cult. 46: 238-240.

Chundawat, B. S. 1990. Arid Fruit Culture, Oxford & IBH Publication, New Delhi.

Gauri Shankar 1985. Phalsa *In:* Fruits of India: Tropical and Sub-Tropical (ed. by Bose, T.K.) Naya Prakash, Calcutta.

Jadhav, Ajay Kumar 2007. Studies on propagation of Phalsa (*Grewia subinaequalis*) by cutting. M.Sc. Ag. Thesis submitted to University of Agriculture Sciences, Dharwad. Karnataka, India

Jauhari, O.S. 1960. Sci. and Culture, 26:114.

Kathrotia, R. K and Singh, S.P. 1995. Adv. In Hort. and Forestry, 4:35-41.

Khurdiya, D. S. 1979. Ph. D. Thesis, IARI, New Delhi.

Madan, T. R and Khurdiya, D. S 1987. J. Oil Tech. Assoc., India, 19:23.

Mahammed Shafat and Chauhan, K.S. 1970. Vegetative propagation of Phalsa. Indian J. Agric. Sci, 40(4):581-586.

Prasad, J. Ram Lallan and Ram, R.A. 1989. Studies on Propagation Technique in Phalsa (*Grewia subinaequalis* DC) through stool layering. Indian J. Hort., 46:189-192.

Roy, S. K., Khurdiya, D. S. and Waskar, D. P. 1995. In: Handbook of Fruit Science and Technology: Production Composition, Storage and Processing (Eds. D.K. Salunkhe and S.S. Kadam), Mareel Dekker, Inc., New York, pp. 553-555.

Salunkhe, D. K. and Desai, B. B. 1984. Phalsa. In. Postharvest Biotechnology of Fruits, Vol. II, Satisfactory Diets (6th Edition), ICMR, New Delhi, pp. 101-126.

Saxena, C. K. 1963. Allahabad Farmer., 37:28-31.

Sharma, B.B. Saxena S.K. and Sharma V.P. 1987. Symposium on Himalayan Horticulture- In the context of Defence Supplies, DRDO, Tezpur, Assam, Oct. 28-30, p.49.

Srivastava, S. R. 1996. Effect of IBA, sugar and Captan on rooting of phalsa *(Grewia subinaequalis* DC.). Recent Horticulture, 3: 1, 52-54.

Singh J. P., Godara, P. S. and Singh, R. P. 1961. Effect of type of wood and palnting dates on the rooting of Phalsa (*Grewia asiatica* L.). Indian J. Hort., 18:46-50.

Singh, D. and Jindal, P.C. 1979. Haryana J. Hort. Sci. , 8:47-48.

Singh, O and Kumar, H. 1974. Response of some growth substances on the propagation of Phalsa (*Grewia asiatica* L.). Indian Jour. of Horticulture, 31;2, 163-164.

Singh, S. P. and Acharya, P. 1999. The Hort. Jour., 12:81-83.

Yadav, U. L. and Rajput, C.B.S., 1969. Anatomical studies of rooting in stem cutting of Phalsa (*Grewia asiatica* L.), Hort.Sci.,1:19-22.

Chapter-20

Gangana [*Grewia tenax* (Forsk.) Fiori.]

J.P. Singh* and R. S. Singh**
**CAZRI, RRS, Jaisalmer, Rajasthan*
***Central Institute for Arid Horticulture, Bikaner-334 006*

Gangana*(Grewia tenax)* is a deciduous tropical shrub widely distributed in arid and semi-arid region. In western Rajasthan, the natural habitat of this shrub is buried pediments, hills and pediplains (Bhandari, 1971). It does best on deep sandy loam, and can grow in very shallow, skeletal, gravelly or clay soils. It has been reported that the species grow abundantly (density = 678 individuals per ha.) in hilly waste lands of Kailana, Jodhpur. Sharma (1997) reported that its efficient root system can penetrate into minute rock crevices. It is extremely drought hardy and can withstand high temperature in hot arid climate. It tolerates frost well but plant growth is slow in arid region. Mostly fruits for sale in the local markets are collected from wild stands. It is underutilized species (Gebauer *et al.*, 2007), the major limitations to the development as a horticultural crop is the lack of detailed information, particularly the genotype, physiology and ecology (Gebauer *et al.*, 2007). Due to its drought hardy and dense root system, it seems promising species in arid region. It has good fodder value for small ruminants. It is naturally grown in range lands, *orans,* and hilly track of Aravalli. It is untapped plant species and its potential should be exploited.

Nutritive Value and Uses

The plant *Grewia tenax* is an example of multipurpose plant species which is the source of food, fodder, fiber, fuel wood, timber and a range of traditional medicines that cure various perilous diseases and have mild antibiotic properties. Ripe fruits are eaten fresh and liked by the inhabitants or left to dry for consumption. The juice is regarded as a good thirst-quencher, during hot summer. Because of high iron, the fruits are used as an iron supplement for anemic children (Gebauer *et al.*, 2007). In Sudan, a local drink is prepared by soaking the fruits over-night, and then they are hand pressed, sieved and sweetened. The fruits consumed by man and animals contain a large amount of iron and can be made into a refreshing drink. Fruit storage can be extended by drying. The dead leaves are eaten, but only while they remain on the plant. Young leaves are consumed by livestock, they are slightly palatable at the end of dry seasons, and have fairly good feed value. Bees visit the flowers for pollen and nectar. The branches are used as firewood, and can be used in charcoal making. The bark is used to make ropes and for binding purposes in house construction. *G. tenax* wood is used in making weapons such as clubs, bows, and arrows and for other general purposes. A mucilaginous bark preparation is used by women against hair vermin. Medicine: In Kenya plant parts are used as a remedy for colds and chest complaints and also as a chief constituent in a typhoid remedy.

Nutritional data presented in table 1 and 2 indicates that the fruits of *Grewia tenax* are rich in nutrients and could be important suppliers to improving the nutritional content in arid region. Pectic substances were found at higher levels in *G. tenex* (13.02%) as compared to *G. villosa* and *G. flavescens*). Variation in mineral composition in Grewia spp. has been reported by Mohammed Elhassan and Yagi , (2010). In Grewia species, good fibre content has noticed by Prasad *et al.* (2010).

Table 1 : Proximate composition, reducing sugar and starch values of *Grewia* spp. (expressed as (%) on DW basis).

	G. tenax	*G. flavescens*	*G. villosa*
Moisture	13±0.17	15±0.19	14±0.13
Fibre	20.5±0.11	42.8±0.14	25.5±0.08
Ash	5.2±0.09	3.4±0.10	4.0±0.12
Protein	7.7±0.12	8.7±0.10	6.7±0.06
Lipids	1.7±0.39	1.3±0.17	1.5±0.13
Carbohydrates	66±0.41	75±0.1.1	84±0.80
Reducing sugars	13.8±0.12	10±0.40	10.4±0.22
Starch	44.4±0.09	38.6±0.60	22.8±0.13

Table 2 : Mineral composition of Grewia spp. (mg/100g dry weight) Source: Mohammed Elhassan and Yagi (2010).

Element	*G. tenax*	*G. flavescens*	*G. villosa*
K	817±0.02	877±0.03	966±0.06
Ca	790.5±0.15	269.8±0.11	536±0.01
Mn (Manganese)	5.1±0.08	0.1±0.07	0.1±0.11
Fe	20.8±0.01	26.9±0.13	29.6±0.05
Cu	1.5±0.07	1.1±0.05	1.2±0.04
Zn	1.9±0.03	1.1±0.01	2.5±0.01

The data are mean values ± (SD) of duplicates

Leaves and twigs are very palatable, and much browsed by the sheep and goats in hot arid region. It has very good coppicing affect after cutting for feed. It is a suitable bush plant for Horti-Silvi-pastoral system in arid region of western Rajasthan (Sharma *et al.,* 1994). The plant preparations are used for the treatment of bone fracture and for bone strengthening and tissue healing. The fruits are used for promoting fertility in females and are considered in special diets for pregnant women and anemic children. The plant is adapted to high temperatures and dry conditions and has deep roots which stabilize sand dunes. The shrubs play effectively for rehabilitation of wastelands. The plant parts are rich in amino acids and mineral elements and contain some pharmacologically active constituents. The plant is identified in trade for its fruits. Plant is also sold as wild species of medicinal and aromatic plant and is direct or indirect source of income for the tribal people (Sharma and Patni, 2012; Ahmed *et al.,* 2011).

Botanical Description

Gangana (*Grewia tenax* Forsk.) also known as Gangaren, Gangeti and belongs to the family Tiliaceae. This much branched deciduous shrub is up to 2.0 m tall; however, under high browse pressure in rangelands remains in short stature. Branches are slender, thinly pubescent, green when young, smooth and ashy-gray when old. Bark smooth, grey, very fibrous so that twigs are hard to break. Leaves are 2.0-4.5 x 1-3 cm, alternate, more or less pubescent and ovate to suborbicular, acute or obtuse at the apex, coarsely dentate. Petiole is 5-8 mm long, slightly thickened at apex pubescent; stipules 4-5 mm long, caducous. Flowers are white, solitary or rarely paired with long stamens and even longer pistils. Fruits usually 2-4- lobed, globose, glabrous, shiny, fleshy; lobes 5-7 mm in diameter, orange yellow or bright red when ripe with one to two seeds in each. It flowers from August to October and fruiting from September-December. The small flowers are in bunches and vascular anatomy of Grewia flowers was studied by Joshi and Diwedi (1979). Fruit is orange-red at maturity, with 1-4 spheroid lobes. Grewia tembensis and G. tenax are virtually indistinguishable in fruit. The specific epithet refers to the plant's tenacious growth habit. Flowering and fruiting occurs during winter season. Fruits are small in size having yellow -orange colour and 2-3 lobes, edible and sweet-sour in taste. The pulp content is low and hard seed is available in the fruit. Seeds of the plant contain 2% fat.

Propagation

Its natural regeneration occurs through dispersal of seeds after being eaten by birds. Seeds excreted by birds germinate during monsoon in suitable habitat. It can be propagated through vegetative means.

Seed propagation

The germination of seed is irregular throughout the year (Sharma *et al.*, 1994) in pasture lands. The germination of seed is very difficult under nursery conditions because of hard seed coat and dormancy. Uniform seed germination has not observed in this plant species. The seedling growth is also slow at initial stage. It is mainly multiplied through seeds.

Winter collected fresh seeds (3-4 seeds per poly bags) are sown in the nursery in mid February to March at 1 cm depth. Before sowing the seeds should be soaked in warm water for 24 hours (Sharma, 1997). Seeds treated with 50% Sulphuric acid solution for 5 minute exhibited maximum germination in a span of 65 days and more over, biomass production was also found to be best in the seedlings obtained Sulphuric acid treated seeds.(Singh *et al.*,1993). Chemical treated and water soaked seeds shown superiority over control but in no case germination could exceed 60%.

Vegetative propagation

Cuttings can also be used for multiplication but success rate is very low under hot arid conditions. Auxins influenced rooting in cutting of Grewia tenax (El-Sidding *et al.*, 2005).

But the prolonged seed dormancy is a typical feature and vegetative propagation is not well characterized for the plant. Micro-propagation by tissue culture techniques may play an effective role for plant conservation. The plant needs phyto-chemical and pharmacological investigations on a wider range which may have great scope in near future. Thus, efforts are needed to conserve, domesticate and cultivate the plant. This paper reviews the plant's medicinal aspects and chemical constituents, also provides brief information of plant bioprospecting and its manifest market values.

Standardization of propagation technique is required for this underutilized crop for quality planting material production for systemic plantations in semi-arid and arid regions.

References

Ahmed, E., Sharif, A., Hussain, S., Malik, A., Hassan Mu, Munawar, M.A., Nagra S.A., Anwar, J., Ashraf, M., Afza, N. and Athar, M. 2011. Phytochemical and antimicrobial studies of *Grewia tenax*. J. Chem. Soc. Pak. 33: 676-681.

Bhandari, M.M. 1971. Flora of Indian Desert. MPS Repros, Jodhpur.

El.-Sidding, K., Gama, P.B.S. and Inanaga, S. 2005. Influence of auxins on rooting of stem cuttings of the small-leaves white cross berry (gudeim), *Grewia tenax* (Forssk.) Fiori. Gezira J. Agric. Sci. 3: 12-23.

Gebauer, J., K. El-Sidding and B.A. El Tahir. 2007. Exploiting the potential of indigenous fruit trees: *Grerwia tenax* (Forssk.) Fiori in Sudan. Genet. Resor. Crop Evol. 54:1701-1708.

Joshi, B.M. and N.K. Dwivedi. 1979. Vascular anatomy of the flower of *Grewia tenax* (Forsk.) Fiori. Proc. Indian Academy Sci. 88: (B part 2) 57-61.

Mohammed Elhassan, G.O. and S.M. Yagi . 2010. Nutritional composition of (*Grewia tenax* (Forsk.) Fiori, G. flavescens Juss and G. villosa Willd.) Fruits. Advance Journal of Food Science and Technology 2: 159-162.

Nidhi Sharma and Vidya Patni 2012. *Grewia tenax* (Forssk.) Fiori.-A traditional medicinal plant with enormous economic prospectives. Asian Journal of Pharmaceutical and Clinical Research 5 (3): 28-32.

Prasad C.V., Rao, K.C., Reddy G.V., Rani, T.S., Yerriswamy B. and Subha MCS. 2010. Characteristic studies of Ligno-Cellulose fabric *Grewia tenax*. Journal of Natural Fibers 7: 194-215.

Sharma, S. K; R. S. Singh, J.C. Tewari and U. Burman 1994. Silvi-pastoral studies in arid and semi-arid degraded lands of western Rajasthan. *In*. Agro-Forestry for degraded lands (Eds. Punjab Singh, P.S. Pathak and M.M. Roy). Oxford and IBH Pub. Co. Ltd., New Delhi , pp.745-754.

Sharma, S.K. 1997. Characterization of common top feed species. *In* : M.S. Yadav, Manjit Singh, S.K. Sharma, J.C. Tiwari and U. Burman (eds.). Silvipastoral Systems in Arid and Semi-arid Ecosystems. CAZRI, Jodhpur. pp. 181-186.

Singh, R.S., J. C. Tewari and S. K. Sharma 1993. Effect of different seed treatment on germination and growth behaviour in *Grewia tenax* Forsk. under nursery conditions. My Forest, 29 (2):151-154.

Chapter-21

Mulberry (*Morus* sp.)

R.S. Singh
Central Institute for Arid Horticulture, Bikaner- 334006

Mulberry (*Morus nigra* , Family : Moraceae) also known as toot, shehtoot is a multipurpose tree. It is found all over India from temperate to tropical regions. Fig and Jackfruit are other important fruit trees of this family Moraceae. *Morus alba* is white mulberry mainly used in Sericulture whereas *Morus nigra* produces black fruits which have tremendous nutritional and medicinal value and are produced for fruit cultivation. Mulberry is a fast growing, deciduous, woody perennial plant, which is mainly grown in the orchards as windbreak due to its dense foliage. It can be grown on wide range of climate and soil conditions (Hare Krishna *et al.*, 2012). It is not commercially cultivated as fruit crop and trees are planted mainly on road sides, wind breaks and in the orchard for getting shade. In most parts of the country particularly southern states, mulberry is grown for its foliage, which is the only food for the silkworm (*Bombyx mori*). Mulberry leaf is a major economic component in sericulture since the quality and quantity of leaf produced per unit area has a direct bearing on cocoon harvest.

Mulberry is one of the rare delicacies. However, it has not been able to attain commercial significance as a fruit crop due to its highly perishable nature. Fruits of red mulberry (*Morus rubra*) are red and slightly acidic and multipurpose Mulberry is very important tree for socio economic point of view in the society. It provides shade at community places (Chandra *et al.*, 1994). It provides fodder, shade and shelters to the animals in grazing lands. It exists as a stray plantation at farmer's house in villages. Its plants are grown in arid region however; it is affected by very low temperature/frost during winters. New flushes starts just after winters over and good flowering/fruiting has been recorded under arid conditions in mulberry genotypes.

Nutritive Value and Uses

Mulberry fruit is nutritious but it is highly perishable. The fruits are consumed fresh or processed into squash, jam, beverages and wines. The fruits are delicate sweet -sour in taste eaten as fresh during summer. Squash from mulberry fruits (CIAH Selection-1) was prepared to develop new antioxidant rich beverage (Hare Krishna *et al.*, 2012). Fruit juice cures sore throat, fever, dyspepsia and melancholia. Its bark is used for paper making in many countries of the world. The wood is also used for making furniture and other packaging items. Wood is light in weight and used for making boxes for packing, spots item, implements. Dried fruits ground to powder make excellent poultry food. Leaves are sweet and fed to cattle and heartwood is used for sport goods, furniture, boat doors etc. Roots have the medicinal value being anthelmintic. The fruits are rich in sugar (about 10 per cent) and iron with only 0.1% acid content.

Botanical Description

Mulberry (*Morus* species) belongs to the family Moraceae. It is believed that origin of mulberry is Asia continent. The species which are found in the country are White mulberry (*Morus alba* L.), Black mulberry (*M. nigra* L.), American Mulberry, Red mulberry (*M. rubra* L.), Hybrid forms exist between *M. alba* and *M. rubra*). The white mulberry is native to eastern and central China. It became naturalized in Europe centuries ago. The tree was introduced into America for silkworm culture in early colonial times and naturalized and hybridized with the native red mulberry. The red or American mulberry is native to eastern United States form Massachusetts to Kansas and down to the gulf coast. The black mulberry is native to western Asia and has been grown for its fruits in Europe since before Roman times.

The white mulberry and to a lesser extent, the red mulberry are quite tolerant of drought, pollution and poor soil. The white mulberry is considered a weed tree in many parts of the country including urban areas. The black mulberry is more fastidious faring less well in cold climates or areas with humid summers. The white mulberry is the coldest hardy of the three species, although this varies from clone to another. The black mulberry is the least cold hardy of the three, although again cold tolerance seems to be depending on the clone.

All the three mulberry species are deciduous trees of varying sizes. White mulberries can grow to 80 ft. and are most variable in form, including drooping and pyramidal shapes. Red mulberries can reach

70 ft in height. The black mulberry is the smallest of the three, sometimes growing to 30 ft in height, but it tends to be a bush if not trained when it is young. The species vary greatly in longevity. Red mulberry trees rarely live more than 75 years, while black mulberries have been known to bear fruit for hundreds of years. The mulberry makes an attractive tree which will bear fruit while still small and young.

The white mulberry is so named for the colour of its buds, rather than the color of its fruit. The thin, glossy, light green leaves are variously lobed even on the same plant. Some are un-lobed while others are glove shaped. Leaves of red mulberry are large and thicker, blunt toothed and often lobed. They are rough on their upper surfaces and pubescent underneath. The smaller black mulberry leaves are similar to those of red mulberry, but with sturdier twigs and fatter buds. White mulberries generally come out in early spring, almost two month before black mulberries. Mulberry bears axillary bearing on new growth of the shoots. Mulberry tree flowers during February –March and the crop matures by May to June in Northern India. The trees are monoecious. The fruits are aggregate of duplets.

Species and Cultivars

More than 100 *Morus* species are found in Genus in the different countries in world. However, its three species are commonly grown for different purposes. In India, there are many species of Morus of which *Morus alba, M.indica, M. serrata* and *M. laevigata* grow wild in the Himalaya. Several varieties have been introduced belonging to *M. multiculis, M. nigra, M. sinensis and M. phillippinensis.* Most of the Indian varieties of mulberry belong to *M. indica.* In Kashmir and Karnataka, it is used to rear silkworms. The total acreage of mulberry in India is around 282,244 ha. Genetic variability has been observed in fruiting type mulberry. In our country, two type mulberries are cultivated for fruits *viz.* green and red colour fruit. However, length of fruit varies according to genotype. In mulberry, Some promising selection have made, however there is no named improved cultivar.

Exotic cultivars

Black Persion : *M. nigra* large black fruit, over an inch long and almost as wide. Juicy with a rich sub acid flavour. The tree is fairy drought resistant once established.

Collier : *M. alba* x *M. rubra.* Medium sized, purplish black fruit, $1^1/_8$ inches long and 3/8 inches in diameter. Flavour sweet with just a trace of tartness. Quality very good. Everbearing. Ripens over a long period. Tree of medium size, spreading. Relatively hardy, very productive.

Downing : The original Downing was a *M. alba* var. multicaulis. Plant grown from seed sown about 1846. The fruit was black with excellent flavour and ripened from June to September. Other varieties have subsequently been sold under the same name.

Illinois everbearing : *M.alba* x *M.rubra*. Originated in Illinois. Introduced in 1958. Black nearly seedless fruit large and very long. Flavour good to very good, very sweet, considered best by many. Matures over along season. Tree vigorous and somewhat dwarfed, extremely hardy and productive.

Kaester : *M.nigra*. Originated in Los Angeles. Introduced in 1971 by Nelson Westree. Large black or deep purple, elongated fruit, $1\text{-}^{1}/_{2}$ inches long and ½ inch in diameter. Flavour very sweet with good sweet/ tart balance. Tree bears heavily.

Pakistan : Originated in Islamabad, Pakistan. Extremely large, ruby red fruit $2^{1/2}$ to $3^{1/2}$ inches long and 3/8 inch in diameter. Flesh firmer than most other named cultivars. Sweet with a fine balance of flavours. Quality excellent. Tree spreading with large heart shaped leaves. Recommended for cultivation in cold areas.

Rivlera : Originated in Vists, California. Elongated, deep purple, black fruits, 1 to 1½ inches in length, ½ inch in diameter. Flesh slightly juicy and very sweet. Very good dessert quality. Ripens over a long period, from April to June.

Russian (Tatarica) : Introduced into Europe from China about 1,500 years ago. Fruit reddish black, of good quality when completely ripe. Tree bushy to 35 ft tall, very hardy and drought resistant. Planted wildy for wind breaks and wild life food.

Tehama (Giant White) : Originated in Tehama county, California. Very large, white coloured, plump fruit, $2^{3}/_{4}$ inches in length and ½ inch wide. Very sweet succulent, melting flesh. Attractive, large- leaved tree. Probably best adopted to mild winter areas.

Wellington : Originated in Geneva. Reddish black, medium sized fruit, 1- $1^{1/4}$ inches long, 3/8 inch in diameter. Form long slender and cylindrical. Flesh soft, of good flavour. Ripens over a period of several weeks. Tree is heavy producer.

Propagation

It is propagated by seeds as well as by vegetative methods. It is commonly multiplied through stem cuttings and budding. Micro-propagation technique has also been attempted to multiply true to type plants on mass scale. For raising rootstocks of mulberry, Russian variety Tatarica is most suitable. This species is drought hardy but sensitive to cold, best suited for raising windbreaks.

Seed propagation

Mulberry can be propagated through seeds. The cross pollination is the rule in mulberry (Reich, 1922), propagation by seeds does not conserve true to type owing to its heterozygous nature. Mulberry seeds remain viable even up to 14 years when kept in dedicator conditions compared to 4 months only in control. Seed propagation is not common practice. Seeds are sown in polybags on a mixture of soil containing one part each of soil, compost and sand. Well grown up seedlings are transplanted in the fields. If plants are raised for rootstocks, it should be budded of suitable variety.

Vegetative methods

Cutting

Various methods of vegetative propagation can be successfully employed for propagation of mulberry. Propagation of mulberry through cuttings is still one of the most cost effective methods for clonal regeneration. Propagation by cuttings is easy. Hard wood cuttings of 25-30 cm length treated by IBA (5g/l) from mature stems with 4-5 buds are planted in nursery in the month of December and January (Fatma and Eylem, 2003). The best rooting percentage was obtained from bunch planting and for January cuttings. Among the three methods examined; bunch planting seems to be the best for the rooting of black mulberry. Root and shoot formation in these cuttings takes place within 3-4 months and during rainy season, these can be transplanted in the fields. Treatment of mulberry cuttings with cane sugar for 24 hrs increased the rooting percentage.

Mulberry cuttings were treated with IBA at 50,100,200 or 400 ppm with and without the addition of 25ppm boron solution. Root production was stimulated by all but highest result was achieved with IBA 200ppm with or without boron (Misra and Jauhari, 1970).

Budding

Mulberry is propagated through budding method. T-budding is found best method during spring season. Method of conventional vegetative propagation by budding or grafting is not practical since it requires special labour power and expensive nursery practice. Also, the success of grafting depends on internal factors like exudation, compatibility, activity of cambium, as well as external conditions like temperature and humidity and characteristics of soils (Reich, 1922).

Air -layering

It is propagated by air layering using root promoting substances. IBA and NAA up to 7,500ppm stimulated rooting in air layers of mulberry (Misra and Jauhari, 1970).

Micro-propagation

Tissue culture techniques can be used for mulberry, even though guaranteed protocols have not yet been reported (Bhau and Wakhlu, 2001). However, success on micro-propagation of mulberry (*Morus alba* L.) through *in vitro* culture of shoot tip and nodal explants has been reported by Mohammad *et al.* (2003). They reported high frequency of sprouting (80%) from nodal and (70%) from shoot tip explants and shoot tip was observed in the primary cultures of *Morus alba* L. on MS medium supplemented with BAP and Kinetin. Bapat *et al,* (1987) observed that axillary buds of mulberry (*Morus indica* L) were encapsulated in alginate and agar to produce individual beads. The beads could be stored at 4°C for 45 days without loss of viability. Amongst the encapsulating agents tested, sodium alginate was found to be a better matrix. Encapsulated buds regenerated complete plantlets on an appropriate medium. This technique would provide an easy and novel propagation system for the elite as well as difficult-to-root species of mulberry.

Mulberry has multiplied through *in vitro* morphogenesis of direct shoot formation, callus induction and subsequent root formation (Singh *et al.,* 2008). The medium composition used for micro- shoot formation was MS medium supplemented with 30g/l sucrose 8 g/l agar and different level of BA (0, 3, 6 and 9 mg/l) alone or in combination of NAA (0, 0.1 and 1 mg/l). *In vitro* morphogenesis in terms of micro-shoot formation, flowering and embryogenic callus formation was remarkably influenced by different media composition and source of explants. The medium containing 6 mg/l BA resulted best treatment for shoot induction. Whereas, no callus or flowering formed on the media composition devoid of BA or NAA. The media composition having 9 mg/l BA + 0.1 mg/l NAA resulted in good flower formation under in vitro condition. Adventitious root formation in micro- shoot or directly on to nodal explants were studied in ½ MS medium supplemented by different auxins such as IBA, NAA and 2,4-D at different levels (0, 0.5, 1 and 2 mg/l). Varied response of rhizogenesis in all genotype was observed under different concentration of different auxins. Better adventitious root formation was observed in all concentration of IBA. Higher concentration of IBA resulted in thick and short roots (~3 cm.) whereas; low concentration (0.5 mg/l) produced longest roots (~5 cm) with auxiliary root (root hair). The results obtained in present study have high significance in a standardization of clonal micro propagation protocol and development of *in vitro* technique for flower and bud differentiation in mulberry and other fruit species.

Quality Standards of Plant Material

There is no quality standards of planting material fix for mulberry. It is cultivated for silk worm rearing as well as fruits production. In general, plants are developed through stem cuttings. Plants should be of well-developed root system, one year old, good foliage for planting in the field under arid conditions.

References

Anonymous, 1996. Shahtoot ki kheti *In* : Rajasthan mein Udayanki phaslein, Directorate of Horticulture, Govt. of Rajasthan, Jaipur, p 47-48.

Bapat, V.A., Minal Mhatre and Rao,P.S. 1987. Propagation of *Morus indica* L. (Mulberry) by escapsulated shoot buds. Plant Cell Report,6(5):393-395.

Bhau, B.S. and Wakhlu, A K 2001. Effect of genotype, explant type and growth regulators on organogenesis in Morus alba. Tissue and organ culture. Plant Cell, 66 : 25-29.

Fatma, K. and Eylem, S. 2003. Rooting of Black Mulberry (*Morus nigra* L.) Hardwood Cuttings. Jour. of Fruit and Ornamental Plant Research,11:53-57.

Hare Krishna, Singh, D., Singh, R.S.; Sharma, S.K. and Parasar, Avinash 2012. Bioactive compound rich squash from Mulberry (Morus sp.) for nutritional security. Crop Improvement, 39: (spl. Issue). 1103-1104.

Misra, A. K and Jauhari, D.S. 1970. Indian Jour. Hort., 27: 141-46.

Mohammad, K.A., Mohammad, F., Mohammad, A. and Singh, S.K. 2003. Micro-propagation of mulberry (*Morus alba* L.) through *in vitro* culture of shoot tip and nodal explants. Plant Tissue culture, 13 (1) : 47-52.

Reich, L. 1992. Uncommon fruits worthy of attention a gardeners guide. Addison-Wesley Publishing Company, Inc. pp. 173-183.

Singh, D., More, T .A. and Awasthi, O.P. , Ram Newas and Srivastava, Shweta 2008. *In vitro* Morphogenesis in Mulberry. *In* Book of abstracts of National Seminar on Opportunities and Challenges of Arid Horticulture Crops for Nutrition and Livelihood held at CIAH, Bikaner, March 8-9, 2008, pp.33.

Chapter-22

Kair (*Capparis decidua* Forsk.)

R. Bhargava and R. S. Singh
Central Institute for Arid Horticulture, Bikaner-334 006

Kair also known as Karil or Ker {*Capparis decidua* (Forsk) Edgew} is an economic desertic plant species, which grows well on sand dunes. The plant is adopted so well under hot arid ecosystem that during extreme of summer when it is difficult to survive by many other vegetation, it produces flowers and fruits. Ker is indigenous to India and growing wildly in arid and semi-arid regions of the country. In India, it is found growing naturally in arid and semiarid tracts of Rajasthan, Gujarat, Haryana, Punjab, Madhya Pradesh, Uttar Pradesh, Delhi and southern part of Deccan Peninsula. Besides India, it is found in tropical Africa, Arabia, Sind, Baluchistan, Socotra, South Persia, Pakistan (Sind), Sudan and Egypt. The natural habitat of ker is pediments plains all over dry regions and its density ranges from 8-44 bushes ha^{-1} (Pareek, 1978). The population density of *Capparis decidua* on different land forms have been reported by Gupta *et al*., (1989) as the population varies from 9-70 bushes/ ha. They recorded highest population density on older alluvial plains (81 plant/ ha) and minimum on the periphery of runs (20 plants/ha). In semi-arid of Pali region, 20-40 bushes /ha was noted in rangeland area (Singh,1993). In recent decade, the natural population of ker bushes has been declined due to increase in population pressure of man and animals and increase in cropping area due to canal irrigation in north-western Rajasthan.

No authentic data on this aspect is available so far, as no systematic plantation of ker has been done. However, as an estimate, ker is spread over 3,540 km^2 pediment plains in Bikaner and Jodhpur districts of Rajasthan and giving an estimated annual production of 70,000 tonnes (Chandra *et al*., 1994).

Nutritive Value and Uses

The immature fruits of ker are used for preparation of vegetable and pickles after processing. The pickles of ker are highly delicious and very popular. The immature dehydrated fruits are also used for preparation of vegetable either alone or mixed with other vegetables and can be used in off-season. Dehydrated ker is one of the important ingredients of '*panchkuta*'; a dehydrated mixture of sangri, ker, kachri, kumat and mango/ kamal kakri. However, mature unripe fruits without processing are unpalatable as these contain high concentration of glucosinolates. The ripe fruits are sweet but acrid in taste and thus not used as table fruit but enjoyed by the children and rural tribes. The flower buds are cooked as potherbs. The fruits are rich source of protein and mineral matter. The protein content in ker (8.6%) is more than guava (8%), grape (1.3%), mango (0.7%) and banana (0.3%). Whereas, Chauhan *et al.* (1986) reported that the unripe fruit of ker had protein 14.88, fat 7.43, ash 5.96, crude fibre 12.32 and digestible carbohydrates 59.41 per cent of dry matter, beta-Carotene, ascorbic acid, calcium, phosphorus, iron, copper, zinc and manganese were 5.4, 120.7, 90.0, 179.0, 3.5, 1.1, 1.6 and 1.9 mg 100 g^{-1}, respectively. Total protein fraction contained albumin 53 per cent, globulin 16 per cent, prolamine 11 per cent and glutelin 12 per cent. Ker fruit had phytic acid 304 mg 100 g^{-1} compared with 54 mg 100 g^{-1} in guava.The unripe fruits are also rich source of proteins (18.6%) and vitamin C (8 mg 100g^{-1} pulp). Duhan *et al.* (1992) reported that among thirteen non-conventional foods; Khejri (*Prosopis cineraria*), pinju (*Capparis decidua*) and kachri (*Cucumis* sp.) contained considerable amounts of protein (15-18%). Among locally available fruits and vegetables of the arid zone of Rajasthan viz., Dansra (*Rhus myserensis*), kachri (*Cucumis callosus*), ker (*Capparis decidua*) and sangri (*Prosopis cineraria*); the ker was found to be the richest source of beta-carotene in fresh and various processed states like blanched, dried and roasted (Chaturvedi and Nagar, 2001).

To assess the nutritional composition of ripe fruits, the fruits were collected from different areas of Bikaner, Jodhpur and Jaisalmer districts of Rajasthan during summer, 2002 and analyzed. The range of the composition of fruit pulp is given in table 1. The pulp content is very low (1.25 pulp: seed). Khera and Singh (1975) reported that fruit pulp is about 55 per cent of total weight, which contains 56 per cent moisture, 8.6 per cent crude protein, 5.02 per cent true protein, 1.8 per cent total sugar, 1.1 per cent reducing sugar, 0.057 per cent P, 1.026 per cent K, 0.055 per cent Ca, 0.055 per cent Mg and 7.81 mg/ 100g ascorbic acid.

Seeds are rich source of oil (20.3%) of which 68.6% is unsaturated fatty acid and 31.4 % saturated acids. The alcoholic extracts of flowers and fruit husk have some antibacterial and antifungal activities. Shushila Rai (1987) has extracted and isolated 15 natural products from *Capparis decidua.*

Table 1 : Nutritional composition of fruit pulp of ker (*C. decidua*).

S.No.	Composition	Value
1.	Moisture	52-60 %
2.	Protein	5.0-8.20 %
3.	Total sugars	1.5-2.30 %
4.	Phosphorus	0.056- 0.059 %
5.	Potassium	1.024-1.030 %
6.	Calcium	0.045-0.058 %
7.	Ascorbic acid	7.50-8.10 mg/ 100 g.

Fruit is useful in cardiac trouble, biliousness, destroying foul breath and discharge urinary purulent. The tender leaves and shoots are used for plastering boils. Top green shoots are grounded to powder and used in affliction of the pain. Tender shoots, when chewed, relieves toothache and the bark is useful in cough, asthama and inflammation and good for analgesic, diaphoretic, akheteric, laxative, anthelmintic, ulcer and boils, vomiting and all inflammations. Root bark is used against intermittent fever and rheumatism (Chundawat, 1990 and Chandra *et al.*, 1994). The fresh young leaves after crushing and soaking in water are strained off and residue is dried and allowed to solidify for application. It is eaten with butter in small quantity, which gives relief from pains of bruise (Gupta *et al.*, 1989).

Ker wood is strong and durable. The ker branches can be used for fencing, which are not affected by white ants and also as fire wood as it burns quickly. Ker bushes are very useful in stabilizing sand dunes. *Capparis decidua* was found to be the best species for shelterbelts to check the movement of sand (Pandey and Rokad, 1992). It is very efficacious plant for checking soil erosion by wind and water. Fertility status of dune sand has been found to be improved near the ker bushes. It affects sharp decline in soil alkalinity (Gupta, *et al.*, 1989). The tender plant part is also eaten by the animals as fodder.

Botanical Description

Ker {*Capparis decidua* (Forsk) Edgew, Syn. *Capaparis aphylla* Roth} belongs to the family Capparidaceae that comprises more than 300 species. The chromosome number is X=7, 8 (a polyploid). The plants are mostly in form of bushes but older plants attained to a size of medium height tree. The study on root system of a mature shrub on rocky site at Bhopalgarh,Jodhpur have shown that roots penetrate up to 4 meter (Singh *et al.*, 2005). Ker plants are densely branched shrubs with green thorny stem and scanty, small caduceus leaves found on young shoots only. The older branches and trunk of trees are brown in colour with slightly splitted bark. The leaves are small, linear oblong, acute, spinous-pointed; petioles are very short; flowers are corymbs that emerge from the old branches or from the short lateral shoots; pedicels are cylindrical; the outer sepals are ciliate, subvolvate and the lower sepals are very saccate, acuminate and the upper being smaller, ovate or oblong. Petals are red and bright scarlet, narrow and oblong. Flowers are hermaphrodite having 13-18 stamens with red filaments, anthers besifixed. Pollen sterlity of about 35 per cent was observed (Rao and Rao, 1977). Ovary gynophore, glabrous. Fruits are berry, ovoid to subglubose and beaked. Generally green to purplish green at immature stage and turn to red or dark pink colour on ripening. The fruit skin is slippery after ripening. The biology of the tree is described, and the medicinal, nutritional, fodder, and economic properties are noted by Srivastava *et al.* (1998).

Under natural population, rich genetic diversity exist with regard to bearing habit, fruit size, colour of fruits and petals, pulp content, etc. but no systematic efforts have been made so far to collect and conserve them and to popularize the promising types. There is a need to identify suitable types with the view to select heavy yielder, large fruit size, high pulp content, rich in protein, proper TSS : acid blend, less acrid, small and soft seeds etc. Some attempts were made by NBPGR, Regional Station, CAZRI, Jodhpur and CIAH, Bikaner to identify promising germplasm of ker from natural population. There is no named improved variety of ker so far its commercial cultivation.

Propagation

Seed

The natural regeneration of ker takes place by seeds and suckers. Though, the seed multiplication is generally not recommended but suckers with some roots can be used as planting materials. Attempts to propagate ker through vegetative means yielded low rate of success so far. Seeds collected from fresh fruits sown in poly bags in the nursery during June-July and germinate after 15-20 days. However, the initial seedling growth is slow in ker plants.

Vegetative propagation

Ker is multiplied through stem cuttings. Vashishtha (1987) suggested that best time for taking cutting is August-September and these cuttings should be treated with 1000 ppm IBA before planting. In later study, Meghwal and Vashishtha (1998) suggested that July-August is the best time for taking the cutting under Jodhpur conditions. Whereas, Shekhawat (1994) claimed best response at 5000 ppm while propagating the ker through cutting. Under mist system of nursery shade net house, Bhargava *et al.* (2002) reported that semi-hardwood cuttings taken in the month of August-September, treated with 7500 ppm IBA + 1000 ppm Thymine in DMSO had given maximum sprouting (50 %). In another experiment, aqueous extracts of *Capparis decidua* cuttings inhibited *Nerium* sp. root growth and cluster bean sprouting. ABA-like substances (at 3.2 µg per g) were observed in *Capparis decidua* aqueous extracts, which may be responsible for inhibition of sprouting and root growth in the other plants (Bhargava *et al.*, 2000).

Micro-propagation

In an investigation on micro-propagation, various concentrations of BA (1, 3 and 5 mg/ l) showed development of multiple shoots and shoot buds in the primary cultures of *Capparis decidua* within 30 days. With nodal explants shoot multiplication was best on medium supplemented with 5 mg l^{-1} BA and after asecond and subsequent subculturing, further multiplication of shoot buds was highest on the medium containing 3 mg/l BA (Bhansali, 1994). Whereas, Deora and Shekhawat (1995) obtained multiple shoots from nodal explants on Murashige and Skoog's (1962) medium + 0.1 mg/litre NAA + 5.0 mg/ litre BAP [benzyladenine] + additives (50 mg/ litre ascorbic acid and 25 mg/litre each of adenine sulfate, L-arginine and citric acid) at 28 ± 2°C, with a 12 h/day photoperiod and 35-40 µ\mol m-2s-1 photon flux density. The shoots were multiplied by (i) subculture of nodal shoot segments onto MS + 0.1 mg/litre IAA + 1.0 mg/litre BAP + additives, and (ii) repeated transfer of original explant onto MS + 0.1 mg/litre IAA + mg/litre BAP + additives, at intervals of 3 weeks. Some 60-70%

of the shoots rooted when pulse treated with 100 mg/litre IBA in half strength MS liquid medium for 4 h, and then transferred into hormone-free half-strength agar-gelled MS basal salt medium. Incubation in the dark at 33 ± 2°C for 6 days favoured root induction. *In vitro* hardened plants were transferred to pots. Bhargava and Saroj (2003) reported that both callus induction and shoot proliferation was best in the medium containing MS = 5 mg/ l BAP + 1.5 mg/ l IBA. However, rooting was not found in this medium, thus the plantlets were sub-cultured in modified MS medium and roots were induced in MS + 0.1 mg/ l IAA + 1.0 mg/ l BAP and 0.1 mg/ l IAA + ½ strength MS media.). Protocol has been standardized for clonal propagation of *Capparis decidua* by using nodes and shoot tips as ex-plant for regeneration (Chahar *et al.*, 2010). Best rooting response (37%) was obtained on MSRS medium supplemented with1.0mg/L. NAA and 1.0mg/L. IBA.

The mass multiplication of elite type of ker should be taken by vegetative means and improving the poor establishment of vegetatively propagated plants.

References

Bhansali, R.R. 1994. In: Advances in Plant Tissue Culture in India (Tandon, P., Ed.) Pub. Pragati Prakashan, Meerut (UP).

Bhargava, R. and Saroj, P.L. 2003. Standardization of protocol for micropropagation of *Capparis decidua*. Annual Report 2002-2003 CIAH, p. 125-126.

Bhargava, R., Saroj, P.L., Neelam Chauhan and Piyush Verma 2002. Development of propagation techniques for *Capparis decidua*. In: CIAH, Annual Report 2001-2002.

Bhargava, R., Vishal Nath and Pareek, O. P. 2000. Note on the role of plant growth inhibitors in sprouting of *Capparis decidua* cuttings, Current Agriculture, 24(1-2): 131-133

Chahar, O.P, Kharb, P.; Ali,S. F.; Batra P. Choudhary,V.K. 2010. Development of protocol on Micro- propagation in Ker (*Capparis deciduas* (Forsk.)Edgew.).World Applied Science Jour., 10(6)695-698.

Chandra, A., Anju Chandra and Gupta, I.C. 1994. Kair In. Arid Fruit Research, Pub.: Scientific Publisher , New Pali Road, Jodhpur. p.299.

Chaturvedi, Y. and Nagar, R. 2001. Levels of beta-carotene and effects of processing on selected fruits and vegetables of the arid zone of India. Plant Foods for Human Nutrition, 56(2):127-132

Chauhan, B.M., Duhan and A., Bhat, C.M. 1986. Nutritional value of ker (*Capparis decidua*) fruit, Journal of Food Science and Technology, 23(2): 106-108

Chundawat, B.S. 1990. Arid fruit culture, Pub. Oxford and IBH Publishing Com., New Delhi.

Deora, N.S. and Shekhawat, N.S. 1995. Micro-propagation of *Capparis decidua* (Forsk.) Edgew. - a tree of arid horticulture. Plant Cell Reports, 15(3-4): 278-281

Duhan, A., Chauhan, B.M. and Punia, D. 1992. Nutritional value of some non-conventional plant foods of India. Plant Foods for Human Nutrition, 42(3): 193-200

Gupta, I.C., Harsh, L.N., Shankarnarayana, K.A. and Sharma, B.D. 1989. Wealth from wastelands. Indian Farming, 38(11): 18-19 & 24.

Khera, A.P. and Singh, J.P. 1975. *Capparis aphylla*- A multipurpose fruit tree for arid region. Indian Horticulture, 20(2):10.

Meghwal, P.R. and Vashishtha B.B.1998. Effect of time of planting and auxins on rooting of cuttings of ker (*Capparis decidua* Forsk.). Annals of Arid Zone,37:401-404.

Pandey, A.N.,and Rokad, M.V. 1992. Sand dune stabilization: an investigation in the Thar Desert of India. Journal of Arid Environments, 22(3): 287-292.

Pareek, O.P.1998. Underutilized fruits- Kair In.: 50 Years of Horticultural Research (Ghosh, S.P.and Bhatnagar, P.S., Eds.) , Division of Horticulture, ICAR, New Delhi, Pp. 95-99.

Rao, B.H. and Rao, P.S.P. 1977. Anther in *Capparis decidua* (Forsk.) Pax. Current Science, 46:19.

Shekhwat, J.S. 1994. Studies on growth, fruit development and quality of ker (*Capparis decidua* (Forsk) Edgew). Ph. D. Thesis submitted to Rajasthan Agricultural University, Bikaner.

Shushila, Rai 1987. Chemical extraction of edible plants of Rajasthan desert with special reference to Capparidaceae. Current Agriculture, 11: 15-23.

Singh, R. S. 1993. Maru Kshetra ka Upyogi phal Kair,Udyan Lok, 5(4):1-2.

Singh,Manjit ,Jindal,S.K. and Srivastava, R. 2005.*Capparis deciduas*-A multi-purpose shrub. In. Shrub of Indian Arid Zone, CAZRI, Jodhpur, pp. 97-104.

Srivastava,A.K; Singh, HK and Pandey, AK 1998. An under exploited minor fruit: kair,. MFP-News. 1998, 8: 4, 7-8.

Vashishtha, B.B. 1987. Vegetative propagation of *Capparis decidua*. Annals of Arid Zone, 26(1&2): 123-124.

Chapter-23

Lasoda (*Cordia myxa* L.)

P. K. Yadav and Vartika Srivastava
S.K. Rajasthan Agricultural University, Bikaner-334 006

Lasoda (*Cordia myxa* L.) known by many names viz., cherry of desert, assyrian palm, sebestene plum, *lasura, lehsua, gonda* and *naruvilli,* is an underutilized, multipurpose fruit tree distributed in arid and semi-arid regions of the world. Arid regions are considered under the category of wastelands which are not being used to their fullest potential. Among the arid region crops, *lasoda* is very suitable crop for cultivation in Rajasthan as it can withstand drought to a very great extent and can tolerate temperature as high as 48 to 50°C during summer months. *Lasoda* thrives well on neglected, low fertile, sandy, gravelly, saline, alkaline soils and hence can be used for forestation of waste lands. However, it grows vigorously in moist sandy loam soils and gives high productivity.

Lasoda is native of north-western India (Stewart and Brandis, 1992) and distributed throughout the country mainly in warmer region up to altitude of 5000 ft., as natural, wild and occasionally cultivated. It is found in a variety of the forests from Rajasthan to moist deciduous forests of the Western Ghats and Tidal forest of Myanmar. In Maharashtra, it grows in moist monsoon forests also. In Rajasthan, its commercial plantations are found in Pali, Sirohi, Udaipur, Jodhpur and Jalore districts. However, it is grown as fruit plant in hot areas of Jaiselmer district of Rajasthan (Chandra and Pareek, 1992). It is a profitable tree for semi- arid parts of Rajasthan (Singh and Vishwa nath, 1991). Its few of the xerophytic characteristics are: deeper root system, deciduous nature, reduced foliage, sunken or covered stomata, waxy coating or hairiness on leaf surface which minimizes the evapo-transpiration and makes plant adaptive to the hostile climate for plantation (Patel and Singh, 2000). Earlier it was treated as under exploited plant. Looking to it vast potential, now it is being cultivated on large scale in semi arid and arid regions.

Nutritive Value and Uses

Its tender fruits are used to make vegetable and mixed pickle. Immature fruits of *lasoda* make a tasty broth which is hot in effect as per ayurveda practitioners. Fruits can make an excellent pickle too which is not too hot. In fact, the preserve is quite effective against indigestion. Ripe fruits are full of vitamins and regular use is supposed to be helpful in growth of hairs. Fruits have profound medicinal value and considered as antihelmentic, diuretic, demulcent, expectorant. An otonecine and saturated polymines (PAS), were found in *Cordia sinensis* and *Cordia myxa* having anti-tumor activity (Wassel *et al.*, 1987). Kernels are used for curing birdlime. The wood is generally used for making furniture, house-posts, beams, agricultural implements, spokes, etc. It is also used for making quality toys, bowls and wooden utensils for kitchen. Since, the wood is light and yet reasonably strong, it is in great demand for ornamental wood work. Antifungal and larvicidal cordiaquinones have been reported from roots of *C. curassavica* (Ioset *et al.*, 2000).

Leaves are used to cover fruits to protect from birds in arid region. Lac insect can also be reared on *lasoda* plants. Ficarra (1995) and Schilcher (1993) worked on leaf preparations of several species of *Cordia* and used in traditional medicine as remedies for osteoarticular diseases. The effects of aqueous, petroleum ether and alcoholic extracts of *C. francisci, C. martinicensis, C. myxa, C. serratifolia* and *C. ulmifolia* leaves were studied in rats using various models of acute and chronic anti-inflammatory, and analgesic activity. (Pundir (1987) reported maximum increase in plant growth parameters during Aug- Sept. than the March-April growth in *Cordia myxa*.). The composition of lasora fruit on dry weight basis is reported by Kaushik and Dwivedi (2004). There is a possibility of its uses in gum, confectionary and processing industries.

Composition of Lasoda fruits (per 100g on dry matter basis).

Ingredients	Quantity (per 100 g)
Moisture	74%
Protein	2g
Fat	2g
Calcium	55mg
Crude Fiber	2g
Carbohydrates	93g
Phosphorus	275mg
Zinc	2.0mg
Manganese	2.0mg
Iron	6.0mg

Botanical Description

Lasoda is a medium sized, broad leaved deciduous tree, its important species is *Cordia myxa* and other species are *C. gharaf, C. macleodii, C. vestita* and *C. wallichis.* It belongs to family Boraginaceae. The fruit is drupe, rounded or pear shaped with sticky pulp in which seeds are embedded. The fruits when immature are green which turn yellow to pink when mature. *Lasoda* is a fairly fast growing tree. It matures in about 50-60 years by then its girth at breast height is about 1-1.5 cm. Its bole (main trunk) is generally straight and cylindrical, attaining a height of nearly 3-4 m. Branches spread in all directions by virtue of which its crown can be trained in to a beautiful inverted dome, like an umbrella. When fully grown-up, total height comes to nearly 10-15 m. Due to these qualities of *lasoda,* it can grow in arid region without any problem after management of some agro-techniques. Plants are susceptible to frost in arid region (Singh *et al.,* 2012). However, after employing management practices affected plants were recovered.

Leaves of the *lasora* are broad, ovate, alternate and stalked with a spread of 7-15 cm × 5-10 cm. In matter of external appearance, these are glabrous above and pubescent below. Young leaves tend to be hairy. Fresh foliage is quite useful as fodder for cattle, more so during grass famines. Leaves contain about 12-16 per cent crude protein and 16-27 per cent crude fiber. Leaf gall formation by mites has been reported in lasoda (Sharma *et al.,* 1998). Pustules formation on leaves in lasoda has been observed.

Lasoda trees flower during March- April, Flower is short, stalked, bisexual and appear in loose corymbose cyme. Inflorescence is generally terminal is usually white in colour. Individual florets are nearly 5 mm in diameter. At places, these are somewhat hairy and white. Being a bisexual plant, both male and female flowers occur on same tree. Calyx part of an independent flower is about 8 mm long and glabrous, but not pubescent. It splits irregularly at the opening of its flower. The filaments are hairy. Anthesis starts with the longitudinal slit at the blossom end of the bud. The slit gradually widens and the petals get unfold. The anthesis occurs between 7:15 pm to 8:45 pm. It reveals that the maximum number of flowers (57.5%) opened during 7:45 pm to 8:00 pm. The percentage of fruit set was found to be very low. Out of 2481 buds observed 2426 dropped in the bud state only (Vashishtha *et al.*, 1985; Pundir, 1987; Duthie, 1960). According to Pundir (1987), fruit setting in *lasoda* was about 33% and application of GA_3 at 100 ppm resulted in highest fruit set (59.96 %).

A wide range of genetic variability was observed in the existing plantations of *lasoda* in western parts of Rajasthan (Samadia, 2007). There is lot of scope to select superior strains of lasora as wide diversity is found in its fruits and other parameters (Kaushik and Dwivedi, 2004). However, there is no named cultivar. In general, two types of plants *viz.*, large fruited and small-fruited are found growing and are sold by nurserymen. Large fruited cultivars have an average fruit weight of 8.55 g whereas small-fruited cultivars have fruit weight of 3 g only. Large fruited types in Gujarat are recognized as Paras Gonda, which in fact is a general term for any fruit variety with big size fruits. Different genotypes were evaluated at CCSHAU, Hisar by Saini *et al.* (2002) and reported variability in height N-S 1.25-4.22 m and E-W 1.0-4.02 m. Kumar *et al.* (2006) observed variation in maturity indices and quality characters in lasoda fruits.

Genetic diversity of *Cordia* species especially *C. myxa* has been collected by NBPGR from Haryana, Gujarat, Madhya Pradesh, Himachal Pradesh and Uttar Pradesh. 134 accessions of various *Cordia* species have been collected. Around 45 accessions of elite germplasm have been collected by NBPGR in collaboration with CCSHAU, Regional Research Station, Bawal from Rewari, Mahendragarh and Bhiwani districts of Haryana and germplasm was established at field gene bank CCSHAU, Regional Research Station, Bawal and NBPGR, Regional Station, Jodhpur. Recently, a promising genotype of lasoda (Thar Bold) has been identified by Saroj and More (2012) and recommended for cultivation.

Germplasm of lasora is being conserved in the field genebank at CCSHAU, Regional Research Station, Bawal (30), CIAH, Bikaner (65), NBPGR Regional Station, Jodhpur (73),SKN College of Agriculture, Jobner, Jaipur and NDUAT, Faizabad. For seed propagation, ripe fruits are collected from the trees and rubbed to remove the flesh. The healthy stones are dried in the shade and kept in the tin containers. The stones can be stored for about one year in the air tight containers kept in a dry place to avoid insect attack. There are 4000-7000 stones per kg. Cryogenic storage of seeds is reported to be beneficial in long term germplasm conservation but when seeds of *lasoda* are stored in liquid nitrogen, they show around 14% decline in viability. Seeds with high recovery were obtained on cryostorage between 6-8% seed moisture content. Seeds of *C. myxa* (24 accessions), *C. crenata* (3 accessions), *C. obliqua* (1 accession) and *C. rothii* (9 accessions) have been cryostored successfully at NBPGR.

Propagation

Seed

Lasoda is commonly propagated by seeds inspite of low seed germination percentage which varies between 25-30 %. However, in *Cordia gharaf*, around 70-80 % of seed germination is observed. For raising seedlings or rootstock, seeds are sown in well-prepared nursery beds or in polythene tubes (25 × 10 cm size) filled with garden soil. It has hard seat coat which does not absorb water easily. Germination is epigeal, 1-2 seedlings may appear from one stone. The seed should be sown in nursery beds or polythene tubes. Seedlings should be picked when the first pair of true leaves appeared. Sowing is done in June-July. If it is sown in nursery bed, then sowing depth of seed should of 2 cm in lines and spaced about 20 cm apart. A seed rate of 80 g/m^2 of nursery area are adopted. Germination starts in about 3-4 weeks and is completed in 6 weeks. The germination percentage 38-50% can be obtained when the seed is treated with H_2SO_4 for 30 minutes. The stumps should be 8-13 mm thick at the root collar with about 4 cm stem and 20-25 cm root. Seedlings should be ready for planting in the field in about one year at commencement of monsoon rains. When *lasoda* is propagated by seeds, directly sown seed take longer period (20-25 days) for germination and show irregular germination. Therefore, pre-treatment of seeds with cow dung for 24 hours and soaking in water is advisable to increase germination percentage and seedlings growth. Watering at 2-3 days interval also induces early and uniform germination (Singh *et al.*, 1986).

Vegetative propagation

Lasoda is multiplied through vegetative methods *viz.* budding and grafting. Micropropagation technique has also been tried to develop true to type plants on large scale.

Raising of Rootstocks

The seeds of *Cordia dichotoma* and *Cordia gharaf* can be used as root stock for budding and grafting but *C. gharaf* is more suitable. The age of the root stock should be 9-12 months however scion should be 3 months old for better success.

Budding

Vegetative propagation in *lasoda* has been uncommon till recent past but present studies have opened a new vista that now it becomes easy to propagate *Cordia dichotoma* by budding. Shield or T-budding can be performed in the month of April-May; during this period success remains very high. T-budding, cleft-grafting and wedge-grafting are also successful (Pundir, 1987; Chovatia and Singh, 1996). T-budding has been found most successful with 96.7% success than the other method. Most suitable period for T-budding is the month of August because plant is most active physiologically during the period of August to September (Chandra *et al.*, 1992; Chovatia and Singh, 1996). Under arid climatic conditions, Patch budding gave better results even on to rootstock grown during current season. In rootstock, N and protein contents were highest during June-July and the carbohydrate content and C:N ratios of both scions and rootstocks were highest in February. Phenolic compounds content was highest in scions and rootstocks in July. Patolia and Singh (1993) also reported ethrel treatment in inducing early sprouting specially in April-May however, it was less successful in the month of July. Trees with small fruits can be transformed to large fruited tress of good quality by top working (Sharma *et al.*, 1983). *Lasoda* can be propagated by side grafting when it is done in the month of July. Sprouting from this scion is observed after 90 days of operation and the success was 6.1%. However, when cleft grafting was done, the success was 30.5%. This finding was observed by Pundir and Porwal (1998).

Top working

In order to improve quality of inferior or small fruited trees, top working technique was standardized at SKN College of Agriculture,

Jobner, Jaipur (Sharma *et al.*, 1983). By this technique five year old trees of small fruited were transformed into large fruited.

Micropropagation

At CIAH, Bikaner, micropropagation was achieved through single node explants. Both type of organogenesis of shoot and root formation was obtained by placing nodal explants on shoot induction medium and subsequently on rooting medium. More than 90% culture were responsive for *in vitro* axillary shoot proliferation on MS medium supplemented with 30g sucrose 8.0g agar and a combination of 2.0 mg. BA and 0.1mg NAA per litre. After culture period of 6-8 week, the nodal explants with microshoot were rooted on MS media supplemented with 30 g sucrose+3 g activated charcoal and different levels of NAA (0.1,3 and 6 mg/l).The season and position of explants on the mother plant was also shown to influence *in vitro* performance of the regeneration in terms of establishment of aseptic culture, percentage of responsive explants and intensity of leaching of growth inhibitory phenolics (Singh *et al.*, 2006). Tissue culture raised plants have planted in field for evaluation.

Quality Standards of Planting Material

Being under utilized fruit tree plant, there is no quality standard fixed for *lasoda* plants however, the plant should be healthy, good foliage and free from diseases, insects and low temperature injury. The plant should be single stem of pencil size thickness of improved genotype used for planting in the field. Proper healing of bud union is required at the time of transplanting. Very old or young plant should not be used for planting. Plants should be procured from either registered or Govt. Nursery. Roots of sapling should not be coiled in container.

References

Chandra, Atul and Pareek, O.P. 1992. *Lasoda* (*Cordia myxa* Roxb.). A potential fruit crop in Jaisalmer district of Western Rajasthan. Agric. Sci. Digest, 12 (1) : 11-12.

Chovatia, R. S. and Singh, S.P. 1996. Propagation of *Cordia dichotoma* Forst. through budding and grafting. Journal of Applied-Horticulture, 2: 1-2, 127-134.

Duthie, J. F. 1960. Flora of Upper Gangetic Plains, Vol. 1:528.

Ficarra,R; Ficarra,P; Tommasini,S; Calabro,ML; Ragusa-S; Barbera,R and Rapisarda,A, 1995. Leaf extracts of some Cordia species: analgesic and anti-inflammatory activities as well as their chromatographic analysis. Farmaco, Edizione-Scientifica., 50: 4, 245-256.

Ioset, J.R., Morston, A., Gupta, M.P. and Hostettmann, K. 2000. Phytochem, 53 : 613-617.

Kaushik, R .A. and Dwivedi, N.K. 2004.Genetic diversity in Lasora, Indian Horticulture, July-Sept. issue, pp. 1 & 27.

Kumar, P.,Yadav, P.K. and Choudhary,M.L. 2006. Studies on physical properties of fruits as indices of maturity in Lasoda (*Cordia dichotoma*). Ind.Jour.Arid Hort., 1(1):74-75.

Patel N.L. and Singh, S.P. 2000. Effect of different trees species on site amelioration. Indian J. of Forestry, 23 (2): 192-196.

Patolia, M.B. and Singh, S.P. 1993. Effect of primary rotting of growth regulators on budding success under different times of budding in sebesten (*Cordia dichotoma* Forst. Advances in Horti. and Forestry 3 : 93-97.

Pundir, J.P.S. 1987. Studies on growth, flowering and fruiting behaviour of Lehsua (*Cordia dichotoma* Frost.) .Ph.D. Thesis submitted to Sukhadia University, Udaipur.

Pundir, J.P.S. and Porwal, R. 1998. Management of Propagation techniques in Cordia (*Cordia dichotoma* forsk F.) under semi arid conditions of Rajasthan. Proceedings of National Seminar on Management of Natural Resources of Rajasthan. pp. 140.

Saini, R.S; Kaushik, R.A. and Singh, S. 2002. Research note on the evaluation of *C.cymxa* Roxb germplasm for vegetative growth character under semi arid conditions Haryana J. Hort. Sci., 31(31-2):62-63.

Samadia, D. K. 2007. Variability and scope of improvement in Lasoda (*Cordia myxa*). Indian Jour. Agro-Forestry, 9:111-115.

Saroj, P.L. and More, T. A. 2012. CIAH Lasoda: The fruit for future, Indian Horticulture, Jan-Feb, 2012, IACR, New Delhi, pp.21-23.

Schilcher,H (ed.) and Phillipson,J.D. (ed.);Loew,D.1993. Hepatotoxic effects of the leaves of Some *Cordia* species. Acta Horticulture, No. 332, 237-242.

Sharma, H.C., Pundir, J.P.S. and Rawat, T.S. 1983. Indian Hort., 28 : 17-18.

Sharma, S.L. Debnath, M and Kant . U. 1998. Histochemical profile of mite induced leaf galls of *Cordia dichotoma* (L.).Journal-of-Phytological-Research, 11: 2, 197-198.

Singh, D., Shukla, A.K., Bhargava, R., Awasthi, O.P. and Meena ,S.R. 2006.Direct organogenesis in single bud ex-plants of Lasoda (*Cordia myxa* Roxb.). Ind. Jour. Arid Horti.,1(1):31-34.

Singh, R. S. and Vishwa Nath, 1991. Shushk kshetra mein Labhkari Lasoda Lagaein, Phal Phool, April-June issue,pp.8-9. ICAR, New Delhi,

Singh, R. S; Vishal Nath and Tewari, J.C. 1986. A promising fruit tree for arid ecosystem. Deco-Mirror, 3 (2): 11-15.

Singh, R.S.; Bhargava, R.; Garima Pal and Sharma, S.K. 2012. Effect of frost on fruit plants and its management in hot arid region. Crop Improvement, 39 (spl. Issue).723-724.

Stewart, J. L and Brandis, D. 1992. *The forest flora of North-West and Central India.* Reprinted by Bisen Singh and Mahendra Pal Singh, New Connaught Place, Dehradun, p. 602.

Vashishtha, B.B., Raja Ram and Singh, R. 1985. *Haryana J. Hort. Sci.,* 14 : 156-159.

Wassel, G., El-Menshawi, B., Saeed, A., Maharan, G. and El-Merzabain, M. 1987. Pharmazie, 42: 709.

Chapter-24

Jamun (*Syzygium cuminii* Skeels.)

Sanjay Singh and A.K. Singh
Central Horticultural Experiment Station, Godhra - 389 340

Jamun fruit holds a dear place in Indian literature and religion. One such tale states that when Rama was exiled from Ayodhya, he lived on nothing but jamun for 14 years. Hindu mythology also likens Krishna's blue skin to the color jamun. Its correlation with Krishna is one reason why the fruit is planted near temples honoring him, even today. Another text states how the Puranas split the cosmos into seven continents. The continent at the center and the one on which humans live was regarded as the island of jamun. At the center of the earth was a single jamun mother tree that produced fruits so large that, upon falling, gushed life-giving rivers for all to enjoy.

The Jamun, *Syzygium cuminii* Skeels. (*Eugenia jambolana*) is a nutritious fruit with a variety of uses. The properties of the species have been extensively studied, particularly with reference to the pulp and seed. Jamun has many valuable properties and virtually every part of the tree has been utilized by both urban and rural dwellers.

Almost all kinds of fruits are grown in India due to the prevalence of varied ecological conditions. Conservation of fruit germplasm is done for current use as well as in the future and comprises entire diversity of primary, secondary and tertiary gene pool. Almost all underutilized fruits were earlier in a state of neglected condition but Pareek (1988) did a comprehensive review of genetic resources activities of arid zone fruits but jamun was not covered under the programme. It is one of the very important crops and needs attention on germplasm conservation in different part of the country. Information regarding area and production of this fruit in India is not available because it is not grown on plantation scale but trees in patches are common in the villages in many states. They can be seen growing in parks, on roadsides, avenues and as windbreak. The production in India is mainly concentrated in the drier states and the produce is collected by the villagers and sold in the local market. The potential production areas for jamun depend on the demand for the products. Its cultivation may spread to arid and semi- arid areas, resource-poor areas and wastelands where other crops cannot be grown.

Besides India, it is also found in Thailand, Philipines, Madagaskar and some other countries. It also occurs in the lower range of Himalayas up to an elevation of 1300 meters and in the Kumaon hills up to 1600 meters. It is widely grown in the larger part of India from the Indo-Gangetic plains in the North to Tamil nadu in the south (Singh *et al.*, 1967). It is one of the most hardy fruit crops and can easily be grown in neglected and wasteland where other fruits plants cannot be grown successfully (Singh and Singh, 2012). The limited work has been done to exploit the genetic resources of jamun. This crop is facing severe genetic erosion and disappearance of several species as a result of activities related to urbanization and intensive agriculture. The genetic diversity of the related wild species is of particular value in search for sources of resistance to physiological races of pathotypes of fungi, bacteria viruses and nematodes besides winter hardiness, resistance to drought and salinity, etc.

Nutritive Value and Uses

The fruit is good source of iron, sugars, minerals, protein and carbohydrate etc. Fully ripe fruits are eaten fresh and can be processed into beverages like jelly, jam, squash, wine, vinegar & pickles. The Jamun fruit has sub- acid spicy flavour and squash is very refreshing drink for quenching the thrust in the summer season. A little quantity of fruit syrup is much useful for curing the diarrhoea. The vinegar prepared from juice extracted from slightly unripe fruit is stomachic, carminative and diuretic, apart from having cooling and digestive properties (Thaper, 1958). Small jamun fruits not suitable for table use were found to be suitable for use in the beverage industry as they contained a high amount of acidity, tannins and anthocyanins (Anonymous, 1986). The volatile oil from the jamun fruits can also be extracted (Vijayanand *et al.*, 2001). Fruits are used as an effective medicine against diabetes, heart and liver trouble (Singh, 2001). The powdered seeds have also reputation of being useful in the treatment of diabetes. Oral administration of an aqueous jamun seed extract for 6 weeks caused a significant decrease in lipids, thiobarbituric acid reactive substances and an increase in catalase and superoxide dismutase in the brain of diabetic rats (Prince *et al*, 2003). Leaf extract of jamun (*Syzygium Cuminii* Skeels) also reduces the radiation induced DNA damage in the cultured human peripheral blood lymphocytes (Jagetia and Baliga, 2002). Jamun seed can be used as concentrate for animals because it is rich in protein, carbohydrates, and calcium (Thaper, 1958). Its wood is used for railway sleepers. The jamun timber is used in buildings, for agricultural implements and for well work, as it resists the action of water (Chundawat, 1990).

The tasty and pleasantly flavoured fruit is mostly used for dessert purposes. The fruit is usually eaten with salt. Jamun squash is a very refreshing drink in summer season. Jamun is used for preparation of wine particularly in Goa. Almost every part of the tree is used for one purpose or other. A tree made of tender leaves and alcoholic extracts of leaves and fruits are used in South America for curing the stomach disorder. Powdered seeds are also very useful for the cure of diabetes. The seed powder of jamun reduces the quantity of sugar in the urine very quickly and permanently. It is used as a lotion for the cure of ring worm. Jamun seeds can be used as a concentrate for animals because it is rich in protein, carbohydrate and calcium. Its wood is used for manufacturing railway sleepers. It is not attacked by wood decaying insects or fungi.

A number of seedling strains with lot of variation in respect of fruit shape and size, pulp colour, TSS, acidity and earliness particularly in Uttar Pradesh, Gujarat ad Maharashtra provide a good scope for selection of better varieties. Lancaster percy and Bose, 1965 noted that F1 hybrid of cross between the Alba variety of water apple (*S.javanicum*) and rose apple (*S.jambos*) had prolific bearing with large fruits than those of the parents. The fruits contained fragrance and sweetness of rose apple. Survey in Pune and Ahmednagar districts of Maharashtra revealed wide variation in fruit weight (3.5 to 16.5 g), pulp content (54.29-85.71 percent), TSS (4.5-17 per cent and acidity (0.16 – 0.55 per cent) from where promising types No.15, 4,14 and 13 were identified (Keskar *et al.*, 1989). Ashraf (1987) reported that fruit shape in jamun varied from round to oblong and base and apex of fruits from flat to pointed. He also observed great variability in physico-chemical characteristic of fruits offering possibility of selecting a variety suitable for fresh market and processing. Small seed size, high pulp content with better chemical qualities is considered ideal characteristics. Singh *et al.* (1999) evaluated eight genotypes of jamun under Faizabad conditions and reported that oblong types had more fruit weight and relatively less seed weight. Among the locally available genotypes of jamun in West Bengal

(Table 1), Selection No.1 (oval shaped large fruit) and Selection No.2 (Cylindrical shaped medium sized fruit) proved to be better on the basis of yield and fruit quality attributes (Kundu *et al*; 2001). A survey was undertaken in the Karnataka state to investigate the nature and extent of variability present in jamun seedling progenies for morphological characters of tree, high degree of variability was observed with regard to plant girth, leaf area, petiole length and leaf length to petiole length ratio (Prabhuraj *et al*; 2002). A survey was carried out in North Goa to identify the elite genotypes of jamun and individual fruit weight ranged from 3.42 to 13.67 g, length from 3.31 to 5.26 cm, girth from 5.21 to 9.82 cm, pulp percentage from58.57 to 84.55, TSS from 12.00 to 26.80%, titratable acidity from 0.59 to 1.63 %, total sugars from 6.87 to 25.31 % and sugar: acid ratio from 15.39 to 27.92 (Devi *et al.*, 2002).

Table 1 : Nutritive value of jamun fruits.

Composition value					
Moisture	84.50-86.4%	Acidity (%)	2.1-2.5	Juice	50-65 %
Protein	0.53-0.65	Pectin	2.3-3.7 %	Calcium (mg)	0.02 %
Fat	0.10	Vitamin A (IU)	74 - 100	Iron(mg)	0.1 %
Calorific value	83 per 100g	Vitamin C	30-40 mg /100g	Total sugars	5.8-6.9 %
T. S. S.	9-11.5 %	Phosphorus	0.01 %	Anthocyanin (OD)	1.8 - 1.9

(***Source :*** Bose *et al.*, 2001, Singh *et al.*, 1999).

Botanical Description

It belongs to the important genus *Syzygium* of the family Myrtaceae. There are about 400 to 500 species of which a few provide edible fruits (Chundawat, 1990).

Syzygium jambos (Rose apple or Safed jamun): The tree is ornamental. The fruits are light yellow- white in colour. They are rose scented and calyx persistent. The trees are medium sized and evergreen. The leaves have very small petiole. The seeds are polyembryonic and it is being grown in Assam, Bihar, Andhra Pradesh, Tamil Nadu, West Bengal, Coastal areas of Maharastra and Gujarat.

Syzygium fruitecosum: The trees are suitable for windbreak. Fruits are edible and small in size.

Syzygium javanica (Water apple): This species is found in South India and West Bengal.

Syzygium densiflora: It is suitable for use as a rootstock for *Syzygium cumnii*. It is resistant to attack of termites.

Syzygium uniflora (Surinam cherry or Pitanga cherry). It is also a small tree and bears small-sized fruits having bright red colour and aromatic flavour. The tree is found in South India.

Syzygium zeylanica: The tree is small and bears edible fruits. It is found in Western Ghats of India.

Jamun is an evergreen tree-attaining medium to large size. The trunk is thick and greyish white in colour. Leaves opposite, simple mostly firm and glossy, elliptic, pinnately veined with lateral veins close together; flowers in a few flowered panicles. Flowers are hermaphrodite, light yellow in colour, borne in the axils of leaves on branchlet, calyx lobes 4, calyx tube not extending beyond summit of ovary, petals 4, white, spreading, stamens many, ovary inferior, 2 celled (Bailey and Bailey , 1978).

New vegetative shoots in the jamun emerged in 2 distinct flushes from February to May and from August to October, the number of inflorescences per shoot was maximum in February flush and least in those produced in October (Mishra and Bajpai, 1975). The maximum anthesis (18.71 to 43.08%) and dehiscence were recorded between 10 am and 12 noon. The pollen fertility was higher in the beginning of the season. The maximum receptivity of stigma was observed one day after anthesis (Mishra and Bajpai, 1975). Honeybees, houseflies and wind do the pollination. The highest level of pollen germination in jamun was obtained in 20 % sucrose solution (Singh, 1978). The maximum fruit set (32.6 to 36.0%) was obtained when pollination was done one day after anthesis and thereafter, a sharp decline was observed in fruit set (Mishra and Bajpai, 1975). Fruit is berry, purplish red, ovoid, edible and polyembryonic.

There are no standard varieties available however; common cultivar grown under North Indian condition is "Ra-Jamun". It produces big sized (length 2.5-3.5 cm) and diameter (1.5 – 2.0 cm), oblong fruit, deep purple colour at fully ripe stage, Juicy and sweet in taste. Fruit ripens in June-July possesses small stone. Another late maturing variety bears a small sized (length 1.5-2.0 cm and diameter 1-1.5 cm), slightly round fruit, deep purple or blackish in colour at full ripe stage. The stone present in these cultivars is comparatively large in size. Fruits ripen in the month of August. The jamun is a fruit, which offers a good scope for selection of better varieties out of the innumerable seedling strains that are existence in this country.

Propagation

Jamun is propagated by seeds and vegetative methods. Seed propagation is very common. The seeds of jamun are recalcitrant which loose oits viability within few days of extraction from the pulp. Seed germination and seedling vigour in jamun has direct relation with the maturity of seeds. However, micro-propagation technique has been tried to multiply large scale planting material.

Seed propagation

The most common method of propagating this fruit is seed. Jamun seeds have no dormancy, hence fresh seeds can be sown (within 10-15 days) 4 – 5cm deep at a distance of 25 cm x 15 cm. The seeds germinate 10-15 days after sowing. Sowing od seeds in pots is preferable than sowing them in beds (Chaturvedi, 1956). The seedlings become ready for transplanting in spring or next monsoon. If the seeds are sown too deep, seedling emergence is delayed and there may be some rotting due to poor aeration. Seeds may also be sown in polythene bags as it facilitate in easy handling of rootstocks and grafted plants. There is occurrence of Poly embryony in jamun up to 20-50 per cent; hence nuceller seedlings may be used to produce true to type plants (Singh and Thakur, 1977). The seeds of jamun took 24 to 61 days for total germination under Bihar conditions (Singh and Thakur, 1977). Studies revealed that seed extraction of jamun after heaping the fruits for a single day was better for getting good quality seeds in comparison to the extraction of seeds immediately after collection (Srimathi *et al.*, 2003). Sashtri *et al* (2001) recorded that large size seeds had higher germination (98-99 %) than small size seeds (79-89%). Sultan Singh and Singhrot (1985) also reported that july-August is the appropriate time of sowing of jaumn seed in nursery bags filled with soil, sand and FYM mixture(1:1:1).

Vegetative propagation

Budding

One year old, 10-12 mm thick rootstocks budded in July-August, showed better success. The scion wood or bud wood is taken from one to two year old vigorously growing shoots. In India, shield, patch and Forkert methods of budding are generally employed. Singh *et. al* (1979) at Basti, Utter Pradesh noted 70% success by patch budding in the month of March. Patch budding was found successful when performed in the month of June under Gujarat conditions (Chovatia and Singh, 2000).

Soft wood grafting

This method was proved to be better for insitu grafting under rain fed conditions. Madalageri *et al;* 1991 reported that soft wood grafting was found successful method for multiplication of jamun under Karnataka conditions. Chovatia and Singh, 2000 recorded 41.67 percent success through soft wood grafting during the month of June under Gujarat conditions.

Inarching: It can be propagated by inarching method but it is not adopted commercially. In the month of June-July, one-year-old rootstocks are inarched with the matching thickness of scion.

Veneer grafting

Bose *et al* (2001) recorded 31% success in veneer grafting when one-year-old seedlings were used as rootstock. The shoots were taken from spring flush and the operation was one in the month of July.

Cutting

It was observed that tip cuttings of *S. javanicum* rooted well under intermittent mist and treatment with IBA at 5000 ppm produced 100% rooting. Bose *et al;* (2001) reported 45% rooting with 100 ppm IBA and 40% with 100 ppm IAA treatments.

Micro-propagation

Micro propagation is a method by which a true-to-type and disease free entire plant can be regenerated from a miniature piece of plant in aseptic condition in artificial growing medium rapidly through out the year. There are several methods of culturing plant tissues such as meristem culture, embryo culture, callus culture, protoplast culture and cell culture.

Yadav *et al* (1990) induced multiple shoots from nodal and shoot tip segments of 10 - 15 days old seedlings of *S.cumini* on modified MS medium supplemented with BA singly and in combination with NAA, IAA or IBA. Excised shoots were placed for root induction on MS medium containing NAA or IBA and then transferred to MS basal medium to complete plantlets .The regenerated plantlets were acclimatized and successfully transferred into the soil. Propagation from elite mother trees is very important. Roy *et al* (1996 a, b) induced multiple shoots from nodal explants of 10 –years –old elite trees of *S. cumini* on MS medium supplemented with 2.5 mg kinetin / l. Explants taken from *in vitro* - proliferated shoots produced further multiple shoots when cultured on the same basal medium also containing 2.5 mg kinetin/l. Repeated subculture resulted in rapid shoot multiplication at an average of 10 shoots per subculture. Jain and Babbar (2000) obtained multiple shoots from the epicotyl segments bearing scaly leaves, excised from *in vitro*- grown seedlings of *S. cuminii,* on MS medium supplemented with different concentrations of BA. An average of 8.6 shoots per explant was produced in 60 days after inoculation, following transfer to fresh medium after 30 days. The shoots were excised and the residual explants were transferred to fresh medium, where they again developed shoots. Rathore *et al,* (2004) attempted for cloning of adult tree of jamun (*S. cumini*). Explants were harvested from rejuvenated shoot sprouts produced by looping of tree. Thus a protocol has been developed to raise plants of *S. cumini* at any time of the year. Somatic ebryogenesis has also been found successful for multiplication of jamun plants (Litz, 1984).

Root stocks

Rootstock selection for vegetative propagation of jamun is important as it controls the vigour and equilibrium between yield and quality. Dwarfing rootstocks are considered to be best because they induce dwarf ness and enable to facilitate easy management of the orchard. Such type of work may be initiated and be standardized for different agro climatic conditions..

Use of Polycontainers

Raising of rootstock in nursery beds and lifting budded plants with earth ball in highly sandy soils is practically not feasible. Transportation of plants from long distance may also cause high mortality particularly under semi arid and arid environment. To reduce the time for raising root stock and to avoid damage during handling and transportation, polyethylene tubes and polyethylene bags may be used on commercial scale. Generally polyethylene bags (40cm x 15cm) and polyethylene tubes (25cm x 10cm) are used for raising the rootstocks. Polyethylene bags with small holes punched in the bottom and sides for drainage and aeration and filled with porous rooting medium or pot mixture for raising rootstocks. Generally 1-2 seeds are sown in each poly bags and then placed in trench bed, so that it can be irrigated easily. Some times coiling of root is become problem, hence root pruner is also used for trimming of roots.

Rejuvenation

The old, unproductive trees or damaged young plants can be rejuvenated easily by top working, employing scion wood or buds from superior clone.

Quality Standards of Planting Material

The appropriate planting material should be used for planting in Jamun. The seedling should be single stem and healthy for vegetative propagation. The sapling should be neither too large nor too small. It should be medium in size with sufficient foliage preferably raised in polybags.Plants raised on older root stock should not be preferred. The height of budding or grafting should not be more than 20-30cm.

References

Anonymous 1986. Research Highlight, 1985, Indian Council of Agricultural Research, New Delhi, p.27.

Ashraf, S.M. 1987. Studies on post harvest technology of jamun fruits. Ph. D. Thesis, Department of Horticulture, NDUAT, Kumarganj, Faizabad.UP.

Bailey, L.H. and Baily, E.J. 1978. Hortus Thitd- A concise dictionary of plants cultivated in the United States and Canada. MacMillan Publishing Co., New York, pp, 457-58.

Bose, T.K., Mitra, S. K. and Sanyal, D. 2001. Fruits: Tropical and Subtropical Vol. II, Pub. Naya Udyog, 206, Bidhan Sarani, Calcutta, India. PP. 643-56.

Chaturvedi, M.B.1956. Jamun is another prized tree. Indian Farming, 5:17-19.

Chovatia, R.S. and Singh, S. P. 2000. Effect of time on budding and grafting success in jamun (*Syzygium cuminii* Skeel.), Indian J. Hort., 57 (3): 255-258.

Chundawat, B.S. 1990. Arid Fruit Culture, Oxford and IBH Publishing Co. Pvt. Ltd., New Delhi, p. 33.

Devi, S.P., Thangam, M., Desai, A.R. and Adsule, P.G. 2002. Indian J. Hort., 59 (2): 153-156.

Jagetia, G.C. and Baliga, M.S. 2002. Toxicology Letters. 132.

Jain, J. and Babbar, S. B. 2000. Plant Cell Respiration. 19: 519-524.

Kadam, S.S. 2001. Indian J. Hort., 58 (1-2): 170-177

Keskar, B.G., Karale, A. R., Dhawale, B. C. and Chaudhary, K. G. 1989. Maharastra J. Hort., 4:117-20.

Kundu, S., Ghosh, D.K. and Maiti, S. C. 2001. Environment and Ecology. 19 (4): 872 - 874.

Litz, R. E. 1984. Hort. Science. 19: 720-722.

Madalageri, M. B., Patil, V. S. and Nalawadi, U. G. 1991. My Forest. 27 (2): 176-178.

Mishra, R.S. and Bajpai, P.N. 1971. Indian J. Hort., 28 (4): 273-277.

Pareek, O.P. 1988. Present status and future needs for genetic resources in arid zone fruits. In: Plant Genetic Resources, An Indian Perspective, Symposium Proceedings, National Bureau of Plant Genetic Resources, New Delhi. pp.320-34.

Prabhuraj, H.S., Swamy, G.S.K., Athani, S.I., Patil, B.R., Hulamani, N.C. and Patil, P. B. 2002. My Forest. 38 (2): 187-189.

Prince, P.S.M., Kamalakkanan, N. and Menon, V.P. 2003. J. Anthropology. 84 (2-3): 205-209.

Roy, P.K., Rahman, M.M. and Roy, S.K. 1996a. Bangladesh Journal of Botany. 25: 159-164.

Roy, P.K., Rahman, M.M., Roy, S.K. and Brumfield, R.G. 1996 b. Mass multiplication of *Syzygium cuminii* from selected elite trees.In Proceedings of XIII International Sympo. on Hort. Economics, Rutger, The State University of New Jersey, Acta Hort., 429: 489- 495.

Rathore,Vinod; Shekhawat,N.S; Singh, R.P,Rathore, J.S.and Dagla,H.R.2004. Cloning of adult trees of jamun (*Syzygium cuminii*). Ind.Jour. of Biotechnology, 3(2) :241-245.

Sasthri, G., Srimathi, P. and Malarkadi, K. 2001. Madras Agril. J., 88 (7-9): 524-526.

Singh, H.L. 1978. Plant Science. 10:151-152.

Singh, I.S. 2001. Indian J. Hort., 58 (1-2): 178-182.

Singh, I.S., Srivastava, A.K. and Singh, V. 1999. J. Applied Hort., 1(1):34 - 37.

Singh, M., Singh, G.N. and Singh, H.L. 1989. Haryana J. Agril. Sciences 18 (1-2): 4-9.

Singh, R.K. and Thakur, S. 1977. Seed germination and seedling growth of jamun (Syzygium (Euginia) cuminii Skeels.). Proceedings of the Bihar Academy, Agril Sci., 25 (1): 139- 142.

Singh, S.K., Krishnamurthy, S. and Katyal, S. L. 1967. Fruit culture in India. ICAR, New Delhi. pp. 255-9.

Singh, U. R., Pandey, I.C. and Prasad, R.S. 1979. Propagation of Jamun, by budding.Punjab Hort. J., 19: 74-75.

Singh, Sanjay and Singh, A.K. 2012. Biodiversity and potential of jamun, chironji, rayan and mahua for sustainable agriculture for tribal region. In. Compendium of winter school held at CIAH, Bikaner 5-25th September,2012 pp.67-73.

Srimathi, P., Ramanadane, T., Malarkodi, K. and Natarajan, K. 2003. Seed extraction in jamun (*Syzygium cuminii* Skeels.).Progressive Hort., 35(2): 221-223.

Thaper, A. R. 1958. Farm Bulletin No. 42, ICAR, pp.255-9.

Vijayanand, P., Rao, L.J.M. and Narsimham 2001. Flavour and Fragrance J., 16 (1):47-49.

Yadav, U, Lal, M. and Jaiswal, V.S. 1990. *In vitro* micro-propagation of tropical fruit trees (*Syzygium cuminii*). Plant Cell Tissue Organ Culture. 21: 87-92.

Chapter-25

Khirni (*Manilkara hexandra* (Roxb.) Dub.)

Sanjay Singh and R. S. Singh
Central Horticultural Experiment Station, Godhra - 389340

Khirni (*Manilkara hexandra*) also known as rayan, is an important multipurpose tree of dry parts of country. It is commercially used as a rootstock for vegetative propagation of sapota in different parts of the country (Singh and Singh, 2012). Seedlings are found in natural state in the field as well forest areas. Its fruit are nutritious, edible and available during summer season at high price because of its low production. At present, the potential of this fruit crop is unexploited. There is no improved cultivar for orchard plantation. It can be planted as wind break on boundary of farm/orchards because of its dense foliage. It is a underexploited fruit tree semi arid to sub tropical regions. Malik *et al.* (2012) reported that Khirni is a socio- economically important fruit species for tribal peoples. They observed a sizable variability in morphological characters and high socio-economic potential of Khirni during survey in Rajasthan, Gujarat, M.P. and Maharashtra state. Its bark, fresh fruits and extracted seeds have high nutritional and medicinal value. Tree provides substantial livelihood support to local inhabitants as collected fresh fruits from natural population fetch good price in local markets. Due to its high market demand, its cultivation should be expanded and elite trees should be conserved at farm site.

Nutritive Value and Uses

Its fruit are sweet, nutritious, yellow in colour at ripe stage. The average fruit weight and pulp weight ranged from 1.25-7.10 g and 0.50-5.90g respectively in different thirty genotypes. Total soluble solids ranged from 21.00-28.00 per cent. Vitamin C and total carotenoid content varied from 16.20-25.64 mg/100g and 4.30-6.50 mg/100g respectively in different genotypes (Singh *et al.*, 2006). The fruit weight ranged from 1.66 to 4.25 g; length from 1.39 to 2.18 cm; diameter from 0.77 to 1.58 cm; pulp weight from 1.28 to 3.33 g; seed weight from 0.21 to 0.92 g in different genotypes (Anonymous, 2007). Total Soluble Solids in different accessions varied from 26.20 to 31.80^{0}B; total sugar from 10.25 to 14.92 % whereas, vitamin C content 9.84 to 19.68 mg/100g at CHES, Godhra. It is mainly used as root stock for sapota propagation through vegetative methods. Its wood is used for timber, fuel and for making small agricultural implements by the tribal people.

Botanical Description

Khirni (*Manilkara hexandra* (Roxb.) Dub.), is medium size tree belongs to the family Sapotaceae. It is woody, much branched evergreen tree suited for semi arid and arid parts of the country. It is highly heterozygous, cross-pollinated fruit tree. The peak period of flowering and fruit set was recorded in the month of November – December and December - January respectively in majority of the genotypes. Peak period of anthesis was recorded between 8-11 AM in all the genotypes. Anther dehiscence commenced after opening of flowers i.e. at 5AM and continued till 2 PM. Peak period of dehiscence was recorded between 10 AM –1 PM in all genotypes. The ripening span in different genotypes was observed from first week of March to first week of June.

A wide variation any genotypes under different climate conditions were observed in fruiting and fruit quality characters in Khirni (Singh *et al.*, 2006). With respect to all horticultural traits studied the genotypes, CPT-1, CPT-3,CPT-4,CPT-6,CPT-8,CPT-12 and CPT -25 were found to be promising. Under evaluation, slow vegetative growth in khirni plants were noted under hot arid conditions. Young plants were also affected by frost/low temperature under arid environment.

At CHES, Vejalpur, Godhra, khirni genotype were evaluated for better yield and fruit quality. Some promising genotypes in khirni have been identified (Anon., 2007). The descriptions of promising genotypes are given below.

GK-1: It was collected from Kharsalia village of Panchamahal district, Gujarat. The peak period of flowering was recorded in the month of November. It ripens in the first week of March and recorded 5.66g fruit weight, 4.93g pulp weight, 21.00 % TSS, 16.20mg/100 vitamin C and 4.30g/100g carotene.

GK-10: It was collected from Parwadi village of Panchamahal district, Gujarat. The peak period of flowering was recorded in the month of December. It ripens in third week of May and recorded 6.64g fruit weight, 5.71g pulp weight, 26.40 % TSS, 24.25mg/100 vitamin C and6.6.06g/100g carotene.

GK-11 It was collected from Bahedia village of Panchamahal district, Gujarat. The peak period of flowering was recorded in the month of December. It ripens in third week of April and recorded 4.67g fruit weight, 4.10 pulp weight, 26.50 % TSS, 25.39mg/100 vitamin C and 6.35g/100g carotene.

Propagation

It is easily multiplied through seeds however, being a cross-pollinated crop and as such seedlings exhibit a wide range of variations in fruiting, yield and quality of fruits (Singh *et al.*, 2006). Further, it can be propagated through soft wood grafting (Singh and Singh, 2012).

Seed

It is propagated traditionally through seeds. Seed takes 15-20 days period for germination. Seed has hard seed coat hence, proper watering in nursery is essential for germination of seeds. Fresh may be used for sowing to raise saplings.

Layering

It can be multiplied through air layers. Gowda *et al.* (2005) observed that maximum rooting percentage (76%) of Khirni air layers was recorded with 5000 mg L^{-1} IBA and the minimum rooting percentage (36%) was recorded in the control shoots at GKVK, UAS, Bangalore. Application of IBA alone resulted in better rooting compared to an IBA+NAA combination. Khirni air layered shoots treated with 10,000 mg L^{-1} IBA recorded the maximum number of primary (9.2), secondary (49.4) and tertiary (34.3) roots.

Soft wood grafting

Khirni is also propagated through soft wood grafting. The technique of soft wood grating in khirni has standardized at CHES, Godhra (Singh and Singh, 2012). In khirni, highest percentage of graft success was noted in March (76.66%) followed by June, July and August where as it was recorded least in the month of February (10.00%).

References

Anonymous, 2007. Natioanl Network project on Underutilized fruits, Annual Progress Report, CISH, Lucknow.

Gowda, Nache V.; Vasanth, K.R. and Shyamalamma, S. 2005. Studies on vegetative propagation of khirni (*Manilkara hexandra* (Roxb) Dub.) by air layering, Acta Horticulture, 727: 85-88.

Malik, S.; Chaudhary., R.; Kumar, S.; Dhariwal, O.; Deshwal, R. Chaudhary, Rekha, 2012. Socio-ecomonic and Horticultural potential of Khirni (*Manilkara hexandra* (Roxb.) Dub.): a promising underutilized species of Indian; Genetic Res. & Crop Evolution, 59(6): 12 55 -1265.

Singh, Sanjay; A.K., Joshi .H. K., Bagle, B.G. and Dhandar, D.G. 2006. Proceedings of National Symposium on Production utilization & Export of Underutilized fruits with commercial Potentialities, BCKVV, West Bengal, Nov, 22-24, 2006. p. 30-33.

Singh, Sanjay and Singh, A.K. 2012. Biodiversity and potential of jamun, chironji, rayan and mahua for sustainable agriculture for tribal region. *In :* Compendium of Winter School, CIAH, Bikaner, pp.69-73.

Chapter-26

Citrus Fruits

Atul Chandra and M. K.Kaul
S.K. Rajasthan Agricultural University, Bikaner-334 006

Cultivation of citrus holds promise in arid irrigated tracts. The major citrus fruits under cultivation in arid regions are acid lime , lemon, Kinnow, Mandarin hybrid, sweet orange and grapefruit .Acid lime and kinnow are widely grown. Citrus fruit are grown through out country for its nutritive and medicinal value (Chadha, 2001) . Citrus is world's leading fruit crop. In India, it is cultivated in the states of Maharashtra, A.P., Punjab, Rajasthan, H.P., U.P., M.P., Karnataka, T.N., J&K., Odisha, Gujarat, Assam, & Sikkim, For obtaining quality plant material, propagation is an important aspect. Citrus is largely propagated by budding except kagzi lime which is raised from seeds because of occurrence of polyembryony and sweet lime which is commercially propagated by stem cuttings. Kagzi lime and lemon can also be propagated by layering. In case of citrus the availability of proper plant material is very much essential. There are many problems related to availability of suitable plant material which includes incidence of different diseases such as Tristeza, greening, canker and other diseases; problem of incompability between scion and rootstock resulting formation of crease at later stage and thus decline of the tree 'availability of rootstocks suitable under high salinity drought conditions, high temperature, low temperature and poor soil fertility, phytosanitation is another aspect which needs proper attention and selection of disease free suitable bud wood. Looking to these difficulties plant propagation aspect in citrus has become very much crucial in deciding the future of the orchards. For the success or failure of citrus orchard in performance, selection of plant material of suitable type is one of the major components.

Nutritive Value and Uses

Citrus fruits are nutritious because of rich vitamin C contents. It is used for fresh consumption and for making several value added products. From fruits about 65% juice can be obtained which can be used for making delicious drinks. Kagzi lime and lemon fruits are used for making pickle, mixed pickle and juice. Squash, RTS, Lemon cordials, Juice concentrate, Pectin, oil, Citric acid , marmalade, Jam, etc value added products are prepared from citrus fruits. Its peel is used for making powder for cleaning utencils. Mandarin, Kinnow fruits are fresh eaten because of sweet sour in taste.Citric acid is mainly prepared commercially from citrus sp.

Botanical Description

Citrus (*Citrus* sp.) belongs to the family Rutaceae. It is evergreen medium to tall tree. Trees are thorny and bears flowers singly or in clusters which bear fruit generally round to oval in shape depending on the species. Flowering takes place during spring season (February - March) and also during July -August in citrus species. The botanical description of prominent acid lime, lemon, mandarin hybrid-Kinnow, Sweet orange and grape fruit cultivars is given as under

Mandarin Hybrid

Kinnow-It is first generation hybrid between King (*Citrus nobilis*) and Willow leaf mandarin (*Citrus deliciosa*).Fruits are medium, globose to oblate, base flattened ,skin deep golden orange when fully mature, rind medium thick, adherent to segments. Segments are not easily separated, very juicy, acidity moderate with fine sugar / acid blend ,seed 12-24 ,matures in January.

Sweet Orange

i) **Malta (common) :** Fruits are medium to large; yellow; base and apex broad and rounded;flavoured with a blend of sweetness and acidity.

ii) **Blood red (Malta) :** Fruits are medium to large; yellow with scarlet blush ;flesh light orange in colour but on ripening turns to deep blood red in colour; juicy ,sweet flavoured with a blend of sweetness and acidity.

iii) **Mosambi :** Fruits are medium to large :spherical, light yellow orange coloured; furrowed ,apex broad with shallow circular ring; difficult to peel ; flesh fine, straw yellow; juicy with low acidity; Seeds 20-22;ripen in Oct-Nov.

iv) **Jaffa :** Fruits are medium to large ;spherical ,light yellow orange coloured ; furrowed ,apex broad with shallow circular ring ; difficult to peel ;flesh straw yellow;juicy with low acidity; seeds 20-22;ripen in Oct-Nov.

v) **Hemlin :** Fruits are medium to large; globose; apex and base smooth; difficult to peel; rich in juice; flavoured with a blend of sweetness and acidity; seeds 1-2.

vi) **Valencia late :** Fruit medium ,slightly oval; skin deep golden yellow; sub acid in taste ,juice abundant , rich flavour, seeds 3-8;ripens during February to March.

Acid Lime

Kagzi Lime-The fruits are slightly oval or round in shape, small in size and have thin and shiny rind and acidic pulp. The skin adheres to the pulp. The seeds are small and have red chalazal caps. The fruits mature during August –February.Sparse fruiting may however, take place round the year.

Pramalini- It is clonal selection of Kagzi lime. It bears fruits in cluster of 3-7.Fruits have 57% juice.It yields 30% more than the normal Kagzi lime.

Vikram –It is also a clonal selection of Kagzi lime. It bears fruits in cluster of 5-10 and give 30-32% more yield over the normal lime.

Chakradhar-It is seedless strain of acid lime. Trees are compact , erect and dense in habit. Fruits are round with thin papery rind and contains 60-65% juice.

Lemon

i) **Eureka** –Fruits are medium, elliptical, oblong, lemon yellow in colour, rind medium, thin with pitted surface , fresh tender, fine, pale greenish and juice is abundant, very acidic with excellent flavor.seeds are few (0-6). It is heavy yielder and starts bearing early.

ii) **Kagzi kalan**-Fruits are small medium, spherical ,yellow with apex slightly nipple like and base rounded , flesh acidic, juicy, seeds(8-13).

iii) **Baramasi**-Fruits are medium, ovoid-globule, lemon chrome coloured, surface, pitted, flesh firm, juicy ,acidic, flavoured. Seeds are none to few.

iv) **Pant lemon-1**-It is a selection of Kagzi Kalan developed at Pantnagar. It is self incompatible. It is resistant to canker and Tristeza disease. Fruits are medium round, smooth with thin rind and juicy.

Grape Fruit

i) **Red Blush** –Fruit small to medium, oblate, peel smooth, glossy and deep yellow having crimson colour in patches at maturity. Deep bright crimson blush in juice vesicles. Mildly acidic and high in TSS, seeds 0-8 , mostly aborted, ripens during last week of November.

ii) **Marsh Seedless**-Fruits are medium to large, round/oblate, light yellow in colour, rind smooth , glossy, flesh fine, juicy, flavored, seeds few. It matures during December-January.

iii) **Duncan**-Fruits are large, oblate to globose, light yellow in colour, apex and base rounded, rind leathery, flesh fine, tender, juicy, highly flavoured and excellent variety . seeds are many. It matures during January-March.

iv) **Ruby Red**- It is bud mutation from Thompson grape fruit. Rind smooth with crimson flesh, with deep red colored.

Besides the botanical descriptions of prominent root stock of citrus viz., rough lemon rangpur lime, Cleopetra mandarin orange is given in Table 1.

Considerable research work has been done on effect of rootstock on performance of different citrus species. Cleopetra has been found the best rootstock for blood red whereas in Mosambi, Pectinifera imparted health and longest life (Chohan, G.S. Personal communication). For Jaffa, budded on jatti khatti gave the highest yield (Mehrotra *et al.*,1984). For Hamlin ,Jatti good rootstock for Marsh Seedless at Abohar based on rootstock trial from 1976-1981. Overall performance of kinnow was found to be the best on Jatti Khatti followed by Karna Khatta in Punjab (Chohan *et al.*,1978). Under rainfed condition at Chethalli (Karnataka) prebearing performance of kinnow was better on rough lemon, Troyer and Rangpur lime than trifoliate orange and cleopatra mandarin (Jalikop *et al.*,1986).

Table 1 : Details and suitability of rootstock

S.No.	Name of Rootstock	Characteristics
1.	Rough lemon	(*C.jambhiri* Lush.)Suitable for Kinnow,sweet orange,lemon and grapefruit.Produce large tree,high yield ,large fruit with low quality,deep rooted,poor cold hardiness,very susceptible to blight, excess soil moisture, phytophthora, Susceptible to drought and tolerant to tristeza.
2.	Rangpur lime	(*C.limonia* Osbeck)Suitable rootstock for sweet orange,mandarin.Trees large and vigorous, susceptible to citrus nematode,greening and sensitive to exocortis.Trees are large and vigorous,tolerant to Tristeza,salt, foot rot, high yield,medium fruit quality better than rough lemon.
3.	Cleopetra mandarin	(*C.reshni* Host,extan.)Suitable rootstock for Blood Red, Kinnow ,Pineapple,Hamlin, Tanggelos and grapefruit. Trees are large, slow growth in nursery, fruit size small,high fruit quality.It is highly tolerant to tristeza, exocortis,salinity and moderately tolerant to alkalinity.It is susceptible to Phytophthora and citrus nematode.
4	Trifoliate Orange	(*Poncirus trifoliate* L. Rafinisqe) It is used for oranges, grapefruit, lemon and mandarin. Trees on it are not vigorous in growth but has better fruit quality and high yield.They have shallow root system consisting of weak lateral root development but abundant fibrous roots. Trees are smaller than standard trees .Resistant to foot rot, tristeza, nematodes and tolerant to phytophthora and have low salt tolerance .It has reasonable drougt and cold tolerance.
5	Pectinifera (*Citrus pectinifera*)	It is suitable rootstock for Mosambi. For the initial one and half year, the tree growth is slow and later it picks up. Leaves are small with obtuse apex and small thorns are present in the leaf axil. Tree attains medium height. Fruits are of small size with broad base. Rind is jacket type ith locules having 6-8 seeds per fruit. The seeds attain maturity in the month of September when the fruits are deep orange in colour. It is very suitable rootstock for Mosambi cultivar and is resistant to delayed incompability and bud creasing.

Characteristics of some selected citrus rootstocks as mentioned by Chadha and Singh(1980)is given in Table 2.

Table 2 : Characteristics of some selected citrus rootstocks.

Rootstock	Horticultural Performance Reaction to									
	Yield	Quality	Plant Vigour	Root rot	Citrus nematode	Tristeza	Excortis	Salt	Drought	Root system
Rangpur lime	G	M	G	MT	S	R	S	R	R	D
Marmalade orange	G	M	G	MT	MT	R	S	R	R	D
Rough lemon	G	L	G	S	S	R	R	T	T	D
Cleopetra Mandarin	M	M	M	T	S	R	R	MT	S	M
Sour Orange	G	G	M	R	T	HS	T	T	MT	D
Sweet lime	M	M	G	S	MT	R	S	T	S	M
Trifoliate orange	L	M	L	R	R	R	H	HS	HS	SH
Troyer citrange	M	G	M	MT	T	MT	S	HS	HS	SH
Carrizo citrange	M	G	M	MT	T	MT	S	HS	HS	SH
Sweet orange	G	G	M	HS	HS	MT	R	S	S	M
Karna Khatta	G	M	M	S	MT	-	T	T	S	D
Nasnaran	M	G	M	S	MT	T	RT	HT	S	M

(G=good. M=moderate, L=poor or low, R=resistant, MT=moderately tolerant, S=susceptible, HS=highly susceptible, D=deep, SH=shallow, - =no information).

The suitable cultivars of different citrus groups being grown in arid areas are given as under :

1. Mandarin hybrid- Kinnow
2. Sweet orange-Common Malta, Blood Red, Mosambi, Hemlin , Jaffa, Washington naval,Valencia Late,Pineapple,Sathgudi,
3. Acid lime-Kagzi lime, Pramalini, Vikram, Chakradhar, Selection -49, PKM-1, Sai Sharbati,Jai Devi,
4. Lemon-Eureka ,Kagzi kalan , Lisbon ,Pant Lemon-1, Baramasi
5. Grape fruit-Red blush ,Duncan, Marsh Seedless. Foster , Ruby Red, Thomson, Saharanpur special

Propagation

In India, citrus tree are propagated both by seeds and vegetative means. Seed propagation is still practiced in case of acid lime and to produce rootstocks for budding (Kalloo *et al.,* 2005). Citrus trees are propagated through budding, grafting, and layering methods. Citrus is now being propagated through micro-propagation techniques.

Seed propagation

Kagzi lime plants are raised generally from seeds because of occurance of polyembryony. Acid lime is propagated commercially through seeds (Chadha,2001). Other citrus fruits if raised by seedlings may not always be true to type (identical to parent type) and may not produce fruit for even 10 or more years. The fruits obtained from seedlings quite often are inferior in quality. Citrus seeds are used for growing rootstocks plant on which the desired cultivars may be budded. However, seeds obtained from edible citrus fruits of grapefruit, sweet orange and mandarin are not recommended for commercial purpose for that rootstock plant such as rough lemon ,sour orange, trifoliate orange, Carrizo, citrange, single citrumelo, Cleopatra mandarin and other lesser known varieties are used. Seeds are abstracted from mature fruit and rinsed thoroughly in water and planted as soon as possible.

Unless citrus seeds are purchased,they should be extracted from mature fruit , rinsed thoroughly in water, and planted as soon as possible. If extracted seeds are not planted immediately after rinsing they should be spread evenly on absorbent paper and placed away from direct sunlight. When the surface moisture has dried, the seeds should be placed in polythene bags and stored at 40-50°F (the refrigerator vegetable drawer works well).

Seeds should be planted at a depth of ¼ to ½ inch in soluble pots or flats containing sterile potting media. Removing the seed coats or soaking seeds in aerated water for about eight hours just prior to planting, can reduce the time required for germinating and seedling emergence. Sunlight ,warm soil temperature and sufficient moisture are required for rapid germination and seedling emergence. Under ideal conditions, emergence will likely occur during a week period beginning 7 to 10 days after planting. The viability of seeds in citrus during storage varies depending upon species and storage conditions. The viability of roygh lemon or Jatti Khatti (*Citrus jambhiri*) seeds were observed up to 131 days without and decrease in germination. However, the germination was quicker in fresh seed compared to older which required double the time (Bhat, 1942).It has been reported in case of Rangpur lime that bigger fruits had more number of bold seeds with higher germination (Desai and Patel,1983). Similarly, seeds from extra large and large acid lime fruits gave better germination and seedling growth (Dharmalingam and Kumar,1987). Storage conditions also influence seed viability as Kagzi lime and Mosambi seeds stored in charcoal powder at 50 per cent relative humidity was better than stored in sealed polythene bags(Bajpai *et al.*, 1963). In another study,seeds of grape fruit stored at room temperature at 5-8°C lost viability after 3 and 4 months,respectively compared to 9 months if these were mixed with saw dust and sealed in polythene bags and stored at 5-8°C. Seeds soaking with growth regulators has improved germination, shoot and root growth. According to Chaudhari and Chakrawar (1981) in Kagzi lime soaking of seed with 40 ppm NAA or GA_3 improved germination, shoot and root growth. Plants should be trained to a single stem (no branches within 6 to 8 inches of the soil).

Vegetative Propagation

Shield or T-budding is most common method of vegetative propagation.Budding is generally done either in spring or in September.

Budding

Budding is usually done when seedling stems are ¼ to 3/8 inch in diameter (about the diameter of a pencil).Budding can be done anytime there is suitable stock on which the bark is slipping when suitable budwood is available. Usually ,the bark is slipping from Feb.-March or September–October (Nijjar,1980). The area to be budded should be pruned clean of thorns and twigs. The preferred budding height is 6 inches above ground level. Lemon and Kinnow are propagated successfully by budding (Fig. 1) method.

Fig. 1 : Budded plants of kinnow.

Selection of budwood

Buds should be collected from healthy, disease free tree of the desired cultivar. This is usually, the bark is slipping from the next to last growth flush (the wood behind the current growth flush) or from the current growth flush after it has begun to harden or mature. Bud wood should be round(not angular as is young wood),relatively straight and have well formed buds in the leaf axils, In a study ,the bud take of kinnow on *C.jambhiri* with better subsequent growth when round buds without wood were used by using angular buds with or without wood (Sharma and Sharma,1986).Often ,the presence of a few longitudinal gray lines on the green bark indicate the proper stage of maturity. Whenever possible , budwood should be discarded and the remaining budwood trimmed to 8-10 inch lengths. Leaves should be cut off leaving a stub of the petiole about 1/8 inch long to protect the bud trimmed.

Preparing the rootstock

A very sharp knife is used to make a vertical cut in a smooth area of the rootstock about 2.5-3.5 cm(1.0-1.5 inches) long through the bark, deeply enough into the wood to be certain the bark has been completely cut. A horizontal cut about 1 cm (0.5 inch) long is made through the bark at the Top (T) or bottom (inverted T) of the vertical cut, again cutting completely through the bark. At the finish of this cross cut, the knife blade is turned slightly twist to open the bark at the T. The point of the knife can be used to lift the bark along the vertical cut if necessary.

Cutting the bud

Cut a bud from the budstick while holding the apical end of the budstick away from you. Start the cut about 1 cm (0.5 inch) above the bud and finish a little less distance below the bud. The knife should be held almost parallel to the axis of the budstick, cutting towards the thumb.

Cut only deep enough to take a thin silver of wood under the bark. The bud should not be scooped out, as this causes too much wood to be taken.

Inserting the bud

Insert the bud shield under the bark flaps of the stock so that the cut surface is flat against the wood. The bud shield should be completely enclosed in the T; if part of it protrudes, cut it off.

Wrapping the bud

Wrap the bud with budding tape (polythene stripes about 1 cm ½ inch) wide and 15-25 cm (6-10 inches)long. Start the wrap below the bud with 3-4 turns, finishing with several turns above the bud. The end is secured beneath the last circular turn of the wrap. Wrapping should be firm without being excessively tight.

Forcing the bud

The wrapping should be removed after 2-3 weeks ,as union with the stock should have occurred by that time, The bud is then forced into growth by lopping the rootstock. Cut about ¾ of the way through the stock, on the same side as the bud about 4-6 cm (1.5-2.5 inch) above it, then push the top over to lay on the ground. The lopped top portion continues nourishing the rootstock top can be removed completely by making a sloping cut(high end on the same side as the bud) about 1 cm (1/2 inch) above the bud.

As the bud grows,it will need to be staked and regular intervals to prevent breakage. Remove all other buds and suckers from the rootstock as they appear. Bud sticks should be labeled as to cultivar, date and bud wood source and used immediately or placed in suitable storage.

The importance of selecting bud wood from disease free trees can not be overemphasized .Several serious virus or virus like, disease can be spread by using infected bud wood. These diseases can kill trees, or make them non productive.

Storing bud wood

Bud wood should be used immediately after its collection as is practical. However, it can be stored for 2-3 months under proper conditions. Bud sticks not used immediately should be placed in sealed polythene bags and stored at between 40-45°F. The vegetable drawer of the refrigerator is usually ideal for bud wood storage. Do not put water or wet packing material into the storage bag as this will cause the bud wood to mold or decay. Inspect bud wood every 2 to 3 weeks for the presence of mold, or excessive moisture inside the bag. Lightly molded should be carefully washed in cold, mild soapy water, rinsed and stored in a clean bag. Bud wood that is excessively moist should be lightly blotted with paper towels. Shriveled, darkened, or heavy mould bud wood should be discarded and any unaffected bud wood should be washed and returned to storage in a clean bag.Stored bud wood should remain moist and cool, but not wet. It is best not to remove bud wood from storage until a couple of hours before its use.

Budding tools

A razor –sharp knife is needed to make smooth, clean cuts necessary for sufficient contact between the scion bud and rootstock during the healing process. The sharper the knife the better the technique would be success better. Knives designed specifically for budding are available at garden supply stores and are usually best, although some pocket knives are suitable. Any blade will become dull with use and require periodic sharpening. Polythene budding tape is used to wrape buds to prevent drying and promote bud union formation .Clear tape allows you to observe the bud during the healing process.

T- budding

T- budding is a relatively simple procedure and is recommended over chip budding for the in experienced budder. Most florida citus trees are propagated by the inverted T budding procedure, but the standard (upright) T bud is equally satisfactory. T budding may be conducted, whenever the rootstock plant has attained suitable size, its bark is slipping (the bark separate easily from the wood underneath) and suitable bud wood is available. These conditions usually occur during periods of growth *i.e.* Feb.to September. All thorns from stems and leaves should be removed from the area to be budded. The preferred budding height is approximately 6 inch above the soil surface.

A very sharp knife is used to make a vertical cut in a smooth area of the rootstock stem bark at the bottom (inverted T) or top(regular) of the vertical cut. The point of the knife can be used to lift the bark along the vertical cut. Remove buds from the bud stick while holding the apical end (tip) of the bud stick away from one self. With the knife blade almost parallel to the axis of the bud wood ,begin the cut about ½ inch above the bud removing a shield shaped piece of bark and wood about ¾ to 1 inch long with a flat, smooth cut surface. Cut only deep enough to remove a thin silver of wood under the bark. The bud should not be scooped out because too much wood will be removed with the bud .Avoid knife blade, or by carefully using the leaf petiole stub as a handle.

The bud should be immediately inserted into the stock, not allowing the cut surface or the bud to dry. Slide the bud shield (the bud with associated bark and wood) under the bark flaps of the rootstock with the cut surface flat against the wood of the rootstock plant. The bud shield should be completely enclosed in the T incision ; if the part of it protrudes beyond the incision , cut it off.

Buds should be wrapped immediately following their insertion into the rootstock wrap buds with budding tape [polythene strips about ½ inch wide by 6-10 inches long]. Begin wrapping below the bud with 3-4 turn and finish with several turns above the bud covering all exposed surface of the bud with tape. The end of the tape is secured beneath the last circular turn. The wrap should be firm without being excessively tight. Wraps should be removed after 14 to 21 days and should not be left on more than 30 days. If a successful union has formed between the bud and the rootstock the bud will be green and show no signs of shriveling or drying. Callus formation should also be evident around the edge of the bud.

Chip budding

Chip budding requires slight more skill than T- budding and is usually done whenever the bark of the rootstock plant is not slipping or has become too thick to T bud. The chip bud is cut while holding the bud stick with the bud stick with the apical end towards the budder. A thin slice of wood with a scion bud is removed by making smooth upwards cut about 1 inch long and just into the wood. A second cut is made at the top of first cut, forming a notch. A chip is removed from the rootstock in a similar manner. Because only two thin lines of a cambial tissue are available for healing on both the scion bud and rootstock, it is important that matching on both sides occur whenever possible. However, if the scion bud is smaller than cambial tissue in a thin layer between the bark and the wood of the tree. This is an area of active cellular growth of the tree. The scion should be wrapped as described for T budding so that all cut edges are completely covered.

Aftercare

After the wrapping has been removed and the union between the bud and stock has occurred,the bud must be "forced" into growth. Naturally occurring plant hormones produced in the upper portion of the rootstock seeding may prevent the scion bud from growing unless the bud is forced. Buds are forced by cutting about 2/3 of the way through the stock, on the same side as the bud and about 1 to ½ inches above it. Then the seeding top is pushed over to lay on the ground. This procedure is known as lopping. The rootstock top will continue to supply the root and developing scion with food and other growth substance during the early stages of scion development. After the scion bud has grown several inches, the lop may be removed by making a cut about ½ inch above the scion. If lopping is not practical, the rootstock top can simply be removed with a slopping cut completely through the rootstock about 1 inch above the scion bud. Rootstock sprouts which form along the main stem, especially in close proximity to the scion bud, should be removed as soon as they develop since they will retard the growth of the developing scion. As the scion grows, it will need to be staked and tied at regular intervals to prevent breaking of the scion. When the nursery tree reaches a height of about 18 to 20 inches, the top should be pinched out to stimulate lateral shot development.

Layering and Cutting

There are several propagation methods for citrus.Layering is a propagation method by which roots are induced to develop from stems while they are still attached to the tree, or cutting can be taken from a citrus trees and rooted. Layering is most commonly used on larger branches, while smaller branches may be used for cutting. Both techniques can be used for citrus , but the resultant plant grow on their own root without the advantages offered by rootstock. Sweet lime is generally propagated by layering or hard wood cuttings (Chadha, 2001).

Arora and Yamdagni (1985) observed improved rooting in leafy cutting in lemon cvs Baramasi , Kagzi kalan and Eureka with application of NAA 200 and 300 ppm at Hisar .

Grafting

Grafting procedure other than budding involve the use of a scion having two or more buds. There are numerous types of graft including whip, cleft ,bridge in arch, stump, side, inlay, bark approach and others. Grafting is most commonly practiced in citrus because it is a more difficult means of propagation compared to budding.

Top working

Top working is the process of changing the top of an established tree from one cultivar to another,or to multiple cultivars ,by budding or grafting . Several procedure may be used when top working citrus trees. These include bark grafting ,cleft grafting and T budding. T budding require less expertise than either of the grafting procedure and will most likely be the method of choice for the orchardist .

To top work a citrus trees by T budding, prune the tree back to leave only a few branches of 2-5 inch diameter ,or smaller. Insert 1-3 buds on the upper side of the remaining scaffold limbs, using the T bud method. Remove unwanted buds and sprouts to insure that only the desired scion buds grow. If limbs are so large that budding would be difficult ,prune back to major scaffold limbs , removing all of the top [caution: severely pruned trees should be whitewashed to prevent sunscald]. After limbs sprout back and mature a bit [6 month or so]

The sprouts can be budded as initially described, using 4-6 of the stronger sprouts on each limb. The same method can be used to bud root sprout that develop when the entire tree top has been frozen back or has declined for other reasons.

Suitable rootstock for different Citrus fruits

Kinnow mandarin	*Citrus jambhiri* L .[jatti khatti]
Grapefruit	
Sweet lime	
Sweet orange [all cultivars except Blood Red]	
Cv. Blood red of sweet orange	Cleopatra for Blood Red and
Cv. Mosambi of sweet orange	Pectinifera for mosabi [In North India]
For salt tolerance	Servenia boxifolia

For heavy clay soil in U.P	Citrus karna is used as rootstock in place *Citrus jambhiri*
For dwarfing	Dragonfly ,microcitrus , R -1633 can be used as Dwarfing rootstock for high density of Orchard .
For cold hardyness	Carrizo ,Troyer Citrumellow rootstock may be used as rootstock in citrus but there is a problem of inverted bottle neck.

Micropropagation

Micrografting is done with microbuds from the indexed elite plant. Grafting buds of a scion cultivar onto seeding rootstock generally propagates citrus tree . This propagation method allow the transmission of virus and mycoplasma diseases including exocortis, psorosis , tristeza, xyloporosis, greening and stubborn (Button, 1977).

1. Enhancement of axillary branches
2. Adventitious shoot formation
3. Somatic embryogenesis

The recovery of disease free stock in cultivars that have over generation been propagated vegetatively.Virus in particular is difficult to eliminate using conventional methods where as micropropagation has made it possible.

In micropropagation , small tissue explants of many plant species can be aseptically cultured under artificial conditions and increased in number by suitable uniforms plants. The technique of multiplying plant is still existing in laboratories in research and developments centers especially for citrus and woody plants.Such standardized technology is mainly available for ornamental and flowering plants in India , in addition to banana and straw berry.

Micropropagation the new concept in vegetative propagation, has many advantages and open immense possibilities in rapid multiplication of disease free plant of superior vigour and productively. Economic feasibility will determine the future of micro propagation. It is extremely important for the will determine the future of micro propagation. It is extremely important for the researches to finetune the micro propagation protocol and to reduce the genetically, morphological and physiological variation among the micro propagated plant.

Since virus diseases of citrus clone are monoembryonic and many commercially desirable cultivars are seedless, hence , micro propagation can be utilized by adopting shoot tip culture technique.

Shoot Tip Culture and Virus Elimination

Nucellar explants or entire unfertilized or abortive ovules have been cultured to produce nucellar plant of some seedless citrus cultivars. Although citrus plant produce from somatic embryos *in-vivo* or *in-vitro* are free from most pathogenic viruses, they are juvenile and thus require an extended growth period before they become productive (Button 1977).

The technique of meristem culture for the production of virus free plant has been unsuccessful with citrus, although multiple shoot formation from apical cultures of citrus has been reported (Kitto and Young, 1981 ; Barlass and Skene, 1982). A modified technique, *in vitro* shoot tip grafting was developed by Murashige *et al.* (1972) and refined by Navarro *et al.* (1975). Aseptically grown, etiolated , 2 week old nucellar seedling are used as rootstock. The seedling are decapitated their cotyledons are removed and the root is shortened. Shoot tips containing the apical meristem and three leaf primordial are isolated from new shoots of a desired scion. An inverted –T incision is made on the epicotyls of the rootstock seedling and the shoot tip is placed in the incision. Grafted plant are allowed to develop aseptically in a nutrient solution for several weeks before being transferred to soil. Navarro *et al.* (1975) reported that the success rate for this technique usually ranged from 30 to 50% and sometime was as high as 90% grafted plant and largely virus free and are not juvenile. This technique is now being used in citrus improvement programs throughout the world to obtain virus free grafting of imported bud material is also effective in reducing the risk of introducing diseases from one country to another (Navarro, 1981 a).

Several factors influence the rate of success of grafting or virus elimination. Successful grafts have been obtained using many different scion cultivars of the commercially grown citrus species, and significant differences in grafting success among these different types have not been noted when appropriate rootstocks were used. Both good and poor rootstock –scion combinations have been identified, but it is not yet known if grafting success *in vitro* is related to graft compatibility *in vivo* (Navarro,1981 a). The frequency of successful grafts increases with the size of the shoot tip, but the percentage of virus-free plants declines, although this depends on the pathogen . Therefore, the shoot tip size chosen reflects a compromise between these two criteria (Navarro *et al.*, 1975; Navarro,1981 a). Rootstocks can be successfully grafted only during a rather narrow range of development (Navarro *et al.*, 1975) and optional conditions may differ for different root stock cultivars(Navarro 1981 a). The cultural conditions of stock plants also affect grafting success. For example,growing rootstock seedlings in continuous darkness appears to be beneficial (Navarro *et al.*,1975). Scion material taken from plants grown in a heated greenhouse yields a higher frequency of pathogen free plants than scions taken from field-grown trees. Scion from *in vitro*-cultured buds produce fewer healthy plants than scions from whole trees (Navarro,1981 a).

Microbudding is also done in citrus. Budsticks 5-10cm long are collected from selected and indexed mother plant. Root stocks can be propagated quickly by tissue culture (Kalloo *et al.*, 2005).

Protocol for *in vitro* shoot tip multiplication of *Citrus jambhiri* and *Citrus sinensis* cv. Mosambi through shoot tip have been tried.Treating the shoot tip with ethanol (70%) for 30 sec. plus mercuric chloride (0.1%) for 8 min was the best sterilizing treatment for both the species.MS medium supplemented with acombination of BAP 1.5mg/l and malt extrct 500mg/laws best for culture establishment and proliferation in case of Citrus jambhiri, where as medium supplemented with a combination of BAP 1.0mg/l and malt extract 500mg/l was found most effective for Citrus sinensis cv.Mosambi (Kour *et al.*, 2009).

Quality Standards for Planting Material

One of the important aspects of successful horticultural development is the use of quality planting material that in turn manifests through production of quality horticultural crops. For successful propagation and nursery stock for subsequent planting in the fields or greenhouses and the procedures followed are taken care of. Quality planting material has more influence on successful horticultural development. For successful propagation and growth of nursery plants, environmental factors such as light, air, water etc., have to be controlled so that congential microclimate is created. This is possible only through the use of greenhouse technology. Different types of greenhouses can be used for the propagation of horticultural crops through cutting, grafting, budding, tissue culture and seeds. Greenhouse propagation of plants confers many advantages with respect to success, quality, labour, space, control of pests and disease, etc. Different types of containers and media suitable to crops can be used, Water quality, Sanitation and control of pests and diseases is relatively easy. Seed propagation and propagation through cuttings, grafting, budding and hardening of tissue culture plants in greenhouses have proved beneficial as compared to propagation under open field conditions. In addition, Green House technology for quality plant material can be used.

A green house is covered structure which protects plants from wind, precipitation and excess solar radiation , temperature extremes and considerable pest and disease attack. Greenhouse technology is the technique of providing favourable environmental conditions to plants. It is rather used to product the plants from adverse climatic conditions by providing optimum conditions of light, temp and CO_2 etc. for the best growth of plants to get better quality and maximum yield. In propagation of citrus plants by methods like rooting of cutting, grafting, rooting and hardening of tissue culture plants can be carried out in green house. Protected cultivation and production of nursery stock is of immense significance for the quality of the end product.

Green house propagation of plants entails greater skills and more attention and care than outdoor propagation. Under this technique propagation is possible throughout the year as green houses provide for manipulation of environmental factors suitable for the purpose. The rate of success is very high viz-a-viz high quality plant material is obtained, In addition to this water requirement is minimum although misting is done periodically.

References

Arora, R.K. and Yamdagni, R. 1985. .Effect of growth regulators on rooting of lemon cuttings with or without leaves. Haryana J.Hort.Sci.,15:77-81.

Bariass, M and Skene, K.G.M.1982. *In vitro* plantlet formation from citrus species and hybrids. Scientia Hort.17 :333-341

Bhat, S.S . 1942. Jambhiri (*Citrus medica* var.limionium) the stock variety for citrus trees. Poona Agric.Coll.Mag., 21:114-117

Button, J. 1977. International exchange of disease free citrus clones by means of tissue culture. Outlook on Agr.9:155-159.

Bajpai, P.N., Trivedi, R.K. and Prasad, A. 1963. Storage of citrus seeds. Sci Cult., 29: 45-46.

Chadha,K.L. and Singh, H. P. 1990 . Citriculture scenario of india. In: Citri culture in North -Western India.Proc. *Citrus* show –cum seminar prospects problems kinnow cultivation (Gill, K. S. ,Kammar, J. S. and Singh, R. eds.)p 21-64.

Chadha, K. L. 2001. Hand Book of Horticulture, ICAR,New Delhi, pp.209-217.

Chaudhari, B.K. and Chkrawar, V.R. 1981. Effect of seed treatment with certain growth regulators on the shoot and root development of Kagzi lime(*Citrus aurantifolia*). J.Maharashtra Agric.Univ., 6: 19-21.

Chohan, G.S., Kumar, H. and Vij, V.K. 1998. Effect of rootstocks n previgour, health, yield and fruit quality in Hamlin co. of sweet orange. Indian J.Horti.45:208-11.

Chohan, G. S .,Sharma,J.N. and Thatai, S.K 1978. Effect of four rootstocks on nigour, yield and quality kinnow mandarin. Indian J. Horti., 35:212-15.

Kalloo, G., Pandey,S .K.,Chakravarti, S.K. and Kaushik,S.K. 2005.Technologies for Prouction of Quality seed and Planting material in Horticultural crops, Tech. Bull.No.71., CPRI, Shimla,pp.77-79.

Kour,Kiran,Bakshi,P.,Kher,Ravi and Chahal,T.S. 2009. Effect of phytohormones on in vitro shoot multiplication of two economically important citrus species. Haryana Jour. Hort. Sci.,38(1&2):6-8.

Kitto, S.L. and Young, M.J. 1981. *In vitro* propagation of Carizo citrange. Hort. Science,16:305-306

Desai, J.B. and Patel,V.K. 1983.Influence of fruit size on seed germination in Rangpur lime(*Citrus limonia* osbeck). Seeds Res., 11:68-70.

Jalikop, S.H., Ravishankar,R. and Srivastava, K. C. 1986. .Preliminary investigations on the evaluation of rootstocks for kinnow mandarins. Punjab Hort.J., 26 :13-17.

Jamed M. Al -khayri and Abdulaze M.Al Bahrany 2001. *In vitro* micro propagation of *Citrus curaufoliar* (lime). Current Science, 81(9) : 1241-46.

Mehrotra, N.K., Jawanda, J.S. and Vij,V.K. 1984. .Evaluation of rootstocks for jaffa orange(*C.sinensis* Osbeck) under arid irrigated conditions of Punjab. Indian J.Horti, 41:29-36.

Murashigae, T., Bitters, W.P., Rangan, T.S., Nauer, E,M., Roistacher, C.N.and holiday,P.B. 1972. A technique of shoot apex grafting and its utilization towards recovering vius free citrus clones. Hort Science 7 :118-119.

Navarro, L. 1982 a. Citrus shoot tip grafting *in vitro* (STG) and its application : a review p 452-456. Proc.Intern.Soc.Citriculture.

Navarro, L., Roisacher, C.N and Murashigae, T. 1975. Improvement of shoot tip gafting in vitro for virus free citrus. J. Amer. Soc. hort. Sci. 100:471-479.

Nijjar , G.S. 1980. *Citrus.* Punjab Agricultural University, Ludhiana, India . p 18-19.

Sharma, K.K. and Sharma, R..C. 1986. Bud take and growth in kinnow mandarin influenced by type of bud mood. Haryana J. Horti Sci., 15:57-58.

❑❑❑

Chapter-27

Kamrakh (*Averrhoa carambola* L.)

Vishal Nath[1] and Bikash Das[2]
[1]*NRC on Litchi, Muzzaffarpur (Bihar)*
[2]*Horticulture & Agroforestry Programme, ICAR Complex for Eastern Region, Ranchi, (Jharkhand)*

Kamrakh or Carambola (*Averrhoa carambola* L.) known as 'Star fruit' is an important fruit of tropical and subtropical regions of the world. However, it is a drought hardy plant and grow in dry land areas. Its fruit pulp is sour-acidic in taste due to oxalic acid content.

Carambola are still in wild form or in homestead gardens and therefore, no improved and recognized varieties and /or systemic cultivation practices for this fruit has been developed so far. It is mainly grown as a fancy fruit in the garden and only limited plants are maintained. The plants prefers warm, moist climate but it can grow on any type of soil with good drainage system.

Nutritive Value and Uses

Carambola fruit is good source of vit. A, B and C together with valuable minerals and iron but, not calcium. The fleshy sweet and sour fruits are eaten in Bangladesh. Ripe carambolas are eaten out-of-hand, sliced and served in salads, or used as garnish on avocado or seafood. The plant parts are credited with medicinal properties *viz.* root extract is used as an antioxidant for poisoning and crushed leaves for curing chicken pox, ring warm and scabies. The fruit gives sour sweet taste which has limited consumption as desert however; it can be used for various preparations like, jam, jelly, pickle, beverages, etc. Carambola juice is served as a cooling beverage. The juice is bottled in India, either with added citric acid (1%) by weight and 0.05 % potassium metabisulphite. Lewis and Groeizaam (1989) reported that in Surinam,the sweet type is used for wine whereas the sour type is used for juice and pulping both are also preserved or consumption as a sweetmeat.Fruit of sweet type are eaten fresh and sour ones can be used for making refreshing drinks, pickling or as substitute for tamarind.

The acid types of carambola have been used to clean and polish metal, especially brass, as they dissolve tarnish and rust. The juice will also bleach rust stains from white cloth. Unripe fruits are used in place of a conventional mordant in dyeing. Carambola wood is white, becoming reddish with age; close-grained, medium-hard. It has been utilized for construction and furniture.

Botanical Description

Carambola or Kamrakh belongs to the family Oxalidaceae. It is native to Indonesia. Kamrakh has vigorous tree growth under transactional cut view of the fruit gives a star shape and therefore, it is also called as star fruit. It is a small tree with drooping branches and bears attractive golden yellow, 3-5 ribbed 10-15 cm long fruit.Watson *et al.* (1988) described the detailed botany of *Averrhoa carambola* including its flowering and fruit development ,fruit characteristics and quality. Du-Xiang Qiou (1998) described that sweet Yangto (*Averrhoa carambola*) has repeated flowering habit (4 times a year) and can be harvested 3 times in a year.It produces as high as 31 tonne / ha yields of fleshy juicy fruit. Flowering and fruiting behaviour of both sweet and sour type carambola varieties have been reported by Nand (1970). Flowering in carambola occurs in 3 flushes, viz. April-May, July- August and September –October. The maximum yield came from the second and third flushes from which fruits were harvested between November and February. Pollination, fruit set and fruit development in carambola has also been reported by Nand (1971). The anthers dehisced as the flowers opened, and the best time for pollen collection is 8-10 a.m. The pollens are 100% fertile and natural pollination by insects' results in a higher percentage of fruit set than various controlled pollination methods. The initial fruit set under natural conditions ranged from 58.9% per panicle in the sour variety to 74.5% per panicle in the sweet. However,16%fruit drop has been reported in carambola till the fruit reached maturity. Fruit growth ceases after 14-15 weeks from the time of flower opening. Salakpetch *et al.*(1990) studied the floral ontogeny, floral cycle and flower bud,sepals were initiated first followed by 5 petal primordia which were initiated alternately and centripetally to the sepals. During sepal elongation, stamen and carpel initials are formed.

The flower cycle of opening and closing lasts for one day. Flowers remained on the plant from one to 16 days. The flowering response to diurnal temperature regimes (30/25,25/20,20/15,30/15 and 34/20°C), photoperiods (8,12,14, 16 h) and water stress was observed in the cultivars Thai Knight and Fwang Tung. Flower initiation was not affected by temperature or photoperiod but a leaf water potential of -2.0 MPa reduced the number of flower buds by 90% compared with the control.

Species and Cultivars

The plus trees of carambola has been identified and marked by the plant explorer in the variability regions.Watson *et al.*(1988) mentioned the fruit quality characteristics of 10 cultivars (Arkin,B 10,Fwang Tung, Maha, Kembangan,Kajang,11-1 and 9-4) in Australia. Wagner *et al.*(1975) carried out preliminary screening tests on fruit from 16 selections and two varieties (Tean Ma and Dah Pon) of *Averroha* for the fresh fruit and processed products markets. Two forms had better flavour and higher ascorbic acid but lower oxalic acid contents.

Campbell (1989) reported that the most important carambola (*Averroha carambola)* cultivar in Florida is Arkin , followed by Golden Star, Fwang Tung, Maha and B10. In Malaysia,B10 is most important, followed by B2 and B17. In Taiwan, cultivars Meeshi, Er-lin, Soft-sih and Cheng-Tsey are commercially grown. Its other species, *Averrhoa bilimbi* produces gherikin like fruit which is used for making pickle and curries.

Propagation

Carambola is mainly propagated by seeds. The vegetative means e.g. budding, grafting and layering are also used for multiplication. Micropropagation method is especially applicable to species in which clonal propagations is needed. Tissue culture technique has been tried in carambola to raise true to type plants.

Seed propagation

Propagation through seed is the most prevalent method of carambola regeneration.In natural habitat, the plants regenerate through seed from the mature fallen fruits.Attempts have been made to standardized the method of seed extraction , strong , time and method of seed sowing, etc. for this crop.

Seed Extraction and Storage

Mature seeds of kamrakh are extracted from mature dried fruits along with pulp. The extracted seed are then in sunshine. It has been observed that drying of kamrakh seed in sun shine give better germination and seedling vigour. Sekiya and Chunha (1999) reported best germination of seed and seedling growth when carambola seed were sown after sun drying with or without aril.

Stephen (1936) mentioned that only a small percentage of seed are fertile in carambola. Therefore of well developed, sound and mature seeds should be done for uniform germination and seedling growth.

Time and Methods of Seed Sowing

Carambola has small seed which should be sown in pot or polythene containers under partial shade after cleaning and drying. Maintenance of optimum moisture and warmth in the growth is essential to ensure high germination of the seeds therefore , the time of seed sowing in kamrakh is of paramount importance. Seed should be sown during June - July or February-March. After about 1.0-1.5 month of germination, the newly emerging seedling should be transplated in nurseries at 25 x 25 cm apart or in the polythene bags of 20 x 10 cm size having sufficient hole in the bottom for aeration . Air root pruining container have been found to increase the proportion of roots in half of the root ball and formation of new root in carambola seedlings in Florida (Marler and Willis, 1996) . Therefore , the seedlings can also be raised in air root pruning container . Marler and Mickelbart (1992) also studied the growth response of containerized *Averrhoa carambola* seedlings to GA_{4+7}, applied to trunks in lanolin paste under greenhouse conditions. Concentration of 0, 250, 500 or 750 mg/litre and an untreated control (no lanolin) were used. Internode length and increase in plant height and trunk cross-sectional area (TCA) did not differ between control and plants treated only with lanolin, but they were mostly enhanced as the GA concentration increased. Increase in TCA was observed in plants treated 500mg GA_3/litre. The average recommended graftable size (7mm) was reached in 47 days in plants that were GA treated and 93 days in control plants, suggesting that GA_3 may be used to shorten nursery time for producing graftable rootstock. For raising the carambola orchard through seed, well grown seedlings after proper hardening and growth up to almost 8-9 month , are finally planted in the well prepared pit at appropriate spacing (6m x 6m) in square system.

Carambola is a cross pollinated species and a lot of variation among seedling population has been noticed by Garner and Chaudhari (1976). They mentioned that due to cross pollination, heterogeneity in seedling population is quite common which result unlike growth, fruiting and fruit quality in spite of delayed bearing in the seedling trees. To obtain true -to-type plants of kamrakh, the vegetative method of regeneration is considered to be the most appropriate.

Vegetative propagation

Carambola can be vegetatively propagated either on their own roots through cutting and morcotage or on the roots of seedling root stocks of carambola or bimbli through inarching (approach grafting) and /or through shield of forkert method of budding on seedling rootstock or through top working (detached method of grafting).

Cutting

Campbell (1970) reported that propagation of carambola through cutting has not been successful in Florida. However, Bora and Das (1998) reported that use of auxin in suitable concentration have been found effective to induce rooting in hard wood cutting of carambola.

Bastos *et al.* (2004) reported that top and base cutting of *A. carambola* soaked in IBA(0,1000, 3000, 5000 or 7000 mg / litre) for 10 second before planting them in wooden boxes containing vermiculite as substrate and maintained in the greenhouse under intermittent mist for 70 days . Survival and rooting capacity and number of medium sized roots per cutting were evaluated .The top cuttings resulted in greater survival (49.26) , rooting (34.84%), and number of medium -sized roots per cutting (20.51)than the base, soft and semi hard wood cutting of carambola in peat:sand or peat:perlite media. Chua and Teoh (1973) mentioned the use of woody cuttings of 1 m length and 3-10 cm diameter for propagating carambola plants. Borah and Das (2000) reported that the basal end of the cuttings treated with exhibited the highest rooting percentage in summer. On the other hand,minimum percentage of survival was observed carambola in autumn.

Layering or Morcottage

Carambola can be successfully propagated by layering. The method involves covering of shoot with soil or suitable medium after treating the shoots to restrict the flow of phloem. Later the roots emerge from the treated portion give rise a seperate entity of plant.The rooted shoots are then severed from the tree below the root and established in a pot or bed. It has been mentioned that layering success in carambola have been improved by wiring the shoots or by etiolation of performed on shoots much above the ground by providing the soil medium around the treated shoots, the process is known as morcottage or air layering. In this method a narrow ring (2-3 cm wide) of bark is removed where root arises from the upper edge of the ringed sections after callus formation .Application of rooting hormones particularly the auxins(IBA,IAA) have been found to improve the formation of fibrous roots which ultimately helps in survival of layers. Campbell (1965) and (1970) reported slow and inconsistent results of layering in florida whereas Singh (1963) reported 100 per cent success in India when the layering was done in January.

Singh (1965) described a morcotting box for aerial propagation of kamrakh and other such fruits which comprises of a divided tin box.Carambola has not been much successful in this method because the amount of callus was abrupt in this method. Later it was concluded that use of rooting hormones coupled with layering at active stage of foliage growth and sap flow contributes towards profuse callus formation and development of finer roots in carambola.The method gave desired success in air layering of carambola under warm-humid climatc of India. Nath (1994) reported that removal of 1.5cm diameter of bark from pre etiolated portion of 1 year old shoots before 40-45 cm of the growing tip and application of 5000 ppm IBA during April gave the maximum success.Treatment with 5000 ppm IBA and layering on 1st April gave the highest percentage of rooting (72.4%), number of primary roots (11.0), root length(7.5cm) and survival of rooted layers (75.6%).

At present,carambola is successfully propagated through air layering during January and June. The technique has been fine tuned and now, a very handy step has been advocated. This includes selection of shoot , ringing of shoots, application of IBA paste (1000 ppm) covering of ringed portion with suitable mixture and wrapping with 400 guage white polythene sheets.Rooting takes place in 1-1.5 months and rooted layers are separted from mother tree after 2.5-3 months and shifted to polythene bags or nursery beds under partial shade.

Banerjee *et al.* (1980) found that ringing of one year old terminal crambola shoots about 25 cm long followed by application of 1000-10,000 ppm., IAA,IBA or NAA in lanolin paste gave higher success when layers were severed 45 days after treatment. IBA at 10,000ppm gave 100% rooting followed by NAA at 10000 ppm (90%). Rooting in the control (ringed and treated with lanolin only) was 20%. Campbell (1989) reported that air layering is used successfully in Asia but has not been successful in Florida because root formation is slow and the percentage rooting is low(20-30%).

Budding and Grafting

Attempts have been made to propagate the kamrakh plants through attach and detach methods of grafting. Among detach methods veneer grafting and shield and forkert method of budding have been found more successful by various workers in India and abroad.

Root stock

Seedling plants of carambola are considered to be most appropriate rootstock due to anatomical and biochemical synchronisation of cambial growth of scion and rootstock with each other. Carambola seedlings are suitable root stock for inarching,budding and veneer and cleft grafting .Six weeks old seedlings are transplanted under highly shade conditions from the beds are removed . Six weeks old seedlings are transplanted in lighty shaded nursery beds at 30×30 cm spacing .To develop into hard and study root stock the shade from the beds are removed after 2-3 months coupled with normal weeding and watering. Carambola has thin bark which lifts easily while budding. One year old root stocks are ucsd for budding with cent per cent success. Seedlings of carambola attaining 0.6-1.3 cm diameter at matrix (10-15 cm from the collar) have been found most suitable for grafting and budding (Garner and Chaudhri, 1976).

Chua and Tech (1973) reported that in Singapore, the seedling rootstock of carambola are raised in polythene bags of 25×30 cm size and 0.06 mm gauge thickness filled with a special mixture called John Innes compost and arranged in nursery trenches.The compost consists of 7 parts of medium loans, 3 parts of coconut peat and 2 parts of coarse sand.The bags are arranged in partial shade and properly watered up to 3 months after sowing the seed. The seedlings are gradually hardened in sunlight before being grafted. The Young seedlings are nourished with NPK mixture (10g/seedling) every month. Six to eight month old seedlings are used for inarching whereas one year old seedlings are suitable for budding and grafting adopting appropriate technique.

Approach grafting

In inarching, attached scions are placed with appropriate rootstock at a junction called "the matrix" with various techniques like spliced approach grafting,toung approach grafting. Once the graft is healed the lower part of the scion twigs and upper portion of rootstock are cut away to seperate new inarched plant. Majumdar(2004) mentioned that approach grafting is successful method of carambola propagation. He mentioned that this method has the problem of operation as the rootstock has to be supported close to scion during the process of grafting and union. For the purpose, seedlings are either raised or shifted to light container and/or some platform is raised where rootstocks are rested during the process. Inarching during June or January has given 80 and 100 per cent success in Nilgiris (Singh, 1963). Similar results were also reported in florida by Cobin and Ruele (1949) Chua and Teoh (1973) and Anon (1950).

Chua and Tech (1973) reported that in Singapore, approach grafting is preferred over budding because latter is done on one year old rootstock where as the former is done on 6-8 months old rootstock The rootstock are raised in polythene bags and shifted on low benches near the selected tree which is to provide the scion. The matrix ,after making the desired cut and attaching both the cuts face and coinciding the cambium layer of either side, is tied with polythene film(100 gauge).After 4-6 weeks the plastic film is removed and graft is separated from the scion parent tree. In this way the grafts become ready to plant -out within a year of sowing the rootstock seed whereas budding or veneer grafting takes almost two years period to develop a carambola plant.

Grafting

Among detach grafting techniques ,carambola plant is propagated through veneer or side and cleft or top grafting apart from top working in special cases. It has reported that in Florida, veneer grafting is considered to be the best method for raising large number of carambola plants in nursery. Selection of scion stick is most important for successful grafting of carambola. Well mature shoots of 6-8 month age having healthy leaves are preferred for scion stick.These selected shoots are defoliated one week prior to grafting which helps in union at matrix and emergence of healthy shoots. The thickness of scion stick is of utmost importance. For maximum success, the thickness of scion stick must coincide with thickness of the rootstock at matrix(0.6-1.3 cm diameter). Cobin and Ruele (1949) also reported successful grafting of cultivar Golden Star on seedling rootstock through cleft method in Florida. They described that in cleft grafting, there are comparatively more cut surfaces which promotes stronger union and straight growth of the sapling.

Budding

Shield budding is the most common method of budding being practiced in carambola for raising true to type plants in Singapore. Rootstock of same species has been advocated by Hayes (1970) for better success in India . The selection of bud wood and stage of rootstock at the time of budding decides the success of shield budding in carambola. Sufficiently mature, but not old or hard shoots having purplish appearance from well grown up superior plants should be selected. Bud of 2-2.5 cm length along with petioles are used for budding. Size of bud can vary under different situation. In this method,non-petiolated bud patches are cut from,one year old mature wood during the rainy season. Chip budding is the most common method of propagation in Florida. Seedlings of selections such as Golden Star and M-18960 are better adapted to calcareous soil than other selections and are suitable for use as rootstocks.

Micropropagation

A number of workers have attempted for micro propagation of carambola in India and abroad. Regeneration of plants *in-vitro* from somatic tissue of *Averrhoa carambola* was first attempted by Litz and Canover(1980). Kantharajah *et al.*(1992) described a technique for micro-propagation using MS medium supplemented with BA and NAA. They used roots as source of explants for regeneration .

Kantharajah *et al.*(1992) described a technique for the clonal micropropagation of caramola. Multiple adventitious shoots were produced on root explants placed on Woody Plant Medium or MS medium supplemented with 2.0 mg BA+0.2 mg NAA/litre. When these shoots were excised from the roots and placed on the same medium a multiplication medium for >7 weeks became chlorotic and eventually died. Chlorosis was avoided if the shoots were transferred to a growth regulator -free basal medium for 10 days. Shoots were successfully rooted on a basal medium containing 1% activated charcoal and were successfully transferred to soil and planted in the greenhouse.The technique provides a commercially practical technique for the clonal propagation of carambola.

Griffis and Litz (1993-94) experimented on organogenesis in tissue cultures of *Averrhoa carambola* L.'Arkin'. They described that callus from youngest fully expanded leaflets of grafted, container -grown carambola cv.Arkin plants was initiated on MS medium supplemented with 1 mg L-cysteine, 30 g sucrose, 3.5 g agar, 1 g gelrite, 0-5mg 2,4-D and 0-2mg

IAA/litre and various concentrations of 6-furfurylaminopurine(kinetin), BA and /or 2 ip (isopentenyladenine). Calluses were subcultured onto shoots induction medium. Shoots arose directly from callus which was initiated on medium supplemented with 5g 2, 4-D and 5 mg 2ip/l and then transferred to medium supplemented with 5 g 2ip/litre alone. These shoots were subcultured onto medium supplemented with 0.5mg IAA and 5.0 mg 2ip/ litre to stimulate axillary bud proliferation. Neither rooting *in-vitro* nor rooting of microcuttings in the greenhouse was successful in this study.

Rashid *et al.*(1992) studied the *in vitro* organogenesis of carambola. They mentioned that adventitious shoots were regenerated from root explants of carambola cv. B10, using MS medium supplemented with IBA at 1.0mg/litre and BA at 0.1-4.0 mg/litre. Maximum proliferation occured with 1.0 MS medium supplemented with 2.0 mg IBA/litre or 2.0 mg IBA+0.1 mg IAA/litre. Amin and Razzaque (1993) from Bangladesh reported that *in vitro* plants of carambola can be regenerated by using callus culture. Hypocotyl and cotyledon explants of aseptically grown seedlings produced callus and subsequently 2.00 mg BA and 0.1-0.5 mg NAA/litre. A low concentration (0.1-0.2 mg/litre) of the auxin along with a comparatively high concentration (0.5-1.0 mg/litre) of the cytokinin was most suitable for regeneration and growth of shoots from calluses of both types of explant.Further multiplication (4-to 6-fold every 4 weeks) of axillary shoots could be achieved using nodal segments of the regenerated shoots were rooted *in vitro* using half-strength MS (both major and minor salts) medium and 0.2 mg IBA/litre. The rooting success after 4 weeks of culture was 74%at 25 degrees under 16-h photoperiod and 100% when cuttings were incubated at 30 degrees in complete darkness for 1 week followed by 3 weeks in usual growth room conditions.

Li *et al.* (1999) from China advocated shoot tip culturing in MS medium supplemented with BA and NAA for *in vitro* regeneration of Carambola where as Khalekuzzaman *et al.*(1995) mentioned the cotyledons culturing in *Averrhoa carambola.* The explants from fully grown cotyledons callus and subsequently adventitious shoots on MS medium with 4.4-13.2 μM BA and 0.54-2.702 μM NAA.Addition of NAA to the medium was essential for organogenesis. The distal part of the cotyledon produced fewer shoots than the part of the cotyledon proximal to the embryo axis, where prolific shoot regeneration occurred. Regenerated shoots were subsequently established in soil. The nodal segments have been used as explants for rapid propagation of *A.carambola* by Liu and Wu (2004). Among cytokinins (1.0mg-1-1), Zeatin

(1.0mg 1-1) was optimal for the plant growth; the proliferation coefficient being as high as 3.7 and the rate of abnormal plants was only 23.3%. For rooting the half-strength MS medium containing 0.2% mg IBA 1-1,0.1mg IAA 1-1 and 2-3% sucrose was the best.The rooting rate was 87.1%.

Li *et al.* (1999) reported a rapid micropropagation method for carambolas cv. Shuimiyangto. They found that the best medium for shoot tip culture was MS (Murashigae & Skoog) supplemented with BA(benzyladenine) 0.5 mg/litre + NAA (0.2mg/litre) under such conditions,rosette buds were formed after 21-22 days .Rosette buds were transferred to half –strength MS supplemented with IBA(0.1mg/ litre)+IAA(0.1mg/litre),placed in the dark for 1 week,then moved to a 16 h photoperiod at 27°C for rooting. After 1 month, rooting rate could reach 94% and the transplant survival rate was 90%.

Roy *et al.* (2007) tried an efficient and reproducible method for rapid clonal propagation of *A.carambola* from aseptically grown seedlings. The result of the present experiment demonstrated a reproducible and efficient regeneration protocol through *in vitro* culture of *Averrhoa carambola.*

Quality Standards of Planting Material

The appropriate planting material of carambola should possess the following standards.

1. The seedling plants should be healthy, having single stem,the height of plant should be between 80-100cm with a collar diameter of 0.8to 1.2 cm.
2. The vegetative propagated plants should be true to type to the mother cultivar. The plants propagated through cuttings/ layering should have well developed roots.The plants propagated through grafting or budding should have proper graft/bud union at the matrix.
3. The height of the budding or grafting should not be more than 20-30 cm.
4. The plants showing incompatibility symptoms (differential vigour of rootstock and scion) should not be selected for planting.
5. Plants raised on older rootstock should not be prefered.

6. The sapling should neither be too large nor too small. It should be medium in size with sufficient foliage preferably raised in polythene bags for easy transportation to distant places.

References

Amin, M,N. and Razzaque,M.A.1993. Regeneration of *Averrhoa carambola* plants in vitro from callus cultures of seedling plants. Journal of Horticultural Science ;68(4):551-556.

Anonymous 1950. .Propagation of fruit plants. Agric.Anim. Husb.U.P.1(3):19-27.

Banerjee, D.P.; Chatterjee, B.K. and Rao,D.P.1980. Propagating carambola (*Averrhoa carambola* L.)by air layering with different concentrations of growth regulators. Indian Agriculturist. 24(3/4):165-168.

Bastos, D.C., Martins, A.B.G., Scaloppi- Junior,E.J., Sarzi,I. and Fatinansi,J.C. 2004. Influence of indole butyric acid on the rooting of top and base cutting of star fruit(Averrhoa carambola L.) under intermittent mist. Revista -Brasileria -de-Fruticularia .26(2),284-286.

Borah, A. and Das, R. P. 2000. Efficacy of indole butyric acid (IBA) on rooting of stem cuttings of some minor fruits of Assam. Haryana Journal of Hort. Science. 29(3/4);196-198.

Borah. A., and Das, R.P 1998. Role of indole butyric acid (IBA) on vegetative growth of rooted cuttings of some minor fruits of Assam .Journal of the Agricultural science Society of North East India.11(2):197-201.

Campbell, C.W. 1965. Propagation and Production systems for carambola. Proceedings-of the Interamerican Society for Tropical Horticulture.;33:66-71.

Campbell, C.W. 1989. Storage and Handling of Florida carambola. Proceedings Interam. Soc. Trop. Horti., 33: 78-82.

Chua, S. E. and Tech, T. S., 1973 Propagation by approach grafting and woody cutting of some tropical fruit and ornamental trees.Singapore, Journal of Primary Industries.1(2):87-95.

Cobin, M. and Ruele, G. A. 1949. Sub tropical crops of minor importance. A R Fla. Agric. Expt.Stat.:245-247.

Du-Xiang Qiou 1998. Cultural techniques for growing sweet yangtao in Hainan province. South China Fruits;27(6):33-34.

Garner, R.J. and Chaudhri, S.A. 1976.The propagation of tropical fruit trees. Horticultural Review No.4. Commonwealth Bureau of Horticulture and Plantation Crops, Kent, FAO:297.

Griffis, J.L., Jr and Litz,R.E. 1993-94. Organogenesis in tissue cultures of *Averrhoa carambola* L. 'Arkin'.Proceedings of the Interamerican Society for Tropical Horticulture.

Hayes, W.B. 1970. Fruit Growing in India . Kitabistan , Allahabad, India.

Kantharajah, A.S., Rechards, G.D. and Dodd, W.A. 1992. Roots as surce of explant for the successful micropropagation of carambola (*Averrhoa carambola* Linn.). Scientia Horticulture, 51(1-2):169-177.

Khalekuzzaman, M., Islam, R .,Reza,M.A. and Jorder,O.I. 1995. Regeneration of plantlets from *in vitro* cultured cotyledons of *Averrhoa carambola* L.(Oxalidaceae). Phytomorphology, 45(1/2):107-111.

Lewis, D. and Groeizaam, M 1989. The cultivatio and utililzation of carambola in Surinam Proceedings of the Interamerican Society for Tropical Horticulture, 33:59.

Li Ji Hong, Shao HanShuang: H uang Guixiu and Fu Shaoping 1999. Experiment of fast propagation of carambola *in vitro*. China Fruits: (4):37.

Liu jian fu and Wu-qing 2004. Rapid *invitro* propagation of *Averrhoa carambola*. Journal of tro-and-subtropical Botany; 12 (3):265-267.

Majumdar, B.C. 2004. Minor fruit crops of India Tropical and sub tropical, Daya Publishing House, New Delhi:28.

Marler, T.E. and Willis, D.1996. Chemical or air root pruning containers improve carambola ,longan and mango seedlong root morphology and initial root growth after transplanting. Journal of Environmental –Horticulture;14(2);47-49.

Marler, T.E.,and Mickelbart, M.V.1992. Application of GA_3+7 to stem enhances carambola seedling growth.Hortscience 27(2):122-123.

Nand, D. 1971 Pollination ,fruit set and fruit development in carambola (*Averrhoa carambola* Linn.) Indian Journal of Horticulture.;28(4):278-284.

Nand, D.1970. Flowering and bearing behaviour of carambola (*Averrhoa carambola* Linn.):Indian Journal of Horticulture.;27(3/4):147-152.

Nath, J.C 1994 Studies on vegetative propagation of carambola (*Averrhoa carambola* Linn.)by airlayering. Horticultural Journal;7(1) :63-65.

Rashid, A, A., Hanim-Ahmad, Aziz, M.A. and Alang Z.C.1992. In vitro organogenesis of carambola. Acta-Horticulturae, (321)579-588.

Roy, P.K., Mamun, A.N.K. and Ahmed,Golam 2007. Rapid multiplication of Averrhoa carambola through *in vitro* culture, *J. Bio-Sci.*,15:175-179.

Salakprtch, S., Turner, D.W. and Dell, B.,1990. Flowering in carambola (*Averrhoa carambola*). Acta- Horticulture: (275):123-129.

Sekiya, R, F, M and Chunha 1999. Preparation ad storage effect on carambola seed germination. Revista Brasileria de Fruticultura, 21(1):57-59.

Singh, M.P. 1965. An improved marcot box for aerial propagation of fruits. Indian J. Hort.11(3):89-93.

Singh, D.1963. Kamrakh Sweet and Vitamin rich Indian Hort. 7(2):8.

Stephen, S.E. 1936. Some tropical fruits No.7 ,The five carner. QD.Agric. J. 45:270-71.

Wagner, C. J., Brrrrrryan, R. E. Knight and R. J., Jr 1975. Carambola selection for commercial production :Proceedings of Florida State Horticultural Society;88: 466-469.

Watson, B, J., George, A.P., Nissen, R. J. and Brown, B. I. 1988. Carambola : astar on the horizon Queensland, Agricultural Journal. 114(1):45-51.

Chapter-28

Chironji (*Buchanania lanzan* Spreng.)

Sanjay Singh and A.K. Singh
Central Horticultural Experiment Station, Godhra - 389 340

Chironji or Charoli (*Buchanania lanzan* Spreng.) is an important fruit tree of agro forestry and social forestry. In the wasteland development and dryland horticulture, it assumes great significance due to its multifarious uses and capacity to withstand adverse environmental conditions. At present, it is growing under natural forest as an underexploited fruit and gives monitory reward to the tribal community of the country and seems to be boon for them. It is valuable tree species found in dry deciduous forest throughout the country excluding eastern Himalayan forests (Singh, 1982).

The area and production of this fruit crop in India is not available because it is not grown on plantation scale. It can be seen growing in forest areas and boundary of fields. The production in India is mainly concentrated in the drier states and the produce is collected by the villagers and sold in the local market. Its cultivation may spread to arid and semi- arid areas, resource-poor areas and wastelands.

A tree of the dry region is found in north, west and central India. Its trees are available in forest area of Uttar Pradesh, Madhya Pradesh, Chhattishgarh, Maharashtra, Bihar, Jharkhand, Orissa, Andhra Pradesh and Gujarat (Singh and Singh, 2012).

Nutritive Value and Uses

The flesh of ripe fruit is very palatable and is largely eaten raw or roasted and the oily kernels are the most important part and are used in preparation of puddings (Singh, 1982). Mesocarp of fruit is edible and cherished by children (Munde *et al.*, 2003). The kernel is highly nutritious and rich in protein and yields sweet oil, which can be used to substitute olive and almond oil. Kernel contains 33.50% oil, 1.90% of which is unsaponifiable. The saponifiable part contained 20.00% of linolic acid. Chironji oil is non-repellant and non-toxic and is suitable for human consumption (Banerjee and Jain, 1988). Hemawathy and Prabhakar (1988) reported that total kernel lipid extracted from chironji was 65.60% of dry kernel and consisted of 90.40% neutral lipids, 3.4% glycolipids and 6.2% phospholipids. Kernel is used in sweets and many food items. Kernal is extracted from fruit after drying which is tedious work and done manually. The wood is used as quality timber for making implements and furniture.

Botanical Description

Chironji (*Buchanania lanzan* Spreng.) belongs to the family Anacardiaceae. It is highly heterozygous, cross-pollinated crop and as such seedlings exhibit a wide range of variations. It is a dicot woody plant of deciduous nature. Flowers are typical of family. Flowering starts in the month of January-February on the well-developed panicles, which have hermaphrodite and pseudo hermaphrodite flowers, and they are located on the annular prominent disc and fruit is drup.

Reproductive biology of the crop was studied at Central Horticultural Experiment Station, Vejalpur, Godhra, Gujarat. It was observed that peak period of panicle emergence and flowering was recorded in the month of January and February, respectively. The peak period of fruit set was recorded in the month of February in majority of genotypes. Highest panicle length (35.11cm) and fruit set per panicle (36.20) were noted in CPT 1. Time taken for complete development of flower ranged from 16-20 days. Peak period of anthesis was recorded between 6-11 am in all the genotypes. None of the genotypes showed anthesis before 4 A.M. and after 2 P.M. Anther dehiscence commenced after opening of flowers i.e. at 7 A.M. and continued till 3 P.M. Peak period of dehiscence was recorded between 8 am -12 noon in all genotypes. The flower diameter varied from 5.12-6.30 mm. The stamen and carpel length varied from 1.97 - 2.12 mm and 1.22 - 1.38 mm, respectively. Pollen viability ranged from 54.55-70.38 %. Pollen germination ranged from 20.00 – 35.00%. Maximum stigma receptivity was recorded on the day of anthesis. Fruit set / panicle was found to be positively and significantly associated with panicle length and it may be observed while selecting elite genotypes. Its hard wood is also used for making agricultural implements and as timber.

There is no standard variety of chironji since a little work has been done to exploit the genetic resources of chironji. This crop is facing severe genetic erosion as a result of activities related to deforestation. A number of seedling strains with lot of variability in fruit weight, total soluble solids, acidity, protein and earliness particularly in Bihar, Uttar Pradesh, Gujarat ad Maharashtra have observed which provide a good scope for selection of better genotypes (Singh and Singh, 2012). A survey was undertaken in Uttar Pradesh to investigate the nature and extent of variability present in chironji seedling progenies for morphological characters of tree and high degree of variability was observed with regard to fruit yield, seed size and chemical composition of the fruit (Rai, 1982). Singh and Chaturvedi, (1983) and Chadha and Sharma (1997) surveyed the chironji growing area and recorded significant differences in morphomatrics and yield attributes. An extensive survey and exploration was carried out in Gujarat by Singh *et al.* (2006). They reported that earliest flowering (first week of February) took place in 'CPT 1', 'CPT 5', 'CPT 10', 'CPT 15', 'CPT 17', 'CPT 19' and 'CPT 22', while the latest (4th week of February) occurred in 'CPT 13'. Maximum panicle length (35.11 cm) was found in 'CPT 1', while 'CPT 18' recorded least panicle length (14.20 cm). Fruits become ready for harvest during the month of April and May. Ripening time varied from third week of April to second week of May in different genotypes. The highest fruit yield per plant was found in 'CPT 7' (28.00 kg/ plant). With respect to all horticultural traits studied the genotypes, 'CPT 1', 'CPT 2', 'CPT 5', 'CPT 6', 'CPT 7', 'CPT 8', 'CPT 12' and 'CPT 30' were found to be promising and genotypes have been planted in the field for further evaluation.

Propagation

Seed propagation

Chironji plants are generally propagated by seed giving a long gestation period (10-15 years) and large variability. Per cent germination in freshly extracted seeds is poor because of hard seed coat on the kernels. Prasad *et al.* (1988) recorded significant differences in the germination percentage of chironji seeds when sown in different months and May was found to be the best time, it was completed within 26 days from the date of sowing. Seed germination of 83.00% within 18 days could be achieved with satisfactory seedling growth by mechanically damaging the stony endocarp before sowing in the month of June, plant survival of 93-96% with satisfactory plant growth was obtained when one year old seedlings were planted in 60x60x60 cm pit filled with red soil and 10 kg of well rotten farmyard manure coupled with proper mulching around the plants during winter and summer season and with or without thatching (Shukla and Solanki, 2000). Seed treatment with Sulphuric acid (5-7%) was also found very suitable for promotion of seed germination in chironji. Seeds can be sown on raised beds or in polythene bags during June-July and it germinates within 25-35 days. The seedlings become ready for grafting after one year from the date of sowing.

Vegetative propagation

Chironji is multiplied by vegetative means and propagation method viz., soft wood grafting techniques has standardized at CHES, Godhra for large scale production of plants.

Soft wood grafting

It is propagated by soft wood grafting. When one-year-old seedlings raised for rootstock commence putting on new growth and the leaves are of light green colour, these are ready for soft wood grafting. Shoots of 3-4 months, which have prominent apical bud, are taken as the scion material. Defoliation of such shoots is done 8-12 days before detaching them from the parent tree for grafting operation. At the time of removal of these shoots, the apical buds should remain intact. The top of the new growth of the stock is cut and the scion is fitted by cleft grafting or splice grafting. The union is tied with 200-gauge polythene strip. If the selection of scion and rootstock is proper, 60-70 % success may be obtained during the month of July-August. (Table 1).

Table 1 : Effect of time of soft wood grafting on bud sprouting and success percentage.

Month	Time taken for bud sprout (days)	Bud sprout (%)	Success (%)
July	24.66	71.00 (57.42)	66.66 (54.70)
August	25.67	64.00 (53.13)	40.33 (39.41)
September	29.33	26.33 (30.85)	24.00 (29.33)
October	31.67	19.00 (25.84)	17.00 (24.35)
November	0.00	0.00 (0.17)	0.00 (0.17)
December	0.00	0.00 (0.17)	0.00 (0.17)
January	0.00	0.00 (0.17)	0.00 (0.17)
February	30.00	4.33 (11.97)	3.33 (10.47)
March	25.00	40.00 (39.23)	34.33(35.85)
April	25.00	35.00 (36.27)	29.00(32.58)
May	27.33	17.67 (24.80)	16.33(23.81)
June	26.67	44.33 (41.73)	39.00(38.65)
CD at 5%	1.29	2.24	1.51

(Figure in parentheses are transformed value)

Soft wood grafting *in situ*

This method was proved to be better for establishing *in situ* Chironji orchard in gravelly soils and drier tracts where mortality of nursery raised grafts was very high. Freshly extracted seeds are sown in the polythene bags for germination. After attaining height of 5-10 cm, it should be directly planted in the field at the desired distance. These vigorous seedlings, attaining pencil thickness after one year are soft wood grafted with scion sticks of the suitable genotypes. Singh *et al.* (2007) advocated that about one year old rootstocks should be planted in the field in the month of June for *in situ* grafting under semi arid ecosystem of western India. Such plants grow very fast and attain a stature in few years. Main advantages of this method are better plant growth and no mortality. Response of charoli strains for soft wood grafting under Bagalkot condition showed significant influence of soft wood grating. The maximum graft success was recorded in strain KRC-14, KRC-26, KRC-46, KRC-47 each recorded 90% success (Alloli *et al.*, 2010).

Chip budding

Chip budding has been reported as one of the successful methods for vegetative propagation of Chironji (Tewari and Bajpai, 2001). In this method of budding, a healthy bud is selected in the axil of 2 months old leaf emerging on new flush. Leaf blade is removed with the help of sharp knife leaving petiole intact. The upper cut is given about 2.0 cm above bud which goes downwards, taking wood portion with it simultaneously up to 1.0-1.5 cm below the bud, then lower cut is given about 1.0 cm below the bud so as to take out bud with wood. The same rectangle cut is made on the rootstock and bud is placed at the juncture. The bud is pressed by hand to remove open space if any and tied tightly with white polythene strip. In case, if cuts on rootstock are wider, at least one side bark of scion and rootstock must be matched properly. The rootstock is detopped about 10 cm above the bud to facilitate bud sprouting. The bud sprouts within a month of budding indicating that the bud establishes vascular connection with rootstock. After union, the top of the seedling is cut a little above the bud union and polythene strips are carefully removed. If the selection of scion and rootstock is proper, appreciable success may be obtained during the month of July-August.

Root cutting

Experimentally, this method has been tried with varying degree of success. Due to poor viability of seeds and slow growth of seedlings, the *Buchnania* species has not so far been raised in plantations despite the high price of the kernels in the market. Rooting in root cuttings is very difficult; however with the use of auxins, appreciable success may be obtained. Two- year old root cuttings having thickness of 3.6-5.5 cm, treated with IAA 1600 ppm recorded 67.66 % rooting (Singh *et al.*, 2002). Root cuttings are found successful, but the method has limited use as it results in serious injury to the mother trees when large number of root cuttings is taken. Moreover, taproot is also absent in the plants that affect the longevity of the plants particularly, in degraded lands.

Rootstocks and Use of Polycontainers

Rootstock selection for vegetative propagation of Chironji is important as it controls the vigour and equilibrium between yield and quality. Dwarfing rootstocks are considered to be best because they induce dwarfness and enable to facilitate easy management of the orchard. Such type of work may be initiated and be standardized for different agro-climatic conditions.

Raising of rootstock in nursery beds and lifting budded plants with earth ball in highly sandy soils is practically not feasible. Transportation of plants from long distance may also cause high mortality particularly under semi arid and arid environment. To reduce the time for raising rootstock and to avoid damage during handling and transportation, polyethylene bags may be used on commercial scale. Generally polyethylene bags (25cm x 15cm) with small holes in bottom filled with porous rooting medium are used for raising the rootstocks. Generally 1-2 seeds are sown in each polythene bags and then placed in trench bed, so that it can be irrigated easily.

Model nurseries should be established in local area to raise and supply of quality plant material.The seedlings should be planted 10 m apart, whereas grafted ones 8 m apart during July- August month in well prepared pits. Healthy and disease free plant should be transplanted in the field.

References

Alloli,T.B., Athani, S.I, Karody, S.V. and Naik,K.R. 2010. "Standardizationof propagation studies in Charoli, In. National Conference on production of Planting Quality Seed and Planting material, held at New Delhi,11-14 March, 2010, pp28.

Banerjee, A. and Jain, M. 1988. Investigations on *Buchanania lanzan* Spreng. seed oil. Fitoterapia, 59 : 406.

Chadha S.K. and Sharma, M.C. 1997. Trends in fruit production in *Buchanania lanzan* trees. Vaniki Sandesh, 21: 1-3.

Hemawathy,J. and Prabhakar, J.V. 1988. Lipid composition of chironji (*Buchanania lanzan* Spreng) kernel. Journal of Food Composition Analysis, 1 : 366-370.

Munde, V.M., Shinde, G.S., Sajindranath, A.K., Prabu, T. and Machewad, P.M. 2003. Correlation and path analysis studies in charoli (*Buchanania lanzan* Spreng). South Indian Horticulture, 50: 517-521.

Prasad R., Chadha S.K. and Parvez J. 1988. Some observations on germination of four useful but difficult forestry seeds. Jour. Trop. Forestry, 4 : 395-398.

Rai, Y.C. 1982. *Buchanania lanzan* Spreng. Studies on methods of propagation and estimation of fruit yield. Indian Forester, 108 : 501-511.93-96.

Shukla, S.K. and Solanki, K.R.2000. Studies on seed germination, plant survival and growth of Chironji (*Buchanania lanzan* Spreng). Journal of Tropical Forestry, 6: 44-49.

Singh, Jagdish, Banerjee, S.K. and Francis, A. 2002. Vegetative propagation of *Buchanania lanzan* Spreng root cuttings. Indian Forester, 128 : 700-704.

Singh, R.P. and Chaturvedi, O.P. 1983. Primary production of deciduous forest at Varanasi. Indian Forester, 109 : 255-260.

Singh, R.V. 1982. Fodder trees of India. Oxford and IBH Publication, New Delhi. pp 663.

Singh, Sanjay, Singh, A.K. and Apparao, V V. 2006. Genetic diversity in Chironji(*Buchanania lanzan* Spreng) under semi arid ecosystem of Gujarat. Indian Journal of Agricultural Sciences 76:695-698.

Singh Sanjay, Singh A.K., VV Apparao, Bagle, B.G. and Dhandar, D.G. 2006. Physio-biochemical changes during fruit development in chironji (*Buchanania lanzan* Spreng). Paper presented in National symposium on underutilized horticultural crops 8-9th June, 2006 at IIHR, Bangalore.pp 78.

Singh, Sanjay, Singh, A K and B G Bagle 2007. Propagating chironji by *in situ* soft wood grafting. Indian Horticulture, May-June, 35-36.

Singh, Sanjay and Singh, A.K. 2012. Biodiversity and potential of jamun, chironji, rayan and mahua for sustainable agriculture for tribal region. In. Compendium of winter school held at CIAH, Bikaner from 5-25th September,2012 pp.67-73.

Tewari R K and Bajpai, C K 2001. Propagation of chironji by chip budding. Agroforestry News Letter, 13: 1-2.

Chapter-29

Salvadora (*Salvadora oleoides*)

J.C. Tewari* and R. S. Singh**
**Central Arid Zone Research Institute, Jodhpur - 342 003*
***Central Institute for Arid Horticulture, Bikaner-334 006*

Salvadora the "Grape of the Desert" and commonly known as *Pilu* or jaal is a popular multipurpose tree (MPTS) found growing luxuriantly in Thar desert and Kutchh region of Gujarat. It is an extremely drought hardy and salinity tolerant plant, capable of growing in hostile weather conditions. *Salvadora* provides life and support to the people living in fragile ecosystem of Thar desert and hence can be rated as a life line tree. The two important species of *Salvadora* i.e. *Salvadora oleoides* Decne and *Salvadora persica* L. are known to exist in India and known by different vernacular names. *S. oleoides* is called jaal, Mitha jal (Rajasthani); Pilu (Sanskrit); Bhapilu (Hindi); Kankhina (Marathi); Pil (Punjabi) and Kalawa (Tamil) while *S. persica* is known as Pilu (Sanskrit), tooth brush tree (English); Chotapilu (Bengali); darakht-I-miswak (Persian); Pilu (Marathi); Varagogu (Telgu) Ughaiputtai (Tamil); goni-mara (Karnataka). *Salvadora* is native of India, East Africa and Arabia. It is found in wild form in extensive areas of Africa particularly, Turkama, Sambura and Marsabit districts of Northern Kenya, Ferli in North Senegal, South Sivani in Egypt and N. E. Tanzania. Salvadora is also of common occurrence in South coastal areas of Pakistan, Sindh and Baluchistan, Saudi Arabia and other middle eastern countries.

In India good diversity of *Salvadora oleoides* trees are found in Jaisalmer, Barmer, Jalore and Pali district of Rajasthan, Bhuj and Khanpur districts of Gujarat, parts of Punjab, Haryana, Bihar, Uttar Pradesh and in Chambal ravines of Madhya Pradesh and to a lesser extent in Karnataka, Tamilnadu and Andhra Pradesh.

In most countries, *Salvadora* is a subsistence tree mostly meeting local demands. Consequently, although it is grown in India, parts of Africa, Arabia and middle east countries, precise data on area and production are lacking. However, in India the annual seed production is estimated to be 47, 000 tonnes having potential for 15, 000 tonnes oil.

It is an extremely drought tolerant crop and can grow even in areas receiving rainfall< 80mm. Dry weather is important for flower initiation and fruit development and regardless of total annual rainfall, *Salvadora* produces more fruit when subjected to fairly long annual dry period. The tree can tolerate temperature range of –3° C to 45° C. *Salvadora persica* plants are however, frost sensitive.

Nutritive Value and Uses

It is an excellent plant for wind breaks and shelter belts in deserts as it makes almost impenetetrable growth thorough root suckers around the stem and seedlings which spring under the shade of canopy. Leaves provide nutritious feed to livestock, particularly camel during lean period of the year. It provides the moisture laden leaves when other vegetation is scarce. Animals foraging upon fresh leaves of *S. oleoides* on ranges get considerable quantities performed water embodied in these leaves including crude protein 9.6 per cent, while in *S. persica* the crude protein is 11 per cent. Ganguli *et al.* (1964) observed that the platability rating of *S. oleoides* was highest after *Ziziphus nummularia.* Twigs and roots of *S. persica* commonly known as *"Meswak"* is used in Saudi Arabia and other middle east countries as an oral hygiene tool for removing unwanted filth bacteria and freshness of gum. The extracts obtained from root and stem extracts are reported to contain chloride, sulphate, theocyanate and nitrate (Darout *et al.*, 2000). The fruits of both the species are sweet and eaten by natives, although it has been reported to produce tingling and ulceration in mouth, when eaten fresh and one at a time. Therefore, they are eaten in handfuls. Fruits are rich in non edible oil (40-50%), which has a lot of industrial use. The bruised root bark of *S. oleoides* has a vesicant effect. The wood is soft and is used for prepairing agricultural implements. Egyptians use it for making coffins. It can be used in klins of bricks if mixed with other wood. Ocassionally it is used as a fuel.

Leaves are used as purgative and as a cure for cough, headaches, joint pains, leprosy, itching mouth boil or ulcer. Its shoot and leaves are used as antidotes for poisons. Fruits are used in biliousness and rheumatism. Dry flowers are used for curing ulcers of intestine. It is used for kidney treatments as well. Seeds of *S. persica* resembles to that of *S. oleoides* in size, colour and contain 30-40 per cent oil which is of great economic significance. The seed fat of *S. persica* is used as a base for ointments (Anon., 1972). It is also employed in dyeing and in the preparation of suppositories. Seeds are 44-46 per cent of the total fruit. Purified oil is used in soap making, candle and detergent industries and can be used as a substitute for coconut oil. Lauric and Myristic acid found in oil is used for saponification. The kernel oil is yellow to green and contains 40-45 per cent oil. Physico-chemical properties are melting point 41°C, specific gravity 0.867, saponification value 223 to 245.2, iodine value 3.6 to 5.5, acid value 0.71 and unsponification matter 0.5 per cent. Proximate composition of *Salvadora persica* seeds and fatty acid composition in seed and kernel are given in table 1.

Table 1 : Chemical and fatty acid composition of seeds of *Salvadora persica* (per cent).

Chemical composition		Fatty acid composition	
Moisture	2.8	Lauric	31.4
Fat	45-46	Myristic	38.7
Albuminoids	18.9	Palmitic	10.4
Carbohydrate	23.5	Stearic	2.2
Fibre	5.8	Oleic	4.8
Ash	3.5	Linoleic	1.2
		Malvic	7.8
		Sterculic	3.5

Singh and Tewari (1995) (Hosamani and Pattanashettar, 2002).

Botanical Description

Salvadora (named after J. Salvador, a Spanish Botanist) belongs to the family Salvadoraceae and genus *Salvadora*. Four-five species are found in warm regions of Africa and Asia of which two species *Salvadora oleoides* Decene and *Salvadora persica* L. are commonly found in India. The somatic chromosome number in both the species is confirmed as 2n=24. Two basic numbers n=12 and 13 are contemplated for the genus (Kumar *et al.*, 2002). *S. oleoides* is an evergreen shrub or small tree with twisted trunk and drooping ash coloured branches and branchlets. Leaves are opposite, linear, lanceolate and flowers are greenish white, sessile, compact, erect, axially panicle, shorter than leaves, calyx cup shaped, divided mid way into four rounded lobes, corolla longer than calyx. Flowering continues for a long period from January to April and the fruits are available for a extended period of three months from April to June. Fruits aromatic drupe, globose usually yellow when ripe and acidic. *S. persica* L. resembles that of *S. oleoides* and is an evergreen shrub 10-15 feet high, branches numerous, spreading branchlets with drooping nature. In Sindh it grows to a large size (30-40') than any of its other habitat. Leaves are 1.5-2.5 by 7.13 inches elliptic, ovate, oblong or oval, opposite, petioled, smooth, polished, obtuse often mucronate, rather thick and fleshy, green or glaucous base narrowed or round. Flowers in November and May. Flowers pedicuate in lax axially and terminate, often nodding panicles, which are longer than the leaves, calyx open, lobes less than half the length of the reflected petals. Fruit a berry, 6-7mm in diameter, globose smooth, red when ripe, supported by persistent yellow cup of the calyx, seed 1-4 mm in diameter, subglobse, smooth, brown. The berries has strong aromatic smell and taste.

Phenological diversity exists in *Salvadora* and tree to tree variations are common in flowering and in maturing fruits, which may reflect either genetic variation or genotype x environmental interactions and or both. Jindal *et al.* (1999) reported considerable variability in fruit colour, seed oil content and seed germination showing potential for improvement of fruit production. Parthenocarpy (seedless fruiting) has been recorded by Balasubramanian and Bole (1993) in *S. persica*. Although there is a wealth of *Salvadora* germplasm across the region where it is naturally grown, no systematic germplasm collection and evaluation has been attempted so far in spite of its future potential (Rao and Tarafdar,1998).

Concerted efforts are needed to collect and categorise the existing variability in *Salvadora*. From the diverse germplasm, plants with fruits having higher sweetness, higher oil content and lower level of compounds like dibenzeyl, thio urea, dibenzyle urea and thioglucoside, glucotropalohin which makes its oil non edible should be selected. It is also equally important that methods and programmes are also developed to systematically evaluate these collections. This should include description of the approved descriptor including the important economic characters wherever relevant.

Although *Salvadora* has commercial potential, no attempts have so far been made to improve it as a crop plant. Investigations of the breeding system by Jindal *et al.*(1999) revealed that apomixis in *Salvadora* was totally absent. Maximum fruit set was with selfing, followed by sibbing and open pollination. Reproductive capacity per inflorescence was very low, causing poor fruit set. For immediate returns, mass selection for genetic improvement for oil content and resistance to disease particularly to witches' broom is advocated in this crop.

There is no specific known variety of *Salvadora* under cultivation. However, there are *Salvadora* trees with very high yield potential in Jaisalmer and Coastal saline Kutchh regions of Gujarat (Rao *et al*, 2001).

Propagation

It is commonly propagated by seed and root suckers. The plant coppices fairly well and regenerates through natural layers.

Seed

Salvadora multiplied through seeds. Natural regeneration and occurrence of pure *albino* seedling during seed germination of *S. oleoides* has been reported to be rare (Mertia and Kunhamu, 2000a). As the seeds have high catalase activity de-pulping the fruit and pre treating the seeds in cold water for 2hr or in hot water for 0.5hr enhances seed germination. Fresh seeds have a germination percentage of 65-70 per cent. The species is known to be viviparous. The problem of vivipary can be overcome in *S. persica* by the application of abscisic acid. Seeds treated by mechanical scarification (partial seed coat removal) have been reported to give 100 per cent germination in *S. oleoides* (Mertia and Kunhamu, 2000b). Extracted seeds should be sown in 23x10cm size polyethylene bags filled with soil, sand, FYM (1:1:1) at a depth of 0.5cm during winter season and six months old seedling can be transplanted during rainy season. One kg of *Salvadora* fruits contains approximately 3400 seeds. Sowing time can also be adjusted depending upon the stage of growth required for transplanting in the field. Seeds usually germinate 7-10 days after sowing. Seedling raised in polybags becomes ready for transplanting at about 11-12 months after seed sowing. In arid region, the best time for planting is with the onset on monsoon. For better establishment of plants, irrigation after every 10-15 days interval should be done for the first two months after planting. However, once the seed germinates frequent watering depending upon the temperature and relative humidity is to be carried out till the seedlings attain a height of 25-30cm.

Vegetative propagation

Although there are no reports of *Salvadora* being propagated vegetatively, but can be raised through air layering, cuttings, inserted in a sand in a glass house for rooting.

Micro-propagation

Salvadora plants can be raised through tissue culture of axillary buds of selected plants (Jindal and Singh, 2005).Clonal propagation using cotyledonary nodes and nodal segment has been reported in *S. persica*. Mathur *et al.* (2002) induced root formation in *in vitro* produced shoots of *S. persica* in one half strength MS medium supplemented with 3.0 mg IBA/litre. The regenerates were transferred to field conditions after acclimatization with a success rate of 60 per cent.

Batra *et al.* (2001) multiplied nodal segments from 2-3 year old plant cultured on MS salts supplemented with cytokinins (BAP, benzyladenine and kinetin at 1.0-7.0 mg/litre) and auxins (IAA, IBA and NAA). Multiple shoot formation was obtained on 4.0 mg BAP/ litre combined with 0.5 mg IAA/litre. Highest frequency of root induction was achieved on the medium having 3.00 mg IBA/litre. Plantlets survived in potting soil and exhibited normal growth under field conditions. Mathur *et al.* (2002) multiplied shoots through cotyledonary nodes excised from 15 to 20 day old seedlings. Maximum shoot proliferation, from single explant was obtained on MS medium incorporated with BAP.

The spacing between trees varies from 4-5 meter. However, Kasera *et al.* (2003) obtained maximum plant height and collar diameter of *S. persica* at a spacing of 5x5m. Planting density of 4x4m is recommended for planting of *S. persica* on highly saline black soil (Rao *et al.*, 2000). Once established the plants do not require any special care. Flowering starts 6-7 year after the establishment of the crop.

There is considerable potential for expansion of both the species of Salvadora in most of the countries where it is presently grown, but efforts to improve the species have been neglected. There is strong need to build up information on the status, occurrence, collection and conservation. For this purpose inventories and biodiversity maps can be prepared.

Conservation of *Salvadora* germplasm is an important pre-requisite to prevent loss of genetic diversity. Raid propagation methods are necessary to produce seedlings and make them freely available to growers. The quality planting material should be free from diseases. Most of the rural people use fruits of *Salvadora oleoides* for fresh consumption. Processing technology is lacking. Simple methods for processing, storage and value addition should be developed. Value addition through product diversification is another area that deserves research attention.

References

Anon., 1972. The Wealth of India- Raw materials, Publication and Information Directorate, Council for Scientific and Industrial Research, New Delhi, India, Vol.9. Pp.193-195.

Balasubramanian, P., Bole, P. V. 1993. Fruiting phenology and seasonality in tropical dry evergreen forest in Pt. Calimere Wildlife Sanctuary. Journal of the Bombay Natural History Society, 90 (2): 163-177.

Batra, A., Mathur, S. and Shekhawat, G. S. 2001. Clonal propagation of *Salvadora persica* L. : an important medicinal plant of arid zone. Journal of Plant Biology, 28 (3): 277-281.

Darout, I. A., Christy, A. A., Skaug, N and Egeberg, P. K. 2000. Identification and quantification of some potentially antimicrobial anionic components in miswak extract. Indian Journal of Pharmacology. 32 (1): 11-14.

Ganguli, B.N., Kaul, R.N. and Nambiar, K.T.N. 1964. Preliminary studies on a few top feed species. Annals of Arid Zone, 3 (2): 33-37.

Hosamani, K. M., Pattanashettar, R. S. 2002. *Salvadora persica* seed oil: a rich source of oil and minor source of cyclopropenoid fatty acids. Journal of Medicinal and Aromatic Plant Sciences. 24 (3): 713-715.

Jindal, S.K., Bhansali, R. Satyavir, R, Faroda, A.S. 1999. Variability for seed oil, seedling characters, their relationship and breeding systems in *Salvadora oleoides*. (ed. Joshi, N.L., Kathju, S ., Kar, A.). In: Recent advances in management of arid ecosystem. Proceedings of a Symposium held in India, March 1997. pp 405-408.

Jindal, S.K.and Singh,Manjit 2005. Salvadora in degraded saline habitat of Arid Region, In. Shrubs of Indian Arid Zone,CAZRI,Jodhpur,pp.83-86.

Kasera, P. K., Prakash, J., Chawan, D. D. 2003. Effects of spacing and irrigation levels on growth and biomass production in *Salvadora persica*. Journal of Tropical Forest Science, 15 (4): 626-629.

Kumar, A., Rao, S. R. and Rathore, T, S. 2002. Cytological investigations in some important tree species of Rajasthan IV. Male meiosis studies in the genus Salvadora L. Cytologia, 67 (2): 105-115.

Mathur, S., Shekhawat, G S. and Batra, A. 2002. An efficient *in vitro* method for mass propagation of *Salvadora persica* via apical meristem. Journal of Plant Biochemistry and Biotechnology, 11 (2) : 125-127.

Mertia, R. S. and Kunhamu, T. K. 2000a. Occurrence of albino seedling in *Salvadora oleoides* Decne. Indian Forester, 126 (12):1349-1350.

Mertia, R. S and Kunhamu, T. K. 2000b. Seed germination trial on *Salvadora oleoides* (Decne). Journal of Tropical Forestry. 16: 1, 50-52; 2

Rao, A.V and Tarafdar, J.C. 1998. Selection of plant species for rehabilitation of gypsum mine spoil in arid zone. Journal of Arid Environments, 39 (4) : 559-567.

Rao, G. G., Babu, V. R., Nath, A. and Kumar, R. 2001. Salt tolerance in *Salvadora persica*: osmotic constituents and growth during immature phase. Indian Journal of Plant Physiology, 6 (2): 131-135.

Rao, G.G., Nayak, A. K., Chinchmalatpure, A. R. and Babu, V. R. 2000. Influence of planting density on growth and development of Salvadora persica grown on highly saline black soils. Indian Journal of Soil Conservation. 2000, 28: 2, 167-174; 6 ref.

Singh, R.S and Tewari, J.C. 1995. Salvadora: A multipurpose tree of Arid Tract. Deco Mirror, 2 (2): 19-22.

Chapter-30

Noni (*Morinda citrifolia* L.)

A. K. Singh
Central Horticultural Experiment Station, Godhra-389 340

N*oni (Morinda citrifolia* L.) is a unique horticultural plant. It has a tremendous demand due to its high medicinal value. Increasing demand has led to commercial farming of the crop. Noni is an ideal plant for mixed farming with other horticultural plantations. Noni is grown as medicinal plant in most of South Indian temples as sacred plant like basil (tulsi). All parts of plants have traditional and modern use, including roots and barks in dye and medicine. Fruit juice is a rich source of vitamin A, C, E, B, B_2, B_6, B_{12}, calcium, iron, niacin, folic acid, phosphorous, magnesium, zinc, copper, chromium, manganese, molybdenum, sodium, potassium and carbohydrates. Noni fruit is a raw source for more than 150 neutraceuticals, required by our body on daily basis. It is good source of anti-oxidants. Fruit juice purifies blood, protects from toxins and pollutants, promotes self healing mechanism and stimulates production of T-cells in immune system. Noni fruit attained significant economic importance world wide through a variety of health and cosmetic products, made from fruit leaves and fruits.

Noni grows well in all types of well drained soil. It requires water during early stages of crop for better growth and development. At the time of selection of land ensure water availability either by bore well or other source of irrigation. Water logging and clay soil are to be avoided for better productivity. Noni is a drought hardy plant, which can be grown in waste land and saline soils (Singh and Rai, 2007).

It is adapted to a wide range of soils. For high yield potential and good plant growth, deep loam and well-drained soils are needed. Such soils also retain sufficient moisture, which is beneficial for the optimum growth and good fruiting. Noni tolerates shallow, saline, sodic and infertile land. Noni withstands and even thrives in flooded condition for longer period of time. Noni is tolerant to extreme salinity in general and is thought to possibly gain nutritional benefit from minerals contained in sea water. Noni can be grown in wide range of climatic conditions; it comes up very well between 20-38^{0}C temperature. It can also be grown in adverse climatic conditions, from dry to humid areas of tropics to sub-tropics zones.

Nutritive Value and Uses

The use of different parts of several medicinal plants to cure specific ailments has been in vogue from ancient times. Uses of Noni are mentioned in Indian ethnopharmacology and ayurvedic medicines (Table 1). Use of plants for curing various ailments are not confined to the doctors only but known to several households as well. There are many interesting and sometimes astonishing things to learn from collectors of medicinal herbs. Spreading and preserving this knowledge on medicinal plants and their uses has become important for human existence. There is a growing tendency all over world to shift from synthetic to herbal product.

Table 1 : Traditional uses of noni for different ailments.

Digestive system	:	Diarrhea, Intestinal Parasites, Indigestion, Stomach Ulcers
Bones & Joints	:	Arthritis, Broken Bones, Sprains
Skin	:	Abscesses, Boils, Blemishes, Wounds, Infections
Internal Disorders	:	Diabetes, High blood pressure, Headache, Kidney and Bladder disorders, Tumors, Fevers
Gender specific	:	Childbirth and pregnancy, Menstrual Cramps, Regulate Menstruation, Prostate Complaints
Eyes	:	Eye infections
Mouth & Throat	:	Sore gums, Sore throat with cough, Gingivitis, Toothache
Chest	:	Cough, Tuberculosis, Asthma, Respiratory Afflictions
Aging	:	Tonic to slow Aging or Anti aging

It is known in Vietnam, Africa and the Caribbean, where it is called the "painkiller tree". Australians call it "cheesefruit" from the overpowering smell. Dyes are extracted from noni roots and young fruits, and all parts *viz.* root, bark, leaf, fruit and flower are used medicinally. In 1992, Dr. Arthur W. Whistler of the National Botanical Garden, Hong Kong, reported that noni was the 'Most widely and commonly used medicinal plant prior to the European era".

A Pandora of miracles, the wonder plant *Morinda citrifolia* L. occupies a pride of place for over 3000 years for its wide range of therapeutic and medicinal uses. Noni is an attractive shade tree with large glossy leaves and is well suited to tropical coastal areas. Perhaps its other common names give some hint as to why it was not more popular in Horticulture. "Rotten Cheese fruit" or "Vomit fruit". There is no doubt that the odour of the fruit is pretty unpleasant. Noni was first found in India, and migrating peoples may have carried it around regions in the Pacific.

In the market a number of value added products are available *viz.* Noni Juice, Ready to serve Beverage (RTS), Squash, Concentrated Liquid, Flavoured Beverage, Fortified Beverage, Mixed or Blended Juice, Diet Noni Juice, Wine, Harbal Drink, Instant protein Noni Drink, Noni pulp: Fruit leather Bar, Sweet pulp Concentrate, Harbal Prash, Mixed Fruit Prash, Puree, Jam, Spread, Mixed Fruit Jam, Powder, Sauce and Chutney and toffee.

Traditional Uses

- Traditional uses of *Morinda citrifolia* in different parts of the world in China, Japan, Hawaii various parts are used to control fever and as a health tonic.
- In Pacific Islands and Hawaii various parts are used for treating eye problems, skin wounds and abscesses, gum and throat problems, respiratory ailments, constipation fever.
- In Marshall Islands various parts are used for treating stomach pains and after delivery.
- In Malaysia, heated leaves applied to the chest relieve coughs, nausea, colic problems.
- Juice from the leaves is taken for arthritis in Philippines.
- The fruit is taken for lumbago, asthma and dysentery in Indo-china.
- Pounded unripe fruit is mixed with salt and applied to cuts and broken bones: ripe fruit is used to draw out pus from an infected boil which is common in Hawaii.
- In Malay the juice of over-ripe fruits are taken to regulate menstrual 110w and ease urinary problems.
- In Malay and Hawaii the fruits are used to make a shampoo and to treat head lice.

- In West Indies hot or wilted leaves are pressed on painful parts of the body: the leaves are poulticed to headaches. Crushed leaves in lard or camphor are placed on the head for head colds and neuralgia. In Ailigandi the leaves and bark are used to treat malaria.
- Indochina and Australian aborigines eat the raw fruit with salt or cooked as a curry.
- Tongans used noni leaves to treat skin problems. Leaves were heated over the fire, ground and mixed with noni fruit, coconut, tamanu, and or kukul nut oil.
- In Tahiti, antioxidant rich leaves were dried out and used internally as a tea to treat a variety of discomforts and to aid in digestion.
- Many other diseases treated with the plant by the islanders include venereal diseases.
- Polynesians used dried out and crushed up non leaves to create a rudimentary form of tea. They used this noni leaf tea to treat a variety of discomforts.
- The fruit and its juices have been used in the treatment of diabetes, heart troubles and high blood pressure, with different portions prescribed for different illnesses. In these days of ozone depletion, Noni is useful in the treatment of skin cancers.
- Folklore of the ancient inhabitants of French Polynesia recounts the healing properties of the Noni leaf. Healers use to place Noni leaves on deep wounds caused by coral shaft cuts in black pearl diving.
- Parts of the stem are used to treat scorpion-fish puncture wounds.
- Children are fed noni leaves as a treatment for vitamin-A deficiency.
- Foetid oil extracted from the syncarp is used in the hair as an insecticide.
- Juice from the fruit is used to ma1e a medicinal drink, aumiki 'awa, as a remedy for tuberculosis.
- Noni seeds are used to make an essential oil.

Therapeutic Studies on Human

Nervous system, Mental Health, Pulmonary, Cardiovascular, Renal, Hepatic, Immune, Blood Related, Metabolic, Healing and Inflammatory, Genito-Urinary, Gynaecological, Gastro-Intestinal, Musculo-Skeletal, Skin.

Noni is native to South East Asia (Indonesia) and Australia. It is now found throughout tropics. In India, it occurs in wild forms in coastal areas and Western Ghats. Noni grows widely throughout the pacific and is one of the most significant sources of medicine among peoples. *Morinda citrifolia* is found in the costal belts of our country.

Botanical Description

Morinda citrifolia belongs to the family Rubiaceae . The *M citrifolia* is bisexual and self pollinating plant. Leaves are opposite, pinnately reticulate veined, and glossy. Blades are membranous, elliptic to elliptic-ovate, 2-45 cm long, 7-5cm wide and glabrous. Petioles are stout, 1.5-2cm long. Stipules are connate or distinct, 1-1.2 cm long, the apex entire of 2-3-lobed.

Flowers are perfect with about 75-90 in ovoid to globose head. Peduncles are 10-30 mm long, calyx a truncated rim. The corolla is white, 5-lobed, the tube greenish white, 7-9 mm long, the lobes oblong deltate, approximately 7 mm long. There are five stamens, scarcely exerted; the style is about 15 mm long.

Fruit is greenish white to pale yellow. They contain a number of seeds about 4 mm long. Morphological features which unite the species into one genus include syncarps or capitula (aggregated fruits) forming from ovaries and parts of conate calyx and tricolporate. The syncarp may be fleshy and each syncarp is probably derived from condensation of umbels of separate flowers. Seeds have a distinct air chamber and can retain viability even after floating in water for months. Seed color is tan dark brown. A medium size fruit contains 100-150seeds. Approximately, 35-40 seed counts make one gram.

The genus *Moridna* has about 80 species (Smith, 1999). There is a relatively high degree of genetic variability of fruit and leaf with in the species. *Morinda citrifolia is* a primary form because of its commercial cultivation for medicinal value. It is in the greatest abundance in pacific region. It bears medium to large fruits. There are no named cultivars. There is wide variation in plant habit, leaf color and size, fruit size and quality, keeping quality, seed color and size. Based on morphological features, several workers have identified different species of genus *Morinda*. Hooker (1880) reported four species viz. *Morinda aspera* wight & Arn., *M. multiflora* Roxb., *M. tinctoria* Roxb. and *M. tomentosa* Heyne ex Roth. Peter (2007) recorded presence of different variants in *Morinda citrifolia* and *Morinda pubescens* and proposed two variants of *Morinda citrifolia* and four variants of *Morinda pubescens* and also recorded three other species *Morinda stenophylla, Morinda tomentosa* and *Morinda umbellate* in South India.

In tribal belts of Dahod, Panchmahals, Gujarat and Jhabua, M.P., *M. tomentosa* (noni) is found. At CHES, Godhra, promising genotypes of *Morinda tomentosa* are being evaluated and it will be boon for its commercial exploitation particularly under dry land conditions after evaluation particularly for phytochemicals.

Propagation

Morinda citrifolia is mainly propagated by seeds, stem or root cuttings (Singh and Rai, 2007). But the disadvantages of producing plants from cuttings are not as strong as seedlings plants. *Morinda tomentosa* can be propagated through air layering successfully during July month. It can be propagated vegetatively through grafting and air layering and for commercial exploitation tissue culture has great potential (Singh and Srivastava, 2010).

The planting stock that maintained as a source for commercial propagation is referred to as a mother block. Therefore, for preservation of pathogen free planting stock, proper isolation, sanitation, periodical inspection and testing and maintenance are very important. As far as possible, mother block should be in the close proximity of the nursery site. The scion shoots should be taken only from bearing plants. There should be proper record and certification of planting stock. Commercial planting stock can be referred as certified stock, if it is grown under supervision of a legally designated agency with prescribed regulations designed to maintain minimum standards of cleanliness and clonal identity.

Seed Collection

Noni fruits are harvested when they start turning white or even they are fully ripened. Collect fruit from plants with desirable qualities like large fruits and vigorous plants. After picking good fruits, let fruit ripen fully until it all turns soft. This may take 3 to 5 days, if semi ripened fruits are collected. The pulp of the noni fruits must be cleaned away from seeds. Average fruit contains 150 to 200 seeds. After pulp is separated from seeds, noni seeds are dried under sun for 2 hours and then placed in dry breezy location in shade to dry for 2 days at ambient relative humidity and temperature. Germination is high in fresh seeds. Seeds remain viable for 1-2 years.

Nursery production

Fresh noni seeds can be sown immediately after extraction from fruits. Seed propagation has certain disadvantages. Dried seeds can also be used to raise seedling. Noni seeds require hot and wet conditions for optimum germination. Noni seeds take 2-6 months for germination. Germination period can be reduced by pre-sowing seed treatment such as hot water (40°C) for 24 hours or acid scarification with sulfuric acid (1%) for 5 minutes. Noni seeds require scarification to reduce the time to germination and increase germination percentage and uniformity (Nelson, 2005). Under optimum conditions, at least 60-70% scarified noni seed germinate. Various seed germination studies indicated that application of Gibberllic Acid 800 ppm followed by 1000 and 1200 ppm induced significant higher germination percentage to the tune of 91.06;87.9 and 80.4 per cent respectively against soaking of seed for 24 hr. in water (78.75%) (Singh and Srivastava, 2010).

Seed treatment with GA_3 (4000 ppm) will reduce germination period. After treatment, seeds were sown in a row on a raised nursery bed. After Noni seeds germinate, growth is slow in beginning. When seedling reached 4 leaf stages, it is time to transplant to a poly bag containing 1.5 kg potting mixture of FYM, soil and sand in equal proportion. Seedlings can be hardened for 2 months. Seedlings are ready for transplanting to main field in 5 to 6 months.

Cuttings

For vegetative propagation, non-hollow soft wood cutting should be preferred for better performance. Rooting hormone is used to improve rooting in cuttings. Cutting should be kept under partial shade and moist conditions for root initiation. The cuttings treated with 4000 ppm of IBA had more number of roots and performed better growth pattern (Singh and Srivastava, 2010).

Layering

Air layering is another method tried for producing quality planting material at Farm level, CARI, A&N Islands (Peter, 2007; Singh and Srivastava, 2010).

Grafting

Grafting is method for producing planting material with better root stocks. The study on grafting showed an encouraging result with 88% survival of plants (Singh and Srivastava, 2010).

Micro-propagation

The commercial *in vitro* propagation protocol has been established for multiplication of elite clones. The explants are collected from apical region nodal explants and young leaves are used for callus initiation. Cent per cent rooting was achieved in vitro using MS media with 1.0mg/l IBA and 0.5mg/l NAA. Cytokinin and Auxins in combination were found effective in vitro induction of shoots from nodal explants (Singh and Srivastava, 2010). To induce callus from leaf explants, 2.0 mg/l 2, 4-D was used in both solid and liquid MS media.

Hardening of Nursery Plants for Planting

The plants raised under growing structures or protected environments require hardening before planting in the field. In fact, hardening is a term applied to any treatment that makes the tissues firm to endure better in the unfavourable environmental conditions. In hot arid region, hardening is done to withstand plants against sunshine and hot winds, hardening may develop protection against freezing. The process of hardening should be such that there is an assurance of high survival and slow growth under the conditions to be expected at the time of transplanting. In Indian conditions, withholding watering followed by shifting of the plants under partial shade is the best and practical method for hardening. A hardened seedling having at least 25-30 cm of wood stem tissue with 5 to 6 months age is used for planting in main field. The square system at 4 m x 4 m of planting is the most popular for nursery raised plants. Irrigation is required after transplanting of seedlings.

Quality Standards of Planting Materials

The supply of quality planting material is pre-requisite for successful commercial cultivation. Production of quality planting materials involves selection of mother plant, which should be high yielding with desirable quality attributes, disease and pest free and should have vigorous, luxuriant growth and must have uniformity in production and quality fruits.

References

Peter, K. V. 2007. Indian Noni: A unique fruit crop with high medicinal value. Under utilized and under exploited horticultural crops. (*ed.* K. V. Peter), NIPA, New Delhi pp 301-312,.

Smith, A.C., 1999. *Morinda* pacific tropical Botanical garden, Hawai, Hawaii, Flora Vitensis Nova., 4:332-34.

Nelson, Scot 2005.Noni seed handling and seedling production, Fruits &Nuts, 10,:1-4.

Singh, D. R. and Rai, R. B. 2007. *Morinda citrifolia* Linn. -An important fruit tree of Andaman and Nicobar Islands, Natural Products Radiance, 6(1):62-65.

Singh, D. R. and Srivastava, R.C. 2010. Methodologies for multiplication of *Morinda citrifolia* L. (Noni).*In.* Abstracts of National Conference on Production of quality seeds and planting material-Health management in Horticultural crops, held at New Delhi from 11-14 March, 2010 pp.201-202.

Chapter-31

West Indian Cherry (*Malphigia puncifolia* L.)

Jasmine Jaya, Swaminathan, A. A. and S. Krishnasamy
Regional Research Station, Tamil Nadu Agricultural University Aruppukottai - 626 707

West Indian Cherry is a nutritive fruit mainly grown in semi dry and tropical parts of the country. It is native to West Indies and Central and South America and distributed in Southern Florida, Hawaii and other tropical and sub tropical regions of the world. West Indian Cherry received considerable attention in the 1950's in its area of distribution because of its extremely high vitamin C content (2000-4600 mg / 100 gm of edible matter). It has been considered to be far superior to guava (*Psidium guajava* L.) as a potential source of vitamin c and it is richer than even the Indian gooseberry (*Phyllanthus emblica* L.), which is reported to contain roughly 800 mg of ascorbic acid per 100 g of edible matter. One fruit the size of a large cherry can furnish a man's daily requirement of this vitamin. Thus the fruit of West Indian Cherry can rightly be called as the "pill of nature's vitamin". The fruit can be utilized for the production of crystalline vitamin C. The fruit is also rich in vitamin A, iron and can be consumed fresh or can be used in preparation of sherbet, ice cream, jelly and baby foods, fruit nectars and soft drinks. The juice can also be used to fortify ascorbic acid content of various other products and fruit juices to improve their nutritive value. It can withstand mild frost and drought. Fruits are good source of vitamin C and bark yields tannin (Chadha, 2001).

West Indian Cherry was described previously under the binomial (*Malphigia glabra* Mill sp.) which belongs to a wild relative inhabitating the West Indies, Tropical America and the low lands of Mexico to

Southern Texas. *Malphigia punicifolia*. L. (Syn.*Malpighia punicifolia*) has been generally approved as the correct botanical name for the West Indian Cherry, which is also, called Barbados cherry, native cherry, garden cherry, French cherry, and acerola in Spanish. West Indian Cherry is native to the Lesser Antilles from St. Croix to Trinidad also Curacao and Margarita and neighboring northern South America as far south as Brazil. It has become naturalized in Cuba Jamaica and Puerto and Rico after cultivation and is commonly grown in dooryards in the Bahamas and Bermuda and to some extent in Central and South America.

Nutritive Value and Uses

It is rich source of minerals, vitamins and used to eat as fresh fruit. The following minerals are present in West Indian Cherry / 100 gm of edible portion.

Constituents	100 gm of edible portion	Constituents	100 gm of edible portion
Calories	59	Moisture	81.9 - 91.10 g
Protein	0.68 - 1.8 g	Ether extract	0.09 - 0.19 g
Fiber	0.6 - 1.2 g	Carbohydrates	6.98 – 14 g
Fat	0.1 - 0.18 g	Ash	0.77 - 0.82 g
Calcium	8.2 - 34.6 g	Phosphorous	16.2 - 37.5 g
Iron	0.17 - 1.11 g	Vitamin A	408 – 1000 IU
Carotene	0.003 - 0.408 mg	Thiamine	0.024 - 0.040 mg
Riboflavin	0.038 - 0.079 mg	Niacin	0.34 - 0.526 mg

West Indian Cherry has 4500 mg /100 g in green, 3300 mg /100 g in medium ripe and 2000 mg /100 g of ascorbic acid in very ripe fruits. The ascorbic acid level of unripe fruits can range upto 4676 mg and are exceeded only by the fruits (rose hips) of *Rosa rugos* Thumb which may have as much as 6977 mg/100g, This constituent varies as much as 25% with the clone, the locality, cultural methods and degree of exposure to sunlight during developmental stages and after harvesting.

The ascorbic acid is not totally destroyed by heat for the jelly may contain 499-1900mg/100g, of the total ascorbic acid in Barbados cherry juice 0.18% is in the bound form. Other constituents include dextrose levulose and a little sucrose.

Botanical Description

West Indian Cherry also known as Barbados cherry is a member of Malphigiaceae family. The West Indian Cherry is a large bushy shrub or small tree attaining upto 20 ft (6m) in height and an equal breadth with more or less erect or spreading and drooping minutely hairy branches. The trunk is short, 4 inches (10cm) in diameter with conspicuous raised white lenticels or breathing pores on its bark.

Its evergreen leaves are elliptic oblong - obovate or narrowly oblanceolate somewhat wavy ¾ to 2¾ inch (2-7cm) long, 3/8 to 1 5/8 in (9.5-40 mm) wide obtuse or rounded at the apex, acute or cuneate at the base; bearing white silky irritating hairs when very young, hairless dark green and glossy when mature.

The flowers occur in a forked cluster in the leaf axil in sessile or short peduncled cymes, have 5 pink, lavender or white spoon shaped fringed petals. The ten erect stamens are thicker and have longer filaments than the others. Three styles point outward with stigmatic area on the inner angle.

The fruits are borne singly or in 2's or 3's in the leaf axils are oblate to round cherry like but more or less obviously 3 lobed. ½ to 1 inch (1.25-2.5 cm) wide; bright red, with thin glossy skin and orange coloured very juicy acid to subacid pulp. The 3 small rounded seeds each have 2 large and 1 small fluted wings thus forming what are generally conceived to be 3 triangular, yellowish, leathery coated corrugated inedible stones.

Generally, two distinct types of West Indian Cherry are found.

1. Pink flowered type

The plant is semi erect with drooping branches. The leaf is large in size elliptic to oblong, light green with entire margin and opposite phyllotaxy. The flowers are pink coloured, borne in clusters in the axils of leaves of both current and past season growth. Two flowers bloom at a time in the axil of leaf. Fruits are larger in size (2.1 cm in diameter and 1.9 cm in length) and weighing 6.2 g each on an average. The fruits are crimson coloured when fully ripe with sub acid taste.

2. White flowered type

The morphological characters of this type are similar to that of pink flowered type. The only variation is in the colour of the flower where it is white. Leaves are smaller in size. Fruits are also smaller in size (1.6 cm in diameter and 1.3 cm in length) and weighing 1 g each on an average. The fruits are orange coloured when fully ripe with sub acid taste.

The first promising selection in Puerto Rico on the bases of fruit size, yield and vitamin content were identified as A1 and B17 but these were later found to be inferior to B15 in ascorbic acid level and productivity. Yields of 10 clones (A1, A2, A4, A10, A21, B2, B9, B15, B17 and K7) were compared over a 2 years period in Puerto Rico and B15 far exceeded the others in both years.

A horticultural variety in St. Croix formerly known as *M.thompsonii* Britton & Small has unusually large leaves and fruits and more abundant flowers than the common strain of cherry.

Florida Sweet

University of Florida's Agricultural Research and Education Centre in Homestead evaluated and selected a superior clone Florida Sweet that was observed to have an upright habit of growth, large fruits, thick skin, apple like semi sweet flavour and high yield.

Propagation

It is multiplied through seed and vegetative methods. It is propagated by cuttings, grafting and air-layering (Chadha, 2001).

Seed propagation

It is propagated through seeds. The seeds should be selected from high yielding clones without cross pollination. The seeds in an individual fruit develop unevenly and only fully developed seeds from ripened fruits germinate satisfactorily. The rate of germination ranges from 5 - 50 per cent. The seeds should be collected from fully ripe fruits, cleaned free of pulp, shade dried and treated with fungicide. Seeds are to be sown in well prepared beds and when the seedlings are about 2-4 months old they are ready for planting.

Vegetative methods

The vegetative propagation methods such as chip budding, shield budding, side grafting and veneer grafting have been reported to be successful (60-90%). It can also be propagated by side veneer or cleft grafts and on trees, which produce inferior fruits.

Layering

It is also propagated by air layering (marcottage). Air layering is best done during spring and summer while the plants are growing and requires 6 to 8weeks for rooting. Marcotting with polythene film with sphagnum moss as medium recorded 80 per cent rooting and good survival.

Cuttings

Leafy hardwood cuttings from healthy plants root within 2 months. Leaf and leafless stem cuttings when treated with IBA 5000 ppm and kept under intermittent mist showed better rooting (85 per cent) while in the leafless cuttings rooting was lesser. Indole butyric Acid (IBA) at 500, 1000 and 1500 ppm was tried in hard and semi hardwood cuttings for vegetative propagation. After 90 days maximum survival (90%) number of leaves (25) number of primary branches (6.6) and number of secondary branches (6.6) were recorded in hardwood cuttings treated with IBA at 1500 ppm. Similarly the maximum number of primary roots (9.37) number of secondary roots (16.37) length of primary roots (22.04 cm) and length of secondary roots 15.10 cm were observed in hardwood cuttings treated with IBA at 1500 ppm. The hardwood cuttings treated with 1500 ppm IBA were most successful for the vegetative propagation of West Indian Cherry.

Micro-propagation

There is a great genetic variability in West Indian Cherry when propagated through seeds. To obtain genetically uniform plants micropropagation was done using nodal segments and shoot apex. Contamination index was lower when shoot apex was used. The genotypes G1, G10, G20, G21 and G27 performed well in half (Murashige and Skoog media) MS and half (Woody Plant media) WPM with 2 mg/1 glycine, 100mg/1 mio inositol, 30g/1 sucrose and 6.5g/1 agar . The NO_3/NH_4 ratio had a strong influence on shoot development. Nodal segments in WPM gave good results with 3.2 shoot per explant. On comparison of genotypes IPA-3 was the best which produced 2.44 shoots per explant.

Recent interest in this fruit as a natural source of vitamin C has not stimulated much interest in new commercial plantings due to its economical incompetence with a much cheaper synthetic ascorbic acid but it continues to be a popular fruit for home garden.

References

Alves, R.E. 1989. Contribution to the study of acerola (*Malphigia glabra* l.) Asexual propagation and texts of nutrients. 65f. *Momografia* - Federal University of the Paraiba Sand, PB.

Alves, R.E. 1992. Culture of acerola in: Donadio, l. C., Martins, A.B.G.and Brave, J.P. Tropical Fruiticultura. Jaboticabal. Fune pp.15-17

Arostegui and W. Pennock. 1956 The Acerola. Pr .Agr. Expt. Sta. Misc. Pub. 15, 10pp.

Asenjo, C.F Frerie D. E and A.R. Guzman. 1946, The High Ascorbic Acid Content of the West Indian Cherry. Science 103:219.

Braga, M.F, Caldas, L.S and M.H. Habe. 1947. Establishment of acerola (*Malphigia glabra*) *in vitro*. Effect of clone and the explants. Brazilian magazine of Fruiticultura cross of the souls. V. 19. N.3. p.335-346.

Chadha,K.L.2001. Hand Book of Horticulture, Indian Council of Agricultural Research, New Delhi.

Couceiro, M. 1985. Course of extension on the culture of acerola. *Recife*. UFRPE, 1985. 45p

Ledin, R. B. 1958. The Barbados or West Indian Cherry. Fla. Agr. Expt. Sta. Bul. 594, 28pp

Lloyd, G and B. Mc Cown. 1980 Commercialization feasible micropropagation of mountain laurel *Latifolia kalmia* by uses of shoot tip culture. Combined proceedings of international plant propagation society. Seattle, v. 30 p.421-427.

Melo, N.F., Okasaki, W.Y., Milk. C.B. and F.Mikloson 1999. Establishment of the culture in vitro of the acerola (*Malphigia emarginata* DC). Ciencia and Agrotechnologia. You Cultivate v. 23 n.1 p. 102-107.

C.C. Bazore, K and M. Bartow 1965. Fruits of Hawaii. University of Hawaii Press. Miller, pp.229

Miyashita, R.K., Nakasone, H. Y and C.H., Lamoureux. 1964. Reproductive Morphology of Acerola (*Malphigia glabra* l) Univ. Hawaii Agr. Expt. Sta. Tech. Bul. pp.63,31

Morton, J. 1987. Barbados cherry *In:* Fruits of Warm Climate. *pp.*204-207 .

Moscoso, C. G. 1956. West Indian Cherry: Richest Known Source of Vitamin C. Econ. Bot. 10((3) 280-294.

Murashige, T. and F. Skoog. 1962. Revised medium will be rapid growth and Bioassays with tobacco tissue cultures. Physiologia Plantarum. Vol..15 p. 473-497

Shanmugavelu, K.G. 1987. West Indian Cherry. Production Technology of Fruit Crops, pp. 399-407.

Singh, D.B and B.L. Attri. 2000. West Indian cherry. Advanced search Jour.Appl.Hort. 2(2) 134-135.Central Agricultural Research Institute, Port Blair -744101

Yamane, G. M and H.M.Y. Nakasone. 1961 Pollination And Fruit Set Studies Of Acerola In Hawaii Amer. Soc. Hort. Sci. Proc. 78: 141-148

Chapter-32

Mahua (*Bassia latifolia* Roxb.)

Sanjay Singh and A.K. Singh
Central Horticultural Experiment Station, Godhra-389 340

Mahua (*Bassia latifolia* Roxb.), is a deciduous, hard woody multipurpose tree species (Venkatesh,1998). It is very hardy and thrives well on rocky, gravelly red soils and also on saline and sodic soils. It can grow even in pockets of soil between crevices of barren rock. Trees are grown on degraded rocky area including salt-affected soils and forests area. Mahua prefers tropical and sub-tropical climate. It can withstand drought admirably. A characteristic tree of the dry region is found in north and central India. It commonly grows in eastern Uttar Pradesh, Madhya Pradesh, Maharashtra, Bihar, Odisha, Andhra Pradesh and Gujarat. In Rajasthan, it is also found growing on the wastelands particularly in southern part of the state. The tree is very well known to rural folk since ages in India.

The area and production of this fruit tree in India is not available because it is not planted in systematic manner but trees in patches are common in the villages in many states. They can be seen growing in parks, on roadsides, avenues and as windbreak. The production in India is mainly concentrated in the drier states and the produce is collected by the villagers and sold in the local market. Its cultivation may spread to arid and semi- arid areas, resource-poor areas and wastelands.

Nutritive Value and Uses

Every part of Mahua yields an economic product of great potential value. Its flowers, fruits and oil obtained from seeds are eaten in various ways. Its wood is used as timber wood (Prasad, 1993). The corolla commonly called mahua flowers is a rich source of sugar containing appreciable amount of vitamins and minerals. Flowers are used for preparation of distilled liquors and potable spirits. Flowers and spent flowers (after fermentation) are also used as feed for livestock. Fruits are eaten as raw or cooked. The flowers are used as vegetable, for making cake, liquor. Flower juice is used in the treatment of enlargement of auxiliary gland, neurotic disorder and taken with cow's milk as an aphrodisiac, in cough and bronchitis. The derived liquor desired from the fermentation of flowers is considered to be a tonic and nutritive. The fruit pulp can be used as source of sugar for alcoholic fermentation, whereas dry fruit husk can be used for preparation of absolute alcohol. Seed is a good source of oil. Amount of oil obtained from seeds of the fruit is higher than many oil seed crops and oil-bearing tree. The oil obtained from kernel is used for edible purpose and permitted for preparation of vegetable oil (Singh *et al.*, 1999). In Mahua oil, linoleic and unsaturated fatty acids are found which are useful for heart patients because it reduces the cholesterol content in blood serum. Mahua oil is used in manufacture of soap, lubricating grease, fatty alcohols and candles. Mahua fat is rich source of palmitic acid, which is used in cosmetic and pharmaceutical industries. Mahua seeds can also be used for preparation of defatted flour, which has great potentiality in bakery products. The saponin obtained after extraction has industrial and commercial application. Cake obtained after extraction of oil is used as manure and has insecticidal properties. It provides quality timber wood for various uses.

Botanical Description

Mahua (*Bassia latifolia,* Roxb. Syn. *Madhuca indica)* belongs to the family Sapotaceae. The trees of *Bassia latifolia and B.longifolia* grow up to an altitude of 1,200 m. The *B. malabarica* grows in Western Ghats from Kanara to Travancore and also in the Himalayas. The trees of *B. butyracea* grow in the Himalayan districts up to an altitude of 4,500m. The two major species of genus *Madhuca* found in India are *Madhuca indica* (*latifolia*) and *Madhuca longifolia* (longifolia). *Madhuca* spp. is highly heterozygous, cross-pollinated fruit crop and as such seedlings exhibit a wide range of variations, which aids in the selection of the superior desirable genotypes. Due to cross pollination and predomination of seed propagation over a large period of time, it gives immense opportunity to locate elite trees having horticultural traits. Wide variations were observed in sweetness, acidity, size, shape (table 1-6) and bearing habits in Mahua under Uttar Pradesh conditions (Singh *et al.*, 1999).

Table 1 : Variability in physical characters of mahua fruits.

Maturity group	Flower wt. (g)	Juice (%)	Pomace (%)	Stamens (%)
Early	1.5 - 2.8	66.4 - 69.6	28.3 – 30.8	0.8 – 1.4
Mid season	1.3 - 2.5	64.5 – 67.7	30.5 – 33.4	1.5 – 2.0
Late	1.2 - 2.6	62.8 – 64.1	34.1 – 35.0	1.3 – 1.8

Table 2 : Variability in chemical characters of fresh mahua flowers.

Maturity group	TSS (%)	Acidity (%)	Total sugar (%)	Vitamin- C (mg/100g)	Vit. A (IU)	Minerals (%)
Early	25.0 – 26.7	0.09 – 1.0	23.8 – 25.5	42.9 – 60.9	637 – 650	1.6 – 1.6
Mid season	23.5 – 25.8	0.11 – 0.12	19.7 – 24.0	49.4 – 57.3	525 – 577	0.6 – 0.7
Late	21.0 – 26.0	0.10 – 0.12	19.9 – 24.0	48.8 – 49.2	528 - 696	0.7 – 0.8

Table 3 : Variability in physical characters of mahua fruits.

Maturity group	Fruit wt. (g)	Husk (%)	Seed (%)
Early	16.8 – 35.3	58.0 – 71.7	28.3 – 42.0
Mid season	19.4 – 31.8	58.3 – 70.2	29.8 – 41.7
Late	18.5 – 20.5	70.8 – 71.9	28.1 – 29.2

Table 4 : Variability in chemical characters of mahua fruits (Singh *et al.*, 1999).

Maturity group	Total sugar (%)	Vit. A (IU)	Vit. C (mg/100g)	Mineral (%)
Early	13.0 – 13.8	586 – 667	47.5 – 59.2	1.6 –2.1
Mid season	11.5 – 13.2	823 – 860	48.3 – 72.5	1.8 – 2.3
Late	11.0 – 11.4	742 - 846	51.2 – 65.5	1.4 – 1.9

Table 5 : Variability in physical characters of mahua seeds (Singh *et al.*,1999).

Maturity group	Seed wt. (g)	Kernel (%)	Shell (%)
Early	7.2 – 10.0	73.7 – 77.2	22.8 –26.3
Mid season	8.1 – 9.5	73.9 – 83.6	16.4 – 26.1
Late	5.2 – 6.0	76.8 – 78.8	21.2 – 23.2

Table 6 : Variability in chemical characters of mahua kernels (Singh *et al.*, 1999).

Maturity group	Oil (%)	Free fatty acids (%)	Protein (%)	Minerals (%)
Early	42 – 49	0.53 – 0.54	23.5 – 23.7	5.3 – 5.7
Mid season	42 – 45	0.59 – 0.62	22.6 – 22.7	3.9 – 6.4
Late	42 – 43	0.97 – 1.20	21.0 – 25.7	4.4 – 4.8

The flowering season extends from February to April. The copious fall of succulent, corollas weave a cream colour carpet on the ground. It is rich in sugar (73%) and next to cane molasses constitute the most important raw material for alcohol fermentation. The yield of 95% alcohol is 405 liters from one ton of dried flowers. The matured fruits fall on the ground in May and July in the North India and August and September in the South India. The orange brown ripe fleshy berry is 25 to 50 mm long and contains one to four shining seeds. The seeds can be separated from the fruit wall by pressing. Drying and decortications yield 70% kernels on the weight of seeds. The trees of *B. latifolia* in north and *B. longifolia* in south are found which bear flowers containing a large amount of sugar and seeds rich in quality oil. There is no improved variety of mahua for systematic orchard. Mahua seed contains 35% oil and 16% protein. There is need to develop new products from Mahua flowers and popularize them not only in domestic market but also in international market.

However, a large variability exists in its fruits. Some selections have been made at N.D. University of Agriculture and Technology, Faizabad (U.P.), which are as follows.

NM- 2: Flowering is taken place in the first week of April. Average number of flowers / fascicles 66, average weight of fresh flower 2.8g; average yield of flower (dry) of 7-year-old grafted plant is 9.9 kg. Fresh flower recorded 25.50 % total sugar, 19.60% reducing sugar, 5.6 % non-reducing sugar, 60.90 mg/ 100 g Vitamin C, 650.70 IU Vitamin A and 0.60 % mineral content. Ripening of fruit took place by second week of June. Average fruit weight 16.80; husk content 58.80 %, total sugar13.60 %, reducing sugar 7.8 %, non- reducing sugar 5.80 %, vitamin C 51.30 mg/ 100 g, vitamin A 586.70 IU and mineral content 1.90 % was recorded. Average seed weight and kernel weight was found to be 7.2 g and 5.50 g respectively. Oil content 47.40 per cent, free fatty acid 0.53 %, protein content 23.60 % and mineral content 5.70 % was recorded.

NM- 4 : Flowering is taken place in the fourth week of April. Average number of flowers / fascicles 54.6, average weight of fresh flower 2.6g and average yield of flower (dry) of 7-year-old grafted plant is 8.5 kg. Fresh flower recorded 24.50 % total sugar, 19.60% reducing sugar, 4.80% non- reducing sugar, 49.20 mg/ 100 g Vitamin C, 528.20 IU Vitamin A and 0.67 % mineral content.

Ripening of fruit took place by third week of June. Average fruit weight 19.50 g, husk content 65.30%, total sugar 11.20%, reducing sugar 8.20 %, non- reducing sugar 3.20%, vitamin C 55.80 mg/ 100 g, vitamin A 860 IU and mineral content 2.1% was recorded. Average seed weight and kernel weight was found to be 6.00 g and 4.70 g respectively. Oil content 45.20 per cent, free fatty acid 0.63%, protein content 25.70% and mineral content 4.40% was recorded.

NM-7 : Flowering is taken place in the second week of April. Average number of flowers / fascicles 64.30, average weight of fresh flower 2.50g and average yield of flower (dry) of 7-year-old grafted plant is 10.00 kg. Fresh flower recorded 22.70 % total sugar, 19.50% reducing sugar, 3.00% non- reducing sugar, 48.50 mg/ 100 g Vitamin C, 564.50 IU Vitamin A and 0.82% mineral content. Ripening of fruit took place by third week of June. Average fruit weight 21.8 g, husk content 55.00 %, total sugar 10.10%, reducing sugar 8.00 %, non- reducing sugar 2.30 %, vitamin C 59.20 mg/ 100 g, vitamin A 620.70 IU and mineral content 2.4 % was recorded. Average seed weight and kernel weight was found to be 9.50 g and 7.20 g respectively. Oil content 46.70 per cent, free fatty acid 0.55 %, protein content 22.40 % and mineral content 4.80 % was recorded.

NM- 9 : Flowering is similar with other type in the last week of April. Average number of flowers / fascicles 54.6, average weight of fresh flower 2.2g and average yield of flower (dry) of 5-year-old grafted plant is 6.80 kg. Fresh flower recorded 21.80 % total sugar, 20.00% reducing sugar, 2.10% non- reducing sugar, 30.20 mg/ 100 g Vitamin C, 802.00 IU Vitamin A and 1.40 % mineral content. Ripening of fruit took place by third week of June. Average fruit weight 16.70 g, husk content 60.00 %, total sugar 15.40 %, reducing sugar 12.20 %, non- reducing sugar 3.70%, vitamin C 72.50 mg/ 100 g, vitamin A 890.00 IU and mineral content 1.80% was recorded. Average seed weight and kernel weight was found to be 5.20 g and 4.0 g, respectively. Oil content 48.50 per cent, free fatty acid 0.48 %, protein content 16.80 % and mineral content 4.10% was recorded.

MH-10 : A rich genetic variability in Mahua has been observed during survey in Gujarat state. At CHES, Godhra, Gujarat out of a large mahua collections, a promising genotype MH-10 has been identified which is suitable for cultivation in tribal belts of MP and Gujarat. It ripen in second week of May; produces 13.5 kg seed and 11.00kg kernels. Further, MH-14, MH-35 and MH-64 has been selected as superior genotypes (Singh and Singh, 2005; Singh *et al.*, 2005).

Propagation

It is propagated through seeds and vegetative methods. Micro-propagation has also been tried for mass multiplication of mahua.

Seed propagation

Mahua plants are generally propagated by seeds giving a long gestation period (15-20 years) and large variability. Seed after 5 to 6 days of storage at room temperature showed 80% germination (Chadha, 2001). For this, seedlings are raised to use as rootstock. Mahua seeds have no dormancy. Seeds can be sown on raised beds or in polythene bags about 30 cm deep during July. Seeds germinate within 10-15 days. The seeds and seedlings become ready for transplanting 4 weeks and 1 year after sowing, respectively. Scion wood should be selected from mother plant for grafting. Polyembryony is found in mahua seeds and 15-20% seeds produce nucellar seedlings (Chadha, 2001).

Vegetative propagation

Soft wood grafting

This method was proved to be better for *in situ* grafting under rain fed conditions. The study revealed that naturally defoliated shoots may be used for grafting in the month of March and April with highest percentage of graft success (Singh *et al.*, 2008). The soft wood grafting recorded maximum success in the month of March (80%), followed by February (70%) under semi- arid conditions of western India.

Table 7 : Effect of time on soft wood grafting in Mahua.

Month	Time taken for bud sprout (days)	Bud sprout (%)	Success (%)	Length of sprout at 150days (cm)
July	30.50	66.35(54.45)	64.20(53.25)	24.20
August	30.00	64.50(53.43)	62.00(51.95)	23.10
September	34.00	9.25(17.67)	08.33(16.74)	18.50
October	32.00	22.10(28.04)	20.10(26.64)	16.00
November	0.00	0.00	0.00	0.00
December	0.00	0.00	0.00	0.00
January	0.00	0.00	0.00	0.00
February	0.00	0.00	0.00	0.00
March	26.00	72.25(58.22)	70.00(56.79)	28.00
April	27.33	67.50(55.24)	65.50(54.03)	27.10
May	28.00	63.50(52.83)	60.00(50.77)	20.10
June	29.00	57.10 (49.08)	55.00 (47.87)	20.0
CD at 5%	2.11	3.35	2.14	1.25

(Figure in parentheses are transformed value)

Wedge grafting

Better graft success is recorded in March/April due to faster establishment of vascular connection with root stock and scion. The success rate was 75% when *in-situ* grafting was done in March on defoliated shoots. March and April month is ideal for grafting of elite genotype under semi arid condition of Gujarat.

Veneer grafting

Veneer grafting was found successful method for multiplication of Mahua on one-year-old seedlings as rootstock. The shoots were taken from spring flush and the operation was done in the month of July and August. July is the ideal time for veneer grafting. Plants become ready for planting in September, just 2 months after grafting (Anon, 2007).

Air layering

It can also be multiplied through air layers. Shukla *et al.* (2002) reported that one year old shoots on mahua trees were ringed and upper end of the cuts were treated with IBA 1000 ppm. The ringed portion after hormonal operation was wrapped with moist moss grass, held in a position by tying it with polythene sheet. Rooted shoots were removed from the mother plant 60 days after the treatment and planted in nursery poly bags (10 x 4 cm size) and filled with soil mixture.

Micro-propagation

Micro-propagation of mahua plants (*Madhuca latifolia* Macb.) was achieved by culturing excised nodes on WPM supplemented with different plant growth regulators. A combination of IBA with BA or Kinetin was found suitable for axillary shoots initiation within a week of culture. Excised micro shoots treated with IBA produced roots. Best regerative response was obtained from explnts collected in May (Bansal and Chhibbar, 2000).

Growing Structures for Nursery Plants

The effect of growing structure of nursery plants has been observed. Mist house is used for multiplication of plants through soft wood grating in which controlled misting is provided to maintain humidity throughout the growing period of plants to avoid loss of water. Net house is used for protection of grafted plants from direct sun light or high temperature during hot summer season provided with micro-sprinkler system. Agro-shade net house with 50 or 75% shedding of nylon net is suitable for growth and development of nursery plants under hot arid region.

Quality Standards of Planting Material

Model nurseries for the local supply of quality plant material should also be established. Dwarfing rootstocks are considered to be best because they induce dwarfness and enable to facilitate easy management of the orchard. Such type of work may be initiated and be standardized for different agro-climatic conditions. Plants should be free from diseases with appropriate height of plant.

References

Anon., 2007. Annual Progress Report on Network project on Underutilized fruits (2006-07). CISH, Lucknow, p25.

Bansal,Y.K. and Chhibber, T. 2000. Micropropagation of *Madhuca latifolia* Macb. Through nodal culture. Plant Biotechnology, 17: 17-20.

Chadha. K. L. 2001. A Handbook of Horticulture, Indian Council of Agricultural Research, New Delhi. pp 228-231.

Prasad, R. 1993. Mahua- The tree of the poor. Indian. Jour.Trop.Foresrty,7:171-179.

Shukla,S.K., Solanki, K.R., Tiwari, R. K. and Rao, G.R. 2002. Propagation of mahua by air layering. ICAR News, A Science & Technology Newsletter. 8(1): 14.

Singh I.S, Srivastava, A.K. and Singh, V. 1999. Improvement of some underutilized fruits through selection. Jour. Appl. Hort., 1(1):34-37.

Singh, Sanjay and Singh, A.K. 2005. Studies on genetic diversity in Mahua (*Bassia latifolia*) under semi arid ecosystem of Gujarat. Indian Jour, Agric. Science, 75:519-523.

Singh, Sanjay, Singh, A.K., Appa Rao,V.V., Bagle, B.G. and Dhandar, D.G. 2005. Gentic divergence in Mahua (*Bassia latifolia*) under semi arid ecosystem of Gujarat. Indian Jour, Plant Genetic Resources, 18(3): 244-249.

Singh, Sanjay, Singh, A.K., Joshi, H. K; Bagle, B.G. and More T.A. 2008. Standardization of time of soft wood grafting in khirni, chironji and mahua under semi arid ecosystem of Western India .In. National Seminar, CIAH, Bikaner, held during 8-9 March,2008, pp. 70

Venkatesh, C. S. 1998. Genetic improvement of multi-purpose tree species. The Intl. Tree crop Jour. 5:109- 124.

Chapter-33

Jojoba [*Simmondsia chinensis* (Link) Schn.]

J. C. Tewari* and R. S. Singh**
**Central Arid Zone Research Institute, Jodhpur-342 003*
***Central Institute for Arid Horticulture, Bikaner- 334 006*

Jojoba (*Simmondsia chinensis* (Link) Schn.) is a non traditional crop in arid and semi-arid areas. The search for a replacement for sperm whale oil has led to the investigation of jojoba, as a potential crop plant, because its seeds contain a wax that is essentially identical to the whale oil. Development of jojoba as an economically viable crop requires improved methods of propagation and culture since it is a dioecious in nature. Jojoba is a new oil-producing industrial crop that has attracted much attention in recent years. Jojoba oil is unique in nature. No other plant is known to produce oil like jojoba. It is remarkable to note that, in spite of the variability in size or where grown, each jojoba seed contains an average of 50% pure oil by volume. Jojoba is native to the semiarid regions of southern Arizona, southern California and north western Mexico. Now Jojoba is being cultivated in many countries like Israel, Egypt, Arizona and Mexico. In Rajasthan (India) it provides a renewable source of unique high-quality oil. It is suitable oil yielding plant for hot arid region (Muthana,1981) because of its adoptive nature to arid environment. Much of the interest in jojoba worldwide is the result of the plant's ability to survive in a harsh desert environment. Jojoba can be grown as an oil-producing cash crop. Jojoba is a drought resistant plant and can be grown on marginal lands without replacing any existing crops. Moreover, for optimum production, the crop needs irrigation, management and a good cultivar.

The arid lands of Rajasthan are all set to usher in a revolution in Indian agriculture, with planned large-scale cultivation of the exotic jojoba plant. Rajasthan government has allotted 110 hectare of wastelands including 70 hectare at Fatehpur in Sikar district and 40 hectare at Dhand in Jaipur district for jojoba plantation. Jojoba is now being grown in an estimated area of about 500 hectare in the country and further, area is being increased every year because of its drought hardy plant characters. The production estimate for this plantation is 25 million tonne. In India, about 90 per cent of the jojoba cultivation is in Rajasthan. From 2005 to 2011, the area under jojoba has increased from 500 to 1000 ha in Rajasthan (Kukkar, 2012). Planting material of improved genotypes has been produced for further expansion by providing subsidy by the sate Government. In India, some parts of Kutch in Gujarat and Jodhpur in Rajasthan intensive cultivation of Jojoba is being done Gujarat State Rural Development Corporation Ltd (GSRDC) has undertaken commercial cultivation of Jojoba at its Munjpur farm in Sami Taluka of Patan district. For almost one year GSRDC is cultivating Jojoba plantations on its 50-hectare farm in Munjpur (fig. 1).

Fig. 1 : View of Jojoba plantation.

Our technocrats believe that jojoba oil shows a lot of promise as an alternative source of fuel. It releases a lot of energy when it burns. The driving force to find such fuel source stems from the need to find an alternative fuel to petroleum oil. In addition to this, the jojoba renewable nature is attractive and can produce less pollution. Jojoba oil appears to be promising, also with promising scope for cultivation in the relatively hot weather. Considering Jojoba's potential as an oilseed feedstock for biodiesel, CJP has honour to establish this untapped resource as alternative source for Bio- Diesel industry of future.

Composition and Uses

Jojoba wax is made up of straight-chain esters of mono-unsaturated long chain fatty acids and fatty alcohols and has an average total carbon chain length of 42 carbons. At room temperature, it is a light yellow liquid and does not become rancid or damaged by prolonged exposure to high temperatures. The wax can be isomerized, hydrogenated, sulphurized, chlorinated or transesterified. The wax and its derivatives have a wide range of industrial uses, mainly in cosmetics in which it is incorporated in formulations for skin care preparations such as lotions, moisturizers, massage oils, and soothing creams. It is also widely used in hair care products, such as shampoos, gels and mousses, and is a very good base for lipstick, makeup and nail products. It is used in the cosmetic, medical, pharmaceutical, food products, manufacturing, and automotive industries (as lubricant) and it is a renewable energy resource. There are many potential uses in pharmaceuticals and in industry (as extenders for plastics, printer's ink, gear-oil additives, various lubricants etc.). After oil expression the meal contains more than 30% protein and can be used as animal feed after detoxification. It can be planted to control desert in arid region. Animals are not easily grazed to this plant. Jojoba (pronounced ho-ho-ba) is being cultivated to provide a renewable source of unique high-quality oil. Jojoba is valuable as a soil conservation and landscape plant for highway shoulders, city parks, and other places that cannot afford much care.

Botanical Description

Jojoba (*Simmondsia chinensis* (Link) Schneider) also known as goat nut, deer nut, pignut, wild hazel, quinine nut, coffee berry and gray box bush belongs to the family Simmondsiaceae . *Simmondsia chinensis* is a leafy, xerophytic, woody evergreen dioecious shrub or small multi-stemmed tree that grows to a height of 0.5–1 m in the wild, occasionally to 6 m tall with taproots to 12 m long. Fully mature shrubs or trees can reach a height of 15 feet with a potential natural life span of 100 to 200 years depending on environmental conditions. It has an extensive and deep root system and requires little care if maximum seed production is not desired. Jojoba is a woody evergreen shrub that typically grows to a height of 10 to 15 ft. The plant develops one or a few long tap roots (up to 40 ft) that can supply water and minerals from far below the soil surface. Leaves opposite, oval or lanceolate, grey green or bluish-green, leathery, oblong, opposite, 2.5–3.5 cm long and contain special tissue with a high concentration of phenol compounds. Jojoba does not shed its leaves with the change in seasons. Flowers apetalous, dioecious; the male flowers are yellow, larger, and occur in clusters with 10–12 stamens per flower; female flowers small, usually solitary in the axils or in clusters at the nodes, pale green with 5 greenish sepals, soft and hairy. Fruits green capsules, ovoid, usually dehiscent with 1-3 seeds. As male plants flower, they should be thinned out to 1 male every 40 ft on the row. As female plants flower, usually in the third year, any slow-growing or unproductive plants are thinned out, leaving 1 female plant every 2 to 3 ft on the row. Seed peanut-sized, dark brown, the endosperm scanty or absent. The jojoba plant is dioecious. The gender of jojoba plants can only be discerned from their flowers. When planted from seeds, jojoba plants can take up to three years to produce flowers. The female plants produce seed from flowers pollinated by the male plants. Seed production is generally limited until the fourth year of growth.

Propagation

Jojoba plantations are established by planting seedlings, rooted cuttings or plantlets produced from tissue culture. Seed propagation is common method. However, now a days vegetatively propagated plants are used for large scale plantation with proper ratio of male plants. It is observed that growth of plants raised through direct seedling is not uniform. Propagation through cuttings has been found to be most successful for uniform growth and good yield.

Sources for Quality Seed and Planting Material

The economic age of Jojoba plant is about 150 years. Therefore, it is essential to use only quality seeds from reliable sources for raising seedlings. The quality seeds can be obtained from the following sources:

i) Central Salt and Marine Chemicals Research Institute (CSIR),Bhavnagar-364002 (Gujarat).

ii) Indian Institute of Petroleum (CSIR), Dehradun -248005, Uttrakhand.

iii) Association of the Rajasthan Jojoba Plantation and Research Project (AJORP), Pant Krishi Bhavan, Jaipur-302005, Rajasthan.

iv) Central Arid Zone Research Institute (CAZRI), Jodhpur-342 003, Rajasthan.

Seed propagation

It is propagated by seeds but it is dioecious in nature. Seedling should be raised in polybags or tray. However, Seeds were sown in tray filled with perlite medium under mist condition for fast germination and saplings growth (Kukkar, 2012). At present, seed propagation is avoided and vegetative methods are used for production of quality planting material.

Germination

Jojoba seeds require no pre-treatment and are usually readily germinate immediately after harvest (Nord and Kadish 1974). They are protected from premature summer germination by a requirement for relatively cool temperature-an optimum 15 to 23°C and slow germination rates. It takes 3 days for the first emergence of the radicle at 20/30 degree°C and 7 days at 10/20°C. Dormancy could be removed in most seeds by breaking the testa the radicular end. Nord and Kadish (1974) reported that Jojoba seeds could germinate at 5-10°C but only after 8hours pre-treatment at 20°C. Germination is hypogeal.

Raising of Plants in Nursery

Jojoba can be seeded or transplanted to the field when the soil temperature reaches 70°F. Low soil temperature may delay emergence by as much as 2 to 3 months. Seeds are sown 1 inch deep, and emergence usually occurs within 20 days. The soil should be kept moist but not wet through emergence. Individual seed can be sown in polybags (25x10cm size) filled with soil mixture. However, for fast germination, it is sown in perlite medium in nursery tray under mist unit and after germination saplings are shifted in to poly bags under Bikaner conditions (Fig.2) (Kukkar, 2012). Jojoba nursery should be prepared either in March or October month. After sowing of seeds, germination starts after 8-10 days and continues up to 20-25 days depending on moisture and climatic conditions. The growth of seedling is very slow and seedlings attain height 25-30cm after 5-6 months of sowing.

Fig. 2 : View of Jojoba nursery

Fig. 2a : Jojoba plants under shade net house

Fig. 2b : View of Jojoba nursery

Vegetative propagation

It can be achieved by layering, grafting, or rooting semi-hardwood cuttings, but the highest number of possible propagules is limited by the size of the plants and time of the year. Cutting material is collected from wild jojoba stands or from field grown plants; the season can have striking effects on rooting potential. Jojoba plants can be successfully propagated from stem cuttings made during spring, summer and to some extent, fall. Variability among jojoba plants may also play a role in rooting potential, although it is not as important as season. Selecting plants with naturally high rooting potential could make it easier to root cuttings and might eliminate the need to collect cutting material at specific times of the year. In general, the use of auxin on jojoba cuttings during periods of high rooting potential promotes adventitious root formation, but during periods of low rooting potential it has no effect or is even slightly inhibitory. In the greenhouse, cutting-grown plants apparently reproductively matured sooner than those grown from seed. If this observation holds true for plants transplanted into the field, earlier fruit production by cutting grown plants would mean earlier return of initial planting and maintenance costs (Clifford and Wesley, 1981). Plants raising through soft wood stem cuttings have been tried. Soft wood cuttings from female plants have taken and planted in well prepared filled with garden soil mixture poly bags. The techniques for planting of cuttings has standardized at Nursery, Daund, Jaipur.

In order to maximize production, it would seem advantageous to plant rooted cuttings from sexed plants which are known high producers or known to have seeds with high oil content. Rooting the cuttings takes a little bit of technical knowhow but it would be worthwhile to have a plantation with 90% to 95% female plants, leaving the 5% or 10% males to produce adequate pollen for all the female flowers. Another alternative is to establish plants from cuttings of known sex. Jojoba can be propagated from softwood stem cuttings taken in the spring or summer (Nord and Kadish, 1974).

Grafting

Grafting of jojoba has many advantages as keeping conditions. Hence, it could be a prime plant species to be the deep tap root of the rootstock which helps to survive introduced for arid land cultivation. It's economic in drought and salinity conditions, propagating plants importance is due to its saturated seed oil waxes that well with previously known sex, propagating unique and recognized for utilization in cosmetics, lubricants and desirable shrubs which will allow predictable plant growth pharmaceuticals etc. Plantations are established by and yield, also shortening juvenile phase of jojoba.

A seedy female jojoba shrub of good traits (seeds of high oil content) was selected as a source of scions and propagated in two dates, April and August during 2010 and 2011 seasons by two grafting methods namely top wedge and veneer grafting. Twelve seedy mature male or non-productive female jojoba shrubs (15 years old) were used as rootstocks. Data showed that grafting practiced in April were significantly higher in success percentage and sprout length than that of August while, grafting in August showed significantly lower number of days for sprouting. Top wedge grafting scored significantly higher success percentage and sprout length. On the other hand, veneer grafting scored significantly lower number of days for sprouting. April was the recommended time for jojoba grafting and top wedge grafting as well (Khattab *et al.*, 2013). Scion of mature wood attained better results compared with those of immature wood.

Micro-propagation

Jojoba seedling explants were cultured on a modified Driver Kuniyuki medium, supplemented with various concentrations of 6-benzyladenine alone and in combination with silver nitrate. Shoot proliferation was successful at all the concentrations tested, with a maximum number of 15.2 shoots per original explant. Shoots produced during the proliferation stage were treated with á-naphthaleneacetic acid, indole-3-butyric acid and indole-3-acetic acid to induce rhizogenesis, reaching 64% rooting in some treatments. When the rooted explants were transferred to the mist system for acclimatization, 90% of them survived and continued to grow after a period of one month (Roussos *et al.*, 1999).

Micro-propagation is highly recommended strategy for obtaining jojoba elite clones. For culture initiation, single-node explants are cultivated on Murashige and Skoog medium (MS) supplemented with Gamborg's vitamins (B5), 11.1 μM BA (N_6-benzyl-adenine), 0.5 μM IBA (indole-3-butyric acid), and 1.4 μM GA_3 (gibberellic acid). Inter nodal and apical cuttings proliferate on MS medium containing B_5 vitamins and 4.4 μM BA. Rooting is achieved on MS medium (half strength mineral salt) amended with B_5 vitamins and 14.7 μM IBA during 7 days and transferred to develop in auxin-free rooting medium. Plantlets are acclimatized using a graduated humidity regime on soil: peat: perlite (5:1:1) substrate. This micro-propagation protocol produces large numbers of uniform plants from selected genotypes of jojoba (Maurizio Lambardi *et al.,* 2012).

References

Clifford, B. Low and Wesley P. Hackett, 1981. Vegetative propagation of Jojoba, *California Agric.* March-April, 1981, pp. 12-13.

Khattab, Magda M. ; Ayman A. Hegazi, Mohamed E. Elsayed and Abd Thi Al Jalal Z. Hassan 2013. Propagation of Jojoba Shrub by Grafting, Jour. of Horticultural Science & Ornamental Plants, 5 (1): 37-40.

Kukkar, Rakesh Raman, 2012. Jojoba ki kheti (Package of Practices). Bulletin, Thar Nurseries Pvt, Ltd., Rani Bazar, Bikaner. pp.18.

Maurizio Lambardi; Elif Aylin Ozudogru and Shri Mohan Jain. 2012. Protocols for Micro-propagation of Selected Economically Important Horticultural Plants, Methods in Mol. Biology, 11013:19-31.

Muthana, K.D 1981. Jojoba- an oil yielding desert shrub. Ind. Farming,31:28-31.

North, E.C. and Kadish,A. 1974. *Simmondsia chinensis* Seed of woody plant in United States,AgricHandbook,Washington DC Forest Service, pp.774-776.

Roussos, P.A.; Tolia-Masiola; A.,Pontikis, C.A. and Kotsias, D. 1999. Rapid multiplication of Jojoba seedlings by *in vitro* culture, Plant Cell, Tissue and Organ Culture, 57(2): 133-137.

Chapter-34

Cactus pear [*Opuntia ficus indica* (L.)Mill.]

R. S. Singh
Central Institute for Arid Horticulture, Bikaner-334 006

Cactus pear (*Opuntia ficus indica* (L.) Mill.) commonly known as Prickly pear or Tuna, which have been identified and established as vegetable, fruit and fodder crop and considered as an important fruit crop for semi-arid and arid regions. In Indian arid and semi- arid regions, it is called *Nagphani* and produces small size fruits in which number of seed is available. It is grown wildly and unexploited. It is thorny plant and generally used for bio –fencing to protect fields from wild animals (Singh, 2000; Pareek *et al.*, 2001) . However, exotic cactus pear produces edible fruits and its tender cladodes can be used to prepare vegetable, salad, pickle and as animal fodder (Singh and Felker, 1998). Owing to the presence of Crassulacean Acid Metabolism (CAM) pathway, cacti have about four to five times more efficiency of water to dry matter conversion than even C_4 plants such as maize (Nobel, 1988). The species which produce edible fruits are *Opuntia ficus indica* (L.) Mill; *O. megacantha* Salm dyck; *O. amyclaea* Tenore; *O.streptacantha* Limaire; *O. robusta* Wandl; *O. lindhaimeri* Engelm . Fruits are rich source of sugar, vitamins and minerals, whereas its tender green cladodes are used as tasty and nutritious vegetable like lettuce and spinach (Rodriguez -Felix and Cantwell, 1988).

The systematic cultivation of *Opuntia* for fruits, vegetable and fodder is on a large scale in Mexico, Chile, Peru, Brazil, Bolivia, Argentina, Italy, South Africa, Israel etc. In recent years, emphasis has been given for intensive research activities and various producer's associations and regional boards have been constituted. The aim is to rationalize the marketing and promote consumption of cactus pear. During the past few years, the economic interest of cactus pears has remarkably increased, especially in arid and semi arid zones. In India, cactus pear has been introduced by the clones procured from Texas, USA only in late eightees and at the moment the area of its cultivation is negligible, but, the crop is now gradually becoming popular in arid and semi-arid regions, because of its high yield potential and export value. This crop has assumed importance because it can provide sustainable farming system in arid and semi arid regions. The suitable ones are those that successfully cope with water shortage, high temperature, and poor soils and easy management to provide food for the subsistence agriculture, in addition to valuable products and by–products (Nefzaoui and El-mourid 2009).

Cactus pear is now the part of the agricultural system of many regions of the world. It is commercially grown in more than 20 countries of world for fruits, vegetables, fodder and for preparation of several products (Russell and Felker, 1987; Nobel, 1988; Barbera *et al.*, 1992). Cactus pear is now being considered as most eco-nomical crop for dry areas of the world (Nefjaoui, 2009).

Looking to its vast industrial and economical value of cactus pear, FAO has established Cactus Network with aims to promote cactus pear as important crop world wide (Pimienta *et al.*, 1993).

Nutritive Value and Uses

The cactus pear has great industrial and commercial value. Prickly pear fruits are nutritious which contain 10-15 per cent carbohydrates, 1.8 per cent crude fibre, 0.5 per cent protein, 60 mg 100 g^{-1} calcium, 30 mg 100g^{-1} Vitamin C and 50 IU Vitamin A along with 85 per cent water. In Mexico, white-fleshed cultivar of cactus pear (*O. amyclaea) i.e.* 'Rayana' is an alternative crop in drought-affected areas.

Prickly pear fruits are used to prepare processed products like jam, jelly, marmalade, syrup, squash etc. Ripe fruits are also used to prepare wine. In South Africa, prickly pear is used for preparation of different processed products. Syrup prepared from red coloured varieties has natural colour and sweetness. Young tender cladodes (*Nopalitos*) are used as green vegetable in Mexico, Chile, Italy, Israel and South Africa. It is also used as *salad*. Nopalitos are rich source of carbohydrate (4.5%), fibre (4-6%), protein (1-2%) and minerals, principally calcium (1%).

It grows abundantly in western prairies of USA and becoming valuable livestock feed. The spring cultivars of *O. lindheimeri* are used as emergency drought fodder for cattle in Texas after burning of spines by propane torches (Russell and Felker, 1987). Cladode produce high coloured butter but had no effect of on flavour and preserving quality. Seed cake of *Opuntia* is also utilized as stock feed. Under limited rainfall conditions, it produces 10 t dry matter ha^{-1} yr^{-1} (Garcia de Cortazar and Nobel, 1992).

Prickly pear has been used for medicinal purpose since prehistoric period. It has antipyretic, anti-inflammatory, anti-diuretic, analgesic and anti-sposmodic properties. Use of nopales to cure diabetes has opened new vista for its utilization in human health care (Hegwood, 1990). Mulas (1993) has observed hypoglycaemic and hypo-cholesteremic action in young cladodes of cactus pear. Seeds are also used in medicines.

It is used as feed for mass rearing of the cochineal insects (*Dactylopius cocus*) to produce red dye carmine acid (Brutsch and Zimmermann, 1993). Prickly pear has high value in cosmetic industry. Nopale based products such as soaps, lotions, shampoos, hair conditioners and moisturising facial creams are now available in the international market (Pimienta *et al.*, 1993).

Being drought hardy, prickly pear plants provide a productive biological barrier and to control desertification (Pimienta *et al.*, 1993; Nefjaoui and El-Mourid,2009). It is a suitable multipurpose crop for marginal lands. Cladodes are a good material for biogas generation either alone or in combination with cattle dung (Uribe *et al.*, 1992). However, studies conducted at CIAE, Bhopal showed that xerophytic plants (*Cereus, Opuntia, Euphorbia*) used as supplemental feed for biogas plant either untreated or treated with 1.0% $Ca(OH)_2$ produced less biogas than the aquatic weed water hyacinth (Anonymous, 1993).

Prickly pears are of high quality and have nutritive value comparable to apple, peach, orange, cherry and apricot (Brutsch, 1984). Fruits are delicious and are highly est-eemed as fresh fruit and also used to prepare juice and syrup. Fleshy fruits are rich source of sugars, vitamins and minerals and contain large number of seeds. Sawaya *et al.* (1983) reported that 100 g edible portion of fruit contains 85.6g water, 0.21g protein, 0.12g fat, 0.02g crude fibre, 0.44g ash, 0.19g pectin, 0.22mg ascorbic acid, 27.6 mg calcium, 27.2 mg magnesium, 0.8 mg sodium, 161.0 mg potassium, 15.4 mg phosphorus, 1.5 mg iron and traces of β carotene.

The nutritive value of nopalitos (cladodes) can easily be compared with vegetables like okra and beans and is a good source of vitamins and minerals, mainly calcium and iron. Nopalitos are a traditional vegetable in Mexico and contain high quantity (92%) of water and carbohydrates with 4-5 per cent fibre, 1-2 per cent protein, 1.0 per cent calcium and 10-15 mg 100 g^{1} vitamin C.

The seeds of Prickly pear contains 16.6% protein, 17.2% fat, 49.6% fibre and 3.0% ash on dry weight basis and have 67.6 mg sodium, 163 mg potassium, 16.2 mg calcium, 74.8 mg magnesium, 152 mg phosphorus, 9.45 mg iron, 1.45 mg zinc, 0.32 mg copper and traces of manganese. Seeds also con-tain sulphur and amino acids (Sawaya *et al.*, 1983).

The edible types of *Opuntia* are native to the western Hemisphere that occur naturally from Alberta, Canada to 40°S latitude in Argentina. (Felker *et. al.*, 1997). However, most of the useful fruit cultivars have their origins in Mexico. The prickly pear cactus has been cultivated in Mexico for the last 9,000 years for food value. Wild thorny *Opuntia* type with thin pads are found in India.

Prickly pear is commercially cultivated in Mexico and semi arid regions of Western South Africa for fruit, vegetable and fodder. The leaves (nopales) and fruit (Tuna), comprise important nutritious and dietary source for Hippancies and Spanish people (Russell and Felker, 1987).

Botanical Description

Prickly pear or cactus pear (*Opuntia ficus indica* (L.) Mill.) belongs to family Cactaceae. Over 300 species of genus *Opuntia* are known in the world. The name *Opuntia* was derived from the word Opuse, an ancient Greek village in the region of Leocrid, which denotes spinous plant. The taxonomy of *Opuntia* is very complicated due to their phenotype which varies greatly according to the ecological condi-tions, polyploidy levels, existence of numerous hybrids, etc. Intensive field studies can recognise and identify the species, their varieties and adaptations reflected in their phenotype. In wild populations, plants located in the middle have been observed to express lower variability in their phenotypic characteristics than those on the periphery probably because of more possibility for out crossing with other species or varieties in the latter. Very few taxonomic studies have been done on this complex genus. Pimienta (1994) described 13 taxa from the commercial plantations but gave only the common names. David Griffiths, established *Opuntia* in early nineties as an important resource for forage and described more than 30 new species in Mexico. The species of the genus *Opuntia* can be divided into three major groups on the basis of their utilization.

The semi arid region of central Mexico is the main Centre for production of Cactus pear in the world (Pimienta, 1994). Cactus pear has an area of 7000 ha in Tunisia, 10,000 ha in Italy (Barbera *et al.*, 1991); 30,000 ha in Brazil, be-sides a large area in Algeria, Chile, Argentina, Mexico (Nobel, 1988), South Africa (Wessels, 1988) and 200 ha in Israel (Nerd *et al.*, 1989). During the last decade, area under cactus pear has been increased manifolds and it plays a strategic role in subsistence agriculture (Pimienta, 1994).

In India, its genotypes are growing wild. However, its cultivation has immense potential in the dry part of the country. In India, commercial plantation of cactus pear is lacking owing to absence of suitable varieties, planting material, agro-techniques and utilization methods. Looking to its potential in dry areas, cactus pear clones have introduced in India from different sources and the research work is in progress at CIAH, Bikaner, NARI, Phaltan, CAZRI, Jodhpur, CSSRI, Karnal to develop the cultural practices.

A large number of Cactus pear cultivars for fruit, vegetable and forage production are available through out the world. The crop was first introduced in Spain and later in northern and southern Africa and Mediterranean region from the Mexico. It has already acquired commercial status in South Africa, Italy, Israel, Spain, USA, Peru, Bolivia, Brazil, Colombia, Argentina and Chile (Russell and Felker, 1987; Barbera *et al.*, 1992). *Opuntia ficus indica* is cultivated as monoculture or intercropped with traditional crops of arid regions in Mexico, Italy, Chile, Israel, Tunisia, South Africa, etc. The main production centres of prickly pear fruit are Mexico and Italy while it is found wild as a forage crop in many countries and provides livestock feed during drought periods.

Cultivated Species

Opuntia ficus indica (L.) Mill.: Plant arborescent, 3-5 m high, with wide glabrous crown; stem 60-150 cm wide, joints obovated, 30-60 cm large, 20-40 cm wide and 19-28 mm thick, dark green, covered by layer of wax; areoles disposed in 8-9 spiral series, piriform, 2-4.5 mm long and approx. 3 mm wide, light brown; glochids yellow, disposed in the middle of the areole, 1-2 mm long, caduceus spines generally absent, rarely even one in few areoles, approx. 1 cm large, greyish, translucent. Flowers 7-9 cm long, orange to yellow pericarpel 2-2.5 times longer than the perianth, tuberculated with approx. 8 spiral series of areoles. Fruit sweet, juicy, edible, 5-10 cm long and 4-8 cm wide, pyriform, slightly sunk unbilicus, yellow, orange, red or purple with abundant fleshy pulp and thin skin. Seed is obovate to discoid, 3-4 mm diameter.

Pinkava *et al.* (1992) mentioned that this species can be hepta-ploid (n=77) or octaploid (n=88). In Mexico, its tender stems are used as vegetable "nopalitos", and the fruits that are very sweet, white or yellow or purpureous are appreciated as dessert of world.

Opuntia robusta **Wendl.:** Prostrated or erect, arboreous or arborescent, glabrous plant with very short trunk joints rounded to obovated, very thick, green glaucous, covered with a layer of wax; areoles 3.6-6 cm apart, about 3 mm diameter with dark wool; glochids numerous, 13 cm long with different colours, base blackish; spines 0 12 (No's), absent in the inferior areoles, unequal, divergent, not differentiated in radial and central, the base reddish spotted black with age. Flower yellowish green, shiny with a purple apex turning orange at the end of the day. Fruit purple or white globous to elliptic with long tubercles and superficial umbilic scar. Seeds discoid to deltoid with wide rill, 4-6 cm diameter and 1-1.5 cm thickness and sweet juice funicle. Three varieties of *Opuntia robusta* a) *O .robusta* var. *robusta* with circular joints and globoid purple fruits, b) *O. robusta* var. *guerrana* with obovoid spiny joints and white fruits, *O. robusta* var. *larreyi* with obovoid spineless joints and purple fruits are common.

Wild Species (subgenus *Nopalea*)

The only species belongs to this group is *Opuntia cochenillifera* (L.) Mill. which has arboreous, 6-7 m high, densely branched and glaborous plants with well defined stem; narrowly oblong joints, 20-30 cm long and 4-7 cm wide with light green colour; areoles disposed in 9 -10 spiral series with yellow wool and glochids; spines generally absent but sometimes present on old joints. Flowers tubular, 5-6 cm long, red purpureous almost do not open at the anthesis; stamens exert, disposed 1-1.5 cm above the perianth, filaments pinkish and anthers yellowish; style exert above the androecium with reddish colour and green stigma has 6-7 lobes generally pollinated by honey birds. Fruit obovate tuberculate, 4-5 cm long, red, spineless. Seed numerous, well shaped, obovate, hilum sunk, lateral sub basal, lateral arillum well developed. It is propagated by seeds or vegetative means in all the tropical and subtropical American countries. In Mexico, it is cultivated on the Atlantic and Pacific coasts. It is also cultivated in Central America, Cuba and Brazil. Its fruits are mainly eaten by birds however, in Mexico, this species is used as a substratum for the breeding of the cochineal insect from which it probably derived its name.

Root System

Opuntia ficus indica is characterized by a shallow and fleshy root system . The root distribution depends on the type of soil and cultural management. Under favourable soil conditions, a tap root penetrating nearly 30 cm into the soil and under drought conditions develops fleshy side roots from the tap roots, thus absorb soil moisture from lower profiles. However, bulk of the absorbing roots are found within a maximum depth of 30 cm and in a spread of 4 to 8 m. Plants supplied with periodical doses of manures and fertilizers develop succulent and unbranched roots while the unmanured plants have more lateral roots which very soon develop a loose exfoliating cork on the surface.

Under drought conditions, the plant develops xeromorphic characteristics in root system. Primarily, either the fine roots are covered with a layer relatively impermeable to water or the roots abscise by a cicatri-zation layer in order to avoid water loss. They can rapidly absorb small quantity of water sup-plied by light rains through "rain roots", which develop within a few hours after a shower and disappear as soon as the soil dries up and decrease shoot transpiration due to high negative potential in root. The root system of *O.ficus indica* is very complex and classified into:

Skeletal roots: A 20 to 30 cm long root which is formed from a primary skeleton of scarcely fibrous root and develops a periderm very soon. The skeletal root keeps abundant active or latent buds from base to apical portion without a regular pattern of distribution. When the root skeleton is kept dry for a period and then remoistened, the absorbing roots are formed from these latent buds within a few hours which quickly respond to the advent of moisture. The initiation of lateral roots is always endogenous from parenchyma cells belonging to the secondary phloem. This fine and fragile mass of roots is formed from short and branched rootlets, covered with root hairs. Abundance of rootlets and root hairs involved in nutrient absorption.

Absorbing roots : These roots are formed within a short period as a response of lateral buds to the advent of moisture. The "rain roots", developed from the hidden latent buds in the cortex of the older roots. These rain roots remain functional till the soil remains moist.

Root spurs : Root spurs develop from the most bulky mass of root clusters (Boke, 1979). Roots developed from these spurs are of two classes i.e. short and fleshy with plenty of root hairs and long cylindrical, similar to the absorbing root system.

Roots developing from areoles : These roots develop when the areoles are in contact with soil. Initially they are without root hairs with a prominent caliptra and epidermal cells forming brat like appendages. The growth of the young roots is very rapid and they become slender with a thick cortex which is covered by many root hairs. All the roots arising from the areoles in due course make up the real root system.

Cladode: Cactus pear bears a succulent stem with green cortex and devoid of foliage leaves.The stem, known as cladode, is typically oblong to spatu-late oblong, 30 to 40 cm long (exceptionally 70-80 cm) and 18-25 cm wide succulent organ. Anatomically, the cladode, is an eustele (in cross section) consisting of skin, cortex (a ring of vascular tissue made from collateral bundles and separated by parenchyma tissue) and pith which forms the major succulent tissue. The skin consists of one layer of epidermal cells and 6 to 7 layers of hypodermal cells with thick primary walls resembling laminar collenchyma tissue. The epider-mal cells are flat, thin and shaped like paving stones. The epider-mis along with thick cells of hypodermis provides an effective mechanical integrity to act as the first line of defence against fungus, bacteria and foraging organisms.

Areoles: The axillary buds in cactus are called areoles which are situated 2 mm below the skin surface. Under appropriate environmental condi-tions, new cladodes, flowers or roots arise from the meristematic tissues of the areoles. In *Opuntia ficus-indica*, the areoles are distributed in a helical way and develop spines rather than leaves. In young cladode, the areole is formed at the base of the podarium which hold a small, subulate, short-lived and caducous green structure corresponding to a leaf. The areoles are formed at the base of the short apical meristem and spines develop from the basal portion (Mauseth, 1984) in variable size and number, often one or two long central spines and some shorter lateral spines. The presence of spines is a special characteristic of the areoles and their morphology has potential taxonomic significance. Spines can be distinguished into true spines and spine hairs (glochids). The number and duration of the spines and glochids in *O. ficus-indica* depends on its type. Usually, spines are present in the first stage of the cladode growth and most of them fall off with rise in temperature. Only few spines remain at the base of cladodes for a long period. Sub family *Opuntioideae* is characterised by rough surface of the glochids and spines . In *O.ficus-indica*, the thorns have a rough and the glochids have smooth surface. The glochids are arranged in clusters of 7 to 12 in the cavities of the areoles.

Thorns in *O. ficus indica* are white and present in groups. In a group one or two thorns are long (1-1.5 cm) and are supported by two smaller thorns. As the tops of the spines harden, the basal meristem continues to lengthen. The surface of the spine hair is covered with calcium carbonate and pectic substances (Buxbaum, 1950). Glochids are grouped in 4 to 6 tight clusters from which it is possible to distinguish two or three longer thorn-like structures. They are hard, sclerified and pointed. Their surface is covered by barbed scales which permit them to lodge in the skin and can not be removed easily. In *Opuntia ficus-indica,* however, the glochids are fastened to the areole during growth which later develops a phellogen in the outer cortical region around the base of each cluster and makes it deciduous which is removed either by touching or by blowing wind.

The spines and trichomes save the plant from being eaten by animals, help to restrict water loss, reduce daytime stem temperature and diminish light interception by the cladode. These have ability to condense water from the air .

Flower: The auxiliary buds in cactus are present in the form of areoles which develop very early in the axils of the leaf primordia. The areole develops into a mass of cells which rapidly form a regular shoot apical meristem consisting tunica, central mother cells, peripheral zones and pith rib meristem. After a certain period the meristem at the centre of the areole depression becomes dormant. The cactus flower consists of a number of perianth segments weakly differentiated as petals, spirally arranged stamens and a pistil of four or more fused carpels enclosed in a floral cup. Its unilocular ovary has parietal placentation and appears embedded at the end of a modified branch. The receptacle contains leaves and perfect areoles which later becomes the peel of the fruit. During the emergence, it is possible to differentiate between vegetative or reproductive bud from its spatial volume. The reproductive bud is more spherical while the vegetative one is flatter. Generally, the ratio of vegetative and reproductive buds is 3 : 1 while 10 per cent cladodes may have both kinds of buds in equal proportion (Sudzuki *et al.*, 1993).

The hermaphrodite actinomorphic flower develops on the upper margin of a 1-3 year old cladode. Both type of buds may develop over the most illuminated plain surface of the cladode. Floral differentiation occurs over a short period of 50 to 60 days after the meristem becomes active till anthesis unlike in other fruits where floral differentiation begins almost a year before anthesis. There are few differences between sepals and petals in the perianth. The sepals are smaller but both are oblong and fused at the base with a brilliant yellow or pink colour. The yellow flowers change colour to orange or pink after fecundation. The pistil has a wide style with a prominently sectioned granular and viscous stigma. The ovary is syncarpous, unilocular, penta-carpellary with parietal placentation. In *Opuntia* sp., the funicle is slender and relatively long. The ovules have three true integuments, where the inner one is beyond the outer. The third integument is a modified part of one of the two normal integuments, which covers the ovule completely. Ovules are circinotropus, because the funicle surrounds the ovule and covers the micropyle like a thick third integument. This structure can be seen in the seeds of *O. ficus-indica,* where it is present with the micropyle pore quite free.

The inner surface of the funicle is covered with short hairs or papillae, especially at the point where the funicle touches the micropyle. The ovules are arranged in two rows in each ventral prominent suture. This situation suggests that the carpel union is margin to margin and the dorsal bundles extend to the ovules .

Pollen grains: The pollen grain has a thick exine which protects it and regulates changes in its volume according to the humidity. The intine is thin and readi-ly adapts itself to changes in size. The patterns of exine have considerable taxonomic and phylogenetic value because the wall may become very thick and complex in structure and the outer layer is sculptured with projecting ridges, spines and granules. The *Opuntia* has a large number of pollen grains which germinate over the glandular epidermal surface of the broad stylar canal after 24 hours of anthesis.

Fruit and seed: Fruit is a simple fleshy berry but since it is developed from an inferior ovary sunk into the receptacle, it is called a false berry. The peel of cactus pear originates in the receptacle and has an epidermis with ephemeral leaves and perfect areoles where the glochids are more prominent than in the cladodes, slender hypodermis and bulky cortex and plenty of mucilage cells without crystals. The pulp is formed from the outgrowth of the trichomes originated in the epidermal cells of the funiculi and the funicular envelope. The fruit size depends on the number of fecundated and aborted seeds (Barbera *et al*, 1994).The fruit often has two kinds of sterile seeds. Lack of viability in the seeds is related with failure of the ovule to germi-nate and the young adventitious embryo .In the latter case, the nucellus continues to grow even after the embryo is formed and at the time the funicle changes into a hard lignified coat. In Chile, many fruits of *O.ficus indica* represent a third kind of non viable seed. The growth of the peel is greater during the first week after flowering whereas the edible portion begins to expand mainly between the fifth to sixth weeks before ripening (Hills, 1995).

Cultivars

Cactus pear has a wide range of varieties. Depending upon the agro-climatic adaptability, the cultivars can be classified into Mexican or American group, Italian group and Argentinean group apart from some cultivars from Israel and Malta. The important cultivars are Amarilla Montesa, Fafayuco, Rosa Pilona, Copena Cardona, Burrona, Cristallina (Mexico); Bianca, Gialla, Rossa (Italy) and Ofer, B.S.-1 (Israel). According to Wessels (1988), South African Prickly pear cultivars may be classified into following three arbitrary groups on the basis of their growth.

1. **Round blue leafed types** which produce fruits of poor quality and are essentially used as Livestock fodder.
2. **Long leafed or tree type** which grows reasonably erect with few small thorns and produces succulent, green, white and khaki co-loured fruits, e.g., Fusicaulus.
3. **Bushy types** which produce spreading, bushy plants of varying form and colour. Some types are vigorous, eg., Nudosa while others have a tendency towards dwarfism. The type produces outstanding yield of fruits of variable shape, size and colour. Fruits are generally elongated to round with green, red, pink or even a mottled red and white colour as in Castillo. Pimienta (1994) hypothesized that the existing commercial fruit cultivars in Mexico have resulted from asexual propagation by mankind from natural tetraploid and octaploid hybrids.Italian cultivars like Rossa, Gialla and Bianca grow errect and produces better quality fruits, are suitable to Indian conditions. Beer sheva Selection-1(BS-1) and Ofer are fruiting type cultivars developed at B.G.U. of the Negev Beer Sheva, (Israel). Fruits of B.S.-1 contain less number of seed while cultivar Ofer produces yellow coloured fruit with sweet pulp. The cladodes of important cultivars have been introduced in India. About 55 exotic clones of cactus pear have been introduced from Texas, Israel, Argentina and Mexico and indigenous types were evaluated under hot arid conditions (Pareek *et al.,* 2003). Variation in number of cladodes per plant and size was observed. Clone 1269, 1270, 1271 and vegetable type 1308 were found suitable to grow under arid conditions.

Propagation

In general, cactus pear is propagated for fruit and biomass production through sexual and asexual means. Vegetative methods are very common and easy for large scale multiplication. Micro-propagation technique has also been employed for diseases free mass multiplication of cactus pear plants.

Seed propagation

Cactus pear is propagated by seed. Plants raised through seeds show variability in growth. Seed viability in cactus pear is longer but the rate of germination is poor. Seed taken from mature fruits germinates within 5-21 days. Young seedlings remain spiny for about a year and then they become thornless. Cactus pear has a hard lignified seed coat which protect the seed by abiotic stress during nursery stage and also checks seed germination. Attempts have been made to improve its germinability (Muratalla *et al.*, 1990) by mechanical breaking and/or after immersion in GA_3, hot water (90-95^{0}C) treatment for 5 to 20 minutes and chemical scarification with concentrated H_2SO_4 followed by soaking in 100 ppm GA_3. Before sowing, seeds should be treated with captan or thiram for escaping fungal diseases. Without treatments, seeds took longer period for germination under nursery conditions.

Seed storage

Cactus pear seeds are small in size but hard seed coat. The cactus seeds can be stored in plastic containers. Long term storage reduces germination rate of seed. However, it can be stored up to nine months under normal conditions. Four months stored seeds showed better germination rate. The short-term storage of seed increases germination rate.

Germination

For better germination, seed should be kept under cool temperatures (13-20^0 C) and diffused light conditions. Seed taken from mature fruits germinates within 5-21 days.

The *opuntia* seed takes time to germinate 25-30 days under hot arid conditions. The germination of seed is also very slow due to hard seed coat. Soaking of seeds in water also enhances the germination. The treatment of seed with $GA_3$100 ppm promotes germination.

Vegetative propagation

For commercial cultivation, prickly pear is only vegetatively propagated using atleast one year old cladodes as planting material. For this, pads of 500-1000 g weight are severed from the mother plant, shade dried and planted directly in the field after treating with fungicide and plant growth regulators. NAA 1000 ppm gave the best results (Mulas *et al.,* 1992). In order to induce more rooting and better growth and development in cactus pear plants, the planting material, i.e. basal portion of cladodes may be dipped for 15 minutes in Indole Butyric Acid (IBA) 1000 ppm and this treatment was better than IAA and NAA. However, 500 ppm NAA and 1000 ppm IAA were more effective than other concentrations of these plant growth regulators employed (Singh and Singh, 2001) under Agra conditions, India. Vegetative propagation, which is widely used, can be performed through the rooting of single or multiple cladodes, small portion of mature cladodes comprising two or more areoles, or by using fruit as propagules.

Size of cladodes

The size of cladode is an important factor for considering the suitability of planting material as it is primarily responsible for the sprouting pattern and further growth especially during first year after planting. Consideration of cladode size is also important for better establishment apart from transportation cost and cost of planting material. The size of cladode or planting material becomes more important factors when its availability is a limiting factor. The planting of 10-15 cm width size cladodes of one year old was found suitable (Singh, 2000). Impact of different size of cladodes on plant establishment and growth behavior has been discussed by several workers (Wessels, 1988; Pimienta, 1990; Inglese *et al.,* 1994). However, very little thrust has been given so far to investigate the propagation aspects in light of type and size of cladodes used for planting. The study on size and effect of planting has been carried out at Agra, India. Singh and Singh (2003a) reported that two piece of a cladode can be planted.

Age of cladodes

Age of cladodes is another important feature related to survival and growth in *Opuntia ficus indica* Mill . Planting of young cladodes is considered inappropriate by some workers (Domigella, 1958). For vigorous growth after planting sufficient quantity of reserve food in planted cladodes is very much useful. The information available on the age of cladodes for planting operations is meagre.

The main limitations with vegetative propagation technique are the availability of sufficient disease-free planting material. A nursery technique to produce more number of cladodes for planting has been developed. The technique involves, cutting of a cladodes into 4-5 pieces and planting in nursery bags/tubes of 25x10 cm size filled with sand clay and FYM in equal proportion. Each piece produces 3-5 daughter cladodes which can be used for planting in next season. The best time of planting is autumn winter or spring season. The cladodes are removed from the mother plant and are allowed to dry in shade for 4-6 weeks before planting in the field. Root primordia originate from phloem cells while the cambium remains dormant .

Season of planting

The planting of cactus differs with other fruit crops. Different workers have reported different time of planting for different locations/ countries. However, in general, planting time vary with latitude and environmental conditions such as rainfall, temperature, humidity, etc. Thus, a lot of variation in time of planting has been reported by the researchers at different places. Under Indian conditions, February - March month (spring season) is suitable time for planting of cladodes (Singh and Singh, 2003).

Method of planting

The method of planting is another important factor which are responsible for survival and growth of plants in Cacti. Consideration of correct row orientation in planting of cladodes is an also important for better growth and production. Many workers have been suggested different methods of planting in Cactus pear (Wessels, 1988; Anon, 1993; Felker *et al.,* 1997). Nobel (1982) observed the orientation of terminal cladode in platy Opuntias in different countries at different latitudes, and found that topographical factors primarily affected the direction of prevailing PAR. The orientation of cladodes planting can be changed to maximise the interception of PAR in order to maximise the growth and production in Cacti. The upright planting of cladodes during spring season is found better for establishment and growth of plant under semi arid conditions (Singh and Singh, 2003). In Nopalitos producing region, planting is ofen done by making a hole with help of small hoe. The cladode plainner surface are oriented east-west in order to allow good growth and prevent sun burn during summer season.According to Singh and Felker,(1998) upright method of planting in cactus is ideal under environmental conditions of Karnal,India. The east-west direction planted cladodes favoured better growth in cactus clone -1270 under hot arid condition (Singh *et al*, 2002).

Spacing in Cactus pear largely depends upon their spread/canopy depending on variety/genotype. Wessels (1988) suggested that cultivars such as Fusicaulus, Director, Guaya, Quit, etc. may be planted with 4 m spacing between plants and 5-6 m spacing between rows. In order types, planting distance may be 3 x 3 m. When planted closer (2 x 4 m), annual pruning must be done to maintain proper health of the plants. Upright planting of cladodes on ridges at 3.0 x 1.5 m spacing has been found suitable under semi-arid conditions (Anon, 1993).

Micro-propagation

Plants can also be raised by tissue culture technique. Mohammed Yasseen *et al.* (1995) successfully pro-duced rooted shoots *in vitro* or *ex vitro* and established the plants in soil.

Although, clonal propagation used traditionally for commercial plantations of cactus pear require large quantity of propagating material. This often poses a practical problem. *In –vitro,* technique have been reported in thousands of species including the member of Cactaceae. The first attempt of *in–vitro* propagation in *Opuntia* spp. has been reported by King (1957). Since then, tissue culture methods have been applied to Cacti for a variety of purposes.

Mohammed Yasseen *et al.* (1995) successfully produced rooted shoots *in vitro* or *ex-vitro* and established the plants in soil. Investigations on growth and development of root system with application of growth promoting substances are very limited. Some *in vitro* experiments (Mouseth and Halperin, 1975; Es'cobar *et al.* 1986) and others on cuttings (Babaeva, 1988) are known but were carried out on species other than *O. ficus indica.* Es'cobar *et al.* (1986) developed a very efficient micro propagation method for *Opuntia amyclaea* by which an average 25,000 plants could be obtained in 100 days from a cladode of 5 cm. However, survival rate of such plants has not been discussed in said study. However, Mulas *et al.* (1992b) carried out experiment on one year old cladode by using IAA, IBA and NAA at the dosages of 250 and 1000 ppm and found that cactus cladode cultivar 'Gialla' treated with 1000 ppm for 15 minutes resulted in slightly better rooting than that of untreated ones.

Khalafalla *et al.* (2007) found that satisfactory rooting percentage (100%) was achieved on both Murashige and Skoog (MS) free-hormone or supplemented with IAA at 0.5 mg/l but, the highest number (15.0) of root per shoot was obtained on MS medium supplemented with 0.5 mg/l IAA. Finti *et al.* (2012) found that *in vitro* conditions, 100% of plantlets rooted and highest number of root (21.05 and 16.2) was obtained, respectively with IBA and IAA. The rooted shoots were easily acclimated and transferred to soil.

Finti *et al*, (2012) observed that young cladode explants containing one areole were cultured on Murashige and Skoog (1962) medium (MS) with adenine sulphate (40 mg/l), monosodium phosphate (50 mg/l), sucrose (50 g/l), phytagel (0.3%) and benzyladenine (BA) at 22.2 μM. Best multiplication rates were obtained with BA at 0.5 mg/l than kinetin. To induce rooting, the proliferated shoots were transferred to a medium containing half MS, sucrose (3%), phytagel (0.3%) and 0.5 mg/l of different auxins. In these culture conditions, 100% of plantlets rooted and highest number of root (21.05 and 16.2) was obtained respectively with indole-3-butyric acid (IBA) and indole-3-acetic acid (IAA). The rooted shoots were easily acclimated and transferred to soil for establishment.

Finti *et al.* (2013) also observed that rooting of *in vitro* generated shoots was achieved more efficiently on half- strength MS basal medium supplemented with 0.5 mg/l of IBA and IAA. Rooting frequencies were in the range from 95 to 100% and the highest mean number of root (19.1) was obtained with IBA for 'Delahia' cultivar. All micro-propagated plants were showed good overall growth.

Quality Standards of Planting Material

For the development of Cactus pear plantations, the emphasis should be given on the establishment of registered nursery to provide healthy rooted cladode from productive orchards. The cladode should be of 1-2 year old, disease free, true to type for better establishment and quality production. There is no quality standards fixed in cactus pear. Before planting, cladode should be kept under partial shade for 15-20 days period which improves healing of cut wound. Appropriate age and weight of cladode also influence the establishment and growth of plant in cactus pear. In case, scarcity of cladode/planting material, pieces can be used for planting after treatment with fungicides.

References

Anonymous, 1993. Final Report of the Project "Genetic selection and improvement of *Opuntia* cultivars for human and animal food on semi-arid lands". Nimbkar Agricultural Research Institute, Phaltan, Maharasthra, India p.93.

Babaeva, SH. R. 1988. Effect of growth regulators on rooting of *Opuntia* cuttings. Introduktsiya-i-Akklimatizai siya Rastemii, 67-69.

Barbera, G., Carimi, F. and Inglese, P. 1991. The flowering of prickly pear (*O. ficus indica*(L.) Mill.) influence removal time and cladode load on yield and fruit ripening. Adv. Hort. Sci., 5: 77-80

Barbera, G., Carimi, F., Inglese, P and Panno, M. 1992. Physical, morphological and chemical changes during fruit development and ripening in three cultivars of prickly pear *(Opuntia ficus indica* (L.) Mill). J. Hort. Sci., 67 307 312.

Barbera, G., Inglese, P. and La Mantia, T. 1994. Influence of seed content on some characteristics of the fruit of cactus pear (*Opuntia ficus indica* Mill). Scientia Hort., 58:161 165.

Boke, N., 1979. Root, glochids and root spurs of *Opuntia arenaria* (Cactaceae). Am. J. Bot., 66(9): 1085-1092.

Brustsch, M.O. 1984. Prickly pear (*Opuntia ficus indica*)cultivation in southern Africa. In: Symposium of agricultural use of the Cactaceae. Prospects and problems. Internatinal Organization for Succulent Plant study, Frankfurth.

Brutsch, M.O. 1979. The prickly pear (*Opuntia ficus indica*) as a potential fruit crop for the drier regions of Ciskei. Crop Prod., 8:131 137.

Brutsch, M.O. and Zimmermann, H.G. 1993. The prickly pear (*Opuntia ficus indica* (Cactaceae) in South Africa: ultization of the naturalized weed, and of the cultivated plants. Econ. Bot., 47(2):154 156.

Buxbaum, F. 1950. Morphology of cacti. Abbey Garden Press. California., USA.

Es'cobar, A.; Villalobos, V.M. and Villegas, M. 1986. *Opuntia* micro-propagation by axillary proliferation. Plant Cell,Tissue and Organ Culture,7:269-277.

Felker, P.; Singh, G.B. and Pareek, O.P. 1997. Opportunities for development of cactus (*Opuntia* spp.) in arid and semi arid regions. Annals of Arid Zone, 36(3):267 278.

Finti, A. El., Boullani, R. El., , Ait Aabd, N., Msanda, F., Serghini, M. A. and Mousadik, A. El 2013. *In vitro* propagation of three Moroccan Prickly pear Cactus Opuntia and plant establishment in soil. Not. Sci. Biol. 5 (1) : 39-44.

Finti, A. El., Boullani, R. El., Ayadi, F. El, Ait Aabd, N. and Mousadik, A. El 2012. Micropropagation *in vitro* of *Opuntia ficus indica* in South of Morocco. Int. J Chem Biochem Sci. 1 : 6-10.

Garcia de Cortazar, V. and Nobel, P.S. 1992. Biomass and fruit production for the prickly pear cactus *Opuntia ficus indica*. J. Am. Soc. Hort. Sci., 117:558 562.

Hegwood, D.A. 1990. Human health discoveries with *Opuntia* sp. (pricly pear). Hort. Sci., 25(12):1515 1516.

Hills, F.S. 1995. Anotomy and Morphology. In: Agro ecology,cultivation and uses of cactus pear. (Barbera, G; Inglese, P. and Pimienta. B.E; eds.) FAO: 28 35.

Inglese, P., Barbera, G. and Carimi, F. (1994). The effect of different amounts of cladode removal on reflowering of cactus pear (*Opuntia ficus indica* (L.) Mill.). J. Hort. Sci., 69:61 65.

Khalafalla, M. M., Abdellatef, E. Mohameed Ahmed, M. M. and Osman, M. G. 2007. Micropropagation of Cactus (*Opuntia ficus indica*) as strategic tool to combat desertification in arid and semi arid regions. Int. J. Sustain. Crop Prod. 2 (4): 1-8.

King, R. M. 1957. Studies on the tissue culture of Cacti. Cactus and Succulent *J.*, 29:102-104.

Mauseth, J.D. and Halperin,W. 1975. Hormonal control of organogenesis in *Opuntia polycantha* (Cactaceae). American. J. Bot.

Mauseth, J.D. 1984. Introduction to Cactus anatomy (Part 7). Epidermis. Cactus and succulent Journal (US), 56:33 37.

Mohammed - Yasseen - Y; Barringer, S.A., Splittstoesser, W.E. and Schnell, R.J. 1995. Rapid propagation of tuna *(Opuntia ficus indica)* and plant establishment in soil. Plant Cell Tissue and Organ Culture, 42 : 117-119.

Mulas, M. 1993. Medicinal properties and yield possibilities of the prickly pear (*Opuntia* spp.) in the Mediterranean environment. Acta Hort., 331 : 79-84.

Mulas, M., Spano, D., Pillizaro, G. an D'Hallevin, G. 1992. Rooting of *Opuntia ficus indica* Mill. in young cladodes. Adv. Hort. Sci., 6: 44-46.

Muratalla, L.A., Barrientos, P.F. and Rodriguez, A.J. 1990. Germinacion y viabilidad de semilla de nopal *Opuntia amyclaea* t. (V 5) y *O. ficus indica* (V 1) y (F 1). In: Memorias de la IV Reunion Nacional y II Congreso Internacional sobre el conocimiento y aprovechamiento del nopal. Zacatecas, Zac. Mexico

Nefzaoui, A. 2009.Cactus: A crop to meet the challenges of climate change in dry areas. Annals of Arid Zone, 48:1-18.

Nefzaoui, A. and El Mourid, M. 2009.Cacti: A key stone crop for the development of marginal lands and to control desertification. Acta Hort.,811: 365-372.

Nerd. A., Karadi, A. and Mizrahi, Y. 1989. Irrigation fertilization and polyethylene covers in prickly pear influence bud development. Hort. Sci., 24:773 775.

Nobel, P.S. 1982. Orientations of terminal cladodes of planty opuntias. Bot. Gaz., 143:219 224.

Nobel, P.S. and Hartstock, T.L. 1984. Physiological responses of *Opuntia ficus indica* to growth temperature. Physiol. Plant., 60:9 8 105.

Nobel, P.S., 1988. Environmental biology of agaves and cacti. Cambridge Univ. Press, New York.

Pareek, O.P. ,Singh, R.S. and Vashishtha, B.B. 2003. Performance of cactus pear (*Opuntia ficus indica* (L.) Mill.) clones under hot arid region of India .J.P.A.C.D.,5:121-130

Pareek,O.P. ,Singh, R.S.; Vishal nath and Vashishtha, B.B. 2001.The Prickly pear, Agrobios (India), Jodhpur,pp71.

Pimienta, B.E. 1994. Prickly pear (*Opuntia* spp.): a valuable fruit crop for the semi arid lands of Mexico. J. Arid Environments, 28:1 11.

Pimienta, B.E., Barbera, G. and Inglese, P. 1993. Cactus pear (*Opuntia* spp. Cactaceae) International Network: An effort for productivity and environmental conservation for arid and semi arid lands. Cactus and Succulent Journal, 65:225 229.

Pimienta, E. 1990. El nopal tunero. University de Guadalejara, Mexico.

Pinkava, D.J.; Parfitt, R.D., Baker, M.A.and Worthington, R.D.1992.Chromosomes number in some cacti of western North America-VI with Nomenclatural changes. Modrano, 39:98-113.

Rodriguez-Felix, A. and Cantwell, M. 1988. Developmental changes in the composition and quality of prickly pear cactus cladodes (nopalitos). Plant Food and Human Nutrition, 38:83-93.

Russell, C.E. and Felker, P. 1987. The prickly pears (*Opuntia* spp. Cactaceae): a source of human and animal food in semi arid regions. Econ. Bot., 41:433 445.

Sawaya, W.N., Khalil, J.K. and Al Mohammad, M.M. 1983. Nutritive value of prickly pear seeds, (*Opuntia ficus indica.*). Plant. Foods Hum. Nutr., 33: 91 97.

Singh R.S., Bhargava R., Sharma B.D. and Pareek O.P. 2002.Effect of depth and directions of cladodes planting on growth in cactus pear. Current Agric.,26: 119-121.

Singh, R.S. and Singh Vijay 2003. Effect of planting seasons on growth performance in Cactus pear genotypes. Progressive Hort.,35: 164-169.

Singh, G.B. and Felker, P. 1998. Cacti: A new world food. Indian Hort., 43 :26-31.

Singh, R. S. and Singh, Vijay 2001.Effect of auxins on rooting and growth behaviour in Cactus pear. Haryana Jour. Hort. Sci.,30: 201-203.

Singh, R.S. 2000. Studies on propagation and growth behaviour in Cactus pear (*Opuntia ficus indica* L. (Mill) as affected by size and age of cladode, season and methods of planting and growth regulators. Ph. D. Thesis submitted to Dr. B.R. Ambedkar University, Agra (U.P.)

Singh, R.S. and Singh, Vijay 2003a. Growth and development influenced by size, age and methods of planting of cladodes in Cactus pear. Jour. Prof. Assoc. Cactus Develop., 5:47-54.

Sudzuki, F, Munoz C. and Berger H., 1993. *El Cultivo de la tuna (Cactus Pear)*. Fac. Ciencias, Agrarias y Forestales U. de Chile.

Uribe, J.M., M.T. Varnero and C. Benavides, 1992. Biomasa de tuna (*Opuntia ficus indica* (L.) Mill.) como acelerador de la digestiion anaerobica de guano de bovino. Simiente. 62(1):14 18.

Wessels, A.B., 1988. Spineless prickly pear. Pub.: Perskor Publication, Johanesburg, South Africa. p 61.

Chapter-35

Marula nut (*Sclerocarya birrea sub* sp. caffra)

R. S. Singh
Central Institute for Arid Horticulture, Bikaner-334 006

Marula nut (*Sclerocarya birrea* subsp. *caffra*) is a highly valued tree in most parts of Southern Africa because of its medicinal and nutritional properties. The marula tree is adapted to the semi-arid conditions that characterize most parts of sub-Saharan Africa and renders them unsuitable for conventional crop agriculture (Moyo Mack, 2009). The unique nutritional properties of marula and its high tolerance to dry conditions provide opportunities for its development into a plantation crop. On the other hand, the demand for marula plant parts, mainly the bark and roots as medicinal remedies, poses a great threat to wild populations. In the long term, the growing demand of marula products in the food, pharmaceutical and cosmetic industries will not be sustainable from wild populations alone. A rich genetic diversity in *Sclerocarya birrea* has reported by Muco *et al.* (2007) in Kenya. Further, variation in plant population occurs in different parts. A study has conducted for sustainable harvest of marula (*Sclerocarya birrea* subsp. *caffra)* in South Africa by Emmanuel *et al.* (2005). Looking to its enormous values, conservation of marula plant is very important. Therefore, this plant is protected in South Africa. It is cultivated for fresh fruit in Arava valley, Israel, where annual average rainfall is about 100 mm and dry climate. In the experimental plots many, branches died due to winter temperatures of -7 ^{0} Celsius. In the spring, however, all plants resumed their growth from the lower parts of the stem. In northern Sotho, marula trees are not cut down when land is cleared for planting and cultivation because of its high value. Its drought resistance make it ideal suited to Namibia, Bostwana, Zambia and Zimbabwe where it is in abundance. Marula fruits abscise when mature but ripen only a few days later. In western arid part of India, it was introduced from Israel and evaluated for growth and development of plant. It is a fast growing fruit tree species but it is susceptible to low temperature/frost during winters in hot arid region (Singh and Bhargava, 2008).

Marula, a underutilized fruit crop should be popularised and improved cultural practices should be developed for different agro-climatic regions. Marula plant is not only yields fruits but also provides timber, firewood, leaf fodder, medicinal value and serve as shelterbelts in arid regions. Owing to its multiple uses, it can be used in different farming systems to meet the basic needs of rural inhabitants of arid region (Singh, 2012).

Nutritive Value and Uses

The marula tree is a multipurpose tree yields edible fruit and nut. Its fruit is very sweet, juicy and aromatic and looks like a small yellow plum. Its fruit is nutritive and the flesh has high vitamin C content. It is also used to prepare jam, juices and alcoholic beverages. Inside the flesh, one or two small tasty nuts are available which are rich in protein. Its seed oil is used as a skin cosmetic. Marula kernels are rich in protein (28-31%), oil (56-61%), magnesium, phosphorus and potassium. Other important nutritional components include trace elements and vitamins such as iron, calcium, copper, zinc, thiamine and nicotinic acid. Kernels can be eaten or oil can be extracted from them for cooking or for use in the cosmetic industry. Fruit and leaves are browsed by livestock and have a variety of medicinal uses, as do the roots and the bark. Fruit and kernels form an important component of the diet of the rural people. Fresh fruit is widely consumed, particularly by children, providing a rich source of vitamin C. Fruits are also collected and processed into juice, alcoholic beverages (wine and beer) and jam, extending the shelf life of the product and prolonging availability and consumption beyond the two to three month fruiting season. The fruits are used to make beer a *marula liquor* is available commercially. The kernel is crushed to remove the seed, which is then pressed to produce oil which can be used as edible oil. The kernels are either eaten raw or roasted, have a delicious taste. The endocarp is oily and edible, occasionally sold on local markets, and contains up to 6% oil. Marula oil contains a similar fatty acid composition to olive oil however; it is 10 times more stable to oxidation. Marula oil have moisturizing properties and used for massage purposes. Its pale, clear, yellowish colour oil is used in cosmetics and lipsticks. Kernels oil is used for coking, as a meat preservative and for skin care, hair applications and massage oil for baby. Marula is one of the most commonly used wild fruit species in South Africa, where 59–77 % of households reported consuming marula fruit between four to five times per week during fruiting season. About 2% of households sell marula products, mainly beer and kernels. In West Africa, marula is one of the ten most used species and sources of income. It is used for both food and medicine. The bark is widely used medicinally to treat diarrohea, fever, malaria and the roots are used for sore eyes. Marula fruit and products, including alcoholic drinks, jams and fruit juice, have been traded locally, regionally and internationally since the Marula commonly occurs in arid and semi-arid areas in the world. Marula wood is used to make kitchen utensils, stool, mortars, bowls, etc.

Botanical Description

Marula nut (*Sclerocarya birrea*) belongs to the family Anacardiaceae. It is drought hardy medium to big tree. The plant is tall, erect, dioecious, deciduous, dense foliage, which grows naturally in northern South Africa and parts of eastern Botswana. Marula generally bears male and female flowers on separate trees. Although, the flowers are usually one sex or the other, occasionally single flowers are both male and female. Although scentless, marula flowers are very attractive to bees and other insects, including flies, hoverflies and, less commonly, wasps. These insects are believed to be the main means of pollination.

There are three recognized subspecies of marula:

- *S. birrea* subsp. *caffra* (Sond.) is the most ubiquitous and occurs in tropical East Africa (Kenya and Tanzania), tropical southern Africa (Angola, Malawi, Mozambique, Zambia and Zimbabwe) and southern Africa (Botswana, Namibia, South Africa and Swaziland). It is also recorded from Madagascar. It is common in savannah areas in South Africa and Zimbabwe. In Ethiopia it is found in northern and southern parts of the country at altitudes of 500–1600 m.
- *S. birrea* subsp. *multifoliolata* (Engl.) occurs in mixed deciduous woodland and wooded grassland in Tanzania and perhaps in southern Kenya.

- *S. birrea* subsp. *birrea* occurs through tropical areas of West, north-east and East Africa. It is found across a range of vegetation types, principally mixed deciduous woodland, wooded grassland, riverine woodlands, bush lands and the open dry savannahs of tropical northern Africa and the Sahelian region.

In the Sahel, marula flowers from January to March, triggered by increasing air humidity at the end of dry season. Fruiting occurs in March and April. Leaf flushing also follows the dry season, when the first rain starts real growth. In Sudan, flowering occurs in January to April and fruiting from April to June. In South Africa, flowering occurs in September to November and fruiting in February to June. The firm green fruits fall from the trees before they are mature and final ripening takes place on the ground, fruits turning yellow and softening. Trees can begin to set seed as early as 5 years old. The flowering/fruiting in seedlings was started after six years of planting. In Arava valley, Israel fruits mature in the month of November –December. Under hot arid conditions of western Rajasthan, India flowering takes place during summer season. Fruits were matured during the month of September-October and fruit characters were studied .Fruit drop has also been observed. The average weight of fruit was 40.12g; size 3.7 x 3.4 cm. but pulp content was low. The average stone weight was 13.4 g and T.S.S. of ripen fruit was recorded 18.0 ^{0}Brix. Its fruit is juicy sour- sweet in taste with good aroma and flavour. The plum sized fruit are covered in a soft leathery, pale green yellow exocarp which encloses the juicy white flesh (Hall, 2002).

Propagation

It is propagated through seeds and vegetative methods. In general seed propagation is not appropriate because of its dioecious nature. Cutting method is easy to produce true to type plants under nursery conditions. For mass multiplication, some work on micro-propagation has been done by Moyo Mack (2009) in *S. birrea* plants .

Seed

The nut is very hard and difficult to germinate without cutting or pre-treatment. Seeds germinated after the operculum is opened which is cut down by saw or scarification. The germination of nut is reported to be takes place during favourable season. Seedlings take longer time for flowering and fruiting than vegetative multiplied plants besides variability in seedlings. Very limited research work has been done on this plant in India since; it is an exotic plant species. Under natural conditions, irregular germination of marula has observed.

Vegetative propagation

It is multiplied vegetative means through stem cuttings. For raising of true to type plants, cutting are suitable methods for multiplication. Vegetative propagation of plant through stem cuttings has become more successful by using growth promoting substances.

Cuttings

The cuttings treated with IBA 1000 ppm gave better results (66.6%) with respect to rooting, survival and growth of plants in marula nut under nursery conditions. Further it was observed that it can be propagated through semi hard wood cuttings under hot arid conditions. Soft wood cutting were sprouted but not survived. In the study conducted at CIAH, Bikaner, low concentration of IBA (1000 ppm) was found effective to multiply marula nut plants through stem cuttings (table 1). February -March is suitable time for planting of cuttings in the nursery for better rooting than July -August.

Table 1 : Effect of Indole Butyric Acid (IBA) on rooting and growth of cuttings in Marula nut.

Treatments	Days taken to sprout	Sprouting (%)	Survival (%)	Number of roots	Length of roots (cm.)	Shoot length (cm.)
IBA 500 ppm	10	86.66	53.33	12.00	31.33	48.00
IBA 1000 ppm	10	93.33	66.66	9.66	38.00	50.00
IBA 2000 ppm	12	93.33	63.33	9.33	50.33	47.33
Dip in Cow urine (1hr)	12	76.66	43.33	8.33	31.50	43.00
Cow urine (12hrs)	15	73.33	46.66	6.66	23.66	41.67
Control	15	66.66	33.33	7.66	23.33	39.67
C.D. at 5%	–	16.23	14.52	4.33	5.79	6.60

Micro-propagation

Plant tissue culture technologies can be useful for *in vitro* multiplication of the plant in the process of domestication and conservation. An ex vitro seed germination study indicated that after-ripening and cold stratification are critical factors. Cold stratification (5 °C) of marula nuts for 14 days improved germination (65%) as compared to non-stratified nuts (32%).

Moyo Mack (2009) achieved direct shoot organogenesis from leaf explants through the induction of nodular meristemoids on Murashige and Skoog (MS) medium and woody plant medium (WPM) supplemented with 6-benzyladenine (BA) in combination with naphthalene acetic acid (NAA), indole-3-butryric acid (IBA) and indole-3-acetic acid (IAA). Induction of nodular meristemoids from 86% of the leaf cultures was achieved on a MS medium with 4.0 mM BA and 1.0 mM NAA. High levels (78–100%) of induction were also achieved on WPM with different concentrations of BA (1.0–4.0 mM) and IBA (1.0–4.0 mM). The highest conversion of nodular meristemoids into shoots on MS initiation medium was only 22% for 4.0 mM BA and 1.0 mM NAA. This was improved to 62% when nodular clusters were cultured in MS liquid medium. The initiation of high numbers of nodular meristemoids per explant provides potential for automated large-scale clonal propagation in bioreactors, in vitro photochemical production and the development of synthetic seed technology, similar to somatic embryogenesis. Plant regeneration through nodule culture has potential for application in mass micro propagation and plant breeding of S. birrea. Adventitious shoot and root induction are important phases in micro propagation. Three auxins (IAA, IBA and NAA) and four aromatic cytokinins (6-benzyladenine, meta-topolin, meta-topolin riboside, and meta-methoxytopolin riboside) were evaluated for their potential to induce adventitious shoot and root formation in *S. birrea* shoots, hypocotyls and epicotyls. Among the evaluated cytokinins, the highest adventitious shoot induction (62%) was achieved on MS medium supplemented with meta-topolin (8.0 mM). The lowest adventitious shoot induction (2.5%) was obtained on MS basal medium containing 2.0 mM meta-methoxytopolin riboside. The highest adventitious shoot induction for hypocotyls was 55% on MS medium supplemented with 8.0 mM meta-topolin. For the tested auxins, IBA induced adventitious rooting in 91% of shoots at a concentration of 4.0 mM after 8 weeks in culture. However, the *in vitro* rooted plants only survived for two weeks when transferred *ex vitro*. A temperature of 25 °C and 16-h photoperiod were optimum for adventitious root induction. *Sclerocarya birrea* leaf extracts had the highest concentration of proanthocyanidins (1.25%). The EC50 values of the extracts in the DPPH free radical scavenging assay ranged from 5.028 to 6.921 mg ml^{-1}, compared to ascorbic acid (6.868 μg ml^{-1}). A dose-dependent linear curve was obtained for all extracts in the ferric-reducing power assay. All the extracts exhibited high antioxidant activity comparable to butylated hydroxytoluene based on the rate of b-carotene bleaching (89.6 to 93.9%). *Sclerocarya birrea* provides a source of secondary metabolites which have potent antioxidant properties and may be beneficial to the health of consumers. *Sclerocarya birrea* young stem and leaf ethanolic extracts exhibited high bioactivity (MIC < 1.0 mg ml^{-1}) against both Gram-positive (*Bacillus subtilis* and *Staphylococcus aureus*) and Gram-negative (*Escherichia coli* and *Klebsiella pneumoniae*) bacteria.

The pharmacological activities observed suggest that *S. birrea* renewable plant parts (leaves and young stems) provide a substantial source of medicinal secondary metabolites. Based on these results, plant part substitution can be a practical conservation strategy for this species.

Quality Standards of Planting Material

There is a tremendous scope of underutilized fruits cultivation in water scarce areas of the country. Under drought conditions, using underutilized plants in horti-pasture and agro- horticultural system are suitable to fulfil the local demand of food, fodder and fuel besides several products of economic uses. The marula tree should be popularized and its genotypes may introduce from available sources. Propagation techniques should also be standardized for large scale production. Planting material should be disease free and of appropriate genotypes.

References

Emmanuel,P.L., Shackleton, C.M. and Boxter, J. S. 2005. Modelling the sustainable harvest of *Sclerocarya birrea* subsp. *caffra* fruits. in the South Africa Lowweld. Forest Ecology and Management, 214: 91-103.

Hall, J.B. 2002. *Sclerocarya birrea.* In. Plant Resources of tropical Africa . Precurser, PROTA Programme ,Wageningen, the Netherlands, pp.127-131.

Moyo, Mack 2009. Micro-propagation and secondary metabolites of *Sclerocarya birrea.* Thesis (Ph.D.)-University of KwaZulu-Natal, Pietermaritzburg,

Muco B.O., Matsumura A, Ishii T and Odee D.W. 2007. Genetic diversity of *Sclerocarya birrea* populations in Kenya. Journal of Arid Environments, 71: 1–11.

Singh, R. S. and Bhargava, R. 2008. Propagation of Marula nut through stem cuttings in arid region. Haryana Jour. Hort. Science, 37 (3&4) 263-264.

Singh, R. S. 2012. Scope and importance of exotic underutilized fruit species under arid ecosystem. In. Compendium of Winter School, CIAH, Bikaner, pp. 11-18.

Chapter-36

Propagation of Less Known Shrubs of Hot Arid Region

J.P. Singh* and V.S. Rathore**
**Central Arid Zone Research Institute, RRS, Jaisalmer*
***CAZRI, Regional Research Station, Bikaner- 334004*

The woody perennial shrubs have been the integral part of the traditional agro forestry system in arid region since ancient times. Inhabitants of the region rely on these species for food, fodder, fuel wood and other day-today requirements since long. Among woody perennials arid shrubs also have a major role in stabilization of drifting sands in the arid ecosystem. Out of about 682 species of Indian arid Zone (Bhandari, 1990), shrubs constitute 70% of the total flora in the Indian desert, yet they cover over 70-80% of desert landscape. Thus shrubs are less in variety, they are ecologically more successful biotypes in the desert environment. The arid shrubs *viz.* Jharberi (*Ziziphus nummularia* (Burm. f.) Wt.), Bawli (*Acacia jacquemontii* Benth.), Phog (*Calligonum polygonoides* L.), Lana (*Haloxylon salicornicum* (Moq.) Bunge), Arni (*Clerodendrum phlomidis* L. f.), Kheep (*Leptadenia pyrotechnica* (Forsk.) Decne), Gangeran (*Grewia tenax* (Forsk.) Fiori), Sania (*Crotalaria burhia* Buch.-Ham.), etc. form the component of natural vegetation cover of western Rajasthan. These species show excellent adaptations to survive under abiotic and biotic stresses of hot arid region. Heavy grazing/ browsing pressure, urbanization, introduction of IGNP coupled with change in cropping pattern brought drastic change in natural vegetation of this region. Combined effect of these factors affected the composition of natural vegetation as well as depletion of important economic plant species of the region (Singh *et al.*, 2002) including native multipurpose shrubs.

In western Rajasthan, some of the shrub species viz., Phog, Lana is among the most threatened species due to over exploitation and their poor natural regeneration in extreme climatic conditions. The species like Phog (*C. polygonoides*), Lana (*H. salicornicum*), however, seed profusely in nature but show limitation in seed germination. In the past, desired research attention was not given to arid shrubs compared to tree species. In recent years need was felt to identify potentially important indigenous shrubs for enhancing the productivity of arid areas particularly under rainfed condition. The plantations of these species on degraded lands particularly on sand dunes not only provide permanent vegetative cover, but also food, fodder, medicine, fuel wood etc. to the inhabitants of the rainfed areas. These shrubs are important for human health because of its medicinal properties.

Few arid shrubs like Lana (*H. salicornicum*) and Phog (*C. polygonoides*), also seem to have potential for ornamental value. Such species can be very well utilized in the gardens and park for ornamental purpose, as they need less water as compared to other conventional ornamental woody species. The introduction of material in suitable habitat will also help in conservation of these shrubs. In the present chapter importance and propagation of some of the less known shrubs of western Rajasthan have been discussed.

Phog (*Calligonum polygonoides* L.) Family : Polygonaceae

It is an important arid shrub of the natural vegetation cover of the Indian Thar Desert. Of the 60 species of this genus only *C. polygonoides* occurs in Indian arid zone. Historically it was the dominant woody shrub of sand dune and sandy plain of the Indian desert (Khan, 1977). It is an indicator species of the sand dune habitat in western Rajasthan and one of the pioneer species, which colonizes deposited sands. It is common on sand dunes, often covering the entire dune; also found in interdunal sandy plains and in agricultural field bunds. It is extremely drought hardy species, which tolerates extremes of temperature, moisture scarcity and strong wind regimes. This nearly leafless much branched shrub attains 1.5 to 3 m height. Young branches are green and shining and stems are modified as phylloclades. Flowers are small, numerous, light pinkish with a pleasant odour. Fruits are oblong and densely clothed with reddish brown bristles. At the time of flowering, it absorbs a very large amount of dew, immediately after absorbing dew the plants begin to flower (Bhandari, 1988). After exploitation for the fuel and charcoal production, this species face extinction and is listed as endangered. Singh

(2004) emphasized that surface vegetation affects the productivity of *C. polygonoides* altering availability of soil resources. Proper management to ensure water availability will be beneficial to enhance the productivity and population of this species under afforestation programme. Conservation of this species is essential to combat desert in arid region.

Nutritive Value and Uses

Food: The flower buds locally called as *Lasson* or *Phogla*, are used by inhabitants particularly during summer months. Usually inhabitants collect flower buds during spring/early summer and utilize during summer after light boiling or frying in ghee and mainly used in preparing traditional dish called *Rayata* by mixing with curd. The preparation is supposed to avoid heat strokes during peak summer. The flower buds are rich source of crude protein (16.87 %), total carbohydrate (71.12 %) with low fibre content (9.07 %). Detailed account of ethnic uses of Phog was dealt by Singh *et al*., (2004) and Beniwal *et al*., (2005).

Fodder: The young green and fleshy branches are browsed by camel in the rainy season by sheep, goat and cattle. Green/dry phylloclades of Phog locally known as *Lasu* having good digestibility is considered as potential top feed (Gaur *et al*., 1982) particularly for camel. Saxena (1988) opined that Phog is good fodder for camel in the arid zone and emphasized that as the twigs contains only 7.4 % crude protein can be mixed with feed materials such as guar or moth husk and fruiting tops of Lana (*Haloxylon salicornicum*). Chemical analysis of shoots made by Bishnoi and Gautam (1991), Dhir *et al*., (1984) and Sharma *et al*., (1984) indicated that it content fairly good amount of nutrients. At flowering stage, it is a good source of nutrients particularly for milching animals.

Medicinal value: The branch decoction after boiling is used as gargle for sore gums. Root decoction mixed with catechu is also used as gargle (Kirtikar and Basu, 1935). The flower buds are supposed to have a cooling effect and are used against sun stroke (Sekhawat, 1961). The juice of branches is also used as an antidote when the milky juice of Aak (*Calotropis procera*) enters into eyes (Bhimaya *et al*., 1961).

Agricultural value: Traditionally its branches are spread as mulch during summer in the crop fields by the farmers (Bhimaya *et al*., 1961). The farmers also claim the higher yield of Bajra (pearl millet) grown in a field, which carried a good stand of Phog plants during the previous year. The farmers also informed the increases of groundnut yield by addition of the foliage of Phog. The farmers also allow growing its seedlings in the crop field and afterwards plough them as green manure.

Other uses: Phog is also used as chief source of domestic fuel wood, live hedge, and building materials for the huts in arid region.

Propagation

Natural regeneration in Phog takes place by self seeding and stumps of young green Phog has good coppicing ability. In natural condition, it bears good fruiting but has limitation in seed germination. Firstly the seeds are covered with sand due to heavy storms and secondly high temperature in rainy season (July –August), affects adversely the seed germination. Bishnoi and Gautam (1991) reported that $20 \pm 2\,^{\circ}C$ is the most optimum temperature for germination and favorable condition it starts after five days of sowing. So winter rains are supposed good for seed germination in natural condition. Phog can be propagated either by direct seed sowing, nursery raised seedlings, or by stem cuttings and layering.

Direct sowing: On sand dunes, seeding at a depth of about 10-15 cm exhibited good germination (Kaul, 1965 b). Penetration of fruits into the sand in the right season and suitable temperature appears to be decisive factors for germination under field conditions of Phog.

Nursery raised seedlings: Nursery tests have shown that Phog seeds have average germination capacity with an average germination of 50 per cent when sown in sand and covered with a layer of 2 cm deep sand mulch. Pretreatment of seeds is not required before sowing, November - December is supposed the best sowing time for good germination in nursery. In nursery condition, 6-7 month old seedling varied from 30-60 cm in height July-August is the best time of transplanting. Transplanted plants show very fast growth during first year and then progressively it slows down.

Stem cutting: Generally the stem cuttings of Phog are planted in July in rainy season. An intensive study on vegetative propagation of Phog was carried out by Kaul, (1965a). Stem cuttings having 0.7-1.2 cm or 1.2-1.7 cm performed better in terms of sprouting and rooting compared to that of 0.5-0.7 cm diameter. Similar trends were observed for rooting of cuttings. Application of 0.8 per cent IBA significantly promoted rooting of stem cuttings. Branch cuttings, irrespective of their diameter, when planted horizontally or in an inverted position failed to produce roots. Some leaves should be retained on them for their easy rooting thus obviating the need for application of a plant regulator which in any large scale-planting programme is not economically feasible (Kaul, 1965a).

Lana (*Haloxylon salicornicum* (Moq.) Bunge) Family: Chenopodiaceae

The species of *Haloxylon* are well adapted to arid environments and develop well fibrous root system. Lana, which is a succulent dwarf chenopod shrub, grows naturally in sandy undulating hummocky plains, dunes, interdunal plains particularly in Bikaner and Jaisalmer districts of western Rajasthan. It grows 1- 1.5 m height with crown spread of 1.0-2.0 m with pale yellow jointed stem and branches Flowers are arranged in alternate pattern on long erect spikes. Seeds are small with perianth attached to it.

In natural habitat, Lana exhibits wide range of diversity in plant types, flowering and fruiting patterns. During flowering stage, a variety of colors in perianth viz., creamy white, light pink, pink, light pinkish white etc. were observed in different plants. Similarly a range in perianth size viz., small, medium and large was also observed at different sites. Germplasm evaluation of different collections of Lana from Bikaner and Jaisalmer districts also showed wide variation in the plant types among and within accessions. Some of them exhibited ornamental look during both vegetative and flowering stages.

Nutritive Value and Uses

Food: In the past, the inhabitants of Lana dominated area utilized its seeds as a article of food. It was much used during times of scarcity and drought. During the famine seeds are used with husk. Some of the villagers still used its seeds for preparing bread especially the old aged persons liked it much.

Fodder : Lana is being used to fed camel, sheep, goats and sometimes to cattle also. The flowering tops called *Fuli* are harvested in the month of November-December and stored for feeding the animals. Its flowering /fruiting tops make useful fodder supplements. It is believed that milk production in cattle increases after feeding its *Fuli* (fruiting tops) along with husk cluster bean and *Lasu* of Phog (*C. polygonoides*).

Fuel wood: Its dried branches are used as fuel wood in the area of its natural occurrence. Inhabitants of the area ranked it as fairly good fuel wood as it gives intense heat and less smoke.

Other uses: Plant ash is used to cure skin diseases. Plant ash is also applied on the body of sheep and goats to kill lice and other external parasites. The natural stands of Lana are also the source of edible mushroom to the inhabitants.

Propagation

Natural regeneration of Lana takes place by seeds under favorable moisture and temperature conditions. In general good regeneration observed during rainfall in January-February. It is propagated through seeds by nursery-raised seedlings and also reported from stem cutting (Kaul *et al.*, 1992). High predation and poor longevity (less than one year) of seeds are reported in *H. saliconicum*.

Nursery raised seedlings: The freshly collected seeds are sown in the polythene bags in the month of Feb.- March. Seed treatment is not required before sowing , however, seeds with perianth have higher rate of germination compared to seeds without perianth. The seedlings are much affected by insects after germination. Much care is required in nursery by applying suitable insecticide just after sowing to prevent the insect attack. The seedlings should be transplanted in July-August.

Stem cutting: Vegetative propagation of Lana by stem cutting is reported by Kaul *et al.* (1992a). They reported that shoot cuttings of the length of around 10 cm are capable of sprouting even without hormonal treatment. There was no additional benefit of treatment with IBA on rooting induction. The rooting induction was greatest in 10-cm shoot. Generally the stem cuttings are planted in July in rainy season.

Kheep (*Leptadenia pyrotechnica* (Forsk.) Decne.) Family: Asclepiadaceae

Kheep is also known as khimp or khimpoli in desert areas. This is an erect, much-branched often leafless, shrub up to 1.5 m high (more in protected sites), with watery sap. Leaves are 2.5-6.5x 0.2-0.3 cm, sessile, linear or linear lanceolate and caduceus. Flowers are greenish yellow in lateral umbellate cymes. Follicles are 7.0-14.0 cm long, terete, lanceolate, tapering to a slender beak, and glabrous. Seeds are 5-7 mm long, ovate lanceolate and glabrous. It is common throughout the western Rajasthan particularly on sandy areas.

Nutritive Value and Uses

Food: The unripe pods known as *Khipoli* are used as vegetable and supposed very tasty and nutritious dish.

Live hedge: It is used as strong live hedge around the field boundaries. It is supposed to check the soil erosion and also the protection from wild animals.

Medicine: It is traditionally used as an expectorant. In the villages, its juice is used to melt/ remove the spines from the body. Its paste is applied on boils and blisters. In the villages, it is believed that the bed of green branches of Kheep is supposed to cure any kind of fever.

Material culture: Since long it is extensively employed for thatching purpose. In western Rajasthan it is used for making huts, brooms and strong cordage /ropes.

Fodder: Generally, it is not used as fodder. However, it is browsed by camel in times of drought. In the Sind Gazetteer it is described as one of the most important fodder for horses, cattle and camel. It is largely used as camel fodder in Sibi.

Fuel wood-: Presently it is one of the species largely used as domestic fuel wood in the western Rajasthan due to shortage of quality fuel wood like Phog (*C. polygonoides*), Khejri (*Prosopis cineraria*) etc

Propagation

Seeds of Kheep are easily dispersed by winds due to their lightweight. It can be established by direct seed sowing in the monsoon. No research work has been done so far for its propagation through vegetative means.

Epilogue

Hot arid region represents a range of diversity in native shrubs, which occur on a wide range of habitats. Some of these plants grow in extreme desertic habitat and survive in harsh climatic conditions. The diverse material of indigenous multipurpose little known shrubs offers a good scope for introduction in the farming system. This will be particularly useful on those areas where some of the species like Phog (*C. polygonoides*), Lana (*H. salicornicum*), Gangeran (*G. tenax*), Kheep (*L. pyrotechnica*) *etc* make the dominant cover even in times of drought. These species with food and fuel wood value also provide the nutritious feed to animals during forage scarcity and drought period. Emphasis should be given introduction of these shrubs in agri-silvi-horti pasture, Agri-horticulture and other alternate land use systems. These shrubs have multipurpose uses and will be a useful introduction in agro forestry system particularly when the systems are now in greater demand of diversification towards side income generating components. Therefore, research attention is required to standardize the propagation techniques of indigenous less known shrubs of hot arid zone.

Research has started at Department of Botany, J.N.V. University Jodhpur, on development of micro propagation methods for cloning and conservation of germplasm of arid shrubs and attempts have been made for generation of know hoe of in vitro culture of *Calligonum, Leptadenia* and *Haloxylon* species. There is also an urgent need for conservation of less known arid shrubs species, which are extensively exploited from their natural habitats.

References

Beniwal, R.K. , J. P, .Singh and V.S. Rathore 2005. Phog (*Calligonum polygonoides* L.) An important arid shrub of rainfed farming system in western Rajasthan. pp 91-96. In: Pratap Narain, Manjit Singh, M.S. Khan and Suresh Kumar (eds.). *Shrubs of Indiana Arid Zone*. Arid Agro-ecosystem Directorate, Central Arid Zone Research Institute, Jodhpur.

Bhandari, M. M. 1988. Life support species for heat tolerance. pp 60-64.In: R. S. Paroda, P. Kapoor, R. K. Arora and Bhag Mal (eds.) Life Support plant Species; Diversity and Conservation. NBPGR, New Delhi.

Bhandari, M. M. 1990. Flora of Indian Desert. (Revised Edition) .MPS Repros, Jodhpur.

Bhimaya, C. P., A. B. Bose and S. P. Malhotra 1961. The human factor in relation to trees and shrubs in a village in arid parts of Rajasthan. Indian Forester 87: 614-617.

Bishnoi, S. and D.D. Gautam. 1991. The reproductive capacity and nutritive value of *Calligonum polygonoides* Linn. Annals of Arid Zone 30: 139-144.

Dhir, R.P., B.K. Sharma and B.K. Dutta. 1984. Mineral nutrient elements in natural vegetation of arid Rajasthan : I Macro elements. Annals of Arid Zone 23: 111-117.

Gaur, D.D., G.R. Purohit and C.S. Mathur. 1982. A note on the formulation of least cost balanced ration for the Indian desert. Annals of Arid Zone 21: 263-265.

Kaul, A. , Suresh Kumar and V. Shankar 1992 . Vegetative propagation of *Haloxylon salicornicum* (Moq.) Bunge. Annals of Arid Zone 31: 227-228.

Kaul, R.N. 1965a. Rooting behaviour of *Calligonum polygonoides* Linn. Annals of Arid Zone 4: 93-95

Kaul, R.N. 1965 b Silvical characteristics of *Calligonum polygonoides* Linn. La-Yaaran 15:91-96.

Khan, T.I. 1977. Conservation of biodiversity in western India. Environmentalist 17, 283-287.

Kirtikar, K. R. and B.D. Basu. 1935. *Indian Medicinal Plants*. IV Volumes. Lalit Mohan Basu, Allahabad.

Saxena, S. K 1988. 'Lasu' an exclusive top feed for camels during drought. Indian-Farming. 1988, 38: 9, 19-20.

Sharma, B.K., R.P. Dhir and B.K. Dutta 1984. Mineral nutrient elements in natural vegetation of arid Rajasthan: II Micro elements. Annals of arid Zone 23: 235-241.

Shekhawat, S. S. 1961. *Calligonum polygonoides* L. -A very useful shrub in Bikaner. Science & Culture 27: 40.

Singh, J.P., R. K. Beniwal, B.B.S. Kapoor and N. D. Yadava. 2002. Herbaceous medicinal plants of the western Rajasthan. pp. 165-187. In: B.B.S. Kapoor, Ahmed Ali, S. K. Mathur and Satish Kaushik (eds.) Advances in Resource Management of the Indian Desert. Madhu Publications, Bikaner.

Singh, J. P., R. K Beniwal, M.L. Soni and B.C. Mondal. 2004. Ethnic uses of Phog (*Calligonum polygonoides* L.) and its germplasm conservation. Range Management and Agroforestry 25: 43-47.

Singh, G. 2004. Influence of soil moisture and nutrient gradient on growth and biomass production of *Calligonum polygonoides* in Indian desert affected by surface vegetation. Journal of Arid Environments 56: 541-558.

Chapter-37

Propagation of Seed Spice Crops

S.K. Malhotra
ADG (Horticulture), ICAR, KAB-II, New Delhi- 110012

Seed spices, an important group of horticultural crops, are defined as vegetable products or mixture thereof, free from extraneous matter, primarily used for flavouring, seasoning and imparting aroma of foods item and beverages. Besides food industry, the seed spices have medicinal properties and thus are used in various pharmaceutical preparations and cosmetics (Malhotra and Vashishtha, 2008). India is known as the home of spices and produces a wide variety of seed spices- coriander, cumin, fennel, fenugreek, ajowan, dill, nigella, celery, anise and caraway. Over the years a good number of high yielding varieties have been developed under AICRP on Spices of ICAR by SAU's and National Research Centre on Seed Spices. India enjoys the status of largest producer of seed spices in the world but the productivity in the country is very low and is 25-30 % of the total potential of the crop varieties. There is a considerable gap between achievable and realized productivity, hitherto the National average yield has not registered the significant rise. Some of the obvious reasons are cultivation of seed spices on traditionally marginal lands with poor management of soil-fertility, irrigation, fertilizer and other biotic and abiotic factors. Furthermore, *Apiaceae* family seed spices - coriander, cumin, fennel, dill, ajowan, celery and anise have inherent weakness of slow germination, resulting in poor crop stand, severe competition from weeds and low productivity. Moreover proper emphasis has not been given on propagation aspect of seed spices for improving the germination and crop stand at field level. Here the mention of propagation of seed spices is followed by the botanical description of these crops.

Nutritive Value and Uses

Seed spices have nutritive value and medicinal properties. Seed spices are commonly used for aroma of food items and garnishing. It has also volatile substances. It's essential oil is used in Ayurvedic and Unani medicines. Seed spices are used for making cosmetics and several other value added products.

Coriander- Its green leaves and fruits have a pleasant aromatic flavour and thus is used in Indian cookery and garnishing foods. It is popular as garnish for salad, sandwiches and cooked dishes as a flavouring to soups. The leaves are used to prepare delicious chutney and curries. The seeds (botanically fruits) are used as condiments in preparation of curry powder, pickling soups and seasonings. Its dried stem/leaves are also used for insect control in storage grains. In the food industry, coriander has big uses in flavouring pastry cake, biscuits, chewing gums, tobacco products and beverages. Its seed are also used as mouth freshener. Because of its medicinal properties, coriander is used in ayurvedic and unani medicine system. Sun dried leaves are also used for flavouring food items in off-season.

Cumin- It is used as condiment and as an essential ingredient in all mixed spices, curry powder, pickle, etc. Seeds are also used for seasoning bakery products such as bread and cakes. It is commonly used in confectionery, meat, sausage and bread manufacturing and as food preservative in food processing. The essential oil is used in perfumery, cosmetics and flavouring beverages.

Cumin is widely used as traditional medicine to treat diarrohea, digestive disorders, colds and fevers and in treatment of wounds. It is also used in veterinary practices.

Fennel- its dried fruits are aromatic and used for flavouring soups, meat, souses, bread rolls, pastries and confectionery. The seeds also have used for flavouring liquors and in the preparations of various types of pickles. The leaves are used in fish sauce and for garnishing, leaf stalks are used both as vegetable and as salad. Chinese, Indian and Egyptian have used it as a condiment. The essential oil is used as flavouring agent in various food items, pickles, perfumes, soaps, liquors, cough drops and liquorice candy. Due to having antimicrobial properties, it is used in cosmetic creams, body lotions, and moisturizers.

Fenugreek- Every part of this crop is utilized as food, fodder, medicine and cosmetics. Its green leaves are used as vegetable and sundried leaves which have aromatic qualities are used as spice, for seasoning of variety of foods in off-season. The seeds are used for making dye and extraction of alkaloids or steroids. Its leaves and seeds are used in cosmetics and hair conditioning. Its seeds are used extensively for medicinal purposes because of its hypoglycemic and hypocolestremic properties. It is used in cattle feed and veterinary medicines.

Botanical Description

Out of the ten main seed spices, the coriander, cumin, fennel, ajowan, dill, celery, anise and caraway belongs to the carrot family *Apiaceae* while fenugreek and nigella are to family *Fabaceace* and *Ranunculaceae,* respectively. The prominent seed spices, their English, Hindi and botanical names, family and centre of origin are depicted in Table 1. The botanical description of seed spices in brief is described here crop-wise (Malhotra, 2004; Malhotra, 2005). The plant part of *Apiaceae* family seed spices crops having economic value is botanically fruit, but here in the text in place of word fruit, seed has been frequently used, as it is perceived easily. Most of seed spices crops bear protandrous flowers and cross pollination occurs through insects except fenugreek which is highly self-pollinated in nature.

Table 1 : Seed spice crops and their centre of origin.

Seed Spices			Family	Centre of origin
(Common name)	(Hindi name)	Botanical name		
Coriander	*Dhania*	*Coriandrum sativum L.*	*Apiaceae*	Mediterranean region
Cumin	*Jeera, Safed Jeera*	*Cuminum cyminumL.*	*Apiaceae*	Mediterranean region
Fennel	*Saunf*	*Foeniculum vulgare Mill.*	*Apiaceae*	South Europe and Mediterranean region
Fenugreek	*Methi* *Kasuri ethi*	*Trigonella foenum-graecum* *Trigonella corniculata L.*	*Fabaceae*	South East Europe and West Asia
Ajowan	*Ajowan*	*Trachyspermum ammi Sprague*	*Apiaceae*	India and Egypt
Dill	*Sowa*	*Anethum graveolens L.* *Anethum sowaKurz.*	*Apiaceae*	Europe, Africa and Asia India
Celery	*Ajmoda, Karnouli*	*Apium graveolens L.*	*Apiaceae*	Mediterranean region
Aniseed	*Vilayati Saunf*	*Pimpenella anisum L.*	*Apiaceae*	Eastern Mediterranean region
Caraway	*Siah Jeera*	*Carum carvi L.*	*Apiaceae*	South Europe and Mediterranean region
Nigella	*Kalongi, Mangrail*	*Nigella sativa L.*	*Ranunculaceae*	Eastern Mediterranean region

Coriander (*Coriandrum sativum* L.) **:** Coriander also known as dhania is an annual aromatic herb, with erect and branching leaf, stem, but the leaves are alternate and pinnatified. Inflorescence is terminal compound umbel, the flowers are zygomorphic and hermaphrodite. Flowers at periphery are bisexual whereas the inner ones are unisexual white or rose coloured with five calyx teeth and a short conical style. Calyx is tubular, adnate to ovate at base with 5 lobes and corolla are 5, spreading, stamens 5 in number and ovary two celled. Fruit is schizocarpic round to globular, yellowish brown at maturity and green when young and ribbed is composed of two concavo-convex mericarps.

Cumin: Cumin also called jeera is a small and slender annual herb, which grows to a height of about 45-50 cm with many branches and linear dark green leaves. The stem has 3-5 primary and 2-3 secondary branches. Flowers white or pink in small compound umbel of 2-3 cm in diameter and one compound umbel contains 4 -7 umbellates. The flowers are regular and bisexual having five sepals, teeth like, subulate, unequal, petals 5, oblong or ovate emarginate, white and often equal. The gynoecium is bicarpellary, syncarpous, ovary inferior and bilocular. The stamens are 5, polyandrous, filaments slender and anthers basifixed. Fruits are oval, 6 mm long and grey brown in colour.

Fennel: It is a stout glabrous herb of 1.5 to 1.8 meters height and is cultivated as an annual crop with a hollow cylindrical stems which have parallel visible veins. Leaves arc stalked, petioles very long, dilated for the greater part of its length into an amplexicant base and memberaneous margins. The flowers of fennel are small, yellow, arranged in compound terminal umbels. The flowers are small ebracteate pedicellate, regular, actinomorphic, complete, bisexual, pentamerous, epigynous. The flower has five sepals, gamosepalous, yellowish green and five petals, polypetalous, valvate, yellow. The stamens are five, polyandrous, alternating the petals, basifixed, introrse and yellow but the gynoecium is bicarpellary, syncarpous, ovary inferior and bilocular. Style extremely short, stigma bilobed, arising from the honey disc (stylopodium) capitate. The fruit cremocarp breaks up into 2 mericarps.

Fenugreek: Fenugreek also known as methi is an annual herb with the height of 45-80 cm. Leaves are light green and are pinnately trifoliate. The flowers are papilionaceous. There are two species of genus *Trigonella,* common methi- *Trigonella foenum-graecum* and kasuri methi- *Trigonella corniculata.* The later one is a slow growing type and remains in rosette condition during most of its vegetative growth period. The flowers are yellow and very showy with sickle shaped pods in bunches bear small sized seeds. The common methi plant bears creamy white flowers solitary or paired and long slender pods with prominent beak. The common methi is quick growing and produces mostly upright shoots, whereas kasuri methi is slow growing initially. Leaves of methi are used as fresh green vegetable as well as dried leaves.

Ajowan: Ajowan is an annual herbaceous plant. The plant is profusely branched with height of 60-90 cm, small, erect with soft fine hair. It has many branched leafy stems, feather like leaves, 2-3 pinnately divided, segments linear, with flowers terminal and compound. The minute greyish white fruits are ovoid in nature. Each mericarp has 5 ridges and odour is similar to thyme.

Dill: Dill or sowa is an annual, occassionally biennial herb, erect in habit, growing to a height of 90-120 cm. Leaves are pinnately compound, having clasping base. The stem is smooth, shiny, hollow, grows upright and bears flat, terminal compound umbels, with numerous yellow flowers, whose small petals are rolled inward. The flowers are regular, bisexual, pentamerous and dichogamous. Fruit consists of two brown, broadly oval, compressed, indehiscent, one seed carpels (mericarps).

Celery: Celery is a herbaceous annual or biennial erect herb growing to a height of 60-90 cm with conspicuous branches bearing well developed leaves on long expended petioles. Stems are branched, angular or fistular, conspicuously jointed. Leaves are radical, pinnate, deeply divided into three segments, once or twice divided toothed at apex. The flowers are small, white in colour and inflorescence is a compound umbel. Calyx teeth are obsolete; petioles 5, ovate, acute with tip inflexed; carpels semiterete, subpentagonal, primary ridges distinct, filiform. The fruit is a schizocarp with 2 mericarps, suborbicular to ellipsoid, 1-2 mm in diameter, aromatic and slightly bitter.

Anise: It is an annual herb with an erect and branching stem. The lower leaves are broadly triangular, incised serrate and the upper are usually trifid. The plant grows to the height of 50-75 cm and has small lacy leaves. The flowers are small and white, arranged in large compound umbels. A single plant produces one to six clusters, each bearing 6 to 10 seeds. The seed is oval in shape about 5mm long with a short stalk laterally compressed and 5 longitudinal ridges on the mericarp. The two carpels containing the seed are hairy, aromatic, greenish brown to greyish brown in colour.

Caraway: The caraway plant is an erect, herbaceous, annual or biennial herb. The plant height varies from 0.5 - 1.0 m. The stem is cylindrical, the leaves are pinnately compound and ultimately segments of lower leaves are lanceolate. Flowers are minute borne in terminally or axillary compound umbels producing cluster of white flowers. The flowers have bracts 1-3, small, linear or none; calyx teeth 5, small or none; petals 5, notched, often enlarged and erect. Fruits are brown, cremocarp, 3-6 mm long, ovoid or oblong, glabrous and laterally flattened.

Nigella: The nigella plant is the small herb of 30-60 cm height. The leaves are 2.5 –3.0 cm long cut into linear lancelet segments; flowers bluish white; seeds trigonous, regulose, angular, small funnel shaped 0.2 cm long and 0.1 cm wide, black in colour, odour slightly aromatic and bitter in taste. Flowers solitary and terminal, beautiful, sepals 5, petaloids, corolla absent, stamens numerous, carpels 5, partially united, fruit a capsule, nectaries many, generally 10, pocket like and epicalyx present.

Propagation in Seed Spices

The major objective of propagation is reproduction of more plants from a source with the qualities like parent. The seed spices crops being annual in nature are propagated through seeds, the simplest and cheapest method of raising the plants. But successful sowing is very important for propagation of the crop and is profusely influenced by certain factors involving purpose of propagation, propagation techniques consisting of propagation methods, sowing depth, plant spacing, and environmental factors. The possible means for improvement of seed germination and crop stand in the field are also suggested.

Purpose of propagation

There are four main purposes of seed spices propagation.

1. **Maintenance of Plant Genetic Resources:** Collection of germplasm and its evaluation is pre-requisite for utilization of plant genetic resources in crop improvement programme. In the endeavor of collection of germplasm, the crop expeditions are done by the scientists and thus rare germplasm materials initially collected in small quantities are propagated with care for evaluation, maintenance and conservation with the ultimate goal to preserve the original characteristic of the plant type. The seed propagation thus eases the process for breeding new plant type and conserving the gene pools. In this context knowledge of seed viability, its storage, time of sowing, factors responsible for germination and care of germinating seedling is essential.

2. **Quality Seed Production:** Once the new varieties are released, it becomes necessary to make available the quality seed to the growers. In the seed multiplication process, the original basic seed goes through the stages of increase and certification. The main stages for raising the propagating material are breeder seed, foundation seed and certified seed. The minimum seed certification standards and necessary guidelines on quality seed production of seed spices have been detailed by Malhotra and Mehta, (2003).

3. **Organic Seed Production:** Seed taken from organically grown, healthy, vigorous plants is organic seed. The organic seed is required for growing an organic product of agricultural crops including vegetable crops, spices and herbs. The seed spices produced organically have its ready market mainly in Germany, The Netherlands, USA and Japan. The organic seed is basic and important requirement as propagating material for raising the organic crop of seed spices crop. The general and specific guidelines for organic production of seed spices have been documented by Malhotra and Vashishtha (2004). In addition to above guidelines, the quality seed production principles are also followed strictly.

4. **Commercial Crop Production:** In the changing paradigm of agricultural production, the seed spices crops have emerged as high value crops and quality seed have become the pre-requisite in successful raising of crops. The goal in propagation in commercial production is to raise and increase the source material for further harvesting the economic part of commercial value which is seed in case of seed spices group of plants. The crop is grown mostly for the seed of seed spices which possesses market demand particularly for whole seed or value added items prepared out of it viz. seed spices essential oil, oleoresin, seed spices water (by product after extraction of essential oil), seed spices powder and curry powder from cumin, coriander, fennel, fenugreek, ajowan, dill, caraway, celery and anise. The seed spices with dual purpose use are coriander, fenugreek, dill and celery, also grown for green leaves which are consumed raw either as vegetable after cooking solely or in mix with other vegetable crops or for garnishing and flowering of salad mixtures.

Propagation Techniques

1. Sowing Method

There are two methods of establishing or propagating seed spices crops -direct or indirect seeding. The seed spices are mostly grown through seeds in the outdoors. The direct seeding is practiced for growing crops of coriander, cumin, fennel, fenugreek celery, caraway, anise, nigella, ajowan and dill, but fennel and celery can be grown through seeds as well as through transplants. The primary propagation method of seed spice crops is enlisted in Table 2.

Direct seeding

Direct method of sowing involves one step in which seeds are sown in the soil directly where they will germinate, grow and go through the entire reproductive cycle. The seeds are easy to handle, transport and store and are thus convenient to use, moreover, seeds are adaptable to mechanization particularly for fenugreek, coriander, fennel and dill. The seed spices crops are easy to propagate through broadcasting, line sowing or dibbling.

The seed spices crops with small size of seeds viz. ajowan, celery, parsley require extra care for maintaining finer tilth of soil and shallow sowing depth. A seed may likely to fail in germination, if not sown at right depth or did not contact with soil to imbibe water. The sowing of seeds outdoor in the field is called direct seeding, is used for most of the seed spices crops. Since the crop is seeded where it will complete its life cycle, thus it is important to choose the site carefully. It should have reach of enough sunlight, water drainage and soil should be free from clods, stones, debris etc.

Indirect seeding

Indirect seeding method involves two step process of propagation of plants in the field. In this method, the seed is sown in nursery, kept till the plants attain seedling stage. Fennel, celery, parsley crops are commonly raised through transplants of about 50- 60 days age, the seeds are sown indoor or outdoor for raising nursery and later transplanted in the main field. Healthy seedlings are transplanted in the well prepared field. The major advantage lying in indirect seeding method are raising and transplantation of healthy seedlings ensuring to the maximum of establishment of plant in the field. Moreover, transplantation method helps in shortening the field growing period. The seedlings can be grown indoors for several weeks during adverse climatic conditions. Though some disadvantages on the part of increase of additional production activity/ costs, being not fit for mechanized planting and propagating material becomes bulky in handling.

Table 2 : Primary propagation methods for seed spices.

S.No.	Seed spices crops	Primary method of propagation
1	Ajowan*	Seed*
2	Anise*	Seed*
3	Caraway*	Seed*
4	Celery*	Seed* and transplants
5	Coriander*	Seed*
6	Cumin*	Seed*
7	Dill*	Seed*
8	Fennel*	Seed* and transplants
9	Fenugreek	Seed
10	Nigella	Seed
11	Parsley*	Seed* and transplants

* Botanically fruits

2. Micro-propagation

Micro-propagation refers to propagation of plants from very small plant parts, tissues grown aseptically in a test tube or container under controlled nutritional and environmental conditions. The propagation of seed spices is rather easy and economic by seed propagation but here micro-propagation can be practiced for *in vitro* propagation of rare plant genetic resource and induction of somatic embryogenesis and plant regeneration from hypocotyle tissues in seed spices crops from callus and production of somaclones. Micro-propagation protocols have been developed and somatic embryogenesis and plant regeneration from hypocotyl tissue has been reported to be successfully transferred in Ajowan (Sehgal and Abbas,1994), dill (Ratnamba and Chopra,1974), celery (Nirmal Babu *et al.*, 1997), aniseed (Sajina *et al.*, 1997a), parsley (Nirmal Babu *et al.*, 2001), fennel (Ravindran *et al.*, 1996), fenugreek (Settu *et al.*, 1997) and coriander (Stephen and Jaybalan, 1998).

Sowing Depth and Spacing

The soil should be worked to a depth of about (25-30 cm) to assure suitable soil condition for root growth. The seeds should be sown at a depth and spacing recommended in the package. If sowing directions are not given, the general rule is that seeds should be planted one and one-half times as deep their diameter. Spacing between seeds should be adopted as the recommended practices. Timing is very important when sowing seeds outdoor. If planted too early, the seeds will fail to germinate or germinate inadequately. If sowing is done late, plants may not mature with in the growing season. It is therefore, necessary to follow the package directions (Table 3). Seeds must be covered lightly after sowing. The fine seeds of ajowan, celery and parsley should be mixed with sand before sowing. It helps in giving a uniform seed spread and in controlling the depth of seeds in furrows during sowing operation.

Table 3 : Sowing depth and spacing for seed spices crops.

Seed spices crops	Maximum sowing depth (in cm)	Row x plant spacing (in cm)	Sowing time
Coriander	2.0-2.5	25-30 x 15	October- November
Cumin	1.0-1.5	22.5-25 x 10	Mid November – First week of December
Fennel	1.5-2.0	45-60 x 20-25	Early season: July (Nursery sowing) :August (Transplanting) Main season: Mid September- First week of October
Fenugreek	2.5-3.5 (Common Methi)1.0-1.5 (Kasuri Methi)	25-30 x 10	Mid October- Mid November
Ajowan	1.0	30 x 20	Early season: July-AugustMain season: October
Dill	1.5-2.0	30-45 x 20	Early season:July-AugustMain season: October
Nigella	1.5-2.0	30 x 15	Mid September- Mid October
Anise	1.0-1.5	30 x 15	October
Celery	0.50-0.75	30 x 15	September- October (Nursery sowing) November- December (Transplanting)
Caraway	1.5	30 x 20	Mid October- Mid November

Environmental factors

Germination involves physiological and biochemical processes, thus environmental conditions must be adequate for seeds to germinate. Such critical factors are moisture, sowing time/temperature, light and air.

Moisture is needed to initiate the enzymatic breakdown of food reserves in seeds during the course of germination. The seed beds should be watered immediately after sowing if initial moisture is less. Excessive moisture is undesirable. Seeds must imbibe moisture to a certain degree for the germination process to be initiated. The seeds will die if allowed to dry after imbibing water or initiating sprouting. Thus it is important to keep the soil surface moist throughout the germination period until seedlings emerge.

Temperature regulates seed germination, because it controls rates of biochemical reactions. The preferable temperature for getting the good germination and time taken to germination in seed spices is given in Table 4. Keeping in view the preferable temperature requirements, the sowing time for different seed spices crops is decided. Similarly the time taken to germination minimum and maximum, gives an idea for paying attention for maintaining optimum moisture till getting he desired level of germination. Most of seeds in seed spices do not need light to germinate, but to fulfill this requirement care should be taken for shallow sowing of celery and parsley. In most of the species, germination is an aerobic process. The low oxygen level of soil air is inhibitory, thus well drainage for good aeration is necessary. Fertilizers are not required until the seeds germinate. The initial addition of organic manure is sufficient to provide nutrients to the new plant; moreover, stored food in the cotyledon is usually adequate for the initial period of germination.

It is therefore, important to note that above environmental factors work together for good germination and must occur in balance to achieve good germination, crop stand and high yield levels in seed spices crops.

Table 4 : Preferable temperature requirement and time taken to germination for seed spices.

S.No.	Seed spices crops	Preferable temperature requirement for germination(^{0}C)	Time taken to germination (days)
1	Ajowan	18-20	10-12
2	Anise	15-20	10-12
3	Caraway	12-15	10-12
4	Celery	10-12	15-20
5	Coriander	14-20	10-12
6	Dill	16-22	8-10
7	Cumin	14-20	8-10
8	Fennel	16-18	8-10
9	Fenugreek	14-18	5-10
10	Nigella	20-25	8-10
11	Parsley	10-12	15-20

Improvement in Germination

The seed once sown under optimum conditions, should ideally germinate within an expected period that is characteristics of the species/variety/type. The uniformity of germination and adequate crop stand in the field involves important field operations in realizing the profitable commercial crop of seed spices. Among seed spices particularly *Apiaceae* family crops have been reported to exhibit poor crop stand due to slow germination, dormancy and sudden shift in climate from optimum to abnormal. To help seeds germination, certain measures can be employed for improvement in following ways.

Overcoming Dormancy

- The seed spices crops have been reported to have less dormancy problem, except for celery, having photo-dormancy and seeds only germinate after fulfillment of light requirement. Thus in case of celery, shallow sowing should be done to allow the contact with sunlight (Desai *et al.,* 1997)
- Furthermore, coriander seed botanically schizocarp should be rubbed for splitting into two parts called mericarp before sowing, not only shall improve germination but also give birth to two plants.

- The soaking of seeds has helped hastening the germination process in most of the seed spices crops. Soaking of seeds in water 8-12 hours in coriander, fennel, ajowan and 18-24 hours in cumin and dill before sowing followed by drying in shade, improves germination by leaching out of chemical inhibitors such as coumarins commonly found in the seeds of *Apiaceae* family seed spices.
- Pre-soaking of seeds in 1% KNO_3 or 1% K_3PO_4 or tri-sodium phosphate or ethyl alcohol or potassium chloride (Acquaah, 1999) before sowing can also activate the enzymes and hormones and thereby can improve the germination. Such primed seeds germinate early and uniformly by initiating physiological processes for quicker germination.
- The sprouted seeds in fenugreek, coriander, fennel, cumin and dill can be used for sowing and obtaining better, early and uniform germination level.
- Direct seeding is extremely difficult when seeds are very tiny as in ajowan, celery and parsley. To improve germination, tiny seeds may be pelleted by coating with clay. Pelleting makes planting easier and spacing more precise.

Use of Plant Growth Substances in Seed Germination

The application of plant growth substances in coriander have considerably helped in improving the seed germination, plant growth, flowering and seed setting.

- Early and higher germination was noticed by soaking coriander seed in solutions of GA_3 (50 ppm) or NAA (20 ppm) or IAA (10 ppm) or warm water for 12 hours. In addition when such soaked seeds were wrapped in polythene sheet or gunny bag cloth, the germination percentage increased, and it was earlier than unwrapped control. The germination percentage was maximum in 50 ppm GA_3 + gunny bag cloth wrapping, followed by 20 ppm NAA+ gunny bag cloth wrapping (Badgujar and Wauhal, 1990).
- Soaking of seeds in cycocel solution at 50-100 ppm improves germination and enhance seedling growth in fenugreek.
- The plant growth substances (IAA, GA_3 and ethephon) were tried by Ortuno *et al.* (1999) and found that treatment of seed with GA (10^{-4} & 10^{-5}M) improved seed germination and

increased the content of diosgenin in 30 days old plants by 19-43% and on a whole plant basis, diosgenin consideration was increased by 77% in 15 days. In another study Alhadi *et al.* (1998) found that under water deficit conditions, use of GA_3 caused change in growth parameters as well as some physiological and biochemical aspect positively in fenugreek.

- Thomas (1994) revealed that florence fennel, seeds soaked for 4 h at 25^0C or for 24 h at 5^0C in GA 4/7 solution (100 mg/1) showed a higher germination percentage and increased rate of seedling emergence compared with untreated dry seeds when sown in compost at 25^0C.
- In Egypt, Omer *et al.* (1997) found that soaking of cumin seeds in ethephon increased the number of germinating seeds and surviving plants and decreased the number of wilted plants. The application (foliar and seed) of plant growth regulators (ethephon, or GA_3) increased seed yield of cumin.
- The celery seeds are reported to possess thermo dormancy resulting in no or slow germination at temperature greater then 25°C. A seed soaking treatment at 10°C using growth regulators GA 4/7, and ethephon at 1000 ppm can over come this dormancy (Thomas, 1990).
- Far red light or sunlight exposure also improves germination percentages when the seeds are very dormant. Therefore it is advisable to sow seeds shallow to enhance light exposure. The thermo dormancy and photo dormancy existing in celery has been reviewed in detail by Desai *et al.* (1997).

The seed spices crops an important group of spices are annual in nature and are propagated through seeds. Owing to inherent weakness of slow germination and further condition of poor crop stand badly affect the yield of seed spices. So far very little attention has been paid on these crops. Thus emphasis is required for adoption of integrated propagation package of seed spices for improving the germination and crop stand by using pelleted seeds, primed seed with nutrients, plant growth regulators application and suitable soil preparation techniques. The yield potential of seed spices can only be realized by following the integrated approach in propagation techniques for meeting the huge domestic as well as export demand. Productivity of seed spices also depends on quality seeds of improved varieties. Intensive research needed to be carried out to work out economically viable cultural practices and measures for improving the germination and crop stand significantly. In seed spices, there is good future scope for investigation of new methods of preparing seeds for sowing with appreciably good germination and crop stand.

References

Acquaah, G. 1999. Horticulture, Principles and Practices. Prentice Hall, Upper Saddle River, New Jersey, USA, pp,281-304.

Alhadi, F.A., Yasseen, B.T. and Jabar, M., 1998 Water stress and gibberellic acid effects on growth of fenugreek plants. Irrigation Sci., 18(4): 185-190.

Badgujar,C.D.and Wauhal, K.N. 1990. Effect of soaking and warpping of seeds on germination of coriander. Spices India 3(10):9-11.

Desai B.B., Kotecha PM and Salunkhe D. K. 1997. 'Seeds Hand Book', Marcel Dekker Inc Pub, New york, USA, pp 280-285.

Malhotra, S.K and Mehta, R.S.2003. Seed Certification Standards for Seed Spices Crops. NRC Seed Spices Folder, March, 2003:1-6.

Malhotra, S.K and Vashishtha, B.B 2008. Package and Practices of Production of Seed Spices, Bulletin, N.R.C.on Seed spices,Ajmer,pp.98.

Malhotra S.K.2004. Minor Seed Spice crops-Breeding Strategy and Important Milestones. National Seminar on New perspectives in Commercial Cultivation, Processing and Marketing of Seed Spices & Medicinal Plants at SKNCOA, Jobner from 25-26, March, 2004. pp 1-2.

Malhotra, S.K 2005. Seed Spices: (Ed.) Sharma, A.K. Breeding Technology of Crop Plants. Yash Publications, Bikaner, pp535-553.

Nirmal Babu,K.,Ravindran,P.N. and Peter,K.V.1997. Protocols for micropropagation of spices and aromatic crops. IISR, Calicut.p35.

Nirmal Babu,K., Minoo,D and Ravindran,P.N. 2001. Biotechnological approaches in seed spices. In : Seed Spices, Production, quality and export. Eds. Agarwal,S., Sastry,EVD & Sharma,R.K. Pointer Pub. Jaipur. p315.

Omer, E.A., Nofal, M.A., Lashin, S.M. and Haggag, W.M.F. 1997. Effect of some growth regulators on the growth parameters and oil content of cumin with disease incidence under two types of soil. Egyptian J. Hor. 24(1) ;29-41.

Ortuno, A., Oncina, R., Botia, J.M., and Rio, J.A., 1999. Regulation of the diosgenin expression in *Trigonella foenum-graecum* plants by different plant growth regulators. Food Chemistry, 65(2): 227-332.

Ratnamba,S.P. and Chopra,R.N.1974. In vitro induction of embroids from hypocotyls and cotyledons of *Anethum graveolens* seedlings. Z.Pflanzenphysiol.73:452-455.

Ravindran,P.N., Peter,K.V., Nirmal Babu,K.,Rema,J., Smsudeen,K., Minoo,D., Geetha, S.P., Sajina, A.,Mini,P.M., Manjula,C. and John,C.Z. 1996. Biotechnological approaches in spice crops- present scenario and future prospects. In Biotechnology for Development. Eds. Das,M.R. and Munayoor,S. Stte committee on Science Tech. & Env. Kerala, pp175-179.

Sajina, A. Geetha,S.P Minoo,D., Rema,J., Nirmal Babu,K., Sadanandan,A.K. and Ravindran,P.N.1997a. Micropropagation of some important herbal spices. In: Biotechnology of spices, Medicinal and Aromatic plants. *Eds.* Edison, S.,Ramana, K.V., Sasikumar,B. Nirmal Babu,K and Santosh,J.E. Indian Society for Spices,Calicut.pp79-86.

Sehgal, C.B. and Abbas,N.S.1994. Somatic embryogenesis and plant regeneration from hypocotyls tissues of *T.ammi*. Phytomorphology, 44(3/4):265-271.

Settu, A., Ranjitha,K and Jaya M.R.1997. In vitro selection for salt tolerance in Trigonella foenum graceum using callus and shoot tip cultures. In: Biotechnology of spices, medicinal and aromatic Plants. (Eds.) Edison,S., Ramana,K.V., Sasikumar,B., Nirmal Babu,K. and Santosh,J.E. Indian Society for Spices, Calicut, pp119-121.

Stephen, R and Jayabalan, N. 1998. *In vitro* flowering and seed setting formation of coriander. Current Science. 74(3):195-198.

Thomas TH. 1990. Hormonal involvement in photoregulation of celery seed dormancy. Monograph Br. Society Plant Growth Regulators. 20:51

Thomas, T.H. 1994. Responses of Florence fennel (*Foeniculum vulgare axoricum)* seeds to light, temperature and gibberellin GA/7. Plant Growth Regulation 14(2) :139-143.

Chapter-38

Propagation of Ornamental Plants of Cold Arid Region

B.P. Sharma
Dr. Y.S. Parmar Univ. of Horticulture and Forestry, Nauni, Solan – 173 230

The propagation of plants has been a fundamental occupation of mankind since civilization begins. Agriculture started some 10,000 years back when ancient people learned plants and grow kind of plants that fulfilled nutritional needs for themselves and their animals (Hartman *et al.*, 1981). As civilization advanced, people added to the variety of plants cultivating not only for additional food crops but also those that provided fibres, medicine, recreational opportunities and beauty. From the great diversity and variation in plant life, it has been possible to select kinds of plants particularly useful to the welfare of humans and their animals. Bio-diversity has attracted the world wide attention because of growing awareness of its importance on the one hand, and the anticipated awareness depletion on the other. About 20 per cent of all species are expected to be lost within 30 years and 50 per cent or more by the end of 21st Century (Myers, 1993). India is one of the 12 mega-diversity countries globally recognized for its rich diversity, since its geographic position extends from sea level to the highest mountain ranges in the world. Also the areas of J&K like Leh and Ladakh are also in the same class. India's biological diversity has been distinguished into 43 major vegetation types including cold deserts. Indian arid zone occupy nearly 12 per cent of total area and covering over 38.7 m ha area located in various states of the country. Out of 38.7 m ha area, only 7.0 m ha area is under cold arid zone. Areas of Kinnaur, Lahaul - Spiti and

Chamba districts come under agro-climatic zone (iv) of H.P. chiefly a mountainous tract and represents the dry temperature and cold deserts of the trans- Himalayas. The general climate is dry temperate, snowy and frigid. Vegetation has shown a great variation, mesophytic on Northern aspects; otherwise it is more xerophytic in dry and arid regions (Sehgal, 2004). The main and immediate causes of bio-diversity losses can be attributed to habitat loss, degradation and fragmentation of ecological niches, resource - exploitation and climatic alternation of these, by for the most important factor is habitat loss due to over grazing, lopping for fire and fuel wood, construction of road and other major developmental projects.

Conservation of floricultural species is very essential from ecological as well as economic point of view for the people of cold arid hill regions. Many floricultural species are suitable to grow in the cold arid areas and important from beautification, medicinal and economic point of view. The flowers are used for extraction of essential oil and perfumes and for preparation of value added products.

Species/Cultivars

Ornamental plants are those plants which are used with the object of commercial utility and beautifying our surroundings as well. Of late these plants have enjoyed a spectacular rise in popularity due to their beauty and capacity to moderate climate control, soil erosion, reduce air and noise pollution and enliven the surroundings. A number of species of flowering plants and their cultivars are available in cold arid region.

Classification of Ornamental Plants

1. Trees
2. Shrubs
3. Bulbs and Climbers
4. Herbaceous perennials
5. Annuals
6. Indoor pot plants
7. Cactii and succulents
8. Palms and cycads

9. Grasses and bamboos
10. Ferns and selaginellas

Propagation

Many genetic resources are so complex and vulnerable that their sustainable management regions specialized knowledge of plants and animals and their interaction in such an environment. The ornamental plants are propagated both by asexual and sexual means. Seed propagation is very common in flowering plants especially in seasonal plants. Several vegetative viz. cuttings, grafting, tubers/bulbs, and suckers are used for multiplication (Randhawa and Mukhopadhay, 1986). Further, micro-propagation technique is also used for mass multiplication of true to type plants. Maximum flowering plants are multiplied by seeds. Flowering plants are also propagated vegetative methods and tissue culture.

1. Trees

A variety of woody perennials have ornamental value, which are propagated either by seeds or vegetative means or both (Sharma and Srivastav, 2004). Ornamental trees are commonly multiplied through seeds or stem cuttings (semi to hard wood). Cuttings are treated with plant growth regulators for rooting and planted under mist conditions.

I. Locust *(Robinia pseudoacacia* Linn.) family Leguminosae

Tree upto 80 ft in height with deeply furrowed dark brown bark and prickly branches; flowers white, very fragrant; leaves green. It is propagated by the following methods:

i) Seeds: Seeds sown in spring season and germinate readily
ii) Cutting
iii) Grafting

II. Holm Oak *(Quercus ilex)* family Fagaceae

Ever green tree to 60 ft in height with large rounded hand. Leaves very variable flowers 1-3 usually provided, acorn, ovate, scales thin, apprised, rarely slightly spreading. It is multiplied through seeds and side grafting method.

III. Ash *(Fraxinus xanthoxyloides)* family Oleaceae

Hardy deciduous ornamental trees; large green leaves; small flowers In panicles either appearing before or with leaves and whitish in showy panicles.

Propagation

i) Seeds: Seeds gathered in fall are sown immediately or stratified, sown in spring and covered with soil; sometime remain dormant until the 2nd year. The varieties and rare kinds are budded in late summer or grafted in spring on seedling of any common species.

IV. Poplar *(Populus ciliata)* family Salicaceae

The poplars are dioecious with both pistillate and staminate flowers naked but form a cup shaped disk and in slender mostly drooping precocious catkins in which subtending scales are toothed, cut or cleft at the apex; flowers a small 2-4 valved caps containing cottony seeds; leaves alternate and stalked mostly broad & fall colour is yellow. It is propagated by hard wood stem cuttings. Cutting is an easy and suitable propagation method for poplar.

V. Willow *(Salix* spp.) family Salicaceae

Deciduous trees planted for foliage and drooping in habit; flowers in catkins, mostly in spring; branches long and pendent, yellowish green sometime blended with brown. It is mainly propagated through seeds and cuttings.

VI. Birch *(Betula utitis)* family Betulaceae

Ornamental deciduous tree, 40-60' trunk with reddish brown bark; branchlets pubescent while young not glandular; leaves ovate, round at the base; acuminate density irregular, serrate, 2-3" long. It is propagated through seeds and cuttings.

VII. Alder *(Alnus nitida)* family Betulaceae

Ornamental woody plants; grown for foliage; leaves alternate, deciduous, short petioled, usually dentate and serrate; flowers monoecious, apetalous, in catkins, staminate ones elongated, each branch with 3 flowers pistillate catkins short each branch with 2 flowers. It is propagated by the following methods.

1. Seeds: Seeds gathered in fall and well dried; sown In spnng with slight covering and kept moist and shady.
2. Suckers
3. Cuttings: Hardwood cutting placed in moist and sandy soil for rooting.

VIII. Maple *(Acer oblongum)* family Aceraceae

Tree 50'; leaves oblong coriaceous, long but bluntly rounded acuminate, reticulate and usually glaucous beneath, 2-7" long, panicle short, pubescent; flowers small, greenish, wings of flowers at it angles.

Propagation

i) Seeds: Seeds should be sown soon after maturity or stratified and sown III
ii) Budding: Budding should be done in summer for rare varieties.
iii) Cuttings: Half ripened greenwood cuttings in summer.

IX. Milk Vetch *(Astragalus. carolinianus)* family Leguminosae

3-10' long plant; leaflets 15-31, glabrous or slightly pubescent, oval, obtuse at apex; flowers yellowish -1" long in a dense spike.

Propagation

i) Seeds: Chiefly by seeds which germinate slowly or by careful division in early spring.

X. Horse Chestnut *(Aesculus indica)* family Hippocastanaceae

Tall tree, 60'; leaflets 5-9, short stalked, obovate - lanceolate finely serrate, acuminate at the base, glabrous, 6-10" long; flowers 1" long with white unequal petals, the upper with yellow blotch & the lower tinged rose. It is multiplied by seeds , side grafting and budding on common species and shrubby form by layering.

XI. Fir *(Abies pindrow)* family Pinaceae

Leaves lanceolate, sessile, persistent for many years on young plants and lower sterile branches, flattened, usually deep green and lustrous above and silvery white beneath from the presence of many rows of stomata, variously notched at the apex, appearing & ranked by a twist at their bases; flower axillary appearing in early spring from buds formed the previous summer on branchlet of the current year.

Propagation

i) Seeds

ii) Side grafting: Using *A. picea* and *A. spinosa* as rootstock.

XII. Cypress *(Cupressus torulosa)* family Pinaceae

Evergreen tree; branchlet very slender, pendulous flattened; leaves rhombic- ovate, spreading at the acute tips, glaucous. Cypress is propagated by the following methods:

i) Seeds: Seeds sown in spring

ii) Cuttings: Cuttings taken from mature wood.

iii) Veneer-grafting

XIII. Spruce *(Picea smithianay)* family Spinaceae

Tree to 150' with wide spreading branches and slender pendulous brachlets; leaves crowded usually thicker than broad, acute, bright green 1-2" long; flower purple in colour. It is commonly propagated by seeds, stem cuttings and grafting method. Cutting is done in oven bottom heat.

2. Shrubs

A large number of shrubs are found in cold arid region out of which some are being listed here. Shrubs are multiplied through seeds and vegetative methods (Hartman *et al.*, 1981). Growth of flowering plants depends on genetic feature as well as environmental conditions besides management practices. The growths of plant propagated by vegetative means are similar and uniform to mother plants.

I. Berbaris *(Berbaris aristata)* family Berbaridaceae

Shrubs to 8'; first year branches yellowish brown, slightly angular, spines mostly simple; leaves elliptic to ovate oblong acute, spineless dentate, occasionally entire 1- 2 W′ long; racemes 10-25 flowers, sessile usually spreading and shoot flowers usually W′ long than stout pedicel, bright red finely purple at blooming.

Propagation

i) Seeds: Seeds broadcast in fall in beds.

ii) Cuttings: Cuttings taken in 1 st-week of June.

II. Cotoneaster *(Cotoneaster microphyllas)* family Rosaceae

Low prostrate shrub; densely branched; leaves cuneate, oblong, acute, shining above, densely pubescent beneath, 1/3" long; flowers usually solitary, calyx pubescent and red in colour.

Propagation

i) Seeds: Mature seeds sown in spring.

ii) Cuttings: Half ripened wood.

iii) Layering: Done in fall.

iv) Grafting: *Cotoneaster-vulgaris* is used as a rootstock.

III. Tick Trefoil *(Desmodium* spp.) family Leguminoceae

Mostly herbs: leaves pinnate with 3-5 leaflets; flowers small and papilionaceous in terminal or axillary racemes in summer, mostly purple.

Propagation

i) Seeds: Seeds sown in February in light sandy soil in pots and placed in warm, close atmosphere.

IV. Horse tail *(Ephedra gerardiana)* family Gnetaceae

Usually low much branched shrubs, often procumbent and sometimes climbing bearing minute, scale like sheathing leaves in distant pair or whorls; flowers dioecious in small egments forming usually peduncled axillary clusters. It is mainly propagated by seeds, suckers and layers.

V. Spindle tree *(Euonymus fimbricatus)* family Celastraceae

Shrubs to 12'; leaves obovate or obovate-oblong, finely and acutely, serrate; flowers white 1/5" across and globose.

Propagation

1. Seeds: Seeds usually stratified and sown in spring.
2. Cuttings: Half ripened wood of fall of this species grow readily under glass or during winter in greenhouse. Varieties are sometimes grafted or budded on stock of their typical species.

VI. Seabuck thorn *(Hippophae rhamnoidesi* family Eleagnaceae

It is found in cold arid parts of the country. Shrub occasionally to 30'; branches gray, usually spiny, winter buds golden brown; leaves linear lanceolate, short petioled, covered on both sides with silvery scales, at maturity glabrescent, $1/_3$" long; flowers appearing before the leaves, very small and yellowish in colour.

Propagation

Seeds: Seeds sown at once are stratified.

Cuttings: Cuttings of mature wood in spring.

Root cuttings, suckers and layers.

VII. Indigofera *(Indigolera gerardiana)* family Leguminosae

Low much branched shrub with silvery conascent branchlets; leaves 1-2" long, with thickish opposite pale white bristly leaflets, 9-17 in number; flowers pale-red, canascent outside, the peduncle raceme 12-20 flowers. It is propagated through seeds and cuttings.

VIII. Juniper *(Juniperus macropoda)* family Pinaceae

Evergreen with branchlets spreading in all directions; leaves either all needle shaped and in 3' s and usually opposite flowers, dioecious, rarely monoecious, staminate yellow consisting of numerous anthers united into oblong catkin, pistillate greenish, minutely globular, with several bracts.

Propagation

1. Seeds: Seeds which usually germinate In 2nd or 3rd year. To hasten their germination, they may be plunged for 3-6 seconds in boiling water.
2. Cuttings: Nearly ripened wood in fall under glass either outdoors or In greenhouse. As a rule, those species with needle-shaped root much more easily than those with scale like leaves and latter are, therefore, mostly increase by side grafting during winter in greenhouse on young potted plants of typical form or allied species.

IX. Honey suckle *(Lonicera spinosa)* family Caprifoliaceae

Rigid, spinescent shrub to 4'; leaves linear oblong, about long; flowers 3-4" long with slender cylindrical tube and spreading limbs, pale bluish red or whitish blooming.

Propagation

i) Seeds: Seeds sown in fall or stratified

ii) Cuttings: Cuttings of ripened wood, also by green wood cuttings under glass in summer.

X. Myricaria *(Myricaria germanica)* family Tamaricaceae

Glabrous under shrub, 4-6" high with upright and hand like branches; leaves minute, bluish green, lanceolate; flowers terminal racemes, light pink, 4-6" long.

Propagation

i) Seeds: Seeds which are very fine and should be only slightly covered

ii) Cuttings: Cuttings of ripened wood in open ground or by green wood cuttings under glass.

XI. *Rhus (Rhus semialata)* family Anacardiaceae

Shrub or flat headed tree to 25'; leaflets 7-13, short stalked, ovate to ovate- oblong, acute rounded at the base 2-6" long; flowers creamy white in large and broad panicles to 12" long.

Propagation

i) Seeds: Seeds sown in autumn or stratified

ii) Cuttings: All species grown from root cuttings, some species may be increased by layers or cuttings of mature woods.

XII. *Rose (Rosa microphyllaj)* family Rosaceae

Much branched spreading shrub 6' high with straight or ascending prickles; leaflets 6-15, elliptic or oblong-elliptic, acute, serrately sharp; flowers pale, pink often solitary, 2-22" across.

Propagation

i) Seeds: True species can be propagated by seeds but as roses are likely to hybridise, only seeds taken from isolated fruits ought to the used. The hips should be gathered as soon as ripe, the seeds washed out and sown at once or stratified and sown in spring.

ii) Cuttings: Cuttings from ripened wood in summer under glass.

XIII. Gooseberry *(Ribes aspestre)* family Saxifragaceae

Upright shrub to 10'; branches with stout spines to 1" long and usually in 3's, obtuse lobes; leaves cordate, 1-2" broad; flowers 1-21, short peduncle, greenish or sometimes reddish flower.

Propagation

i) Seeds: Seeds which germinate readily.

ii) Cuttings: Hardwood cuttings in autumn or by greenwood cuttings in summer under glass,

iii) Mound layering: In summer sometimes practiced.

iv) Budding/Grafting: Usually resorted to if quick propagation of rare varieties is required.

XIV. Rose tree *(Rhododendron campanulatum* Don.) family Ericaceae

Shrub attaining 16 feet height; leaves elliptic to elliptic-oblong, usually rounded at both ends, 3-6" long; flowers in cluster, pale purple or almost white with few purple spots 2" across.

Propagation

i) Seeds: Seeds sown in early spring in frames and pans in sandy peat, without covering and kept moist and shady.

ii) Grafting: Named cultivars can be grafted on common species. Usually by Veneer grafting in autumn in the greenhouse on potted stock.

iii) Cuttings: Mature wood cuttings (2-3" long) can also be used with a heel in late summer and placed in sand under glass.

iv) Layering: Layers usually require two years to root strongly they are made in spring and budded part in closed moss.

3. Bulbous plants

Botanically bulb is underground good storage structures. Some plants have e.g. bulbs, corms. tuber and rhizomes. These plant structures are used for further multiplication. All bulbs create new bulbs as part of their reproduction cycle, although some produce more bulbs than others. Bulbus plants are multiplied through bulbs (Hartman and Kester, 1986). After a few years in the ground, clumps of bulbs usually need to be dug up and replanted into refurbished soil. When you lift the clump, you will find anything from a few to hundreds more bulbs than you planted. Bulbs are stored properly in dark, cool places or in cold storage.

I. *Anemone (Anemone obtusilova)* family Ranunculaceae

Turfted, glabrate to hairy, perennial herbs 10-30 ern tall; radical leaves orbicular, deeply cordate; flowers yellow or white twiged with blue near the base or occasionally blue to bluish-violet. It is multiplied through seeds and root cutting/division.

II. *Ranunculus (Ranunculusfalcatus)* family Ranunculaceae

Erect, glabrous or slightly woody, annual herbs; leaves all radical, 3-lobed; flowering stems solitary or more 3-8 cm high, l-flowered, longer than the leaves flowers yellow in colour. It is a beautiful flowering plant and multiplied through tubers, and fresh seeds.

III. Lotus *(Lotus corniculatus)* family Nympbaceae

Perennial, prostrate, ascending and decumbent, leaves 5-folialate; leaflets oblong, nearly sessile two at the base of the leaf petiole and three at the top; flowers 1-7, in a long peduncle axillary umbels, yellow in colour. Multiplication is done by the following methods:

i) Bulbs/Tubers

ii) Cuttings: In early spring.

iii) Seeds: Seeds sown in fall or spring.

IV. Iris *(Iris kemaonensis)* family Iridaceae

10-30 cm high, stems tufted; root stock creeping; leaves linear, 10-30 em long; flowers solitary, spathe 1-2 flowered, pedicel very short; perianth tube very long dark lilac to purplish blue. It is propagated through tubers and division of stem.

V. Allium *(Allium* spp.) family Alliaceae

Bulbs cylindrical to ovoid, outer coats coriaceous; leaves 2-6, broadly linear to lanceolate, falcate, fistular; flowering scapes slender to 30 cm long; flowers in umbels of different colours. It is commonly multiplied by seeds and bulbils.

1. Seasonal flowers (annuals, biennials and perennials)

I. Viola *(Viola bijlora)* family Violaceae

Erect or decumbent, glabrous upto 30 cm high leaves reniform to broadly ovate, cordate, serrate glabrous to hirsute, stepules present or absent; flowers of different colours, 2-4 cm across may be spreading. It is propagated by Seeds and Division: If plant is large.

II. Diantthis *(Dianthis anatolicus)* family Caryophyllaceae

Tufted, erect perennial herb, 15-20 cm tall; leaves linear with serrulate margin; flowers solitary to many of different shades. It is propagated by seeds.

III. Gypsophylla *(Gypsophylla cerastoides)* family Caryophyllaceae

Decumbent, spreading, pubescent, perennial herb 10-20 cm tall; leaves obovate; flowers white, streaked with purple in many flowered lax cymes. It is multiplied by seeds.

IV. Linum *iLinum perenue)* family Linaceae

Perennial herb upto 75 cm tall; branches ascending glabrous; leaves alternate, linear; flowers blue in 3-5 flowered cymes. Linum is propagated by seeds.

V. Balsam*(Impatieris amplexicaulis)* family Balsaminaceae

Plants are 15 cm-2.5 m in height; leaves long, acuminate, alternate or opposite, whorled, crenate to dentate; flowers of pink in umble like clusters. Lip spotted with different types of spurs. Seasonal plant is mainly multiplied by seeds by raising nursery.

VI. Aster *(Aster indamellus)* family Asteraceae

Perennial herb plants upto 60 cm tall; leaves villous, oblong to ovate, entire or toothed; heads of different shades in terminal corymbs or solitary.

Propagation

i) Seeds: Mostly sown in spring

ii) Cuttings: Softwood cuttings in sprint

VII. Chrysanthemum *(Chrysanthemum pyrethroides)* family Asteraceae

Erect woody, pubescent to wooly, branched herbs, 25-35 cm tall; leaves pubescent to woody, linear - oblong pinnatified, redial leaves petioled; heads solitary terminal, obtuse; involucral bracts ovate-oblong; ray floret white, disk florets yellow.

Propagation

i) Seeds

ii) Cuttings: Lateral cuttings in fall or early spring

VIII. Senecio *(Senecio krascheninnikoviiy)* family Asteraceae

Erect, glabrous, branched herb; leaves petioled or pinnatifid; heads yellow, long peduncled. The annual plant is multiplied by seeds and perennial by division.

IX. Campanula *(Campanula alsinoides)* family Campanulaceae

Sub erect to erect, glabrous herbs, 10-40 cm tall; leaves thick, entire long petioled; flowers solitary andterminal. It is propagated by seeds during spring season.

X. Amaranthus *(Amarantltus retroflexus)* family Amaranthaceae

Erect, simple and tall plants; stems stout; leave ovate to rhomboit, obtuse to sub-acute; flowers are in axillary panicles, male and female flower are intermixed. It is propagated by seeds.

XI. Kochia *(Kochia stellaris)* family Chenopodiaceae

Erect, diffusely branched, tomentose; leaves small flat, linear, acute; flowers greenish yellow, clustered in branched leafy spikes. It is multiplied by seeds.

XII. Salvia *(Salvia rubicola)* family Labiatae

Tall, robust, viscidly hairy herbs 50-60 cm high; leaves petioled, ovate-oblong, apex acute, margins crenate; flowering whorls few, yellow in colour. Salvia is multiplied by seeds and climbers during spring season.

I. Jasmine *(Jasminum humiles)* family Oleaceae

Deciduous or evergreen glabrous climber, 1.0-2.0 m tall; leaves alternate, variable in size, leaflets 3-5, coriaceaous, dark green above, paler beneath; flower in terminal corymbose cymes. It is propagated through stem cuttings during spring- summer season.

II. *Convolvulus (Convolvulus arvensis)* family Convolvulaceae

Prostrate, creeping, slender, glabrous perennial herb with milky juice; flowers white or lilac, 1-3 flowered cyme; leaves ovate to lanceolate. Cutting is suitable for its multiplication in cold arid region.

References

Bailey, L.H. 1935. The Standard Cyclopedia of Horticulture, Vol. I, II, III, The Mac Millan Company, New York, USA.

Hartman, H. T., Flocker, W.J. and Kofanek, A.M. 1981. Growth development and utilization of cultivated plants. *Plant Science.* Engle wood Cliffs, Prentice Hall, New Zealand.

Hartman, H.T. and Kester, D.E. 1986. Plant propagation: Principles and Practices, Prentice Hall of India Pvt. Ltd., New Delhi.

Myers, N. 1993. Ambio. 22: 74-79.

Randhawa, G.S. and Mukhopadhay, A. 1986.Floriculture in India, Allied Publishers Pvt. Ltd., New Delhi.

Sehgal, R.N. 2004. A Final Technical Report on Development of Data Base on Plant Resources in the state of Himachal Pradesh. Deptt. of Tree Improvement and Genetic Resources, Dr. Y.S. Parmar University of Horticulture and Forestry, Nauni-Solan.H.P.

Shrama, R. R. and Srivastav, M. 2004. Plant Propagation and Nursery Management, IBDC, Lucknow, pp. 306- 322.

Chapter-39

Guggul [*Commiphora wightii* (Arnott.) Bhand.]

J.P. Singh and R. S. Singh
Central Arid Zone Research Institute, Regional Research Station, Jaisalmer

Guggul (*Commiphora wightii* Arnott.) commonly known as guggul, is a much branched shrub yielding gum-resin generally obtained by incision of the bark. Its plant is drought hardy and can be grown in wasteland, gravelly, desertic soils of arid parts of the country. This resin, normally referred as guggul, has been shown to have diverse medicinal properties like lowering of lipid activity anti-inflammatory activity , diuretic, expectorant, diaphoretic activity, etc (Atal *et al.*,1975). Its description and uses is mentioned in ancient literatures Ayurvedic treatise Charak (1000 B.C.), Sushruta Samhita (600B.C.). It has a historical use in folk medicine and today it 's highly commercially important target species .In the last few decades there has been a sharp decline in natural population of this species. There is hardly any commercial cultivation of this species. However, at Mangliavas near Ajmer in Guggul Herbal Farm, more than 15,000 plants have been planted on the rocky site and the material so established is being used for research purpose on tapping of guggul and other related aspects.

It is endemic to dry arid rocky areas and piedmonts of Rajasthan and Gujarat which are the natural habitat of this species. It occurs in rock crevices on sandy-loam soils. Usually it prefers rocky slopes where rain water is quickly drained away (Bhandari, 2005). It is one of the species in priority medicinal plants being promoted by National Medicinal Plants Board (AYUSH). It is reported that Guggul has Rs. 60-80 million approximate domestic sales.

Composition and Uses

Plant produces oleo-gum resin which is used for ayurvedic medicines since ancient time. Gugulipoid has a long history of use in Ayurveda. The Atharva Veda, one of the four well-known Holy Scriptures (Vedas) of the Hindus, is the earliest reference to the medicinal and therapeutic properties of guggul. Detailed descriptions regarding the actions ,uses, and indications as well as the various Nighantus (medical lexicons) were written between the 12th and 14th centuries A.D. that were based on the Ayurvedic literature.

The list of traditional uses for guggul is extensive. It has been indicated for healing bone fracture to inflammation, arthritis, cardiovascular conditions, obesity and lipid disorders. Several other external and internal uses for guggul have been described in folklore and ethnomedicine as well. Although, several external and internal uses for guggul have been described in folklore and ethnomedicine as well. This shrub is highly useful plant for human health point of view (Singh and Sharma, 2005). Although, several therapeutic uses were indicated for guggul, the Indian gum resin was mainly used for treating various types of Arthritis. Ayurvedic physicians extensively used guggul for treating for Arthritis and related conditions for centuries. It generally grows in association with trees like 'Thor', 'Mitha jal' and'Kair'. The rate of regeneration of the tree also increases with other trees. Age of the *Commiphora* tree can be determined by the increment rate of its diameter.

It has a strong insect repellant and also has properties that reduce cholesterol in humans. It is used as laxative, aphrodisiac, alternative, anthelmintic, biliousness, ulcers, fistula, piles, pyrorrhoea, chronic tonsillitis, pharangitis, chronic bronchitis, chronic nasal catarrh, laryngitis, phthisis, urinary troubles, asthma, cough and cold, tumors, leucoderma, lowering blood cholesterol. The gum is used as an incense to fumigate rooms. Besides aromatic properties, such fumigation has antiseptic effects. It can be grown as headge as goats can eats its leaves.

Botanical Description

Guggul (*Commiphora wightii* Arnott.) belongs to the family Burseraceae. *Commiphora* is a genus of about 185 species (Mahr,1998).Out of these only 4 species are found in India, they are :*C. wightii, C. stocksiana, C. beryal* and *C. agallocha*. *C. wightii* is branched, dwarf bush with crooked with knotty branches ending in sharp spines.

The centres of origin of *Commiphora spp.* are the arid regions of south Africa, namely, Somalia, Medagascar, Botswana and Pakistan .It is also found in the mountains of Niger, Tanzania, Welo, Yemen, Arab Republic and Namibia. In Indian sub-continent, it is found in Rajasthan, Gujarat, Madhya Pradesh, Karnataka and other low rainfall areas of south India. It is common flora in Thar desert. In Rajasthan, it is widely distributed in the regions of Jaisalmer and Ajmer; whereas, moderately in Jodhpur, Barmer and Bikaner districts (Kumar and Shankar,1982).

Commiphora wightii is a much branched shrub, 1-2 m high, with silvery, paper-like bark peelings; young parts glandular, pubescent, the branches knotty and crooked, usually ending in sharp spines. Leaves 1-3-foliate; leaflets 20 x 8 mm, rhomboid-ovate, serrate near the apex, entire near the base. Flowers are solitary or in fascicles of 2-3; almost sessile. Fruit a drupe, 6-8 mm in diam., ovoid, acute, readily splitting into two, red when ripe. Guggul excludes a resin sap out of incisions that are made in its bark. The resin has been used for centuries as a part of India's traditional medicine called Ayurveda.

In natural habitats, most plants of *C.wightii* are females, where isolated or in groups .Plants maintain their sex in the subsequent years. Seed set in females was not dependent upon the presence of pollen. Hand pollination experiments and embryological studies have confirmed the occurrence of non psuedogamous apomixes, nucellar polyembryony and autonomous endosperm. Anthers from male and bisexual plants, though produce functional pollen grains, the pollen tubes fail to grow beyond the proximal one third of the style (Gupta *et al.*,1998).

The plants starts to set leaves on the onset of monsoon .The leaves are of leathery texture. In all the marked plants, leaves started falling with the onset of winter, but in some plants they continue to stay on the plants till the onset of dry summer season (only 3- 4 plants).

Our studies showed that days required for buds to become flowers ranged from 2 to 7 days with a mean 4-10 days. Days required by the plants from flower age to fruit initiation ranged from 2 to 8 with a mean of 20 and standard deviation of 1.30.Variability in morphological character of plant, branching has been observed in Beriganga area, Jodhpur.

Guggul plant flowers twice in a year. First period starts at the end of March and continues till the end of May while second period starts from the last week of October to mid –December. Flowering is asynchronous. On an average, 4.1 days required for the buds to develop into flowers, and 5.2 days for flowers to initiate fruiting

Flowers of both the flowering periods bear fruits. Out of every 1.8 flowers only 1 is able to set fruit. Percentage of fruiting is 57.50 per cent. Fruit is an ovoid drupe, of approximately 0.7 cm length and 0.5 cm width .only one seed per fruit was observed which had approximate dimensions of 0.7 × 0.4 cm, and 100-seed weight of 4.3. Fruits turn reddish on ripening. Ripening takes about 1.0 to 1.5 months.

Flowers from about 50 plants were observed under stereomicroscope and dissected to determine the sex. All the flowers studied were female flowers having conspicuous gynoecium. Stamens were present in rudimentary form Gupta *et al.* (1996) have also reported presence of predominantly large number of female individuals. Female plants set fruits irrespective of presence or absence of pollen by non-psuedogamous apomixes.

Propagation

It is commonly multiplied by seeds and vegetative methods. Due to poor germination of seed, now a days, stem cuttings is used for plant multiplication in guggal. For mass multiplication, tissue culture technique has also been tried. It has been found that the seeds routinely sown give a germination per cent of up to 5%. However, heavy seds are reported to give more than 65 % germination. Guggul can be propagated through micro-propagation (Singh *et al.*, 1997).

Seed propagation

Seed is the major propagation source in nature. Monsoon season creates conducive atmosphere for germination. The temperature during monsoon ranges from maximum 30-37°C to minimum 20-25°C and relative humidity is high. Mature seeds are washed, along with clay and soil, to the crevasses between rocks and germinate there. Raising of seedling through seeds is problematic in semi arid conditions at Pali. Various seed treatment were tried to enhance germination and growth of seedling. Seeds treated with potassium nitrate (KNO_3) 0.25% solution exhibited maximum germination followed by treatment with H_2SO_4 (50% conc.) for 5-10 minutes and seed soaking in water for 8 hours. The maximum germination (33.3%) was observed even after pre-treatment with KNO_3 solution. Maximum average seedling height (19.5cm) was recorded in case of treatment when seeds were soaked in water for 8 hours.(Singh *et al.*,1998).

Seed germination is very poor (5%) but seedlings produce healthier and robust plants (Dalal *et al.*, 1989). Seedlings can withstand drought and high wind velocity due to better root system, which is a usual phenomenon in the desert. For such areas seed-raised plants should be preferred. Dalal *et al.*(1989) reported that mechanical scarification(with sand paper) of the seed coat an keeping the scarified seeds under running water for 24 hours is essential for good seed germination. Seedlings from such seeds were raised in polyethylene bags and the hardened seedlings were successfully transferred to the field.

Another way of germination is between 'Thor' (*Euphorbia cauducifolia*),which has a property of retaining moisture within its periphery . Mature seeds along with slit are washed in between 'Thor' plants are also said to retard termite infestation ,which cause major damage to *Commiphora* plants (Tajuddin *et al.*,1994).

During experiments large amount of seed set and their germination below guggul was observed where the plants were irrigated regularly. The overall growth of such plants were poor than those raised using cuttings. Besides being washed along with clay and soil, another way of seed dispersal is by birds. Depending upon the frugivore diversity the rate of seed dispersal ranges from 9 to 66 per cent .Some of the species which help in dispersal are: lesser vasa parrot (*Coracopsis nigra*), ant (*Aphaenogaster swammerdami*) and one lemur species.

The seeds germinate poorly due to coat hardness. Germination of seeds stored at ambient temperatures, ranging from 24-30 C, is maximum .Seeds collected from around anthills germinate better than those collected from under the tree and germination decreases after 12 months of storage with no germination after 18 months. Dark coloured seeds give better germination than lighter ones. At Nakoda (Rajasthan)mature fruits produce 2 types of seeds: black and white with only the black ones being viable(36.25%). Seeds sometimes produce 2,3and 4 seedlings also. The best time of collection is mid-February to mid March and of sowing is 20-30June.

Cuttings

Commiphora wightii can be successfully propagated vegetatively by stem cuttings .Cuttings of 20-25 cm long and 1.5-2.0 cm diameter can be planted at a depth of 15 cm, in nursery beds in the month of June. Plant growth substances can be used to enhance rooting in cuttings. Once rooted, the cuttings can be transplanted in the fields during monsoon. Semi hard wood cuttings of guggul plant were found suitable for planting in semi arid region.

The studies conducted on vegetative propagation using auxins (IBA, IAA and NAA) at concentrations 2000, 4000 and 6000 ppm, showed that dip of cutting for few seconds in 5000 ppm IBA solution was enough to initiate good growth of cuttings (table 1). The results were much better when cuttings were left at room temperature for two days before hormone treatment and putting in nursery bag. Beginning of March has been identified as the best month for raising plants of guggul using cuttings.

Table 1 : Sprouting in cuttings of guggal in Nursery during March month.

Treatments	Days for Sprouting	% of sprouting up to 90 days
Fresh cutting	60	30
Fresh cutting + 5000ppm IBA	60	100
Two days old cutting	30	100
Two days old cutting+ 5000ppm IBA	15	100

Air layering in *C. wightii* using solution at 500 and 1000 ppm also gives good rooting. Increase in number of shoots and better regeneration can be obtained by coppice re-growth.

Almost 100% establishment of cutting has been observed with planting during summer when cuttings do not have leaves. about 80% establishment of different germplasm of cuttings was observed when these were planted during the summer at Anand. But when cutting were made during early monsoon (1st july), and planted in the nursery, 62% cutting got established (Dalal *et al.*,1989). When 49 days old, July planted cuttings were examined, numerous fibrous roots were observed and many of them were as long as 15cm or more. At the Udaipur centre, cutting of different genotypes collected and planted after the first fortnight of July, showed 80% establishment. It has been reported that the best season for planting the cuttings of guggul at Mangaliavas (Ajmer) is June and 80 - 90% sprouting was observed (Anonymous, 1982).

Dalal *et al.* (1989) suggested that cuttings be collected during summer and planted under irrigation within three to five days to get good establishment. In addition, air-layering within three to five days to get good establishment. In addition, air–layering was found to be a faster method and is 100% successful for producing rooted plants.

It was observed at the Udaipur center that cutting obtained from semi-woody stems of 15-20 mm thickness established well. These cuttings show sprouting within seven to ten days but the root develops after 15 to 20 days. The cuttings should be planted in a well- manured nursery 1.0-1.25 feet apart between rows as well as plant to plant. The cutting could also be planted in well manure soils in polyethylene bags. After one year, they should be transplanted in the field. Periodically, irrigation should be provided in the nursery.

Micro-propagation

For *in vitro* propagation, media requirements of *Commiphora* is very specific. It can be micro-propagated through forced axillary branching on MS medium supplemented with benzyl adenine (BA) and kinetin. Shoot formation is initiated using MS medium supplemented with 17.8μM BA,18.6 μM Kinetin ,100 mg glutamine and 10 mg thiamine HCl/l and 0.3 per cent activated charcoal. Shoots are rooted by treating them with both IAA and IBA for 24 hours in darkness and transferring to low salt basal medium with activated charcoal (Barve and Mehta,1993).The regenerated plants established well in soil.

Somatic embryogenesis can be achieved by sowing immature zygotic embryos (2-3mm)on B_5 medium containing a low concentration of 2,4,5,-T, kinetin, IBA, IAA or NAA. Somatic embryos of explants supplemented with 2,4,5-T and kinetin are white and of various Shapes; Whereas ,those of medium supplemented with IBA and kinetin are green and well developed (Singh *et .al.*1997).In a report where cell suspension cultures and mobilized cells were tested for guggulusterone production ,it was observed that, in suspension cultures was 4.5 to 5 times lower than in leaves of intact plant whereas ,in immobilized cells it was 2.5 times lower.

Quality Standards of Planting Material

There is no quality standard fixed for guggul plants. Healthy seedlings can be planted. However, well rooted cuttings of appropriate size (1-1.5feet height) plants raised from cuttings should be used for planting in the field. The plants should be disease free and not very old for better survival and establishment.

References

Anon, 1982. Guggulu and its cultivation, Parishad Samachar, CCRAS, New Delhi, 4:1-7.

Atal, C. K., Gupta, O.P. and Asfaq, S.M. 1975. *Commiphora mukul*: Source of Guggul in Indian system of medicine. Economic Botany, 29: 208-218.

Bhandari, M.M. 2005. Shrubs-The Savious of Desert Ecosystem. *In: Shrubs of Indian Arid Zone* (*eds*. P. Narain, M. Singh, M.S. Khan and S. Kumar). CAZRI, Jodhpur. pp. 5-13.

Brave, D.M. and Mehta, A.R. 1993. Clonal propagation of mature elite tree of *Commiphora wightii,* Plant Cell Tissue and Organ Culture, 35:237-244.

Dalal, KC, Patel, N.A, Upadhyay,N.V,; Patel,D.H, Patel, R.B., Hirpara, B.V., Patel, M.C., Patel, P.B. and Gupta, R.. 1989. Germplasm Collection and therapeutic, botanical, Agricultural, chemical and clinical aspect of plants- Guggal. AICRP on Medicinal &Aromatic Plants, Workshop held at Faizabad,pp.67-79.

Gupta, P., Shivanna, K.R. and Ram, H.Y.M. 1996. Apomixis and polyembryony in guggul plant, *Commiphora wightii*. Annals of Botany 78(1): 67-72.

Gupta, P., Shivanna, K.R. and Ram, H.Y.M. 1998. Pollen pistil interaction in a non psedogamous apomict, *Commiphora wightii*. Annals of Botany, 81(5): 589-594.

Kumar, S. and Vinod Shanker. 1982. Medicinal Plants of Indian Desert : Commiphora wightii (Arnott.) Bhand. Arid Environments, 5: 1-11.

Mahr, D.L.1998. A brief visit to succulent commiphora of Namibia, Aloe, 35(3&4): 72-75.

Singh, A.K., Suri, S.S., Ramawat, K.G. and Sonie, K.C. 1997. Somatic embryogenesis from immature zygotic embryos of Commiphora wightii, a woody medicinal plant. Gartenbauwissenchaft, 62(1): 44-47.

Singh, R.H. and Sharma G.G. 2005. Arid Shrubs for human health *In: Shrubs of Indian Arid Zone.* (P. Narain, M. Singh, M.S. Khan and S. Kumar (eds.); CAZRI, Jodhpur. Pp. 39-43.

Singh, R.S.; Tewari, J.C. and Pandey, Manish 1998. Effect of seed treatment on germination, survival and seedling growth of Guggul (C. *wightii*.Arnott.) under nursery conditions. J.Trop. Forestry, 14: 21-26.

Tajjudin, A.K., Duhan, S.P.S., Ram, D. and Thakur, R.S. 1994. E.II. Indian Bulletin (*Commiphora wightii*). Current Research in Medicinal and Aromatic Plants, 16(3): 75-86.

Chapter-40

Senna (*Cassia angustifolia* Vahl .)

R.S. Singh* and J. P. Singh**
**Central Institute for Arid Horticulture, Bikaner-334 006*
***Central Arid Zone Research Institute, Regional Research Station, RRS, Jaisalmer*

Senna (*Cassia angustifolia* Vahl.) popularly called as 'Sonamukhi' is widely used medicinal herb in Ayurveda, Unani and also in Allopathic system of medicine. It is well known for its purgative activity (Bentley and Trimen, 1999). The plant is known by various names in common conversation likes Sennai, Sona, Sonamukhi, Sennamukki, Sonpar, etc. Presently, there is a growing demand of senna dry leaves and processed products in the national as well as in international markets. In recent years, its cultivation has gained momentum in arid and semi- arid parts of Rajasthan and Gujarat due to its diversified adaptability. It has also been widely adopted by the farmers because of its low cost of cultivation with yield stability. Its main characteristics *viz.*, drought hardy, perenniality, non-palatability to animals has also attracted the farmers for its popularity. Moreover, the senna crop is eco-friendly as it can be utilized for wasteland, sand dune stabilization and desert development (Bohra and Sankhla, 1997; Tripathi, 1999).

Senna is a native of Yeman and Hadramaunt province of Saudia Arabia. Its sporadic distribution is also reported in parts of Sindh (Pakistan), Mundra and Anjar in Kachchh region of Western Gujarat (Gupta, 1974). In India, it was introduced by Arab traders in Tirunelveli district of Tamil Nadu during mid eighteenth century; hence it is also called as Tinnevelly senna (Bentley and Trimen, 1999). It is now naturalized in the Tirunelvelly and Ramnathpuram district of Tamil Nadu.

Senna is widely grown in dry area in Arabian countries for medicinal uses in constipation. In Arava valley of Israel, it is also grown as dryland medicinal crop. At present India is main producer and exporter of Senna leaves and pods in the world market. The annual income to the tune of Rs. 300 million is obtained from the export of senna (Maiti and Kumar, 2001). From Tuticorin, it is exported around 6000-8000 tonne of senna leaves a year, earning of Rs.30-40 crores. In India, Senna is being grown in the states of Tamil Nadu, Andhra Pradesh, Rajasthan, Gujarat, Maharashtra, Karnataka, West Bengal, Tripura, in about 25,000 ha area (Maiti and Kumar, 2001). It can be grown successfully under rainfed conditions. In Rajasthan, its cultivation is extended in the district of Barmer, Jaisalmer, Bikaner, Nagaur, Pali, Jodhpur, Churu, Sikar, Jalore and Sri Ganganagar (Bohra and Sankhla, 1997; Lal and Prajapati, 1999 ; Tomer *et al.*, 2000).

Composition and Uses

The senna leaves and pods contain sennosides A,B,C,D, rhein, aloe-amine, kaempferein and iso-rhein in free and compound (glycosides) forms (Gupta and Pareek, 1995). Sennoside A and B are present in larger amount and account for 60 per cent biological activity of the crude drug. Gupta *et al.* (1977) reported that young immature leaves and pods contain more sennosides than mature ones (Table-1). It has been found that leaves of fresh plants contain maximum sennosides (3.58%) as compared to older plants. Flowers contain maximum sennosides content of 5.18%, pods 3.84% and stem contain 1.18% (Sharma and Azam, 2001). Ilangovan *et al.* (1991) reported that leaf sennoside content increased with raising P application rate. The leaves and pods of plants under rainfed conditions had slightly higher sennoside content than plants under irrigation. They also found that wider spacing in senna plant favoured more sennoside content than those at the closer spacing.

Table 1 : Sennosides content of leaves and pods at different growth stages (Gupta *et al.*, 1977).

S.No.	Material	Stage of plant Growth	Sennoside content (%)
1.	Immature leaves	Vegetative	6.93
2.	Mature leaves	Vegetative	3.92
3.	Mature leaves	Flower-bud	3.06
4.	Mature leaves	Full-bloom	3.98
5.	Mature leaves	Maximum pod bearing	2.92
6.	Pods	3-5 days old*	11.92
7.	Pods	5-10 days old	8.30
8.	Pods	10-15 days old	8.75
9.	Pods	Full grown and mature	3.0 to 4.0

* Stage after fall of petals in a flower.

Senna is a well known drug in the Ayurveda and Unani system of medicine and widely used as purgative. The utility of this plant as laxative was first recognized by early Arabian physician and recorded in Drugs of Plant Origin in 1600 BC. (Gupta and Pareek, 1995). Senna stimulates the muscular coat of the intestine and produces purgation; it is therefore one of the most useful purgatives especially in cases of habitual constipation. Shiva (1999) mentions that if it is not taken with aromatic like ginger, cloves, dill, fennel, coriander orange peel or liquorices, it may cause nausea and gripping. Its leaves and fruits are also used as cathartic, febrifuge, in splenetic enlargements, anemia, typhoid, cholera, biliousness, jaundice, gout, rheumatism, tumors, foul breath and bronchitis and probably in leprosy (Anonymous, 1992). It is also used for certain skin diseases and the powdered leaves in vinegar are applied to wounds and burns and to remove pimples. It can be used as an anthelmintic, for removing intestinal worms (Shiva, 1999). In India, senna is also used in several household preparations such as decoction, powder, syrup, infusion etc. In Kerala, Parumbikulam tribals used the roots of senna for joints pain. For this 10-20 g root powder is mixed in 100 ml coconut oil and paste is applied on painful joints (mottuvali) one time per day for 7 days before bath (Rajendran and Mehrotra, 1996). In Europe, leaves are preferred as ingredient of herbal tea (Maiti and Kumar, 2001).

Senna plants can be also utilized for developing wastelands in arid region as it does not require much irrigation and care. The leaves alongwith henna are used to dye the hair black (Anonymous, 1992). It can also be grown under different farming system as ground storey crop or as inter crop in dry land areas.

Botanical Description

Senna botanically known as *Cassia angustifolia* Vahl. belongs to Caesalpiniaceae subgroup of family Leguminosae. Some taxonomists recently changed *C. angustifolia* Vahl into *Senna alexandrina* Miller (Hussain and Kapoor, 1990). Several species are utilized under the plant name senna. As different native species are used for the same purpose in different countries. Some taxonomists had clubbed the two into a common taxa under *Cassia senna* Linn.. Fairbairn and Shestra (1967) reported that surface markings on the seed coat and stomatal index of leaves are distinct and remarkably consistent in the two species to maintain their separate identity. Upadhyaya and Singh (1989) reported that phytochemically *C. angustifolia*Vahl and *C. acutifolia* Del. do not have distinct identity and these result support the reduction by Brenan [See Kew Bulletin (1958) 13:231] of these two species to *Cassia senna* var. senna.

Senna is a small, perennial and branched under shrub upto 1-2 m tall. It grows taller when left uncoppiced. Its stem is erect, smooth and pale-green to light-brown with long spreading branches. It bears compound pinnate leaves; having 4-8 pairs of shortly stalked leaflets. Leaflets are narrowly lanceolate or obovate lanceolate, 2-5x1.5 cm, pale green to bluish green in colour, entire, glabrous or very slightly pubescent beneanth. Senna plants usually commence flowering at 60-70 days after sowing. Flowers are brilliant yellow coloured, showy arranged on 30-45 cm long axillary or subterminal racemes. Pods are flat, thin 3.5-7.0 cm long and 0.5-2.5 cm broad, broadly oblong, pale green in the beginning and change to greenish brown and dark brown on maturity and after drying. Pods contain 5-8 obovate-oblong, yellow to off cream coloured dark brown seeds; 14000 to 35000 seeds weigh to a kg.

Variation in morphological traits in senna genotypes like height of plants from 34.6 to 45.7cm, and fresh leaves yield/plant 3.38 to 6.54g have observed during trials at CAZRI, Jodhpur (Singh *et al.*, 2005). Crop improvement is essential to achieve higher quality production of leaves and pods. Use of different doses of gamma rays (10 to 50 kR) also could not create significant variably for improvement of desirable traits in senna (Jindal *et al.*, 2005).

Plants shed their leaves with the onset of cold weather in North India. Senna is predominantly a self pollinated species but percentage of out crossing is fairly large and pollination is done by beetles (Gupta and Pareek, 1995).

Species and Cultivars

There are several species under the plant name senna but two species of *Cassia* viz. *C. angustifolia* Vahl and *C. acutifolia* Del. are well recognized for cultivation. The following varieties are recognized for cultivation in different parts of the country.

Tinunevelly senna- It is a semi spreading type and very popular in Tamil Nadu. Tinnevelly senna is a very fine variety and this is commonly exported from Tuticorin, South India.

Sona- This variety was released by Central Institute for Medicinal and Aromatic Plants (CIMAP) Lucknow in 1997. It is 115-120 cm tall, leafy and highly branched type. It flowers after 100-115 days and produces a large number of pods per plant. Pods are 5.0 cm long and 1.25 to 1.75 cm wide. It is suitable for cultivation in northern India. Dry leaves and pods yield is 11.0 and 4.0 q/ha, respectively while sennoside content averages 3.5 per cent (Singh *et al.*, 1997).

ALFT 2- It is a late flowering type and identified by the All India Coordinated Research Project on Medicinal and Aromatic Plants at GAU, Anand Centre through selection. It produces higher foliage yield.

Senna KKM-1: A superior genotype released from TNAU is a high yielding variety and is a selection from Tenkulam local. It is suitable for cultivation as rainfed crop.

In addition to this, farmers of the semi-arid and arid regions of western Rajasthan grow local varieties seeds.

Senna requires warm, bright sunshine and dry weather conditions. It is very sensitive to heavy rainfall and waterlogging conditions. It can grows in places where the average minimum and maximum temperature ranges between 10° to 42° C. In arid areas 250-400 mm rainfall is sufficient for a good crop. While in other areas rainfall of 600-700 mm per year is considered enough for good growth. It is observed that quality of senna leaves and pods deteriorates with heavy rainfall, cool climate and formation of dew during growth period.

Propagation

It is grown mainly through seeds since it is a herbaceous plant. Seed propagation is easy and cheaper method hence, other vegetative method has not tried so far. Biotechnological approaches has also tried for multiplication of plants in other species. For a rainfed crop, the land is prepared after the first monsoon rains and for an irrigated crop in December after the end of north-east monsoon. In Tamil Nadu, two distinct seasons are recognized for senna cultivation. The first start with the onset of south-east monsoon in September-October which is rainfed while second is sown in February. The field is ploughed 2-3 times, harrowed once or twice and than levelled for uniform germination of seeds. As crop cannot survive under water logging situation, therefore good drainage system in the field is very important aspect. It has been reported that application of Methyl parathion dust 5% (25 kg/ha) at last ploughing and levelling of land protect the young plants from attack of cutworms.

Seed sowing

Senna crop is grown both as rainfed as well as irrigated crop. Its sowing time varies from place to place. In western India June-July is optimal time for seed sowing. Optimum moisture in the soil is essential for uniform and early germination of seeds. July-August is the suitable time for sowing for rainfed crops in arid region. Lal and Prajapati (1999) mentioned that in the undulated land and sand dunes of Rajasthan, the seeds may be dry sown even in the month of June. In this condition seeds would germinate as soon as rains occur. In south India, it is sown in September-October after paddy harvesting as residual moisture available for germination of the seed.

For rainfed condition the ideal seed rate is 10-25 kg/ha (Lal and Prajapati, 1999). For rainfed crop, 27 kg/ha seed is required when seeds are sown by broadcasting. About 15 kg seed is needed for sowing under irrigated conditions (Gupta, 1971).But lower seed rate i.e. 7 to 15 kg/ha has been recommended for irrigated condition. For good germination seed should be of good quality and free from diseases. If the crop is grown in line sowing under irrigated condition then seed rate can be reduced. It is also reported that 6 kg/ha seed is sufficient if the seeds are sown by dibbing method in line seeding (Maiti and Kumar, 2001).

Senna seeds are reported to have dormancy which inhibits seed germination. Scarification of seed either with sand or by lightly pounding in a mortar with sand facilitates germination (Anonymous, 1992). Pre-sowing treatments enhance the germination and ensure a uniform growth of the crop. Before sowing seeds are soaked in water for 10-12 hrs. Seeds can also be soaked in liquid cowdung for 24 hrs and later liquid-mud treatment for 6 days may be given to hasten germination. Treatment of seed with Captan or Thiram or Agroson G.N.(2.5-3.0 g/kg seed) is effective in preventing the seedlings from damping off diseases.

The treated seeds are sown soon after the rains. Seeding is done through broadcasting of seed. In western India, line sowing at 45x30 cm spacing produced maximum yield. Line sowing and proper spacing provide ease in intercultural operations. Seed should be sown at 1-2 cm depth for proper germination. Deeper placed seeds either fail to germinate or produce uneven germination. After sowing light irrigation is required for quick germination and a good plant stand. Germination of seed takes 3-7 days depending upon availability of soil moisture. A plant population of 70,000 to 75,000 plants/ha is found optimum for maximum yield (Pareek *et al.*, 1983).

Generally, in north-west India, it is grown as rainfed crop. If rains are irregular or fails in early stage of growth, then it is advised to provide 1-2 irrigation for proper plant growth and yield. When crop is grown under irrigated condition, two irrigations are very critical; first after sowing and second at 30 days after sowing if soil moisture is inadequate (Maiti and Kumar, 2001).

Leaves are harvested when they are fully grown, thick and bluish in colour. The pods are collected when still green, and not considered good when they turn brown (Anonymous, 1992). It is reported that when crop is grown under irrigated conditions, three cuts are taken to get maximum foliage yield. The first harvest is taken after 90 days of sowing and subsequently second and third harvests at 150 days and 210 days (Maiti and Kumar, 2001). Under rainfed conditions, crop is harvested after 120-150 days after sowing. It is also reported that the crop which is harvested for leaves shows lower yield of pods. For seed production purpose, the pods are collected in February-March, when the pods are having brown colour. Seeds of this period show high percentage of germination. Pods yield was significantly correlated with plant height, number of branches and number of pods set/raceme at 100 days. Plant height and length of leaves exhibited the highest positive direct effects on leaves yield (Shankarnarayanan, 1995).

Leaves and pods yields in senna varies with rainfed and irrigated conditions and also with the region. In rainfed condition 600-700 kg dry leaves per ha can be obtained. In irrigated conditions 1500-2000 kg dry leaves and 700-1000 kg pods/ha can be harvested (Maiti and Kumar, 2001). A good senna crop can yield 300-400 kg seed per hectare. The crop can give about Rs. 5000-7000 net profit per ha of land.

Although a considerable research work on senna has been done but more emphasis should be given on its varietal improvement and production technologies especially for arid region and wasteland. The crop may help in the development of wasteland and establishing of sand dunes in arid ecosystem. For its commercial production, good quality seeds of improved variety should be made available to the farmers. In view of this, intensive research is needed on development of new variety having high yielding characters of good quality leaves and pods. Seed production technology is also needed to fulfill the seed demand. The main problem of farmers for this crop is marketing system of produce. Senna may be useful in rainfed farming system particularly for poor marginal soils and upliftment of the rural peoples as it is export oriented crop.

References

Anonymous 1992. The Wealth of India. Vol. 3., C.S.I.R., New Delhi.

Bentley, R. and Henny Trimen (1999). Medicinal Plants. Vol. II. Asiatic Pub. House, Delhi.

Bohra, M. K. and Sankhala, P. S. 1997. Senna- a cash crop for arid region. Vaniki Sandesh 21: 3

Fairbairn, J.W. and Shestra, A. B. 1967. The taxonomic validity of *Cassia acutifolia* and *C. angustifolia*. Lloydia 34: 67-72.

Gupta, R. 1971. Senna has a growing export market. Indian Farming, 21(8): 29-32.

Gupta, R. 1974. Wild growing senna (*Cassia angustifolia*) from Kutch, Gujarat. Current Science, 43: 89.

Gupta, R. and Pareek, S. K. 1995. Senna. pp. 325-336. In. K.L. Chadha and R. Gupta (eds.) *Advances in Horticulture* Vol. 11- Medicinal and Aromatic Plants. Malhotra Pub. House, New Delhi.

Gupta, R, Srivastava, V. K. and Maheshwari, M. L. 1977 . Sennosides content in senna leaves and pods as influenced by growth and nutrition. Indian J. Pharm. 39: 109-111.

Hussain, T. and Kapoor, S. L. 1990. Ennumeration of Legumes in India. N.B.R.I., Lucknow.

Ilangovan, R. ; Subbiah, R. and Natarajan, S. 1990. Effect of spacing, nitrogen and phosphorus on certain growth parameters of senna (*Cassia angustifolia* Vahl.). South-Indian Horticulture 38: 53-54.

Jindal, S. K, Manjit Singh and Sivadassan,Rekha 2005.Induced mutations in Senna, J. Arid Legume,2(!):120-123.

Lal. S. and Prajapati, N. D. 1999. Senna- The Money Minting plant in drylands. AGRO-INDIA. April 1999, pp. 14-15.

Maiti, S.M. and Kumar, D. 2001. Cultivation of Senna. NRC for Medicinal and Aromatic Plants, Anand, Gujarat.

Pareek, S. K.; Srivastava, V. K.; Maheshwari, M. L.; Mandal, S. and Gupta, R. 1983. Investigation in agronomic parameters of senna (*Cassia angustifolia* Vahl.) as grown in north-western India. Indian J. Trop. Agric. 1: 139-144.

Rajendran, S. M. and Mehrotra, B. N. 1996. Unrecorded medicinal uses of plants among Parambikulan tribals, Kerala, India. pp. 181-183. In S. K. Jain (ed.) Ethnobotany in Human Welfare. Deep Pub. New Delhi.

Shankarnaraynan, R. 1995. Evaluation of senna. Madras Agric. J., 82:12.

Sharma, A. K. and Azam, M. M. 2001. Senna : A potential industrial crop for desert. p. 84. Symposium Abstract- Impact of Human Activities on Thar Dersert Environment. Feb. 15-17, 2001. CAZRI, Jodhpur.

Sharma, B. B. L. and Prajapati, P. 2001. Senna : A new commercial crop for the arid region. pp 221-226. *In:* B. B. S. Kapoor; K. K. Singh; Ahmed Ali and R. K. Gehlot (eds.) An Assessment of Natural Resources the Indian Desert. Madhu Publications, Bikaner.

Shiva, Alka 1999. Medicinal Plants First-Aid Remedies their Sources and Methods of Uses. Surya International Publication, Dehradun.

Singh, Manjit, Jindal, K. and Sivadassan, Rekha 2005.Evaluation of senna genotypes for dry biomass and related traits under arid conditions, J. Arid Legume,2 (1):123-126.

Singh, S. P.; Sharma. J. R.; Misra, H. O.; Lal, R. K. ; Gupta, M. M. and Tajuddin A. D. 1997. Development of new variety Sona of senna (*Cassia angustifolia* Vahl.). Journal of Medicinal and Aromatic Plant Sciences, 19: 446.

Tomar, V. K. S.; Singh, S. P.; Singh, K.; Singh, A. K. and Kumar, S. 2000. Senna Ki Kheti, Phal-Phool, October- December : 33 & 36,

Tripathi, Y.C. 1999. *Cassia angustifolia*- a versatile medicinal crop. International Tree Crops J., 10(2): 121-129.

Upadhyaya, S. K. and Singh, V. 1989. Phytochemical identity of *Cassia angustifolia* Vahl. and *C. acutifolia* Del. National Academy Science letters, 12: 181-182.

Chapter-41

Henna (*Lawsonia inermis* L.)

R. S. Singh
Central Institute for Arid Horticulture, Bikaner-334 006

Henna (*Lawsonia inermis* L.) is a multipurpose shrub also known as mehandi. It is grown through out the country as a hedge plant. In India, it was once grown extensively in Punjab and Gujarat. However, gradually major cultivation has shifted to the arid fringes of Rajasthan with the Pali district having the maximum area under henna cultivation. It is taken as an annual ratoon crop under rain fed conditions. Presently there are over 22,000 ha of henna plantations in the Pali district nearly 70 percent occurring in the Sojat area which produces around 20000 tonnes dry leaves worth Rs.40 crores annually. Sojat city is world famous for its mehandi or henna (Heena) powder.

Farmers in this region have produced henna for generations, and regard it as a desirable crop, as well as part of their heritage. Much of this area is hilly, with small creeks and some rivers. Henna requires little or no fertilizer, survives on rainwater, and requires little specialized labor. Both men and women work the fields. Women are often responsible for weeding and harvesting; men take care of soil preparation and transportation. A family can often manage their own cultivation and harvest. Henna will survive a drought season that would damage other crops, and is therefore highly attractive to farmers (Khandelwal *et al.*, 2002).

In Gujarat, mehandi is also cultivated as cash crop in Kachchh region. Mehandi is cultivated in low rainfall area on marginal and salt affected soils. The cultivation of henna in such problematic areas will not only improve socio-economic conditions of the farmers but will also improve soil fertility by controlling soil erosion through wind and water in Kachchh region (Vyas, 2001). The henna plants can grow any type of soil from loam to clay loam and can tolerate alkalinity in the soil (Lodha, 2003). Henna (*Lawsonia* sp.) is a rainfed crop commercially cultivated in Rajasthan on 32,084 ha (2001-02) for its leaves, which are important source of a natural dye. The area of heena cultivation has increased to41,149ha with production of 17,511MT in Rajasthan (2010-11). A study on assessment of the profitability and economic viability of henna cultivation based on the data collected from a sample of 100 farmers during 2003-04 in the Pali district was carried out. Based on the factors like net present value, internal rate of return, benefit cost ratio, henna cultivation practiced by the farmers has been found to be a financially viable proposition. The financial viability parameters have been found to be more sensitive to changes in prices than cost (Chand and Jangid, 2007).

Besides this, leaf cuttings are also collected from existing farm and field boundary plantations in Gujarat and Madhya Pradesh. In Uttar Pradesh, henna is grown for its scented flowers used in the essence industry.

Composition and Uses

This common garden species is one of the oldest sources of plant dye known to man. Henna leaves are used as dye. Dried henna leaves contain about 9% moisture, 15% ash, 5-10%tannin, and 25-33% total soluble solids with the dye compound, lawsone (2-hydroxy-1,4-napthoquinone), being the major active principle. The dye produces a reddish orange colour in aqueous solution and shows wide variations in concentration up to over 2% on dry weight basis. Other chemical constituents of leaf include Gallic acid, glucose, mannose, flavones glycosides, fat, resin and mucilage etc. People in the Middle East and Asia have been using the leaves for hair dying and staining of palm, sole of feet and nails. In addition of colour development henna is found to check hair lice and inhibit growth of many dermatophytes, including *Pityrosporum ovale* which causes dandruff. In view of this hypoallergenic nature and other beneficial attributes, henna leaf extract is rapidly becoming an important constituent of herbal cosmetic products like conditioning hair oils, shampoos and other skin care formulations.

Henna has several medicinal uses owing to its antipyretic, analgesic, immune stimulatory and diverse anti microbial activities (Mohammed, 1996). It is an important ingredient of ayurvedic and unani medicines, and is used as a folk medicine in the Middle East and Asia. Almost all parts of the plant are useful. While leaf paste or decoctions help treat or prevent various skin disorders like boils, burns, bruises, skin inflammation, sore throat and ulcers in the mouth, the flowers are used as a soporific. The fruit is considered an emmenagogue and is found to contain diosgenin saponin. Henna is also effective in the treatment of diseases like leprosy, herpes, jaundice, gonorrhea and bronchitis. Roots are bitter, refrigerant, depurative, diuretic, emmenogogue and abortifacient. It is useful in burning sensation, dipsia, leprosy, skin diseases, amenorrhoea and dysmenorrhoea. It is also useful in premature graying of hair. Leaves are bitter, astringent, acrid, refrigerant, diuretic, expectorant, anti-inflammatory, constipating, febrifuge and liver tonic. It is useful in wounds, ulcers, burning sensations, inflammations, scabies, leprosy, leucoderma, boils, amenorrhoea, dysmenorrhoea, fever, ophthalmia, haemorrhages, splenopathy and strangury. Flowers are cardiotonic, refrigerant, febrifuge and tonic. it is useful in cephalagia, burning sensation, insomnia and fever. Leaves are useful in wounds ulcers strangury cough, bronchitis, dysentery etc. It lowers body temperature to soothe headaches, fevers and burning feet. Besides cosmetics and medicinal applications, henna could prove useful in agriculture as a bio protectant or bio pesticide against crop diseases and pests. Henna leaf extract exhibits strong fungi toxicity (Tripathi *et al.*, 1978). It is also grown as hedge plant for good topiary work in the garden. The dried twigs of heena plants are used for fuel and fencing purposes. Fragrant flowers are used for extraction essential oil which used as scent.

Botanical Description

Henna (*Lowsonia inermis* L.) is a perennial shrub native to North Africa and South west Asia. Mehandi belongs to the family Lythraceae. It is a much-branched glabrous shrub or small tree 2-6 m in height, which may be spiny. Bark grayish-brown, unarmed when young, older plants with spine-tipped branchlets. Young branches quadrangular, green but turn red with age. Leaves opposite, entire, subsessile, elliptic to broadly lanceolate, 1.5-5 x 0.5-2 cm, glabrous, acuminate; veins on the upper surface depressed. Flowers small, white or reddish in colour, numerous; in large pyramidal terminal cymes, fragrant, 1 cm across, 4 petals crumpled in the bud. Calyx with 2-mm tube and 3-mm spread lobes; petals orbicular to obovate, white or red; stamens 8, inserted in pairs on the rim of the calyx tube; ovary 4 celled, style up to 5 mm long, erect. Fruits small, brown, globose capsules 4-8 mm in diameter, many-seeded, opening irregularly, split into 4 sections, with a persistent style. Seeds are 3 mm across, angular, with thick seed coat. The specific epithet means unarmed or without spines. The plant tolerates heavy pruning and has quick regeneration ability. Every year pruning is done to harvest leaves. Ratton crop can also be harvested after winter rains.

According to Cowen (1970), the genus of this plant was named *Lawsonia* to honour John Lawson, the traveler. Plant finds importance due to presence of orange -red dye in the leaves and essential oil in the flowers.

Propagation

It is commonly propagated by seeds and vegetative means *viz.* stem cuttings (Lodha, 2003). Cuttings are used for growing hedges but seedlings are transplanted for commercial cultivation due to their better survival rate in the field. At least one lakh nursery raised seedlings are needed for transplanting one hectare. However, developing a good seedling nursery of henna requires intensive care and management.

Seed

Regeneration from seed is also possible but seeds are not always viable. Seeds are pretreated by steeping in water for 3-7 days with daily changing of the water; they are then placed in small heaps and kept moist and warm for a few days. The swollen, well-drained seeds, with a soft seed coat, are mixed with sand and a pinch of the mixture is put on top of the area to be sown. Germination takes place after 2-3 weeks, at a rate of 70%. Seedlings require 4-5 months in the nursery, at which time they will be about 40 cm tall; they are lifted, cut back to about 15 cm and transplanted into the field. Henna seedlings are hardy and can survive with little watering, but the planting site should be well prepared by ploughing and incorporation of manure or compost.

Henna seeds have a seed coat and take lot of time to germinate, with less than 20 percent germination. Therefore, seeds are treated to break the hard seed coat. Nursery growers normally soak the seeds in water before sowing for a period seven to ten days, with frequent changing of water, and achieve around 60-70 percent germination. Hot water treatment and acid treatment are also suggested. The seeds after conditioning are mixed with fine sand and sown broadcast in well prepared 3 X 10 m^2 nursery beds at the rate of one kg of seeds per bed. Sowing is done during March when temperature condition are optimal for germination (25-30°C). In about 10-15 days, the seeds germinate completely. About 10kg seed is found sufficient to obtain seedlings for planting one hectare. The manure and fertilizer schedule followed in the nursery beds is 10 tonnes FYM+60kg N+40Kg P_2 O_5 per hectare. Nitrogen is normally applied in split doses, escaping the basal dose, as and when required to boost seedling growth.

Henna seedlings are very small and extremely sensitive to water stress and hot desiccating winds for about 30 days between emergence and the onset of the rapid growth phase. During this period, the beds are watered regularly, once every two days, by the check basin method and preferably during the night. Later, the irrigation frequency is decreased to once every three to five days depending on the weather. Further, the beds are kept free from weeds by hand picking. The nursery is also protected from hot, dry winds by providing windbreakers around its perimeter. Propagation of henna is done mainly by seeds for commercial cultivations and by stem cutting for hedge plantation as this gives flexibility in time of planting and uniformity of plant type (Singh,1994; Singh and Gupta,1998).

The seeds have a xeromorphic structure. A study on depth of sowing demonstrated that seeds on soil surface germination best (88.8%), whereas those sown 2.5 cm deep do not germination.

Henna nursery is also susceptible to a number of diseases such as damping off, bacterial leaf spot and leaf blight. Suitable control measures are necessary to keep the nurseries healthy. In Rajasthan, dodder infestation is also a significant threat; on its occurrence, the affected seedlings are either eliminated along with the parasite by uprooting or are saved by pruning the affected plants parts, which affect the seedling quantity. As the fruiting bodies and seeds of higher plants parasite resemble those of henna, it is difficult to detect it in the seed lot. Therefore, to prevent its occurrence, the seeds must be obtained from clean fields. To further ensure three to four months, the seedlings reach 30-40 cm height and are robust enough for transplantation in the field.

Vegetative propagation

Cuttings

L. inermis is easy to propagate from cuttings. It is propagated through seeds or stem cuttings during favourable season (Lodha,2003). Vegetative propagation was attempted in February-March in net-house having 50 per cent shade using IAA, NAA and IBA singly and in combination during 2001. Percentage of rooting over the treatments varied from 53.3 to 100 percent, with a mean of 90.7 per cent. It was possible to get cent per cent rooting by using IBA or IAA. Best performance for all the observed characters except for number and length of roots per cutting was given by 7000 ppm Indole Acetic Acid (IAA). Results for vegetative propagation of henna during different months without use of any hormone are presented. Rooting per cent ranged from 33.3 to 100 per cent with a mean of 72.8 per cent. Cent per cent rooting was seen in the month of March, July and August. Maximum numbers of root per cutting was observed in March (9.3), length of roots per cutting in November (18.8 cm) and fresh and dry weight of roots per cutting in January (0.98 g and 0.30 g, respectively). But overall best performance was shown in the month of July, which showed 100 per cent rooting and values for all other characters were at par with the highest values.

There is variation in seed germination in different accessions. In general, it is not difficult to raise nursery, even though initial soaking, proper sowing depth, maintenance of moisture during initial germination stages are some of the critical points. Henna is a easy to root species and hence vegetative propagation is most suitable method for multiplication.

Micro-propagation

Henna is multiplied through tissue culture techniques for raising true to type plants on large scale. Raj Bhansali (2005) observed that bud proliferation and their multiplication were initiated from apical and axillary explants of nodal buds and seed explants tissues of *L. inermis* within 8-10 days of inoculation onto MMS basal medium supplemented with BA, KN and NAA. Among the different cytokinins tested, BA+KN were the most effective for shoot proliferation and multiplication. The maximum shoot proliferation was observed both in apical and axillary meristems cultured on MMS medium supplemented with 1 mg/l BA, 0.5 mg/l KN and 0.1 mg/l NAA within 4 weeks of cultured. The shoots proliferated in large number as miniature plants from dark callus tissues or nodal buds. There was no sign of shoot proliferation when explants were cultured in media devoid of cytokinin. At higher concentrations of BAP or KN the rate of shoot proliferation declined. The study also revealed that the continuous light (24hr) was more conducive to higher rate of shoot multiplication than (16hr) photoperiod. Proliferation of shoots from callus of apical bud of *Lawsonia inermis* on MMS medium supplemented with 1mg/l BAP , 0.5mg/lKN , 0.1mg/l NAA and 3% sucrose was archived. The rate of multiplication was high and stable with regular subculture after every 25 days.

According to Singh *et al.* (2012), attempts were made to develop micro propagation method for cloning of henna plant. The explants of

3-4cm. in length with 2-3 nodes were initially washed with tap water and then surface sterilized with 0.1% $Hgcl_2$. Surface sterilized explants were inoculated on MS medium. The nodal explants show bud breaking on MS +BAP (3.33μM) medium. After optimizing the condition for the culture initiation from explants the in vitro produced shoots with part of mother plant were sub cultured on fresh medium supplemented with different cytokinin BAP, Kin(4.44+2.32μM) medium. The repeated sub culturing was performed at regular intervals of 3-4 weeks. *In vitro* regenerated shoots were excised individually and cultured on MS medium with one fourth, half strength of MS salts supplemented with IBA .The roots were developed within 10 days. By this way, large scale planting material can be produced.

Quality Standards for Planting Material

There is no standard for planting material in henna crop. However, plants should be true to type and disease free. Generally, seedlings are used for transplanting in the field, which may be of optimum age and height of 30-40cm.

References

Chand, Khem and Jangid, B.L. 2007. Economic Viability of Henna in Semi-arid Rajasthan, Agricultural Economics Research Review, 20 : 137-146.

Khandelwal, S.K.; Gupta, N.K.; Sahu, M. P. 2002. Effect of plant growth regulators on growth, yield and essential oil production of henna *(Lawsonia inermis* L.*)*. The Journal of Hort. Sci. and Biotechnology , 77(1): 67-72.

Lodha, R.M. 2003. Cultivation of Heena in Rajasthan, Natural Product Radiance,2(6):306-309.

Mohammed, A. 1996. Chemical and Medicinal evaluation of *Lasonia inermis* (Heena) . Hamdard Medicines, 39:43-48.

Raj Bhansali, R. 2005, Micro-propagation of Mehandi (*Lasonia inermis*). Heena, Cultivation, Improvement and Trade. CAZRI, Jodhpur, pp. 45-47.

Singh, K.C. and Gupta J.P. 1998. *Lawsonia alba*- cash crop for arid zone wastelands, ICAR News, 4(3):15-16.

Singh, R.S. 1994. Pashchami Rajasthan mein Mehandi ki kheti. Udyan Lok, 6(1):22-24.

Singh,S., Rajora, R. and Sharma, N. K. 2012.*In vitro* clonal propagation and regeneration of *Lawsonia inermis*, Int. J.Biotech. Res., 2(2):1-7.

Tripathi,R.D., Srivastava, H.S. and Dixit, S.N. 1978. A fungitoxic principle from leaves of *Lawsonia inermis*, Lam. Experientia, 34:51-52.

Vyas, S. P. 2001. Heena- A cash crop for arid Gujarat. Den News, CAZRI, Jodhpur, 5:2-3.

Chapter-42

Khejri (*Prosopis cineraria*)

J. C .Tewari* and R. S. Singh**
**Central Arid Zone Research Institute, Jodhpur - 342 003*
***Central Institute for Arid Horticulture, Bikaner- 334 006*

Khejri (*Prosopis cineraria*), also known as *shami* (Sanskrit), *janti* and *jand* (Delhi, Haryana, Punjab), *sumi* (Gujarat), *banni* (Karnataka), *kandi* (Sindh) is a leguminous tree species of arid and semi-arid regions. It not only tolerates the extreme edapho-climatic conditions of these regions but also bears flowers and fruits and that too during the driest period from February to May (Pareek, 2002). Khejri is a tree having multipurpose uses, *viz.* nutritious rich pods and fodder and valuable fuel, besides its favourable effects on ecology and soil fertility(Arya *et al.*,1991). It is an important perennial component compatible to almost any companion crop in various agri-horti-forestry systems of production in the hot arid ecosystem (Harsh and Tewari, 1998). The crops such as bajra, guar, moth, mung, cowpea and ground nut during kharif and wheat, mustard, gram, *etc.* during rabi season are grown between the natural plantations of khejri trees without adverse effect on growth of crops. Almost all the vegetable crops grow better between the widespread natural this leguminous tree. Also the fruits such as ber, aonla, bael, kinnow, lime, etc. have been observed to perform better in association of khejri trees. Excellent growth and development of desert plant species of commercial significance such as jharber, ker, lasora, pilu, rohida, kumat, phog, kheep and grasses like sewan, dhaman, *etc.* has been observed in association with khejri trees (Mann and Saxena,1980; Samadia,2007, Samadia *et al.*, 2009).

The tree is found extensively in semi-arid and arid parts of Rajasthan, Haryana, Gujarat, Punjab, Delhi and Southern states also.

Naturally occurring of khejri tree has been reported from Afghanistan, Iran, Pakistan and India (Puri *et al.*, 1964).

The distribution pattern of khejri from Aravali hills to Thar of Jaisalmer is varied. All arid land forms except hills and saline depression receiving an annual rainfall of 150-500mm have good population of khejri tree. Optimum density of its tree population is, however, confined in areas having 300-400 mm rainfall. Because of its capacity to avail perched water and to absorb moisture from rains through its foliage, it can grow in extremely arid tract (100-150 mm rain fall). Plant density varying in north India from 5 plants per ha. in extremely hot areas like Pokaran, Jaisalmer to 150 plants per ha in Sri Ganganagar which is arid command area. In Pali- Sirohi-Jalore region the density of plants vary from 10 to 250 per ha.

In Rajasthan state, the extremes of high (March - July) and low temperature (December - January) conditions alongwith other abiotic environmental stresses in the hot arid region limiting the choice of crops, cultivars and quality of produce and productivity. Horticultural exploitation with native crop species is generally found to be potential and stable in arid and semi-arid areas that receive approximately 150–500 mm of rainfall annually, which is concentrated between the months of June - September during the monsoon season. Out of 9 - 12 mean rainy days/annum in the arid areas, much of this rain often comes in only a few (1–3) heavy rainfall events, which can be erratic and unpredictable. The extreme high temperatures conditions during the kharif season from July to October can ranged between 30 to 46^0 C, prolonged period of drought in combination with sandy soils are create an environment where few rainfed crops can survive.

Nutritional Value and Uses

The khejri tree is a source of nutritious fodder (*loong*) for animals in the arid region. Its green leaves contain 11.9-18.0% crude protein, 13-22% crude fibre, 43.5% nitrogen free extract, 6-8% ash, 2.9% ether extract, 2.1% calcium and phosphorus. From a mature (20-25 year age) tree about 25-30 kg loong can be harvested every year. A full grown seedling tree at an age of 20 years yields 10-15 kg tender pods (sangri) in a year. Tender pods are extensively used as nutritious vegetable both fresh as well as after dehydration. These pods are rich in crude protein (18%), carbohydrates (56%) and minerals such as phosphorus (0.4%), calcium (0.4%) and iron (0.2%). The fully ripe pods are edible and it can be used for animal feed. They contain 9-14% crude protein and 6-16% sugars.

The ripen pods (khokha) can also be powdered and used for preparation of bakery items such as biscuits and cookies.

Khejri is an important flora of Thar Desert (Bhandari, 1978). It readily perpetuates itself from the dispersed seeds consequently giving rise to a wide range of variability with respect to agro-morphological traits in the natural stands. Natural population of khejri is abundant in arid and semi-arid region of north-western India (Pareek and Nath, 1997; Dwivedi *et al.*, 1997; Samadia *et al.*, 2002). The density of khejri varies depending upon rainfall pattern, range management and the farming system. Utilization systems determine their population density on non-culturable wastelands and barren lands. In general, the density of seedling population ranges from 15 - 165 trees/hectare from arid to semi arid agro-climate of Rajasthan state. In Bikaner district, the density varies from 10 - 55 trees/hectare. The elite khejri has identified in naturally growing belt of Rajasthan keeping in view to horticultural traits (Vishalnath *et al.*,1998).

Khejri plants have long been used by mankind in many ways. In desert, khejri is called, kalpa taru because of its several utility. It has vast socio-economic importance in arid region. Khejri tree is worship by a community of people for Godess of snake *i.e.* Goga ji in western Rajasthan. The parts of tree like bark, flowers etc. are also used for medicinal purposes. Its nutritive green leaves contain 11.9 crude protein, 17.50 % crude fibre, 4% phosphorus, 2.1% calcium and 8.1 % ash with high calorific value (Bhandari, *et al.* 1979). Its leaves are considered to be good fodder for animals. The leaves (fodder) is called 'loong' which is obtained every year by lopping of trees.

Nutritionally pods of khejri have 18% crude protein, 2% fat, 56% carbohydrates, 26% crude fibre, 0.4% phosphorus, 0.4% calcium and 0.2% iron. Khejri pods (green) i.e. Sangri which are used as vegetable both fresh as well as after dehydration. It is used for making mixed vegetable with other local plant material i.e. called Punch kutta a mixture of five species. Pods are used for pickle preparation. Its ripe pod is also edible and is used to prepare flour. It contains 9-14% crude protein and 6-16% sugars. Lights brown to chocholate colour pods on ripening are eaten raw by children in rural areas. Twigs/Branches are used as fire wood and for fencing of fields. The wood is suitable for cart-wheel, agricultural implements, tools handles, doors, windows, furniture and other smell turnery articles besides fuel wood. Khejri wood is hard and durable.

Botanical Description

Khejri (*Prosopis cineraria* (L.) Druce) belongs to sub family Mimosaceae of Leguminosae family. It is most common medium size much branched slow growing, evergreen, woody tree of Indian desert. Its foliage is dark green even during hot summer season. Trunk is dark gray, bark splited with deep and rough surface and thorns of 3-6 mm size are found on the stem and slender glabrous branches in scattered manner. The plant has deep and extensive root system and can withstand drought for many years.

A wide range of variability in horticultural traits of this species has been observed in Indian arid zone (Vishalnath *et al.*,1997). Wide variation is observed in quality traits of pods such as taste (bitter, acrid, sweet), tenderness (hard, semi-hard, less tender, tender), type (round, flat, flat-round or straight, wrinkled), fibre content (fibrous, less fibrous, fibreless), colour (green, light green, dark green), length (6-39 cm), thickness (0.9-3.5 cm), seed number (2-30/ pod), seed weight (20-50 mg).

The new foliar growth, flowering and pods formation are found during extreme dry months (March- June) when all other plants in desert may be seen as leafless. Pinnae mostly 2 pairs with a gland between each pair opposite, 2.5-8.0 cm long, leaflets 8-12 in pair oblong and murconate at apex, unequal sides, base rounded and very oblique 5-12 x 4 mm leaflet size.

Flowering starts during February-March and the pods are formed in March-April. Yellow colour flowers in slender spike 7-11 cm long and they are mostly found solitary or in terminal panicles. Calyx 5x1.5 mm long cup shaped fantly 5 toothed corolla yellow 3mm long, recurred stamen filament 3 mm long. It is a cross-pollinated crop and natural regeneration takes place through seeds. Arya *et al.* (1991) found pods of khejri in Shekhawati region to be 12-15 cm in length and 5-8 mm in width containing 2-20 brown or black seeds. A wide range of variability in horticultural traits of this species has been observed in Indian arid zone (Vishalnath *et al.*,1998). Pods are formed in bunches, light to dark green, straight 10-20 cm long, cylindric having 10-15 seeds of dull brown colour at maturity (Bhandari, 1978). .

Intensive study was made to assess the existing variability within a population of *Prosopis cineraria* trees having horticulturally important traits (Samadia *et al.*, 2002).

Owing to natural perpetuation through seeds and cross pollination behaviour, high degree of variability is observed with respect to vegetative growth, flowering and fruiting, pod yield and quality attributes. This provides an excellent opportunity to identify elite plants that bear quality pods or any other traits of commercial significance.

Earlier, there was no named variety for uniform and high quality sangri yields under the traditional system of production and there is no systematic orcharding for its commercial cultivation. However, a kherji variety "Thar Shobha" has been released at Institutes' level. All the seedling population so far available in the country is as scattered seedling plantations by the growers or forest plantations. Traditionally, due to edible quality characters of khejri pods at tender stages, the rural people collect it for fresh consumption as cooked vegetable or dehydrated tender pod for off-season use. There are no quality standard characters in the pods. So far, what-ever produce of sangri is available and the same is being harvested as per requirements and used it by the desert dwellers. In the past, due to non-availability of crop production practices in khejri, no much progress has been made for the varietal improvement and it was first of all started at CIAH, Bikaner and now, Thar Shobha (Khejri Selection - 1) was recommended in 2007 (More *et al.*, 2008: Samadia *et al.*, 2009) and popularized for commercial orcharding.

Thar Shobha is the first variety to develop plantations for uniform, better quality and high sangri production on commercial scale. A five year old grafted plant yield to harvest of 4.25 kg sangri and 6.27 kg loong in a year with improved production technology. On the basis of characterization of Thar Shobha over the seasons under varying situations, the important characteristic on foliage and pods are summarized. The tender pods at marketable stages are 13.1 - 20.2 cm in length, 0.17 - 0.42 cm width, 0.97 - 1.75 g weight and seeds not separable in the stages of vegetable use. Tender pods are light green in colour, straight, roundish-flat, soft and sweet. The ripen dried pods are 13.5 - 22.5 cm in length, 0.49 - 0.65 cm width, 1.65 - 2.15 g weight and number of seeds ranged from 11.15 - 16.85 per pod. The weight of 100 seeds is 5.02 g. Plants of this variety are vigorous in growth and thornless in nature(More *et al*, 2008). At the age of eight years, the total bio-mass production from Thar Shobha is more than double in comparison to the seedling trees of same age. The variety tolerates extremes of high (> 48^0 C) and low temperature (- 4^0 C) and aridity conditions in Thar Desert. It is a drought tolerant and therefore, it has high potential of production under rainfed crop cultivation.

Propagation

Khejri was mainly regenerated by seeds but now the technology has been developed for its fast multiplication. The vegetative methods of propagation is tried to develop true to type plants because of wide range of variability in seed propagated trees of khejri in arid region. The air layering methods gives very little success. However, rooting in cuttings are very difficult even after treatment with growth substances. The treatment with 6000 ppm IBA produced maximum (32%) sprouts. However establishment of sprouted cuttings were not encouraging (Pareek *et al.*, 1997) The tree readily produces root suckers on removal of main trunk. The tree has high regeneration capacity in arid region.

Both patch and shield budding are found successful (76%) method of propagation in *Prosopis cineraria*. The better results were obtained when budding was done either in morning or evening hours (Pareek, 2002).

Propagation is usually made by seeds, but the trees coppice easily and may be managed that way. The trees stand lopping and pruning.

Seed propagation

The wide spread distribution of khejri is solely by seed propagation. The natural dispersion of seeds resulted into self-sown seedlings as well as nursery raised seedling plantations has been protected and promoted by the inhabitants as a lifeline tree of desert eco-system. The ripen pods with full of seeds falling down in June and eaten by small animals (goat and sheep) and through their litter its spreads everywhere. The seeds do not show any dormancy and germinates quickly. The monsoon rains in June-August helps in easy germination of seeds. The young seedlings grown up very fast and develop strong tap-root system within the rainy season. The plants under protection grow very well and took shape in 4 - 5 years. Where as the unprotected saplings drastically and regularly bruised by animals and its above ground growth hampered but with well developed root system for the survival.

Freshly harvested and non-infested seeds should be sown in polyethylene tubes (30´10 cm) filled with mixture of FYM and sand in 1:1 ratio in the nursery. Prior to sowing, seeds should be soaked in water containing 2 % common salt for two hours, and after that dry it for few minutes and treated with fungicide like Captan or Bavistin @2g./ kg seed. The seedlings raised under open conditions were with better height, collar and stem diameter than those under the shade net. Irrigation by rose can was better than by mini sprinklers. The seedlings of February and June month sown nursery are ready for field planting during July and October, respectively and gave very high (80 - 90 %) field establishment. Since, the success in budding and subsequent rate of sprouting of scion depends on the vigour of rootstock and the desirable thickness of rootstock, it is therefore, proper care is required in the saplings either developed *in situ* or raised in nursery followed by field planting.

To raise true-to-type plantations of the selected elite trees vegetative propagation technique is required. The vegetatively propagated plants would also have precocious flowering and pod production, i.e. 3-4 years after planting, as against the long juvenile phase of 10-15 years observed in khejri seedling trees. Attempts have been made from time to time to multiply khejri trees by cuttings, air layering and tissue culture. But the results were not encouraging. Also the plants raised by cuttings and air layers lack strong taproots and thus may not survive in the extreme arid agro-climate. Realizing the potentiality of khejri and its significance as a horticultural crop, the systematic technological interventions for its exploitation were initiated at CIAH at Bikaner with the objectives such as identification of elite plant types for use as scion material studies; to develop simple vegetative propagation technique, and to identify rootstocks and standardize nursery technique for fast, easier and mass multiplication or *in situ* establishment techniques.

Vegetative methods

Khejri can be multiplied through vegetative methods to raise true to type plants. It can be multiplied through stem cutting, budding and grafting.

Cuttings

It is the simple and low cost method of vegetative propagation. Propagation by cuttings could be a suitable technique for rapid mass multiplication of khejri, but it was not so successful to raise plants easily. Rooting in cuttings was low (Pareek *et al.*, 1997). Further, low percentage of rooting and survival in cuttings have also reported by Arya *et al*,(1994) Propagation by cuttings is difficult, as often occurs with *Prosopis* spp.

Air layering

At CIAH, studies on layering were done on one year old shoots emerged after lopping the khejri trees in December and treated with lanolin paste. The success obtained in air layering was very low and gave 35 % rooting in June. The method is difficult, slow and plant establishment in the field is also very poor. Attempts were also made to develop true-to-type plants through air layering in khejri (Rawat *et al.*, 1982-83; Solanki *et al.*, 1986).

Budding

Khejri trees can be successfully propagated by patch budding or bud grafting technique (Pareek, 2001), and therefore, systematic initiatives were taken at CIAH from the year 2001 (Pareek, 2002; Pareek and Purohit, 2002)). For this, its selected elite tree, having a record of producing good quality edible pods, were pruned during December to induce new growth bearing juvenile scion buds from March. The scion buds were patch budded in the month of March - April and July - October on *P. cineraria* seedlings rootstock both under nursery raised and *in situ* planted seedlings. On an average, 76 per cent success was obtained. The extent of scion bud success, its growth in terms of length, diameter and linear growth depended on the stem diameter of the rootstock, environmental conditions and management after bud intake. The patch budding technique was also found to be most potential for *in situ* establishment of khejri plantations on 12 - 24 months old rootstocks as well as top-working the wild seedling trees of over 3 year age. The identified mother tree of the variety should be pruned during the month of November - December to induce juvenile buds suitable for budding. On *P. cineraria* rootstock, budding success depended on diameter of the stem. It appears, for best results stem diameter of the rootstock should be more than one cm. In that case, it may be possible to even achieve nearly cent per cent budding success. Bud sprouting took 18 to 40 days and all the sprouted scion buds survived (Samadia,2007; Samadia *et al.*, 2009).

Wild seedlings of *P. cineraria* and *P. juliflora* and seedlings raised from the seeds of *P. alba* and *P. nigra* were used in the studies as rootstock to graft the buds taken from the selected mother tree of *P. cineraria* growing at the research farm of CIAH, Bikaner. The possibility to develop inter-specific graft combinations of *Prosopis cineraria* with the other species such as *P. juliflora, P. alba* and *P. nigra* were also explored. The graft combinations of the Asian *P. cineraria* with any of the three species of *Prosopis* section *Algarobia* were incompatible. The species within the section *Algarobia* were compatible with each other. Budding success was poorer when *P. cineraria* scion buds were grafted on the *Prosopis* species of South American origin (*P. alba* and *P. nigra)* than on its own roots. There was no graft success on *P. juliflora* rootstocks. As high as 88.9 % successful grafts in *P. cineraria* and *P. juliflora* and 72.2 % in *P. alba* were obtained by patch budding on their own rootstocks. In *P. nigra,* 83.3 % successful grafts were obtained by budding on *P. juliflora* rootstock. Thus by taking bud-wood from the selected mother tree of these species (considering the attributes such as high yield of good quality edible pods, leaf fodder, timber, fuel-wood, etc.) and then patch budding on the identified rootstock, their true-to-type plantations can be raised. The quality standard of the existing wild plantations can also be upgraded with respect to these attributes by top-working with the bud-wood from elite mother trees (Pareek *et al.*, 2006).

Quality Standards of Planting Material

The planting material should be of improved variety and of appropriate age for planting in the field. It is a slow growing tree hence, healthy plants should be selected. Plant should be free from disease and insects infestation.

References

Arya, S., O.P.Toky, R.P.Bist and R.Singh, 1991. *Prosopis cineraria*- Promising multipurpose tree for arid lands. Agro-forestry Today, 3: 13-14.

Arya, S., R. Tomar and O. P. Toky 1994. Effect of plant age and auxin treatment on rooting response in stem cuttings of *Prosopis cineraria*. Journal of Arid Environments, 27(1): 99-103.

Bhandari, D.S., H.M. Govil and A. Hussain 1979. Chemical composition and nutritive value of khejri (*P. cineraria*) tree leaves. Ann. Arid Zone, 18: 170-173.

Bhandari, M.M., 1978. Flora of Indian desert. Scientific Pub., Jodhpur, p. 152.

Dwivedi, N.K., D.C. Bhandari and S.K. Verma 1997. Collecting *Prosopis cineraria* germplasm in western India. IPGRI Newsletter for Asia, The Pacific and Oceania, 22: 19.

Harsh, L.N. and J.C. Tewari 1998. *Prosopis* in the arid regions of India: some important aspects of research and development. In: *Prosopis* species in the arid and semi-arid zones of India (Ed. J.C. Tewari, N.M. Pasiecznik., L.N. Harsh and P.J.C. Harris). Conference Proceedings, Central Arid Zone Research Institute, Jodhpur (India). November 21-23, 1993; pp.5 -10

Mann, H.S. and Saxena, S.K. 1980. Khejri (*Prosopis cineraria*) In the Indian desert- its role in Agroforestry. CAZRI Monograph 11. CAZRI, Jodhpur.

More, T. A., D. K. Samadia, O. P. Awasthi and S. S. Hiwale 2008. Varieties and Hybrids of CIAH. Technical Bulletin - CIAH/TECH/PUB. No., 30: 1-11 pp.

of khejri seedlings and *in-situ* budded plants. Annals of Arid Zone, 42 (1): 1-5.

Pareek, O. P. 2001. Vegetative propagation of khejri. ICAR News, 7(4): 12.

Pareek, O. P. 2002. Top-working wild khejri trees. Indian Horticulture, 47 (1): 11-12.

Pareek, O. P. and A. K. Purohit 2002. Patch budding in khejri (*Prosopis cineraria*). Indian Journal of Horticulture, 59: 89-94.

Pareek, O. P. and Vishal nath 1997. Variability in horticultural traits of khejri in Thar Desert. IPGRI Newsletter for Asia, the Pacific and Oceania, 24: 23-25.

Pareek, O. P., A. K. Purohit and D. K. Samadia 2006. Graft compatibility among the Asian and Algaobia *Prosopis* species. Indian Journal of Arid Horticulture, 1 (2): 1-5.

Pareek, O.P. 2002. Top working wild khejri trees. Indian Horti., 11-12.

Pareek, O. P, Vishal Nath, Singh R.S. and.Sharma B.D 1997. Regeneration of *Prosopis cineraria* through cuttings. In. National Seminar on Propagation and Nursery Management in Horticulture, GAU, Anand.

Puri, G.S.; S.K.Jain, S.K.Mukherjee, Shanti Sarup and N.N.Kotwal 1964. Flora of Rajasthan. Proc. Bot. Survey of India, 19: 1-159.

Rawat, T. S., H. C. Sharma and J. P. S Pundir 1982-83. Air-layering–a success in propagation of khejri (*Prosopis cineraria* L.). Udyanika, 5 (1&2): 43-44.

Samadia, D. K. 2007. Sangri utpadan ke leai khejri ki bagwani. Technical bulletin (Hindi) - CIAH/TECH/Pub. No., 20: 1-32 p.

Samadia, D. K., A. K. Purohit and O. P. Pareek 2002. Genetic diversity in vegetable type khejri. Indian Journal of Agro-forestry, 4 (2): 132-134.

Samadia, D. K., O. P. Pareek, A. K. Purohit and T. A. More 2009. Technological advancement for transformation of *Prosopis cineraria* in to a horticultural crop. In. Abstract International conference on nurturing arid zones for people and the environment: Issue and agenda for the 21st century. Nov. 24-28, 2009; Jodhpur (India), pp. 144-146.

Shanker, V.; M.K.Dadhich and S.K. Saxena 1976. Effect of khejri tree (*P. cineraria*) on the productivity of range grasses growing in its vicinity. Forage Res. 2: 91-96.

Solanki, K. R., N. L. Kackar and S. K. Jindal 1986. Air layering in *Prosopis cineraria* (L.) Mac Bride. Indian Forester, 112 (3): 202-207.

Vishal Nath, R.S.Singh and N.K.Dwivedi 1998. Variation in some horticultural traits in *Prosopis cineraria* in Rajasthan. *In*. Proceeding on *Prosopis* spp. in the arid and semi-arid zones of India. The Prosopis Society of India and H.D.R.A., U.K., pp. 121-124.

Chapter-43

Indian Aloe (*Aloe barbadensis* Mill.)

R. S. Singh
Central Institute for Arid Horticulture, Bikaner-334 006

Indian Aloe (*Aloe barbadensis* Mill.) is commonly known as *Ghritkumari in* Sanskrit and Bangla, *Ghikwar* or *Ghikanwar* in Hindi, *Aloe* in English, *Ghiquara* in Urdu, *Darkhtesibra* in Persian, *Sabbarat* in Arbi, *Guarpatha* in Rajasthani, *Koryed, Korkanta* and *Elwa* in Marathi, *Kunwargandal* in Punjabi, *Lolesara* and *Katthaligida* in Kannada, *Kattavala* in Malyalam, *Sirukanthli, Sirukattashai, Kattazhai, Sorttukathali* and *Chirukattali* in Tamil, *Kalawand, Manjikattali, Pinngorisht Kalvand* and *Chinnakalabanda* in Telgu, *Kumari* and *Mushaboro* in Oriya and *Musabbar* in Kashmiri. Because of its nutritional qualities, anti-oxidant and medicinal values, it is often called as 'The Miracle Plant', 'The Natural Healer', 'The Health Plant' and by many other names.

The word 'Aloe' is derived from the Arabic word 'Alloeh', which means 'bitter and shiny substance'. Aloe is native to North Africa but it is found abundantly throughout the warmer parts of the world. It is found in wild forms in southern and eastern Africa and the shores of the Red Sea. It is cultivated in West Indies, Italy, Malta, Sicily; it has been naturalized in Barbados and has been found in the Zapata region of Texas (USA). It is reported to grow wild on the islands of Cyprus, Malta, Sicily, the Canary Cape and Cape Verde. In India, 4 species are reported to occur and of these *A. barbadensis* is most widely naturalized (Anon.,1962). Its spread could be explained by Sumerian's extensive trade with Egypt, from where it reached rest of the world.

Aloe is one of the most popular natural occurring plants with super therapeutic uses. The healing properties of Aloe have been exploited for centuries. In fact, records of its medicinal uses date back over 4000 years. The earliest reference to Aloe is found in the form of *Sumerian Clay Tablets*, in one of the oldest civilizations the Sumerian civilization located along the rivers 'Tigris' and 'Euphrates'- that dates back to 2200-1700 BC, where it is mentioned as a laxative. The Egyptians prepared its concoctions for internal and external healing. Aloe was used by ancient Persian healers. It was also used extensively by ancient Indian civilizations for its cathartic, emmemagognic (promotion of menstrual flow) and anti-helminthic (de-worming) properties.

The studies on clinical applications were started since the inclusion of Aloe in the United States in Pharmacopoeia in 1820. George Ebers in 1862 first discovered its antiquity in an ancient Egyptian papyrus about 3500 B.C., which was in fact a collection of herbal remedies. Collins and Collins (1935) first described *Aloe vera* in the treatment of radio-dermatitis. Wright (1936) described the use of Aloe in the treatment of radiation ulcers. Rowe *et al.*,(1940) noted that Aloe helped heal burns faster than any other treatment available at that time. Lushbaugh and Hale (1953) reported further evidence of improved healing in radiation dermatitis and ulcers.

In India, few research organizations *viz.* NBPGR, New Delhi, CIMAP, Lucknow, NBRI, Lucknow, Directorate on Medicinal and Aromatic Plants, Boriavi, Anand and CAZRI, Jodhpur, etc. are doing the work on collection, conservation and evaluation of Aloe germplasm while CIAH, Bikaner initiated work on production technology of *Aloe barbadensis* under hot arid eco-system. Many NGOs and private sectors are also involved in production and processing of this crop in India.

Nutritive Value and Uses

The active principle of Aloe is a mixture of glycosides called 'aloin' varies in different species of Aloe. The name aloin was given to a crude material from which barbaloin could be isolated. Barbaloin is an anthracene glycoside. Chemically, it is a 10-glycopyranosyl derivative of aloe emodine anthrone [10 (1-deoxyglucosyl) aloe-emodin anthrone]. $C_{21}H_{22}O_9.H_2O$, MP-148^0-99^0 (andhyd.)], which is a pale yellow, crystalline glycoside soluble in water. The other constituents include 'iso barbaloin', 'Aloe-emodin' (a hydrolytic product of barbaloin), resins and some water-soluble substances. The percentage of aloin may vary from 10-30%. The percentage of barbaloin in Indian aloes population varies from 5.53 to 22.76 per cent of dry weight of the exudates. The characteristic odour of gel is due to the traces of an essential oil (Anon., 1959).

The phytochemical analysis of Aloe reveals the presence of aloin, aloe embodin, monosaccharides, glycosides and sterols triterpenes etc. Pulp/juice i.e. water-soluble fraction contains about 17 amino acids, D-mannose and uranic acid etc. The aloin group of glycosides is the major constituents of Aloe juice or pulp. Besides, some resins and aromatic oils are also found in it. The pharmacopoeia recognizes three types of aloes i.e. *Curacao* aloes (obtained from *A. barbadensis*), *socotrine* aloes (obtained from *A. perryi*) and cape aloe (obtained from *A. ferox*) (Pareek, 1990).

The gel prepared form the leaf sap of Aloe contains as many as 75 known ingredients. The rich cocktail of nutritional elements, whose combined action and balance produce a more powerful effect together than, would be expected from the addition of the individual components. According to Chopra *et al.* (1982), Aloe contains a glycoside group named aloin, which is the active constituent of the drug. It is maximum (30%) in *Curacao* aloes, lesser in *socotrine* and only 10% in cape aloes. Dried juice of aloe leaves is known as 'aloes'. It is also called *Kumarisar* or *Mussabar*.

The minerals including magnesium, manganese, zinc, copper, chromium, calcium, sodium, potassium and iron are also found in considerable amount. Aloe contains 20 amino acids with all the 8 essential amino acids that are not synthesized by human body. The long chain polysaccharide sugars are present in Aloe, which act on the immune system of the body to boost its effect. Among the anthroquinones, the most important one is '*Aloe-emodin*', but altogether, they are strong painkillers and are known to possess anti-bacterial and virucidal activity. In their pure forms, they are very powerful laxatives. Besides these, Aloe leaves also contain salicylic acid, saponins, lignin, etc.

The plant produces at least 6 antiseptic agents such as lupeol, salicylic acid, urea nitrogen, cinnamonic acid, phenols, and sulphur. All of these substances are recognized as antiseptics because they kill or control mould, bacteria, fungi, and viruses. Lupeol and salicylic acid present in the juice are two very effective painkiller. It contains at least three anti-inflammatory fatty acids, cholesterol, campersterol and B-sitosterol (plant sterols). These are highly effective in treatment for burns, cuts, scrapes, abrasions, allergic reactions, rheumatoid arthritis, rheumatic fever, acid indigestion, ulcers, plus many inflammatory conditions of the digestive system and other internal organs, including the stomach, small intestine, colon, liver, kidney, and pancreas. B--sitosterol is also a powerful anti-cholestromatic, which helps to lower harmful cholesterol levels, helping to explain its many benefits for heart patients. About 23 polypeptides (immune stimulators) are present in Aloe juice, which helps to control a broad spectrum of immune system diseases and disorders. The polypeptids plus the anti-tumor agents, Aloe emodin and Aloe lectins, are now also used in treatment of cancer.

Food

The leaves of Aloe are eaten as vegetable throughout the year. Pickle made by small pieces of leaf pad is a common preparation in western Rajasthan. Flowers are also used for pickle preparation. Sharma and Goel, 2002 as well as Anon., 2003; standardized the recipes of various Aloe products viz., vegetable, pickles, *laddu*, jam, squash, biscuits and *churna* by using sensory evaluation technique. Fresh fleshy leaf pad is used as green *salad* and helpful in treatment of indigestion and constipation. The immature flower stalks that are completely free from bitter content, are also used for vegetable purpose. Aloe based soft drinks are very popular in California, Japan, Korea and some European countries.

The juice of Aloe is not stable due to action of enzyme cellulose and microorganisms and undergoes degradation within few hours of harvesting. An efficient preservation process for Aloe juice is essential (Chakraborty and Mukhopadhayay, 2005).

The leaves of *A. barbadensis* contains protein (3.2 g), fibre (15.2 g), iron (9.4 g) and phosphorus (198.93 mg), which were remarkably high as compared to several vegetables (Sharma and Pareek, 2003). The leaf of Aloe is a treasure house of vitamins, minerals, enzymes, amino acids, sterols and anthroquinones. It contains a wide range of other useful metabolic substances. Aloe is one of the few plant sources in the world of Vitamin B_{12}. It also contains a good amount of Vitamin C, E and a-carotene - a precursor of Vitamin A.

Medicine

Aloe is being used for curing several ailments for over centuries. In *Ayurvedic* medicines, it is considered as purgative, fattening tonic, aphrodisiac, anthelmintic and alexipharmic. Kirtikar and Basu (1989) reported that fresh juice of the leaves is cathartic and cooling.

The anti-inflammatory and local anesthetic qualities of aloe gel have made it a household name. The juice of fresh leaf quit popular as elio, is a household remedy for intestinal worms in children and anti-dysenteric. Its local application on painful inflammations of the body and in chronic ulcers is also useful. Squash made from extract of Aloe is used as tonic. The strained mixture of juice and rose water is useful as eyewash for catarrhal and purulent of opthalmia and other eye disorders.

Regular use of pulp is also useful in abdominal tumours, dropsy, piles, sciatica, rheumatism and retention of urine in fever. The pulp mixed with honey and turmeric powder is also recommended in cough and cold. It is mentioned that Aloe extract is useful in curing of kidney ailments, to develop the mammary glands, to enhance sexual excitement, treatment of tuberculosis, cancer and HIV etc. but proper clinical studies are needed to verify and to quantify the dose for not only these diseases but also in several other uses. It has been reported that for ulcers; use of 50 ml Aloe juice, for burns and eczema application of gel liberally twice a day and for stimulation of appetite use of 5 drops of tincture in water before meal (Anon., 1996). Nandkarni (1976) also stated various preparations of *Aloe barbadensis* like confection, lotion and juice are useful remedies for curing various diseases. It is well established that Aloe can cure cancer, digestive, kidney problems and even act against HIV (Chakraborty and Mukhopadhyay, 2005).

In arid region of Rajasthan, it is traditionally used by the inhabitants to cure many diseases and ailments viz., fever, asthma, constipation, flatulence, fracture, inflammation, piles, rheumatism, boils, ringworm and tumors (Kumar *et al.*, 2005). The juice and pulp of Aloe leaves are also used in veterinary medicines. Feeding of about 250- 500 g mixture of boiled Aloe leaf pad + *gur* daily helps in relief of body swelling, fever and dullness in buffalos.

Cosmetics

Apart from bactericidal, virucidal and fungicidal properties, the Aloe gel is also used as natural cleaner and moisturizer. Due to its moisturizing, softening and healing actions on skin, it has become an important cosmetic blend. Maximum gel production is utilized in cosmetic industries. In India, private sectors are coming forward for preparation of herbal shampoo, body lotions, and moisturizers and now there is high demand of raw materials in form of gel paste. Now, it is familiar ingredient in range of health care and cosmetic products. In Israel, Aloe of large sized pad is grown in Arava Valley and maximum Aloe gel is being used in cosmetics industries.

Ornamental Value

The plants of Aloe is very attractive and the leaves are arranged in such a way that it resembles with the lotus petals. Aloe plant can be used as pot plant and also occupy premium place in landscaping as 'Desert Garden' and 'Rock Gardens'. It can be used as an indoor plant too.

Botanical Description

The genus Aloe (family: Liliaceae) has over 350 or 200 species widely distributed throughout the world, ranging from tiny stem less little plants to tree like plants in south-west Africa, growing several feet in height. Out of 275 species of Aloe, 42 belong to Madagascar region (Africa), 12-15 to Arabian Peninsula and rest are distributed over tropical South Africa. In India, only 4 species are reported to occur and of these *Aloe barbadensis* is the most widely naturalized (Anon., 1962). Five species of *Aloe* are considered to possess therapeutic properties viz., *Aloe barbadensis* (Syn. *A. vera*), *A. arborescens, A. saponaria, A. ferro* and *A. perryi* (Anon, 1959). A number of species have been introduced in India of which *Aloe barbadensis* has become naturalized almost in all parts of the country. It is the most prominently used and studied in most parts of the world, while Japanese have done significant research on *A. arborensis* and *A. saponaria*. *A. ferro* and *A. perryi* are used in some parts of Africa. Other species also occur as wild/ornamental at some places. Four types of *Aloe* are official in Indian Pharmacopoeia i.e. i) *Curacao* Aloe obtained from *A. barbadensis;* ii) *Socotrine* Aloe obtained from *A. perryi;* iii) *Zanziabar Aloe;* and iv) *Cap* Aloe from *A. ferox* and its hybrids (Anon., 2002). It is pertinent to mention that though there are about 350 species in genus Aloe mentioned in the literature but no distinguishing characters have been given to identify them. The existence of 350 species is also under question.

Aloe barbadensis is xerophytic, succulent and cylindrical perennial under-shrub reaching the height of 30 to 150 cm. Numerous stolons emerge from the base of simple, soft woody stem. It has thick fibrous root system, which produces large basal leaves. The thick, fleshy, mucilaginous leaves called pads are sessile, densely crowded on the short stem with wide dilated base. The leaves are of spreading nature than ascending type, tapering to a blunt point. The leaf margins are set with hard, distant and hooked prickles. The leaf surface is quite smooth and shining, dark glaucous green or pale green on both surfaces. Sometimes, mottled leaves are also found in aloe plants (Bentley and Trimen, 2000). The vascular bundles in the leaf are isolated and form a line parallel with the epidermis within the mesophyll. Each bundle has a pericycle formed of large thin walled cells filled with yellowish fluid. The mature leaves are 20-24 inches in height and 4-6 inches across the base, tapering to a point at the end. The full grown plant produces a single unbranched flowering stalk (scape)/inflorescence of 50-100 cm long that is topped by a cluster of bright yellow, red purple or pale stripped flowers. Flowers are bisexual with epi-calyx, corolla and six stamens (basifixed, diadelphous) protogynous; ovary superior, trilocular with axile placentation, buds in the inflorescence is dense racemes terminating the scapes. The flowers are present for most of the year, growing in a long raceme at the top of the scape (Bentley and Trimen, 2000). The fruits are triangular capsules containing numerous seeds. Most of the species are male sterile, with scarcely any fertile pollen, hence plant does not produce many viable seeds.

However, under natural population, a lot of variability particularly in morphological characters like leaf colour, size, thickness, spinyness and presence of spots; growth habit of leaves; taste of leaf flesh; colour of inflorescence etc. were observed. Moreover, these characters are also influenced much by environmental conditions. Therefore, morphometric parameters; bio-chemical and cyto-genetical characterization are essentially required for proper cataloguing these variable types.

Although, the commercial cultivation of Aloe is still in its primary stages but in the recent past, its cultivation in the semi-arid and arid regions of Rajasthan, Gujarat, Andhra Pradesh, Madhya Pradesh, Maharastra, Tamil Nadu and Karnataka is fast catching up. However, the *A. barbadensis* which is grown in north-western India mainly for vegetable purpose can grow well under irrigated conditions, since only thick and leathery leaf pads are used for the purpose (Vishalnath and Singh, 1999). In Rajasthan, Aloe is very popular in Bikaner region and forms an integral part of kitchen gardens and also grown on a limited scale in Alwar, Rajasthan, (Saroj *et al.*, 2004), Satnapalli (Andhra Pradesh), Anand, Banaskatha, Rajpipla and Kachchh region (Gujarat) and Kangayam (Tamil Nadu). Now, its cultivation is being extended on large scale by the farmers. Though, the exact data about area and production of Aloe is not available, but as an estimate, about 20 ha area is under Aloe cultivation in western parts of Rajasthan only.

Propagation

Aloe is generally propagated through suckers; arising near the base of the plant from rhizome. Propagation by stem cutting is also possible on limited scale (Pareek *et al.*, 1999). It can be multiplied by micro-propagation techniques.

Suckers

Aloe is commonly multiplied through suckers having 4-6 leaves (Saroj *et al.*, 2004) for better establishment. Suckers can be transplanted throughout the year but July-August is most suitable time for arid region.

Rooted suckers of appropriate size should be detached carefully from mother plants.

Micro-propagation

Limited work on micro-propagation in Aloe has been done so far. The tissue culture techniques can be employed for fast multiplication and production of quality seedlings. Different concentration of BAP (2, 5 , 10 mg/l) when used in MS medium with 2mg/l or 5 mg/l BAP exhibited similar response while BAP at 10mg/l showed poor shoot formation. Saroha *et al.* (2005) obtained multiple shoots in 12 days when 2 mg/l BAP was added with 0.18 mg/l NAA in *A. vera*. The number of multiple shoots ranged from 4 -15 per explants.

Micro-propagation protocol for vegetable type Aloe (*Aloe barbadensis* Mill.) has been developed using shoot tip explants. The protocol was standardized for multiple micro-sucker proliferation, induction of *in vitro* adventitious roots and acclimatization of rooted micro-sucker from *in vitro* environment to greenhouse condition. Maximum proliferation of micro-suckers was obtained with combined supplementation of Benzyl adenine (BA) 2 mg + Thidiazuran (TDZ) 0.5 mg + 0.1 mg Naphthalene acetic acid (NAA) to per litre basal Murashige & Skoog (MS) medium containing 0.8% agar and 3 % sucrose. The maximum rootable micro-sucker (~ 2 m long) was achieved after transfer of culture to hormone free medium for two weeks. The elongated micro-suckers were rooted successfully under *in vitro* condition. The better quality adventitious root formation was observed with NAA 0.5 mg/l supplemented to ½ MS medium (Singh *et al.,* 2009). The regenerated plants were successfully acclimatized under Hi-tech greenhouse and maintained in portrays for further growth and development.

The cultivation of Aloe on scientific line is yet to be standardized for different agro-ecological situations. For cultivation of Aloe, the soil preparation need not to be very deep, as the fibrous root system of Aloe does not penetrate to greater depths. Before planting, the field should be well leveled and made free from boulders and roots of perennial shrubs and trees. Well rotten FYM or compost (15-20 t/ha) should be mixed during field preparation for improving initial soil fertility status of the field. The healthy suckers with roots having 4-5 leaves are separated from the mother plant and directly planted in the field at a spacing of 60 x 60 cm. Healthy suckers of 4-6 leaves are considered good for planting (Pareek *et al.,* 1999a).The treatment of the base of the suckers by fungicide before planting is also recommended.

The closer planting of 50 x 40 cm and 40 x 30 cm under low fertility and water stress situations (Anon., 2002) and 30x30 cm in sandy soils of north-western Rajasthan (Pareek and Mohan, 1995) were also suggested but spacing less than 60 x 60 cm is not advisable for growing of vegetable type Aloe. Under arid conditions, Suckers should be planted in July-August in order to get better survival and growth of the plants. However, under irrigated condition, planting can be done any time round the year except in winter months. After planting of suckers, the soil around the root zone must be firmly pressed followed by irrigation (Saroj *et. al.*, 2003).

Quality Standards of Planting Material

Aloe is an underutilized crop and there is no standard fix for planting material. There is no improved variety of this crop. For the mass multiplication of Aloe, emphasis should be given on the establishment of registered nursery to provide healthy disease free suckers. The sucker should have 4-6 leaves, well rooted, disease free, true to type for better establishment and quality production of leaves.

References

Anonymous, 1959. *Wealth of India.* CSIR, New Delhi, Vol. 1 A. pp. 191-193.

Anonymous, 1962. *Wealth of India*, Raw Materials. Vol. VI. CSIR, New Delhi, pp. 439-441

Anonymous, 1996. *Encyclopedia of Medicinal Plants.* Pub. Darling Kindersley Ltd.. London. p.57.

Anonymous, 2002. Aloe *(Aloe vera)* and its cultivation in India, CIMAP, Lucknow, Pp. 1-15.

Anonymous, 2003. Cultivation of *Aloe vera*, NRC on Medicinal and Aromatic Plants, Anand, Gujarat, pp. 1-6.

Bentley, R. and Trimen, H. 2000. Medicinal Plants. 4:228-306. Asiatic Publishing House, New Delhi-110092.

Chakraborty, A and Mukhopadhyay, Mamta 2005. Preserving the goodness of Aloe using super critical carbon di-oxide. In. Proceeding on Botanical Products Seminar and Expo-2005, Natural Product Laboratories, Jaipur, pp.199-122.

Chopra, I.C.; Abrol, B.K and Handa, K.L. 1982. Medicinal plants of the arid zones. Part I. Today and Tomorrow's Printers and Publishers, New Delhi. Pp. 18-19.

Collins, C.E. and Collins, C. (1935). Roentgen dermatitis treated with fresh whole leaf of *Aloe vera*. Am. J. Roentgenology and Radiation Therapy, 33.

Kirtikar, K.R. and Basu, B.D. 1989. Indian Medicinal Plants. Vol. IV (2nd ed.). Pub. Lalit Mohan Basu, Allahabad, India.

Kumar, S, Azam, M.M. and Parveen, F. 2005. *Aloe vera* in the Indian Arid Zone In: Shrubs of Indian Arid Zone (Eds. P. Narain, M. Singh, M.S. Khan and S. Kumar). Arid Agro-ecosystem Directorate, Central Arid Zone Research Institute, Jodhpur.73-78.

Lushbaugh, C. and Hale, D. 1953. Experimental acute radiodermatitis following beta irradiation. V. Histopathological study of the mode of action of therapy with *Aloe vera*. Cancer, 6: 690-698.

Nandkarni, K.M. 1976. *Indian Meteria Medica* Vol. I (3rd ed.). Pub. Bombay, Popular Prakashan Private Limited. Pp. 73-74.

Pareek, G.N. 1990. Sudhanidhi Vanoshadhi Ratnakar. Pub. Dhanvantri Office, Aligarh. pp. 64 and 73.

Pareek, O. P. and Inder Mohan 1995. Indian Aloe: Wealth from Thar Desert. Indian Horticulture (Oct. -Dec., 1995). Pp. cover II and 29.

Pareek, O. P.; Vishal Nath; Singh, R. S. and Bhargava, R. 1999. Grow Indian Aloe in Arid region., Intensive Agric, 37(5-6): 21-23.

Pareek, O. P.; Sharma, B. D.; Vishal nath; Singh, R. S. and Bhargava, R. 1999a. Effect of Nitrogen and Phosphorus fertilizers and organic manures on growth and yield of Indian Aloe (*A. barbadensis* Mill.). Annals of Arid Zone, 38(1): 85-86.

Rowe, T.D., Lovell, B.K. and Parks, L.M. 1940. Further observations on the use of *Aloe vera* leaf in the treatment of third-degree x-ray reactions. J. Am. Pharm. Assoc. 30: 266-269.

Saroj, P.L.; Dhandar, D.G.; Sharma, B.D. and Bhargava, R. 2003. Evaluation of ber (*Ziziphus mauritiana* Lamk.) based cropping systems for hot arid ecosystem of north-western Rajasthan. Paper presented in 6th Agricultural Science Congress at IISS, Bhopal, February 13-15, 2003.

Saroj, P.L.; Dhandar, D.G. and Singh, R.S. 2004. Indian Aloe, Technical Bulletin, CIAH, Bikaner, pp.30.

Saroha,V.; R.C.Yadav and Neelam R.Yadav 2005. High frequency shoot generation in *Aloe vera* using BAP and TDZ combination of growth regulators. In. National Symposium on Biotechnological Interventions for Improvement of Horticultural crops: Issues and strategies, Kerala Agri.University, Thrissur, held on 10-12 Jan., 2005 pp.41.

Sharma, R. and Goel, M. 2002. Utilization of local plants guarpatha (*Aloe barbadensis*) by women residing in Bikaner city (Rajasthan). In. Proceeding of NSI, XXXIII Annual Meeting, December 1-2, 2002, NIN, Hyderabad.

Sharma, S. and Pareek, Navita 2003 Nutritional composition of *Aloe vera* (Guarpatha)- a dryland product. Paper presented in IX th Asian Congress of Nutrition, New Delhi, Feb. 23-27, 2003.

Singh, D.; Sivalingam, P.N.; Bhargava,R. and Meena ,S.R. 2009. Micro-propagation of vegetable type Aloe (*Aloe barbadensis* Mill.). Indian Jour. Arid Hort., 4(2):61-64.

Vishal Nath and Singh, R. S. 1999. Sushka Kshetra Mein Guarpatha Ki Kheti. Krishi Vistar Samiksha, 9(2): 24-27.

Wright, C. 1936. *Aloe vera* in treatment of roentgen ulcers and telangiectagis. J. Am. Medical. Assoc. 103: 1363-1364.

Chapter-44

Drumstick (*Moringa oleifera* Lam.)

K. Krishnakumary and K. V. Suresh Babu
College of Horticulture, Kerala Agricultural University, Vellanikkara-680 656

Moringa (*Moringa oleifera* Lam.) commonly known as drumstick or sehjna is a multipurpose tree of India and grown intensively in the tropics and sub-tropics. It is a vitamin rich, mineral packed and nutritious vegetable crop. The name drumstick is derived from the pod shape resembling the slender and curved stick used for beating drum. Fruits are also resembled with the siliqua of radish and hence the name is "radish tree". Ancient Indian literature mentioned that moringa is an interesting plant due to its widespread uses in socio-economic for food, medicine, agriculture and industry. This is a backyard tree for daily use by inhabitants of South India. It is grown for its tender pods and also leaves and flowers. The pods are very popular vegetable in South Indian cuisine and valued for their distinctly inviting flavour. Flowers are also used for culinary purposes (Chadha, 2001). Due to its wide adaptability to varied soil and climatic conditions and easiness in cultivation, it is becoming popular in resource constrainted arid and semi-arid areas of north-western part of India and its cultivation is highly remunerative. Presently, it is the most widely known perennial drought tolerant vegetable (Palada, 1996) for its highly nutritious pods (Ramachandran, 1980) grown at about 380.5 km^2 producing 1.1 to 1.3 million tonnes of tender fruits (Rajangam *et al.*, 2001). Drumstick is having not only every part of plant value for food, but also is enriched with highly nutritive minerals and vitamins. Except its edible pod, not much exploitation has been done in drumstick in India yet, though seed oil is utilized as a lubricator in delicate machineries, manufacturing of perfumes and hairdressings and seed powder as a water clarifier. In recent years, drumstick has been established as a stable commercial crop in southern part of India.

Moringa oleifera is a drought tolerant species mainly growing in semi-arid tropical and subtropical areas with an annual rainfall of above 750 mm. It is adaptable to wide range of agro-climatic and soil conditions (4.5 - 8.5 pH). It can not tolerate water logging conditions. It can be a potential crop for diversification of agriculture in the arid and semi-arid areas of north-western India under limited rains or irrigation water. The tree is fast growing and found to grow about 3 - 4 m in one year in arid areas when receiving less than 400 mm rainfall. It grows well in dry sandy soils but is highly sensitive to low temperature (<1^0 C) and frost conditions.

Nutritional Value and Uses

It is used for fresh vegetable and a number of value added products can be made if the availability of moringa is in excess. They are pickle, dehydrated moringa, powder, and fleshy mesocarp powder. Moringa plant has been well documented for its medicinal importance for a long time. The stem bark, root bark, fruit, flowers, leaves, seeds and gum are widely used in Indian folk medicine. The powder of seeds and roots are also used as a spice. All of the parts of the moringa can be used in a variety of ways as food. It is full of nutrients and vitamins and is good as food for humans as well as fodder for animals. The tree provides lots of leafy material that is useful when it is in cropping system models. The leaves, especially young shoots are eaten as greens, salads, vegetable curries and pickles. The leaves can be eaten fresh, cooked or stored as dried powder for many months without refrigeration and reportedly without loss of nutritional value. The leaves are considered to offer great potential for those who are nutritionally at risk and may be regarded as a protein and calcium supplement. The flowers are said to make a vegetable *rayta*.

Analysis of 100g pods gave the values viz. moisture-86.9g, protein-205g, carbohydrate-3.7g and crude fibre-4.8g. Fruits are a rich source of phosphorus (110 mg), calcium (30 mg), carotene (110ug), vitamin C (120mg) and niacin (0.2mg) (Gopalan *et al.*,1989). Leaves are rich in carotene and ascorbic acid. 100g of leaves contain protein- 6.7g, carbohydrates – 12.5g, minerals – 2.3g, carotene- 11300 I U, vitamin A-210 ug, ascorbic acid – 220mg, calcium – 440mg, phosphorous- 70mg, magnesium – 42mg – 259mg, nicotinic acid – 0.8 mg and tocopherol 7.4 mg. The essential aminoacids presents in the leaf protein (g/16gN) are arginine 6, histidine 2, lysine 4.3, phenyl alanine 6.4, methionine 2 and threonine 4.9. The biological value and digestibility coefficient of leaf protein (at 5% level of protein intake) are 41% and 77%, respectively.

Dehydration technologies are available nowdays to preserve the seasonal fruits of this crop. Selected moringa fruits after washing are cut into pieces and exposed to blanching by steam for five minutes and sulphated (blanched cut pieces are dipped in 0.1% potassium metabisulphite solution for 30 minutes). They are loaded uniformly in a tray and dried in a solar drier. The initial and final moisture in blanched and sulphated fruit cuttings were 80 and 13.6%, respectively. The dried moringa can be reconstituted by cooking in boiled water for 20 minutes.

Now, appropriate technology is available to extract the pulp from the boiled fruits and subsequently dried and powdered. This drumstick powder can be mixed in several curry preparations.

In African countries like Kenya, Tanzaniya and Uganda, moringa is grown exclusively for leaves in large areas of land. The leaves after being dried are powdered and exported to European countries for extraction of vitamins and preparation of vegetable soups.

Seed oil of this crop known as Ben or Behen oil has been in much demand for use in soap industry and high value lubrication oil in watches, computers etc. Oil is clear, odourless and do not turn rancid. It is also edible and useful in the manufacture of perfumes and hair oil. The seed cake is used as manure and as water purifying agent as it can be used as organic substitute for water purifying chemicals such as alum (aluminium sulphate).

The young fruits have the general characteristics of a succulent vegetable but are rather high in protein and pods are cooked as a vegetable. In the country moringa has long been cultivated for its edible pods and today these are exported to consumers in Asia and Europe. Tender and mature pods or other parts (leaves, flower buds) of plant are usually harvested at proper marketable stage. In arid region, fruiting may peak in between March - April from non-affected plants from frost injury and non-pruned in the winter season. The main season of harvesting pods is from September - November. Pods of sufficient length and girth are harvested before they develop fibre. The harvest period extends for 2-3 months and each tree bears 250-300 fruit depending on the management practices in semi arid region. Flower buds harvested in the month of April–June has good market for vegetable *rayta.*

Botanical Description

Drumstick belongs to the family Moringaceae. The genus *Moringa* includes 12 species distributed in sub-Himalayan range of India, Sri Lanka, North-eastern and South-western Africa, Madagascar and

Arabia. The best known and most widely distributed species is *Moringa oleifera* (2n=28). Good amount of variability is available in the naturally occurring seedling population both in perennial and annual types. The reported extent of variability in the seedling population of annual type moringa was documented in southern India. Among the nine traits studied, number of flowers per inflorescence (19-126), fruit weight (25 - 231g) and number of fruit per plant (1.0-155.0) exhibited wide range of variability. *M. concanensis* Nimmo is a small tree resembling *M. oleifera* and is found in Rajasthan, dry hills of Konkan, Andhra Pradesh and Tamil Nadu. It can be easily distinguished by the bi-pinnate leaves, some what longer than those of *M. oleifera* and with pinkish yellow flowers. Drumstick is a perennial tree with a fragile stem and grows to a height of even more than 10m. In North India, the tree sheds its leaves in December-January and new ones appear in February-March. Under hot arid conditions, plants are affected by frost or low temperature during winter season.

Flowering and Fruiting

In South India, flowers and fruits appear twice in a year. Anthesis has been reported to commence as early as 4.30 am and continue till 6.30 am with peak at 5.30 am in association with a temperature range of 27.3- 29.3°C and relative humidity of 68-78 %. Anther dehiscence is reported to occur from 4 am to 6.30 am. The anther of longest stamen dehisces first followed by stamens in descending order of filament length. The stigma becomes receptive a day before flower opening and continue to be receptive on the day of opening. The crop is cross pollinated and entomophilous. Honey bees are the pollinating agents. Pollen viability is 70-80 % at anther dehiscence. There is no problem in pod formation.

Varieties

There are few named varieties and mostly perennial types such as Jaffna, Chavakacheri, Chem Murungai, Kadukkal Murungai and Yazhpanam Murungai and Pal Murungai which are grown in South India. Chem Murunga (red tipped pods) which flowers throughout the year is a medium sized tree bearing long pods and is a heavy yielder. There are two distinct type of drumstick found in Thiruneveli district of Tamil Nadu with name Palmurungai and Punamurungai. Horticulture College and Research Institute (TNAU), Periyakulam has an assemblage of 85 accessions. The collections include perennial and annual types with varying fruit and bearing characters, dwarfing stature and tolerant to drought, insect pests and diseases. Another variety, Kattu muruga released from Tamil Nadu is a pure line selection and is an annual type. The yield is 400-500 pods per plant.

Among annual types drumstick, PKM-1 and PKM-2 are most popular for cultivation and PKM-2 registered highest pod yield (44.25kg) per plant followed by PKM-1(36.7kg) (Kader mohideen and Shamugavelu,1982). PKM-1 variety of drumstick, developed at Periyakullam, TNAU, is a pure line selection from the population generated by continuous selfing of seed moringa type for six generations. Average yield is 33kg/year (200-275 pods) having 65-70 cm length and 6-5cm width and 150g weight. Every year after the harvest is over, trees are to be cut back to about 1 m from ground (Kathiresam *et al.,* 2001).

PKM-2 is a hybrid derivative of the crosses between MP-31 and MP-28 from Periyakullam (TNAU). Three to four pods are formed in cluster since flowers occur in cluster. The tree yields on an average 220 pods weighing 62kg with an yield of 98 tonnes/ha.

KM-1 (Kuduminmalai-1) is a variety developed from Anna Pannai. It is a selection from annual type and propagated purely by seeds. The tree bears 400-500 pods/tree. The plants can be ratooned for 2-3 years by cutting the trunk at a height of 1metre and can be replaced by fresh seedling after 3years.

As far as drumstick is concerned fruit length is an important character which decides the consumer attraction. It is understood from the fact that generally in the west Indian markets, the lengthier fruits are cut into pieces in the market itself for easy packing, where as such practices are not prevalent in southern markets. Similarly, small sized fruits are also not preferred by the consumers due to its non-appealing nature. In the present study, PKM-2 recorded the lengthiest fruit of 123.1cm followed by MO-3.

A large number of collections (41) both annual and perennial types of moringa have been made at CHES, Godhra, Gujarat and promising lines have been identified. Moringa lines were evaluated for pod quality and yield under arid conditions (Anon., 2012). The varieties PKM-2, PKM-1 and MO-1 had recorded the highest number of fruits per plant, fruit size, pulp weight and yield per plant coupled with earliness under semiarid conditions of western India.

Propagation

It is mainly propagated by seeds and stem cutting (limb cutting). Perennial types are mostly propagated through vegetative methods.

Seed

Seeds do not undergo dormancy and can be sown as soon as they are extracted. The annual types are propagated by seed. The seeds may be directly dibbled in the pit to ensure accelerated and faster growth of the seedling or nursery raised seedlings can be planted. Seeds germinate 7-9 days after sowing in poly bags (Chadha, 2001).The best time of sowing, the seeds is July-August for arid and semi-arid areas. However, onset of rainy season (June in case of south India) is the ideal time of sowing. A spacing of 2.5 × 2.5 m in between rows and plants should be adopted with intensive cultivation and irrigation facilities, thus maintaining a plant population of 1600 plants/ha. Pits of 60×60×60 cm in size are appropriate. The seeds germinate within two weeks after sowing *in situ*. About 1.0 kg seed is enough for one hectare planting. A spacing of 4x4 m is found to be ideal for ratoon crops of annual type varieties under resource constraints and limited rains or irrigation water in the arid areas.

Cuttings

In perennial types, elite trees are cut down leaving a stump with a 90 -100cm head from which 2 to 3 branches are allowed to grow. From these shoots, cuttings of about 90 cm long and 4 cm diameter are selected and used as planting material. One year old stem cuttings of 1.0-1.35m long and 14-16cm.diameter are used for vegetative propagation (Chadha, 2001). The limb cuttings are planted *in situ* during the rainy season in well prepared pits of 60×60×60 cm at a spacing of 4 × 4 m. The monsoon rains during the period facilitate easy rooting and further growth. While planting, one-third of the cutting should be kept inside the pit.

Ratooning

After harvesting all pods of seedling moringa, tree is cut at one meter height from ground for ratoooning. It is a common practice in moringa. Ratooned stem comes to bearing pod within 4-5 months if all cultural practices are adopted. The perennial types can be retained for ratoon for 15 years.

Quality Standards of Planting Material

Seed of improved varieties of drumstick may be used for sowing of annual types, however, stem cutting is the best option for multiplication of perennial types. Always use fresh seed for sowing either in nursery or directly in the field. Transplanting of seedlings of proper age and height should be done for better survivality.

References

Anonymous, 2012. Annual Report, Central Institute for Arid Horticulture, Bikaner, India.

Chadha, K. L 2001.Drumstick, Hand Book of Horticulture, ICAR, New Delhi, pp.396-398.

Gopalan, C. B. M., Ramashastri, B. V., and Balasubramanium, S. C. 1989. Nutritive value of Indian foods, ICMR, Hyderabad.pp 49-70.

Kathiresam,G., Manoharam, M.L. and Velusamy, R. 2001. Package of practices for drumstick "PKM- 1", Indian Horti., p 49-50.

Kader mohideen, M and Shanmugavelu, KG 1982. Studies on the performance of Annual drumstick *(Moringa pterigosperma.* Gaertn) at Coimbatore condition. South Indian Hort., 30(2): 95-99.

Palada, MC. 1996. Moringa (*Moringa oleifera* Lam.): A versatile tree crop with horticultural potential in the subtropical United States. Hort Science, 31(5): 794-797.

Rajangam, J. 2001. Status of Production and Utilisation of Moringa in Southern India. In "Development potential for Moringa products", October 29th - November 2nd, 2001, Dar es Salaam, Tanzania

Ramachandran, C, Peter, KV, and Gopalakrishnan, PK .1980. Drumstick (*Moringa oleifera*): A multipurpose Indian vegetable. Econ. Bot., 34: 276-283.

Chapter-45

Management of Diseases and Insects in Nursery Plants

S. K. Maheshwari and Hare Krishna
Central Institute for Arid Horticulture, Bikaner -334 006

Nursery is a place where plants are grown, nurtured and sold out. Generally, various commercial crop growers require a good quality saplings or grafts of genuine type.

Diseases and insects are as much a problem in growing plants in the nursery as they are in growing any other crop. It is always better to monitor the nursery continually with respect to disease and insects damage and select the proper control measures or pesticides for identified pests. Horticultural nurseries are heterogenous and especially complex agricultural systems, which present formidable challenges to management of diseases and pests. Moreover, nursery plants transported interstate and internationally can serve as important vectors for pathogens and insects that threaten both agriculture and forestry.

Growing healthy, marketable plants is the primary objective of nursery growers and one of the key challenges for each crop is effective management of diseases. Plants are continually challenged by pathogens agents that enter the nursery in many ways. Some are brought on infected plants from suppliers or through exchange between growers, others with infested soil, equipment, and clothing, and yet others from windblown fungal spores, aerosols containing bacteria, and insects capable of transmitting viruses and phytoplasmas. It is neither practical nor wholly desirable to attempt total elimination of pests. Many beneficial organisms are destroyed in such efforts, and a lack of beneficial organisms can lead to an explosive recolonization of the nursery beds with insects. Therefore, it is imperative to check the buildup of insects in nursery by combining preventive measures with curative methods, and using chemical, biological and cultural control.

Disease Management in Nursery Plants

Nursery plants suffer from several types of diseases which are the product of the interaction between the causal organism, host, and certain environmental factors. Plant pathogens causing diseases in plants may be grouped into the following main categories: fungi; bacteria; viruses and viroids; mycoplasma-like organisms; protozoa; nematodes; parasitic higher plants. Infections involving the foliage leaf (e.g., blight and leaf spot) and drying of growing apical portion or shoots reduce photosynthetic area and hence photosynthetic output, which, in turn, reduces plant growth. Various approaches have been proposed to protect plants from harmful afflictions. Specific pathogens may infect only certain plant species or cultivars, or specific organs or tissue, which varies with the stage of development of the plant. The pathogenic fungi most likely to cause disease development during propagation are species of *Pythium, Phytophthora, Fusarium, Cylindrocladium, Thielaviopsis, Sclerotinia, Rhizoctonia,* and *Botrytis.* These are all soil-borne or aerial organisms *(Botrytis)* that infect plant roots, stems, crowns, or foliage. *Phytophthora, Pythium,* and *Rhizoctonia* are readily disseminated in surface water. However, some tolerant plants or varieties have certain specific heritable characteristics that enable them to allow the pathogen to multiply without any adverse effects.

1. Leaf spots, Blights and Blotches

(Septoria, Alternaria, Cercospora and *Colletotrichum Phytophthora* spp.)

Leaf spots are tan to brown, local lesions with distinct borders on leaves. When the spots become numerous and begin to touch each other the disease is called a blight or blotch or less commonly, scorch.

Symptoms

It appears as dead areas scattered over the leaf surface; the size, color, and shape can vary; usually have definite margins; can result in substantial premature defoliation and appears as necrotic areas. They can appear as blotchy, dead areas or as discrete leaf spots. They require water on the leaf surface for infection.

Management

a) Field sanitation

b) Use of bio-agents (*Trichoderma harzianum, Bacillus subtilis*) and fungicidal sprays (mancozeb, chlorothalonil, thiophanate-methyl and copper compounds).

c) Use of tolerant stock for planting material.

d) Cultural practices to lessen the inoculum distribution and reduce chance of infection.

2) Vascular disease (Wilt)

(*Verticillium, Fusarium and Ralstonia* spp.)

Symptoms

Loss of rigidity and wilting, water stress; yellowing of foliage, drooping of plant parts; often followed by premature defoliation; gradual dieback may be evident. Flagging may occur when stem becomes symptomatic at a time. Infected plants may develop diagnostic brown, greenish, or yellow streaks in the vascular tissues. Since the pathogens grow systemically in the plant; usually results in plant death.

Management

a) Collection and destruction of infected plants

b) Drenching of carbendazim (0.1%).

c) Uprooting of wilted plants.

d) Proper watering in nursery.

3) Damping-off

(*Pythium, Rhizoctonia* and *Phytophthora* spp.)

Probably the best-known nursery disease is damping-off. This is a disease of young seedlings which resulting in reduced seed germination and poor stand of seedlings. 25-75% losses are caused annually due to this disease (Gupta and Paul, 2001). This disease attacks the plants belonging to families Solanaceae, Crucifereae and Leguminosae.

Symptoms

The infection takes place before the hypocotyls have broken the seed coat or the radical and the plumule emerge out of seeds, the seedlings disintegrate before they come out of soil surface. This is called pre-emergence damping-off which results in poor seed germination. Post-emergence is characterized by stem of the seedling becomes constricted just above the surface of the germination substrate, and then the seedling falls over and dies. There is often (but not always) damage to the plant beneath the soil surface. The infection results in lesion formation on the collar region giving a pinched appearance.

Damping off in tomato

Management

a) High plant density, overwatering and heavy shade favour the spread of the disease and should be avoided.

b) Seed treatment with antagonist fungal culture of fluorescent Pseudomonads @ (10 g/kg seeds) or Thiram (2-3 g/kg of seed) and soil drenching with Dithane M 45 (0.2%) or Bavistin (0.1%) and drenching with neem-based like Azadirachtin (0.2%) or neem oil (0.2%) affords protection against the disease.

c) Good control of *Pythium indicum* in tomato seedlings by using bio-agents such as *Trichoderma viride* and *T. harzianum* (Krishnamoorthy and Bhaskaran, 1990). In cauliflower, the disease (*Rhizoctonia solani*) was reduced when suspension of *Streptomyces arenae* was used (Kundu and Nandi, 1993).

d) The nursery should be regularly inspected for the disease affected seedlings.

Such seedlings should be removed and destroyed.

e) Summer ploughing to kill the soil borne inocula of the pathogens.

f) Proper drainage in nursery.

4. Citrus greening

Symptoms

Characteristic symptoms include yellowing of leaf veins and adjacent tissues, followed by premature defoliation, dieback of twigs, decay of feeder rootlets and lateral roots. Infected fruits are small, lopsided with bitter taste. Such fruits may fall prematurely; while those that remain on the tree do not colour properly, remaining green on the shaded side.

Citrus greening

Management

a) Routinely scout citrus for signs of greening so that infected trees can be removed.

b) Spraying with a foliar insecticide to kill any psyllids.

c) Focus Integrated Pest Management (IPM) efforts on using disease-free nursery trees, reducing infection sources (inoculum) and suppressing Asian citrus psyllid.

5. Citrus tristeza virus (CTV)

There are numerous strains of the virus causing several diseases, "quick decline" of sweet orange and mandarin varieties on sour orange rootstock. It spreads by brown citrus aphid (*Toxoptera citricida*).

Symptoms

Severe strains of the virus cause the development of pits in the trunk and branches, resulting in a range of symptoms from deep ropy furrowing to fine pitting. Fine pitting has a more adverse effect on tree health.

Management

a) Removal of early infected plants from nursery beds.

b) Aphids can be controlled by application of Carbofuran (1 kg a.i./ha) in the nursery bed at the time of sowing seeds.

6. Papaya Ring Spot Virus

Papaya ring spot virus (PRSV) infection is a global threat to papaya (*Carica papaya* L.) cultivation.

Symptoms

The earliest symptoms on papaya are a yellowing and vein-clearing of the young leaves. This is followed by a very conspicuous yellow mottling of the leaves and sometimes severe blistering and leaf distortion. Dark-green streaks and rings also appear in the leafstalks and stems.

Management

a) Removal of early infected plants from nursery beds and use of virus-free seedlings.

b) Aphids can be controlled by application of Carbofuran (1 kg a.i./ha) in the nursery bed at the time of sowing seeds followed by 2-3 foliar sprays of Phosphamidon (0.05%) at an interval of 10 days starting from 15-20 days after sowing.

c) Border crop of banana reduced aphid population in papaya plantation (Sharma, *et al.*, 2010)

d) Use of integrated management including application of insecticides like Dimethoate for reduction of aphid population and use of resistant cultivars (Sharma, *et al.*, 2010).

7. Sooty mold disease: (*Capnodium* spp.) of citrus

Symptoms

Sooty mold causes an unsightly blackening over the surface of leaves. These fungi live on honeydew excreted by some insects such as aphids, psylla, and white flies. When insect secretions of honeydew are abundant, the surface of leaves may have nearly a continuous coating of the black, tissue-paper thin layer of sooty mold.

Sooty mold on citrus leaf

Management

a) Collection and destruction of infected seedlings.

b) Control of sooty mold is related to insects producing honeydew.

c) Another important consideration can be ant management. Ants will protect honeydew-producing insects from predators and parasites in order to harvest the honeydew.

d) In many cases, predators and parasites are sufficiently abundant and quickly begin feeding on and reducing populations of scale insects, aphids, psyllids, whiteflies, or mealybugs once ants have been eliminated. If populations fail to decline, apply horticultural oils, neem oil, or insecticidal soap to suppress the problem insects.

8. Crown gall disease: (*Agrobacterium tumefaciens* and *A. vitis*)

The disease is common in tree fruit nurseries and eventually occurs on large numbers of trees in the orchards. It can affect various plant genera and cultivars (fruit trees, tubers and grapevines). The disease is called crown gall because the galls typically are on the plant's crown at or below the soil line along the main root.

Crown gall disease in grapevine

Symptoms

The disease is recognized by wart-like swellings or galls on the roots and crown. Young galls are light in color. They become dark and hard with age. When the galls are numerous, they may disrupt the flow of water and nutrients. Galls on grapevines are usually elongate, erumpent ridges of tissue bursting. Once a gall is initiated, the bacteria play no further part in its growth and development. In this respect, crown gall has been called a plant cancer.

Management

a) Discard diseased plants.

b) Minimize wounding; disinfect pruning tools between plants.

c) Plant only disease-free stock. (Plant resistant varieties if available).

d) Use of copper oxychloride (0.25%) gave best results in reducing tumor incidence.

e) Soil fumigation especially in nurseries

9. Bacterial leaf spot (*Xanthomonas* spp.) of pomegranate

This bacterial disease affects pomegranate and other fruit crops most seriously in nursery. Bacterial spot is more common and severe when trees are growing in light, sandy soils and when nitrogen is insufficient for normal tree growth.

Symptoms

On the leaves, first as small, angular spots that is purple, black, or brown in color. The mature spots remain angular. Often the spots are most numerous at the tip ends and along the midribs of the leaves. While the individual spots are small, several may join. The infected areas may drop out giving the infected leaves a shot hole appearance. Infected leaves eventually turn yellow and drop.

Management

a) Maximum use of resistant cultivars is the most effective control measure.

b) Nurserymen are well aware of the degree of susceptibility of the cultivars they sell and they can provide good information for specific areas.

c) Orchard management programs should be designed to maintain good vigor.

d) Sanitary measures.

e) There are some chemicals (streptomycin or copper oxychloride) for reducing disease.

10. Downy Mildew of cabbage: *(Perenospora parasitica)*:

The disease is very serious in nursery and it can also appear in field planting. High humidity, fog, drizzling rains, and heavy dew favour the disease development and spread.

Symptoms

The first symptom observed are small, light green-yellow lesions on the upper leaf surface, later showing on the undersurface. The spots turn yellow as they enlarge. During periods of high humidity, a grayish white moldy growth is developed on the undersurface of the leaf. Later the leaf may become papery and die. Cabbage heads develop sunken black spots. Though, some plants are infected at the seedling stage, the symptoms does not become apparent until near harvest.

Management

a) All the weeds serving as alternate host to the fungus should be destroyed.

b) The crop should be irrigated judiciously to avoid periods of high humidity.

c) Spraying the seedlings in the nursery beds with Copper oxychloride (0.3%) is effective in controlling the disease. The first spray should be given as soon as the seedlings appear. Subsequent sprayings are given at weekly intervals until the plants are transplanted in the field.

d) Use of bio- agents (*Bacillus subtilis* and *Trichoderma harzianum*).

Insect Management in Nursery Plants

Insects and vertebrate pests

Insects, such as aphids, mealy bugs, thrips, shoot borers and fruit flies actively seek out the plant host by migrating (flying, walking, crawling). Insects indirectly affect photosynthesis by reducing leaf area through devouring the leaf lamina and gradual defoliation. Photosynthates are reduced; thereby, impacting growth and yield of the plants, severely. Chewing insects such as cutworms reduce crop stand by killing plants. Insects may bore through fruits in the field. Root-feeding insects (e.g., wire worms and white grub) can kill plants from below the ground level. Insects act as vectors of disease. Rodents can be controlled, though seldom eliminated, with poisoned bait placed throughout the nursery. Rabbits can be discouraged with repellents sprayed onto the base of plants. Deer are much more difficult to control, although some benefit is claimed for repellents. Fencing the nursery can prevent deer from winter browsing in the production fields.

Insects like mango hoppers, mealy Bug and stone borer, aphid and leaf hopper cause considerable damage during nursery stage of horticultural crops.

1. Mango Hopper: (*Amritodus atkinsoni* and *Idiscopus clypealis*)

Mango hoppers are the most serious pests of the mango nursery.

Nature of damage

a) Mango hopper nymphs and adults suck cell sap from the tender shoots during rainy session.

b) Infested leaves of mango shrivel, turn brown & ultimately fall down.

c) Mango hoppers secrete honeydew like substance which encourage the growth of black sooty mold which severely affects photosynthetic activity of leaves.

d) They affect the marketable quality of seedlings in the nursery.

Shoot damage by mango leaf hopper

Management

a) Spray 0.05% Malathion 50 EC or 0.04% Phosphamidon 85 EC in Oct- Nov.

b) Fogging with mixture of Malathion and Diesel in 1.5: 8.5 proportions.

c) At higher level of hopper density (>4 hopper/ shoot), spray 0.05% Imidacloprid, while at lower density (<4 hopper/shoot), spray of Azadirachtin @ 2ml/liter.

d) Use of natural enemies of hoppers like *Irantha armiper.*

e) White fungus, *Verticillum lacanii* is also an effective bio-agent.

2. Mango Mealy Bug: (*Drosicha maniferae*)

It affects some fruit plants in nursery like mango, guava, papaya and lemon. This insect covered with ashy-white mealy powder and crawling up or down tree-trunks or on ground the tree-base.

Nature of damage

Nymphs suck sap from tender shoots and leaves which dry and drop. The nymphs developing into males undergo some sort of pupation and transform themselves into winged adults.

Mango mealy-bug

Management

a) Soil solarization and deep plough during summer to expose egg masses to the sun and heat and get killed.

b) The application of a soil insecticide in the same area when hatching begins or is expected, so that the just hatched nymphs may be poisoned.

c) Application of 5 cm wide sticky bands around stem/tree trunk in mid December prevents nymphs from crawling up the seedling.

d) Incorporation of insecticide in the sticky band is increase the effectiveness of band.

e) Spray 2.5 ml Methyl Parathion in 10 liters of water.

3. Mango Stone Borer: *(Stemochetus mangiferae)*

As regards the extent of loss, sometimes even 100 per cent of the fruits are infested.

Nature of damage

a) Mango stone borer attacks soft fleshy varieties of mango.

b) Stone borer grub on hatching, create tunnel in a zigzag manner through the pulp, endocarp and seed coat and reaches the cotyledons and destroys them.

c) Stone borer adult also feeds on the developing seeds and it may hasten the maturity of infested fruits.

d) Stone borer in mango destroy the germination viability of seed/ stone in mango nursery.

Infested fruit by mango stone borer

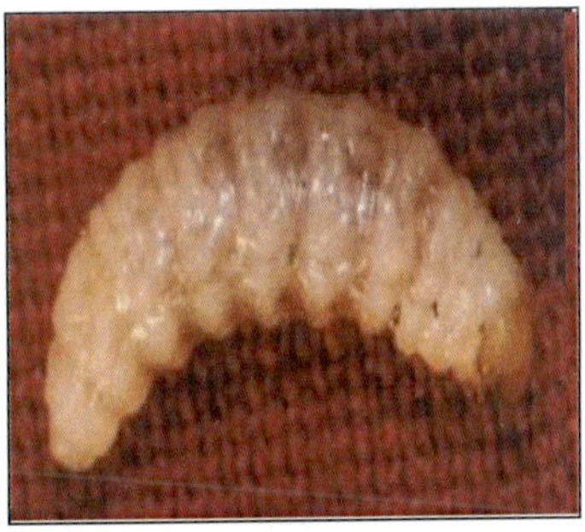

Grub of mango stone borer

Management

a) Collect and destroy fallen fruits and keep the mother plant block clean.

b) Raking soil around tree in October-November and March is helpful.

c) Spraying 0.05% Diazinon on trunk of mango tree and primary branches in October-November.

d) Dipping the fruits in hot water at 50±5 ^{0}C for two hours for killing the insect inside the seed stone.

4. Brinjal shoot borer: (*Leucinodes arbonalis*)

It is a typical cosmopolitan pest causing serious damage to brinjal cultivation and limits the yield to very low level. It is a small larva that bores inside shoots, resulting in the withering of the shoots. It also bores into and feeds on young and maturing fruit. Damage from the FSB starts at the nursery stage and continues until harvest. Losses range from 50-70%. The young larvae of the pest bore in to petioles and midribs of large leaves and tender shoots causing shoot tips to wilt and later they bore in to flower buds and fruits.

Management

a) Indiscriminate use of insecticides leads to an accumulation of pesticide residues in the produce, alongwith the annihilation of beneficial insects, leading to pest resurgence.

b) Collect and destroy affected shoots.

c) Sex pheromones can be used to trap male moths.

d) Dip the roots of the seedlings for 3 hours in imidacloprid solution prepared at 1ml/lit before transplanting.

e) Apply neem cake at 250kg/ha at 30 days after transplanting.

f) Spray of profenophos 0.05 per cent or carbaryl 50WP at 2g/lit starting from one month after transplanting at 15 days interval.

g) Soil application of carbofuran 3G at 30kg/ha 10 days after transplanting is also very effective.

h) Cultivation of brinjal under protected cultivation (net house condition) is found to reduce the incidence of shoot and fruit borer.

5. Flea Beetles in tomato seedlings: (*Epitrix hirtipennis*)

Nature of damage

Flea beetles are common pests of seedling tomatoes in most areas. Adult beetles chew small holes in leaves. High populations of flea beetles feeding on plants younger than the 4- to 5-leaf stage can result in stand loss, especially under hot, windy springtime conditions when the injured plants are desiccated.

Management

a) Cultural control - An overwintering population might exist; if possible, rotate tomatoes with a non-host crop. Replanting rows near borders that have been heavily damaged is an option.

b) Organically acceptable methods- Sprays of pyrethrin are organically acceptable methods for managing these pests.

c) Monitoring and treatment decisions- Monitor fields for flea beetles soon after transplanting or after the plants emerge. Consider a spot treatment in rows that have 5-10% of seedlings or young plants with heavy flea beetle feeding.

6. Citrus Psylla: (*Diaphorina citri*)

Nature of damage

Both the adults and nymphs of this insect suck the sap from the tender parts of the buds, leaves, branches and inject a toxic substance into them. The nymphs excrete white crystalline waxy pellets on which black sooty mould may develop. In case of severe attack, leaves get distorted, curled up and ultimately fall resulting in complete defoliation of the plant. This pest acts as a vector for spreading the 'citrus greening' disease.

Management

a) A number of natural enemies such as species of coccinellids and ladybird beetles attack the nymphs of citrus psylla.

b) Spraying the plants with Phosphamidon (0.025%).

c) Also application of Monocrotophos (0.025%) or Malathion (0.03%) or Dimethoate (1.5%) is useful during March.

7. Citrus leaf miner: (*Phyllocnistis citrella*)

Nature of damage

This insect causes damages both in nursery and in grown up stages of the citrus plant. If the larvae attack tender leaves and feed in the epidermal layers of the leaf by making serpentine mines in which air gets trapped and gives them silvery appearance. The leaves of fruit trees show white glistening irregularly-zigzag galleries within the leaf lamina (Pradhan, 1994). The affected leaves turn pale yellow, get deformed, distorted and crumpled. The whole plant suffers and also becomes prone to the attack of citrus canker disease (Pradhan, 1994).

Management

a) To keep the pest population under check.

b) Pruning of all the affected parts during winter should be done.

c) Spraying the plants with Methyl Demeton and Phosphamidon @ each 0.03% at the emergence of new leaves is highly beneficial.

d) Application of Phorate 10 G (2.5 kg a.i./ha) applied one day before planting is effective in reducing the larval population of leaf miner.

8. Lemon Butterfly: (*Papilio demoleus*)

The lemon butterfly is one of the fewer species of this group which is quite harmful in its larval stage. It is the most destructive pest in nurseries. The caterpillars feed on the young foliage at the nursery stage and also on young flushes of grown up trees.

Nature of damage

The caterpillar feed voraciously on leaf lamina leaving behind only the midrib. In case of severe infestation, entire tree is defoliated. The fully grown caterpillars are green in colour. Due to its capability for rapid population growth under favourable circumstances, it is considered a serious potential threat (Eastwood *et al.*, 2006). The caterpillars can completely defoliate young citrus trees (below 2 feet) and devastate citrus nurseries.

Caterpillars devouring lemon leaves.

Management

a) Hand/mechanical picking of the larvae and pupal stage.

b) Spraying with Monocrotophos (1.5 ml/litre of water) is effective in managing this insect.

9. Anar butterfly: (*Virachola isocrates*)

Cultivation of high yielding varieties of pomegranate with intensive care and management under irrigated condition led to certain severe insect-pest problems. Among them, infestation by Anar butterfly results in reduction of pomegranate fruit yield and heavy loss for growers.

Nature of damage

Its damage is noticed in thirty to fifty days old fruits of pomegranate and the adults prefer thirty to fifty days old fruits for egg laying and most of the eggs are laid on the calyx portion of the individual fruits. Fourth and fifth instar larvae caused the maximum damage to fruits as the larva bores directly into the fruits. Infestation sometimes is so severe that up to 90 % of the fruits, may become wormy. The major constraint in increasing export potential is the quality of pomegranate fruit in terms of size, colour, freedom from blemishes and pesticide residue levels.

Management

a) Covering the 30-50 days old fruits with bags of butter paper in small scale.

b) Collection and destruction of all infested fruits from large compact areas in the beginning of the fruiting seasons.

c) Clipping off calyx cup of flowers immediately after pollination will help to reduce the egg load on the fruits and damage level.

d) Clipping off calyx cup immediately after pollination and two rounds of spraying with neem oil at 3 per cent also controlled pest infestation effectively.

e) Two sprays of Emamectin benzoate 5 SG at the rate of 0.25 g./l or Spinosad 45 SC at the rate of 0.20ml/l has recorded highest reduction in the fruit damage.

References

Eastwood, R.; Boyce, S. L.; Farrell, B. D. 2006. "The Provenance of Old World Swallowtail Butterflies, *Papilio demoleus* (Lepidoptera: Papilionidae), Recently Discovered in the New World" (abstract). Annals of the Entomological Society of America (Entomological Society of America) 99 (1): 164–168.

Gupta, V. K. and Paul, Y. S.2001. Diseases of Vegetable Crops. Kalyani Publishers, New Delhi, 277 pp.

Krishnamoorthy, A. S. and Bhaskaran, R. 1990. Biological control of damping –off disease of tomato caused by *Pythium indicum*. J. Biological Control 4: 52-54.

Kundu, P. K. and Nandi, B. 1993. Inhibition of *Rhizoctonia solani by Streptomyces arenae* in soil. J. of Mycopatho. Research 31: 35-42.

Mehrotra, R. S. 1991.. Plant Pathology. Tata McGraw-Hill Publishing Company Ltd., New Delhi, p. 771.

Pradhan, S. 1994. Insect Pests of Crops. Published by National Book Trust India, New Delhi, 200p.

Ray, P. K. 2012. Plant Nursery Management: How to Start and Operate a Plant Nursery (Paperback). Scientific Publisher, Jodhpur, 407 p.

Sharma, S. K., Zote, K. K., Kadam, U. M., Tomar, S.P.S. , Dhale, M.G., Sonawane, A.U. 2010. Integrated management of papaya ringspot virus. Acta Horticulture 851: 473-480.

Singh, R. S. 1991. Plant Diseases. Oxford & IBH Publishing Co. Pvt. Ltd., New Delhi, p. 619.